Geophysical Monograph Series

Geophysical Monograph Series

Geophysical Monograph 217

Deep Earth

Physics and Chemistry of the Lower Mantle and Core

Hidenori Terasaki
Rebecca A. Fischer

Editors

This Work is a co-publication between the American Geophysical Union and John Wiley and Sons, Inc.

WILEY

Published under the aegis of the AGU Publications Committee

Brooks Hanson, Director of Publications
Robert van der Hilst, Chair, Publications Committee

CONTENTS

CONTRIBUTORS

Yohei Amaike
Department of Earth Science
Graduate School of Science
Tohoku University
Sendai, Japan

Daniele Antonangeli
Institut de Minéralogie, de Physique des Matériaux
et de Cosmochimie (IMPMC), UMR CNRS 7590
Sorbonne Universités – UPMC
Paris, France

Andrew J. Campbell
Department of the Geophysical Sciences
University of Chicago
Chicago, Illinois, USA

Razvan Caracas
CNRS, Ecole Normale Supérieure de Lyon
Université Claude Bernard Lyon 1
Laboratoire de Géologie de Lyon
Lyon, France

Bin Chen
Hawaii Institute of Geophysics and Planetology
School of Ocean and Earth Science and Technology
University of Hawai'i at Mānoa
Honolulu, Hawaii, USA

Patrick Cordier
UMET, Unité Matériaux et Transformations
ENSCL, CNRS, Université de Lille
Lille, France

Haruhiko Dekura
Geodynamics Research Center
Ehime University
Matsuyama, Ehime Prefecture, Japan

Susannah M. Dorfman
Department of Earth and Environmental Sciences
Michigan State University
East Lansing, MI, USA

Thomas S. Duffy
Department of Geosciences
Princeton University
Princeton, New Jersey, USA

Rebecca A. Fischer
Department of the Geophysical Sciences
University of Chicago
Chicago, Illinois, USA

Daniel J. Frost
Bayerisches Geoinsitut
University Bayreuth
Bayreuth, Germany

George Helffrich
Earth-Life Science Institute (ELSI)
Tokyo Institute of Technology
Tokyo, Japan

John W. Hernlund
Earth-Life Science Institute
Tokyo Institute of Technology
Meguro, Japan

Kei Hirose
Earth-Life Science Institute (ELSI)
Tokyo Institute of Technology
Meguro, Tokyo, Japan

Laboratory of Ocean-Earth Life Evolution Research
Agency for Marine-Earth Science and Technology
Yokosuka, Kanagawa, Japan

Hiroki Ichikawa
Geodynamics Research Center
Ehime University
Matsuyama, Ehime Prefecture, Japan

Earth-Life Science Institute
Tokyo Institute of Technology
Meguro, Tokyo, Japan

Seth A. Jacobson
Bayerisches Geoinstitut
University of Bayreuth
Bayreuth, Germany

Observatoire de la Côte d'Azur
Nice, France

Seiji Kamada
Department of Earth Science
Graduate School of Science
Tohoku University
Sendai, Japan

Abby Kavner
Earth, Planetary, and Space Sciences Department
University of California
Los Angeles, California, USA

Kenji Kawai
Department of Earth Science and Astronomy
Graduate School of Arts and Sciences
University of Tokyo
Meguro, Japan

Stéphane Labrosse
Univ Lyon, Ens de Lyon, Université Lyon 1
CNRS, UMR 5276 LGL-TPE, F-69342
Lyon, France

Jie Li
Department of Earth and Environmental Sciences
University of Michigan
Ann Arbor, Michigan, USA

Izumi Mashino
Department of Earth Science
Graduate School of Science
Tohoku University
Sendai, Japan

William F. McDonough
Department of Geology
University of Maryland
College Park, Maryland, USA

Sébastien Merkel
UMET, Unité Matériaux et Transformations
ENSCL, CNRS, Université de Lille
Lille, France

Caitlin A. Murphy
Geophysical Laboratory
Carnegie Institution of Washington
Washington, D.C., USA

Robert Myhill
Bayerisches Geoinsitut
University Bayreuth
Bayreuth, Germany

Itaru Ohira
Department of Earth Science
Graduate School of Science
Tohoku University
Sendai, Japan

Eiji Ohtani
Department of Earth Science
Graduate School of Science
Tohoku University, Sendai, Japan

V. S. Sobolev Institute of Geology and Mineralogy
Siberian Branch, Russian Academy of Sciences
Novosibirsk, Russia

Emma S. G. Rainey
Earth, Planetary, and Space Sciences Department
University of California
Los Angeles, California, USA

The Johns Hopkins University Applied Physics Lab,
Laurel, Maryland, USA

Simon A. T. Redfern
Department of Earth Sciences
University of Cambridge,
Cambridge, UK

Kevin Righter
NASA Johnson Space Center
Houston, Texas, USA

David C. Rubie
Bayerisches Geoinstitut
University of Bayreuth
Bayreuth, Germany

Hidenori Terasaki
Department of Earth and Space Science
Osaka University,
Toyonaka, Japan

Taku Tsuchiya
Geodynamics Research Center
Ehime University
Matsuyama, Ehime Prefecture, Japan

Earth-Life Science Institute
Tokyo Institute of Technology
Meguro, Tokyo, Japan

Xianlong Wang
Geodynamics Research Center
Ehime University
Matsuyama, Ehime Prefecture, Japan

Earth-Life Science Institute
Tokyo Institute of Technology
Meguro, Tokyo, Japan

June K. Wicks
Department of Geosciences
Princeton University
Princeton, New Jersey, USA

PREFACE

The physics and chemistry of the lower mantle and core are responsible not only for the seismic, compositional, and thermal structure of the deep Earth but also for dynamics throughout the entire Earth, from the deep interior to the surface. Structure and dynamics are closely linked to the formation and evolution of Earth's interior.

The interpretations of seismic signatures of the core and lower mantle have long been uncertain or under debate because we could not obtain data experimentally at relevant pressure-temperature conditions. In recent years, we have reached a stage where we can perform measurements at the conditions of the center part of Earth using state-of-the-art techniques, and many reports on the physical and chemical properties of the deep Earth have come out very recently. Novel theoretical models have been complementary to this breakthrough. These new inputs enable us to compare directly with results of precise geophysical observations. As a result, views of the deep Earth have been significantly advanced. Therefore, our knowledge and information of the Earth's interior have been greatly updated recently, and now is a good time to summarize those issues.

This volume includes contributions from mineral/rock physics, geophysics, and geochemistry that relate to these properties. The volume consists five parts (thermal structure of the lower mantle and core; structure, anisotropy, and plasticity of deep Earth materials; physical properties of the deep interior; chemistry and phase relations in the lower mantle and core; and volatiles in the deep Earth) and each part contains three to eight chapters. The idea of this book is based on active discussions in the mineral and rock physics session "Chemistry and Physics of Earth's Lower Mantle and Core" at the American Geophysical Union Fall 2013 meeting.

The proposed volume will be a valuable resource for researchers and students who study Earth's interior to obtain the latest information on deep Earth properties, materials, and behavior. The topics of this volume are multidisciplinary, and therefore this volume could be of interest to researchers and students from a wide variety of fields in the Earth sciences. We have chosen to structure the book to contain many shorter chapters, rather than fewer longer chapters, in order to cover a wide variety of topics and incorporate the views of many experienced authors.

Finally, we would like to acknowledge the chapter authors and reviewers for their valuable contributions and also thank AGU/Wiley editorial staff (especially Rituparna Bose and Mary Grace Hammond) for their excellent support.

Hidenori Terasaki and Rebecca A. Fischer

Part I
Thermal Strucure of Deep Earth

1

Melting of Fe Alloys and the Thermal Structure of the Core

Rebecca A. Fischer

ABSTRACT

The temperature of the Earth's core has significant implications in many areas of geophysics, including applications to Earth's heat flow, core composition, age of the inner core, and energetics of the geodynamo. The temperature of the core at the inner core boundary is equal to the melting temperature of the core's Fe-rich alloy at the inner core boundary pressure. This chapter is a review of experimental results on melting temperatures of iron and Fe-rich alloys at core conditions that can thus be used to infer core temperatures. Large discrepancies exist between published melting curves for pure iron at high pressures, with better agreement on the melting behavior of Fe-light element alloys. The addition of silicon causes a small melting point depression in iron, while oxygen and especially sulfur cause larger melting point depressions. The inner core boundary temperature likely falls in the range 5150–6200 K, depending on the identity of the light element(s) in the core, which leads to a core-mantle boundary temperature of 3850–4600 K for an adiabatic outer core. The most significant sources of uncertainties in the core's thermal structure include the core's composition, phase diagram, and Grüneisen parameter.

1.1. INTRODUCTION

The Earth's core consists primarily of iron-nickel alloy. The presence of several weight percent of one or more lighter elements such as S, Si, O, C, or H is implied by the core's density, and these light elements depress the melting point of the core relative to pure iron [e.g., *Birch*, 1952; *Poirier*, 1994]. The thermal structure of the core plays a key role in many deep Earth properties. It affects the magnitude of the temperature difference across the thermal boundary layer at the base of the mantle, heat flow on Earth, and the cooling rate of the core. Faster cooling rates would imply a younger inner core, while slower cooling would imply an older inner core. The age

of the inner core corresponds to the onset of compositional convection in the outer core due to the preferential expulsion of light elements during inner core crystallization. The core's temperature structure is also linked to thermal convection in the outer core, with these two types of convection driving the dynamo responsible for Earth's magnetic field [e.g., *Lister and Buffett*, 1995]. The vigor of thermal convection in the Earth's core depends on both its thermal structure and its thermal conductivity. Recent studies on the thermal conductivity of Fe and Fe-rich alloys at core conditions have revealed a higher thermal conductivity of core materials than previously thought [e.g., *Pozzo et al.*, 2012; *Seagle et al.*, 2013], implying that higher core temperatures and/or stronger compositional convection are required to power the dynamo.

Knowledge of the core's temperature would inform our understanding of these processes and put tighter constraints

Department of the Geophysical Sciences, University of Chicago, Chicago, Illinois, USA

Deep Earth: Physics and Chemistry of the Lower Mantle and Core, Geophysical Monograph 217, First Edition.
Edited by Hidenori Terasaki and Rebecca A. Fischer.
© 2016 American Geophysical Union. Published 2016 by John Wiley & Sons, Inc.

on the abundances of light elements in the Earth's core, since thermal expansion affects the quantity of light elements needed to match the observed density. The temperature at the inner core boundary (ICB) is equal to the liquidus temperature of the core alloy at that pressure (~330 GPa), since that is the temperature at which the solid inner core is crystallizing from the liquid outer core. Therefore knowledge of the ICB temperature could be combined with measurements of phase diagrams at high pressures and temperatures (P and T) to constrain the identities of the core's light elements. However, the thermal structure of the core is poorly understood.

This chapter reviews the available experimental constraints on the core's temperature. Measuring melting of iron and Fe-rich alloys at core conditions presents significant experimental challenges, leading to discrepancies between studies. Extrapolating melting curves to the ICB pressure provides information about the ICB temperature. Adiabats can be calculated through these P-T points up to the core-mantle boundary (CMB) pressure to determine the thermal structure of the outer core.

1.2. METHODS FOR DETERMINATION OF MELTING

Melting experiments relevant to the Earth's core require the generation of simultaneous extreme pressures and temperatures. This is commonly accomplished through the use of a laser-heated diamond anvil cell, which is capable of reaching inner core conditions. A sample is embedded in a soft, inert pressure-transmitting medium and compressed between two diamond anvils, then heated with an infrared laser until melted. Pressure is typically monitored using an X-ray standard in the sample chamber whose equation of state is well known or by ruby fluorescence or diamond Raman spectroscopy, whose signals shift systematically with pressure. Temperature is measured spectroradiometrically by fitting the thermal emission to the Planck function (see *Salamat et al.* [2014] for a recent review of diamond anvil cell methodology).

While techniques for generating and measuring extreme P-T conditions in the diamond anvil cell are relatively well established, there is disagreement over the best method for detecting a melt signal. Some studies rely on "speckling," a qualitative detection of movement in the sample visualized by shining a second (visible) laser onto the laser-heated spot during the experiment [e.g., *Boehler*, 1993]. This movement is thought to be due to convection of the molten sample, though it has recently been proposed that rapid recrystallization of the sample at subsolidus conditions can cause this apparent motion [*Anzellini et al.*, 2013; *Lord et al.*, 2014a]. Other methods rely on discontinuities in physical properties upon melting, such as a change in the emissivity-temperature relationship

[*Campbell*, 2008; *Fischer and Campbell*, 2010] or in the laser power-temperature relationship [e.g., *Lord et al.*, 2009]. These methods have the advantage of not requiring a synchrotron X-ray source but provide no structural information about subsolidus phases. Synchrotron-based techniques include the use of X-ray diffraction to detect diffuse scattering from the melt and/or disappearance of crystalline diffraction [e.g., *Anzellini et al.*, 2013; *Campbell et al.*, 2007; *Fischer et al.*, 2012, 2013] or, less commonly, time domain synchrotron Mössbauer spectroscopy [*Jackson et al.*, 2013].

In addition to diamond anvil cell methods, the multianvil press has also been used for melting experiments, with analysis of recovered samples used to detect melting [e.g., *Fei and Brosh*, 2014; *Fei et al.*, 2000]. Previously multianvil press experiments were limited in pressure to ~25 GPa, but recent advances in sintered diamond anvils [e.g., *Yamazaki et al.*, 2012] may allow for higher-pressure melting experiments in the multianvil press in the future. Until recently, shock wave experiments were the standard technique for melting experiments at core conditions [e.g., *Brown and McQueen*, 1986]. They provide a reliable method for reaching core pressures and temperatures, with melting determined from discontinuities in the sound velocity-pressure relationship. However, temperatures in shock experiments are frequently calculated thermodynamically [e.g., *Brown and McQueen*, 1986; *Nguyen and Holmes*, 2004] due to difficulties with direct measurements, making them less accurate, and improvements in diamond cell methods have facilitated the access of core conditions by static methods. Additionally, melting curves can be calculated using ab initio methods [e.g., *Alfè et al.*, 2002] or thermodynamic modeling [e.g., *Fei and Brosh*, 2014].

1.3. RESULTS ON MELTING OF IRON

Due to its extreme importance to our understanding of the thermal structure of the core, the melting behavior of iron at high pressures has been investigated by many research groups using all of the techniques discussed in Section 1.2. Despite such a large number of results using a variety of methods, there remains no consensus on the melting curve of pure iron at core pressures. Figure 1.1 illustrates some of the many previous results on iron melting obtained using diamond anvil cell [*Anzellini et al.*, 2013; *Boehler*, 1993; *Jackson et al.*, 2013; *Ma et al.*, 2004; *Saxena et al.*, 1994; *Shen et al.*, 2004; *Williams et al.*, 1987], ab initio [*Alfè*, 2009; *Alfè et al.*, 2002; *Anderson et al.*, 2003; *Belonoshko et al.*, 2000; *Laio et al.*, 2000; *Sola and Alfè*, 2009], and shock wave [*Ahrens et al.*, 2002; *Brown and McQueen*, 1986; *Nguyen and Holmes*, 2004; *Yoo et al.*, 1993] methods. Below ~50 GPa there is fairly good agreement over the iron melting curve. In the ~50–200 GPa range, readily accessible in the laser-heated diamond anvil

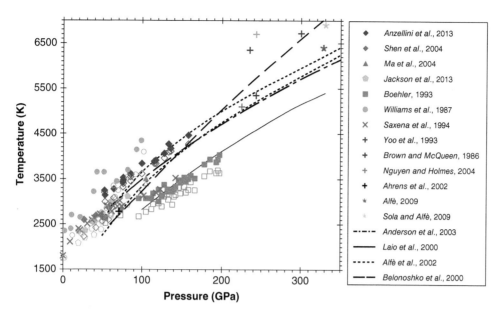

Figure 1.1 Selection of the literature results on Fe melting illustrating range of discrepancy in the literature. Open symbols are lower bounds; filled symbols are upper bounds. Symbols are color coded by study (e.g., teal circles and + symbols are all from *Williams et al.* [1987]), with shape indicating the method used to detect melting. Diamonds: X-ray diffuse scattering. Triangles: loss of X-ray signal. Pentagons: Mössbauer spectroscopy. Squares: visual observation of motion ("speckling"). Circles: changes in sample appearance. × symbols: discontinuity in laser power-temperature relationship. + symbols: shock wave methods. Curves and * symbols: ab initio methods. The lower and upper curves of *Alfè et al.* [2002] were calculated with and without free energy corrections, respectively.

cell, Fe melting curves vary by over 1000 K. Where the shock Hugoniot crosses the melting curve at ~240 GPa, reported shock melting temperatures vary by ~1500 K. At the inner core boundary pressure of 330 GPa, ab initio calculations of Fe melting vary by over 1500 K.

The causes of these discrepancies remain unclear. Among diamond anvil cell studies, it appears that the method used to detect melting is one of the main sources of variation [e.g., *Anzellini et al.*, 2013; *Jackson et al.*, 2013], though there is no clear consensus as to which method should be the most reliable. Uncertainties in radiometric temperature measurements, pressure calibrations, and possible chemical reactions at these extreme conditions may also play a role in the discrepancy. Among shock wave experiments, large uncertainties in temperature determination may explain some of the variability [e.g., *Brown and McQueen*, 1986]. A systematic offset is seen between shock studies in which temperatures are calculated thermodynamically [*Brown and McQueen*, 1986; *Nguyen and Holmes*, 2004] and those in which they are measured spectroradiometrically [*Williams et al.*, 1987; *Yoo et al.*, 1993] (Figure 1.1). Studies in which temperatures are calculated thermodynamically give systematically lower melting temperatures, in better agreement with the static diamond anvil cell results.

The melting point of iron at 330 GPa may be taken as an upper bound on the temperature at the inner core boundary, neglecting the effects of an alloying light element. However, the experimental studies represented in Figure 1.1 reported ICB temperatures ranging from 4850 K [*Boehler*, 1993] to 7600 K [*Williams et al.*, 1987]. *Anzellini et al.* [2013] suggested that earlier diamond cell studies that identified melting at lower temperatures may have actually been identifying the onset of fast recrystallization. The authors point to the agreement between several more recent studies (diffuse scattering results of *Anzellini et al.* [2013], some shock wave studies [*Brown and McQueen*, 1986; *Nguyen and Holmes*, 2004], and some ab initio studies [e.g., *Alfè*, 2009; *Alfè et al.*, 2002]) as evidence of progress toward a consensus on the Fe melting curve. Additionally, coincidence of fast recrystallization with melting temperatures reported using the speckling technique was also reported by *Lord et al.* [2014a] on a different material. Therefore, in this discussion, the melting curve of *Anzellini et al.* [2013] will be used as the reference for pure iron, though uncertainties remain. For example, lower melting temperatures for iron were reported by other shock wave [*Ahrens et al.*, 2002] and diamond cell studies, including those using identical melting criteria [*Shen et al.*, 2004]. Despite decades of effort to measure the melting curve of iron, further work remains necessary.

1.4. RESULTS ON MELTING OF IRON-RICH ALLOYS

The core's density implies the presence of several weight percent of one or more light elements [e.g., *Birch*, 1952], which lowers the melting point of iron. Therefore it is important to consider the effects of these light elements on the melting of iron at core conditions. Surprisingly, the eutectic melting behavior in more complex multicomponent systems involving iron and one or more light elements is often more well-understood than the end-member case.

Melting in Fe-rich iron-silicon alloys has been the subject of several previous studies using X-ray diffuse scattering [*Fischer et al.*, 2012, 2013; *Morard et al.*, 2011], laser power-temperature discontinuities [*Asanuma et al.*, 2010; *Fischer et al.*, 2013], and morphology of recovered samples [*Asanuma et al.*, 2010]. Figure 1.2a illustrates melting results in the Fe-FeSi system on compositions ranging from 9 to 18 wt % Si up to 140 GPa. Since the addition of this amount of silicon must depress the melting point of iron, the melting results in Figure 1.2a lend support to the higher reported melting curves of pure Fe [i.e., *Anzellini et al.*, 2013; *Williams et al.*, 1987]. Taking the Fe melting curve of *Anzellini et al.* [2013] as a reference, the addition of Si causes a melting point depression of approximately 0–400 K at 100 GPa. At higher pressures than this, the eutectic temperature appears to stop increasing with pressure, but there are very few Fe-Si melting data at these conditions. This pressure approximately corresponds to a transition from a face-centered-cubic (fcc) + B2 to a hexagonal close-packed (hcp) + B2 subsolidus phase assemblage [*Fischer et al.*, 2013], which may explain the change in slope of the eutectic temperature with increasing pressure. This behavior could imply that the melting point depression caused by silicon increases at higher pressures; however, additional melting data on Fe-Si alloys at higher pressures are needed to clarify this. The scatter in Figure 1.2a is due in part to changes in the subsolidus phase assemblage being melted, which may be fcc + B2, hcp + B2, DO$_3$ only, or B2 only below ~100 GPa, depending on pressure and composition. However, all compositions shown in Figure 1.2a should melt eutectically from an hcp + B2 mixture above ~100 GPa [*Fischer et al.*, 2013].

Figure 1.2b summarizes melting results in the Fe-FeO system obtained using the "speckle" method [*Boehler*, 1993] and the disappearance of crystalline X-ray diffraction peaks [*Seagle et al.*, 2008] as melting criteria up to 140 GPa. The data of *Seagle et al.* [2008] and *Boehler* [1993] are compatible within uncertainty. It is interesting to note that these two different melting criteria indicate approximately the same melting temperatures in the Fe-FeO system, while they give very different results on

pure Fe (Section 1.3). Oxygen causes a deeper melting point depression in iron than silicon does. At 100 GPa, the experimental results shown in Figure 1.2b demonstrate that the Fe-FeO eutectic temperature is approximately 700–1100 K lower than the Fe melting curve of *Anzellini et al.* [2013]. Thermodynamic calculations of *Komabayashi* [2014] using an ideal solution model for the Fe-FeO eutectic are compatible with the upper end of this range. Changes in the subsolidus crystal structure of FeO at higher pressures [*Ozawa et al.*, 2011] could change the melting behavior, but melting data in the Fe-FeO system are not yet available at these conditions (P > 240 GPa).

Previous results on the Fe-Fe$_3$S melting curve exhibit a remarkable degree of agreement between studies that used a variety of different melting criteria. Figure 1.2c shows some of these results, in which melting was determined from the disappearance of crystalline X-ray diffraction [*Campbell et al.*, 2007; *Kamada et al.*, 2012], the appearance of diffuse X-ray scattering [*Morard et al.*, 2008, 2011], and scanning electron microscope observations of recovered samples [*Chudinovskikh and Boehler*, 2007]. From 21 to ~240 GPa in the Fe-rich side of the Fe-S system, the subsolidus phase assemblage is a mixture of Fe and Fe$_3$S [*Fei et al.*, 2000; *Kamada et al.*, 2010], though Fe$_3$S decomposition at higher pressures could change the slope of the melting curve [*Ozawa et al.*, 2013]. The Fe-Fe$_3$S melting curve shows a small change in slope at ~60 GPa, where the fcc-hcp phase boundary in iron intersects the Fe-Fe$_3$S melting curve [*Morard et al.*, 2011]; effects of this transition are not well resolved in the Fe-FeO system, likely due to scatter in the data. Eutectic melting in the Fe-Fe$_3$S system occurs ~900–1200 K lower than the melting point of iron [*Anzellini et al.*, 2013] at 100 GPa.

Melting in the Fe-Fe$_3$C system was investigated by *Lord et al.* [2009] to 70 GPa using the laser power-temperature discontinuity method and X-ray radiography (Figure 1.2d). New multianvil press experiments and calculations in the Fe-C system by *Fei and Brosh* [2014] have reproduced the Fe-Fe$_3$C eutectic melting curve of *Lord et al.* [2009]. The Fe-Fe$_3$C eutectic is 600–800 K lower than the Fe melting curve of *Anzellini et al.* [2013] at 70 GPa. The Fe-Fe$_3$C melting curve has approximately the same slope as the Fe melting curve of *Anzellini et al.* [2013] at these pressures, so the melting point depression at 100 GPa is also expected to be ~600–800 K. It is likely that Fe$_7$C$_3$ will replace Fe$_3$C as the stable carbide along the eutectic at core conditions [*Fei and Brosh*, 2014; *Lord et al.*, 2009], which could change the melting behavior in this system at higher pressures. Carbon in the core is discussed further in Chapter 22.

Melting in the Fe-H system has not been studied nearly as thoroughly as other Fe-light element systems, due at least in part to technical challenges. It has only been studied up to 20 GPa [*Sakamaki et al.*, 2009] and only in the presence of excess H, which is likely a different

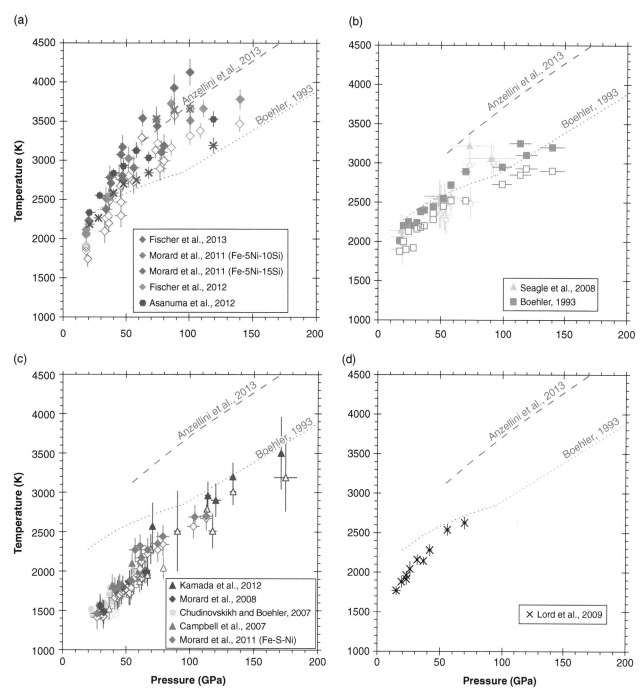

Figure 1.2 Solidus temperatures in (a) Fe-FeSi, (b) Fe-FeO, (c) Fe-Fe$_3$S, and (d) Fe-Fe$_3$C binary systems (some data also contain Ni). Open symbols (and × symbols from *Asanuma et al.* [2010]) are lower bounds; filled symbols are upper bounds. Dashed and dotted grey lines show high and low melting curves for pure Fe for comparison. Symbols are color coded by study, with shape indicating the method used to detect melting. Hexagons: scanning electron microscope imaging of recovered sample texture. Other symbol shapes are as in Figure 1.1.

eutectic from that relevant to the Earth's core composition. Other aspects of hydrogen in the core are reviewed in Chapter 20.

Melting in ternary systems has not been extensively studied at core pressures. *Terasaki et al.* [2011] and *Huang*

et al. [2010] measured melting on Fe-rich compositions in the Fe-O-S system using static and shock methods, respectively. *Terasaki et al.* [2011] report a eutectic melting point depression of ~950–1200 K relative to the Fe melting curve of *Anzellini et al.* [2013] at 100 GPa, indistinguishable from results on the Fe-S system without oxygen present. *Huang et al.* [2010] report a smaller melting point depression of 300–800 K, which is incompatible with results on Fe-S and Fe-O melting, since the addition of a small amount of a second light element is unlikely to increase the melting point.

Nickel has only a minor effect on the melting temperatures of iron-light element alloys, though many of the data on the effects of nickel were obtained far below core pressures. Comparing the melting data of *Morard et al.* [2011] on Fe-Si-Ni alloys to results without nickel (Figure 1.2a) reveals no resolvable effect of 5% Ni on the melting temperature, though this is difficult to assess given the scatter in the Fe-Si(-Ni) data. *Lord et al.* [2014b] reported similar melting temperatures of Fe-Si-Ni alloys. The shock wave study of *Zhang et al.* [2014] reported melting temperatures for Fe-9Ni-10Si at 168 and 206 GPa that are compatible with the Ni-free data in Figure 1.2a. *Zhang and Fei* [2008] investigated the Fe-S-Ni system at ~20 GPa and found that the addition of 5% Ni lowers the melting point by 50–100 K, compatible with the findings of *Stewart et al.* [2007]. *Morard et al.* [2011] found no resolvable effect of 5% Ni in an Fe-S-Ni alloy at higher pressures (Figure 1.2c). *Urakawa et al.* [1987] reported a 100 K melting point depression from the addition of 10% Ni to the Fe-O-S system at 6–15 GPa. *Rohrbach et al.* [2014] found an ~50 K melting point depression when 5% Ni is added to the Fe-C system at 10 GPa. New results on the melting of pure Ni [*Lord et al.*, 2014a] show that its melting curve is very similar to the melting curve of pure Fe reported by *Anzellini et al.* [2013], as opposed to being significantly lower as previously thought, so a deep melting point depression is not required.

1.5. APPLICATION TO THE THERMAL STRUCTURE OF THE CORE

Eutectic melting point depressions caused by different light elements at 100 GPa relative to the Fe melting curve of *Anzellini et al.* [2013] (Section 1.4) are summarized in Table 1.1. They can be used to extrapolate the melting curves of these Fe-light element alloys to the inner core boundary pressure by subtracting this depression from the melting curve of iron, which has been determined to higher pressures and thus requires less extrapolation. Extrapolation of these melting curves assumes that any subsolidus phase changes at higher pressures do not significantly affect the slopes of the melting curves, which is necessary given the pressure ranges of the data. *Anzellini et al.* [2013] report a melting temperature of pure iron of 6200 ± 500 K at 330 GPa based on an extrapolation of their data, which are compatible with the shock data of *Nguyen and Holmes* [2004] and *Brown and McQueen* [1986] and the ab initio study of *Alfè* [2009]. The uncertainties involved are likely larger than this,

Table 1.1 Melting point depressions at 100 GPa (relative to the Fe melting curve of *Anzellini et al.* [2013]) and estimated 330 GPa eutectic melting temperatures for various iron-light element binary systems.

System	Melting point depression (K)	Melting temperature at ICB (K)
Fe	0	6200 ± 500
Fe-Si	0–400	6000 ± 500
Fe-O	700–1100	5300 ± 500
Fe-S	900–1200	5150 ± 500
Fe-C	600–800	5500 ± 500

based on the disagreement between studies at lower pressures (Figure 1.1) and uncertainties inherent in this extrapolation. The melting temperatures at 330 GPa for different binary systems can be calculated by assuming that melting point depressions at 100 GPa can approximate those at 330 GPa. These results are listed in Table 1.1. Since the melting temperatures are compared to the same melting curve of Fe at both 100 and 330 GPa, the melting temperatures listed in Table 1.1 are approximately independent of the choice of Fe melting curve if the melting curves have similar slopes. However, the melting point depressions reported in Table 1.1 do depend on the choice of Fe melting curve.

These results can be compared to a simple extrapolation of the data obtained by fitting the Simon equation to all of the data shown in Figure 1.2 for each composition. However, these fits are very sensitive to the choice of the reference pressure and temperature, which is a large source of uncertainty. Upper and lower bounds were fit separately, resulting in a range of melting temperatures (except in the Fe-C system). This method predicts melting temperatures at 330 GPa of 5250–5600 K (Fe-Si), 4200–4650 K (Fe-O), 4500–5300 K (Fe-S), and 5100 K (Fe-C). These values are generally lower than those listed in Table 1.1 but compatible within uncertainty. The exception is the Fe-O system, in which the data from *Boehler* [1993] imply a significantly shallower slope than the melting curve of *Anzellini et al.* [2013] (though the more recent data of *Seagle et al.* [2008] do not), resulting in a lower melting temperature when extrapolated directly.

It is important to note that the melting temperatures at 330 GPa listed in Table 1.1 are eutectic temperatures. Therefore, they represent lower bounds on Earth's inner core boundary temperature, since the inner core is crystallizing from the liquid outer core along the liquidus, not at the eutectic temperature. Improved understanding of how the liquidus temperature evolves with pressure, temperature, and composition in these systems would produce a better estimate of the ICB temperature. However, the presence of significant quantities of multiple light elements in the core could decrease the melting temperature relative to those of binary systems.

In this discussion the eutectic melting temperatures of the binary Fe-light element alloys listed in Table 1.1 are taken as approximations of the ICB temperature.

These melting temperatures (Table 1.1) therefore represent points to which the core's geotherm may be anchored at 330 GPa for different compositions. The inner core is believed to be relatively isothermal, with a uniform temperature to within ~100 K [*Brown and Shankland*, 1981; *Pozzo et al.*, 2014]. The convecting outer core is mostly adiabatic [e.g., *Birch*, 1952]. Thermally stratified layers have been suggested based on recent calculations of high thermal conductivity of Fe [*Pozzo et al.*, 2012], though thermal conductivity is compositionally dependent [*Seagle et al.*, 2013]. In particular, a thermochemical boundary layer at the base of the outer core [e.g., *Gubbins et al.*, 2008] and a stratified layer at the top of the outer core [e.g., *Buffett*, 2014] have both been proposed, with the bulk of the outer core considered to be adiabatic. The magnitude of possible deviations from adiabaticity is poorly constrained.

Assuming that the temperature gradient in the outer core can be approximated as adiabatic, its temperature profile may be calculated from

$$\gamma(\rho) = \frac{\partial \ln T}{\partial \ln \rho},$$

where γ is the Grüneisen parameter, a function of density ρ. Different equations of state will therefore produce

core adiabats with different slopes. Equations of state of possible core components are reviewed in Chapter 10. Here a recent thermal equation of state of hcp iron [*Dewaele et al.*, 2006] is taken as representative of Fe-rich alloys for geotherm calculations. Using the thermal equation-of-state parameters for iron from *Anderson* [1998], for example, would change the temperature at 135 GPa by ~250 K. This calculation is primarily sensitive to the value of the Grüneisen parameter at core conditions and less sensitive to the Debye temperature and isothermal equation-of-state parameters, such as the bulk modulus, its pressure derivative, and the density at 1 bar.

Figure 1.3 illustrates some example core adiabats anchored to the 330 GPa melting points of each binary system listed in Table 1.1. Inner core boundary temperatures of 5150–6200 K translate into core-mantle boundary temperatures of 3850–4600 K along an adiabat. Extrapolating transition zone temperatures along an adiabat to the core-mantle boundary predicts a mantle temperature of 2500–2800 K approaching the CMB [*Lay et al.*, 2008], implying a temperature contrast across the thermal boundary layer of 1050–2100 K. Assuming a mantle thermal conductivity of 10 W/m·K in a 200-km-thick thermal boundary layer [*Lay et al.*, 2008], this temperature contrast leads to a heat flux q of 0.05–0.11 W/m², corresponding to a heat flow of 8–15 TW. Only the highest end of this range (corresponding to the highest core temperature) is compatible with recent calculations of

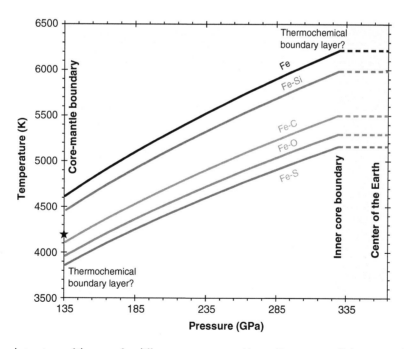

Figure 1.3 Thermal structure of the core for different core compositions. Outer core adiabats are calculated from the equation of state of Fe [*Dewaele et al.*, 2006] and anchored to 330 GPa melting points for pure Fe and Fe-Si, Fe-C, Fe-O, and Fe-S binary systems (Table 1.1). Small departures from adiabaticity may occur at the top and bottom of the outer core. The inner core is approximated as isothermal. Black star at 135 GPa is the mantle solidus [*Andrault et al.*, 2011; *Fiquet et al.*, 2010].

the core's thermal conductivity [e.g., *Pozzo et al.*, 2012], though experiments have shown that iron's thermal conductivity has a significant compositional dependence [*Seagle et al.*, 2013]. Based on estimates of the degree of partial melt at the base of the mantle, which range from zero to ~25% [e.g., *Rost et al.*, 2005], these high temperatures may be incompatible with measurements of mantle melting temperatures, though there are discrepancies between studies. Reported mantle solidus temperatures range from 3600 K [*Nomura et al.*, 2014] to 4200 K [*Andrault et al.*, 2011; *Fiquet et al.*, 2010] at 135 GPa (Figure 1.3), with liquidus temperatures ranging from 4700 K [*Andrault et al.*, 2011] to ~5400 K [*Fiquet et al.*, 2010].

An Fe-Si core implies the highest core temperatures, followed by Fe-C, then Fe-O, with an Fe-S core implying the lowest temperatures. The small difference between the melting curves of pure Fe and Fe-Si (Figure 1.2a) suggests a narrow phase loop between the solidus and liquidus in the Fe-Si system, so that an Fe-Si core would likely drive compositional convection less efficiently than an Fe-O or Fe-S core would. An Fe-Si core may therefore require a larger thermal convection contribution to maintain the geodynamo relative to an Fe-O or Fe-S core. The width of this phase loop between the solidus and liquidus must be compatible with the density contrast between the inner and outer core, offering another clue into the identity of the core's light element(s). The Fe-Si melting curve is steeper than those of Fe-O and Fe-S (Figure 1.2). This implies that for a given core cooling rate an Fe-Si core would crystallize more slowly than an Fe-O or Fe-S core would [*Fischer et al.*, 2013]. Therefore, a core with silicon as the dominant light element would have an older inner core than one rich in oxygen or sulfur, implying an earlier onset of compositional convection if silicon is the dominant light element.

1.6. CONCLUSIONS

Geotherms in the Earth's outer core can be calculated as an adiabat if the melting temperature of the core alloy at 330 GPa and its equation of state are known. The temperature in Earth's outer core likely ranges from 3850–4600 K at the CMB to 5150–6200 K at the ICB, but large uncertainties remain. Despite decades of research and its significance in geophysics, controversy still surrounds the melting temperature of pure iron at Earth's core conditions, though a consensus may be starting to form. However, the pressure evolution of eutectic temperatures in the Fe-Si, Fe-O, and Fe-S systems is better understood, largely due to the development of synchrotron X-ray diffraction techniques for in situ detection of melting. Silicon causes a small melting point depression in iron, while oxygen and sulfur cause larger melting point depressions, with implications for the

core's evolution. For example, an Fe-Si core would drive compositional convection less efficiently and crystallize more slowly than an Fe-O or Fe-S core. Knowledge of these melting curves, combined with information about densities, seismic velocities, solid-melt partitioning, geochemistry, and other independent constraints on properties of various light element candidates (as discussed in Parts II–V of this monograph), offers clues to the core's composition.

Future studies aiming to further our understanding of the temperature of the core should focus on how melting temperatures of Fe-rich systems vary with pressure and composition, especially in ternary systems, and on the Grüneisen parameters of (liquid) Fe-rich alloys at core conditions. It is especially important to focus on measurements of liquidus temperatures since the inner core is crystallizing along the liquidus. Melting measurements are needed at higher pressures, both to reduce uncertainties inherent in extrapolations and to identify the effects of subsolidus phase changes on melting.

ACKNOWLEDGMENTS

I thank Andrew J. Campbell for helpful discussions, editor Hidenori Terasaki for handling this manuscript, and two anonymous reviewers for constructive comments. This work was supported by an American Dissertation Fellowship from the American Association for University Women to R.A.F. and by the National Science Foundation grant EAR-1427123 to A.J.C.

REFERENCES

Ahrens, T. J., K. G. Holland, and G. Q. Chen (2002), Phase diagram of iron, revised-core temperatures, *Geophys. Res. Lett.*, *29*, 1150, doi:10.1029/2001GL014350.

Alfè, D. (2009), Temperature of the inner-core boundary of the Earth: Melting of iron at high pressure from first-principles coexistence simulations, *Phys. Rev. B*, *79*, 060101, doi:10.1103/PhysRevB.79.060101.

Alfè, D., G. D. Price, and M. J. Gillan (2002), Iron under Earth's core conditions: Liquid-state thermodynamics and high-pressure melting curve from *ab initio* calculations, *Phys. Rev. B*, *65*, 165 118, doi:10.1103/PhysRevB.65.165118.

Anderson, O. L. (1998), The Grüneisen parameter for iron at outer core conditions and the resulting conductive heat and power in the core, *Phys. Earth Planet. Inter.*, *109*, 179–197, doi:10.1016/S0031-9201(98)00123-X.

Anderson, O. L., D. G. Isaak, and V. E. Nelson (2003), The high-pressure melting temperature of hexagonal close-packed iron determined from thermal physics, *J. Phys. Chem. Solids*, *64*, 2125–2131, doi:10.1016/S0022-3697(03)00112-4.

Andrault, D., N. Bolfan-Casanova, G. Lo Nigro, M. A. Bouhifd, G. Garbarino, and M. Mezouar (2011), Solidus and liquidus profiles of chondritic mantle: Implication for melting of the Earth across its history, *Earth Planet. Sci. Lett.*, *304*, 251–259, doi:10.1016/j.epsl.2011.02.006.

Anzellini, S., A. Dewaele, M. Mezouar, P. Loubeyre, and G. Morard (2013), Melting of iron at Earth's inner core boundary based on fast X-ray diffraction, *Science*, *340*, 464–466, doi:10.1126/science.1233514.

Asanuma, H., E. Ohtani, T. Sakai, H. Terasaki, S. Kamada, T. Kondo, and T. Kikegawa (2010), Melting of iron–silicon alloy up to the core–mantle boundary pressure: Implications to the thermal structure of Earth's core, *Phys. Chem. Minerals*, *37*, 353–359, doi:10.1007/s00269-009-0338-7.

Belonoshko, A. B., R. Ahuja, and B. Johansson (2000), Quasi-*ab initio* molecular dynamic study of Fe melting, *Phys. Rev. Lett.*, *84*, 3638–3641, doi:10.1103/PhysRevLett.84.3638.

Birch, F. (1952), Elasticity and constitution of the Earth's interior, *J. Geophys. Res.*, *57*, 227–286, doi:10.1029/JZ057i002p00227.

Boehler, R. (1993), Temperatures in the Earth's core from melting-point measurements of iron at high static pressures, *Nature*, *363*, 534–536, doi:10.1038/363534a0.

Brown, J. M., and R. G. McQueen (1986), Phase transitions, Grüneisen parameter, and elasticity for shocked iron between 77 GPa and 400 GPa, *J. Geophys. Res.*, *91*, 7485–7494, doi:10.1029/JB091iB07p07485.

Brown, J. M., and T. J. Shankland (1981), Thermodynamic parameters in the Earth as determined from seismic profiles, *Geophys. J. R. Astron. Soc.*, *66*, 579–596, doi:10.1111/j.1365-246X.1981.tb04891.x.

Buffett, B. (2014), Geomagnetic fluctuations reveal stable stratification at the top of the Earth's core, *Nature*, *507*, 484–487, doi:10.1038/nature13122.

Campbell, A. J. (2008), Measurement of temperature distributions across laser heated samples by multispectral imaging radiometry, *Rev. Sci. Instrum.*, *79*, 015108, doi:10.1063/1.2827513.

Campbell, A. J., C. T. Seagle, D. L. Heinz, G. Shen, and V. B. Prakapenka (2007), Partial melting in the iron–sulfur system at high pressure: A synchrotron X-ray diffraction study, *Phys. Earth Planet. Inter.*, *162*, 119–128, doi:10.1016/j.pepi.2007.04.001.

Chudinovskikh, L., and R. Boehler (2007), Eutectic melting in the system Fe-S to 44 GPa, *Earth Planet. Sci. Lett.*, *257*, 97–103, doi:10.1016/j.epsl.2007.02.024.

Dewaele, A., P. Loubeyre, F. Occelli, M. Mezouar, P. I. Dorogokupets, and M. Torrent (2006), Quasihydrostatic equation of state of iron above 2 Mbar, *Phys. Rev. Lett.*, *97*, 215504, doi:10.1103/PhysRevLett.97.215504.

Fei, Y., and E. Brosh (2014), Experimental study and thermodynamic calculations of phase relations in the Fe–C system at high pressure, *Earth Planet. Sci. Lett.*, *408*, 155–162, doi:10.1016/j.epsl.2014.09.044.

Fei, Y., J. Li, C. M. Bertka, and C. T. Prewitt (2000), Structure type and bulk modulus of Fe_3S, a new iron-sulfur compound, *Am. Mineral.*, *85*, 1830–1833.

Fiquet, G., A. L. Auzende, J. Siebert, A. Corgne, H. Bureau, H. Ozawa, and G. Garbarino (2010), Melting of peridotite to 140 gigapascals, *Science*, *329*, 1516–1518, doi:10.1126/science.1192448.

Fischer, R. A., and A. J. Campbell (2010), High-pressure melting of wüstite, *Am. Mineral.*, *95*, 1473–1477, doi:10.2138/am.2010.3463.

Fischer, R. A., A. J. Campbell, R. Caracas, D. M. Reaman, P. Dera, and V. B. Prakapenka (2012), Equation of state and phase diagram of Fe-16Si alloy as a candidate component of Earth's core, *Earth Planet. Sci. Lett.*, *357–358*, 268–276, doi:10.1016/j.epsl.2012.09.022.

Fischer, R. A., A. J. Campbell, D. M. Reaman, N. A. Miller, D. L. Heinz, P. Dera, and V. B. Prakapenka (2013), Phase relations in the Fe–FeSi system at high pressures and temperatures, *Earth Planet. Sci. Lett.*, *373*, 54–64, doi:10.1016/j.epsl.2013.04.035.

Gubbins, D., G. Masters, and F. Nimmo (2008), A thermochemical boundary layer at the base of Earth's outer core and independent estimate of core heat flux, *Geophys. J. Int.*, *174*, 1007–1018, doi:10.1111/j.1365-246X.2008.03879.x.

Huang, H., X. Hu, F. Jing, L. Cai, Q. Shen, Z. Gong, and H. Liu (2010), Melting behavior of Fe–O–S at high pressure: A discussion on the melting depression induced by O and S, *J. Geophys. Res.*, *115*, B05207, doi:10.1019/2009JB006514.

Jackson, J. M., W. Sturhahn, M. Lerche, J. Zhao, T. S. Toellner, E. E. Alp, S. V. Sinogeikin, J. D. Bass, C. A. Murphy, and J. K. Wicks (2013), Melting of compressed iron by monitoring atomic dynamics, *Earth Planet. Sci. Lett.*, *362*, 143–150, doi:10.1016/j.epsl.2012.11.048.

Kamada, S., H. Terasaki, E. Ohtani, T. Sakai, T. Kikegawa, Y. Ohishi, N. Hirao, N. Sata, and T. Kondo (2010), Phase relationships of the Fe–FeS system in conditions up to the Earth's outer core, *Earth Planet. Sci. Lett.*, *294*, 94–100, doi:10.1016/j.epsl.2010.03.011.

Kamada, S., E. Ohtani, H. Terasaki, T. Sakai, M. Miyahara, Y. Ohishi, and N. Hirao (2012), Melting relationships in the $Fe–Fe_3S$ system up to the outer core conditions, *Earth Planet. Sci. Lett.*, *359–360*, 26–33, doi:10.1016/j.epsl.2012.09.038.

Komabayashi, T. (2014), Thermodynamics of melting relations in the system Fe–FeO at high pressure: Implications for oxygen in the Earth's core, *J. Geophys. Res.*, *119*, 4164–4177, doi:10.1002/2014JB010980.

Laio, A., S. Bernard, G. L. Chiarotti, S. Scandolo, and E. Tosatti (2000), Physics of iron at Earth's core conditions, *Science*, *287*, 1027–1030, doi:10.1126/science.287.5455.1027.

Lay, T., J. Hernlund, and B. A. Buffett (2008), Core–mantle boundary heat flow, *Nature Geosci.*, *1*, 25–32, doi:10.1038/ngeo.2007.44.

Lister, J. R., and B. A. Buffett (1995), The strength and efficiency of thermal and compositional convection in the geodynamo, *Phys. Earth Planet. Inter.*, *91*, 17–30, doi:10.1016/0031-9201(95)03042-U.

Lord, O. T., M. J. Walter, R. Dasgupta, D. Walker, and S. M. Clark (2009), Melting in the Fe–C system to 70 GPa, *Earth Planet. Sci. Lett.*, *284*, 157–167, doi:10.1016/j.epsl.2009.04.017.

Lord, O. T., I. G. Wood, D. P. Dobson, L. Vočadlo, W. Wang, A. R. Thomson, E. T. H. Wann, G. Morard, M. Mezouar, and M. J. Walter (2014a), The melting curve of Ni to 1 Mbar, *Earth Planet. Sci. Lett.*, *408*, 226–236, doi:10.1016/j.epsl.2014.09.046.

Lord, O. T., E. T. H. Wann, S. A. Hunt, A. M. Walker, J. Santangeli, M. J. Walter, D. P. Dobson, I. G. Wood, L. Vočadlo, G. Morard, and M. Mezouar (2014b), The NiSi melting curve to 70 GPa, *Phys. Earth Planet. Inter.*, *233*, 13–23, doi:10.1016/j.pepi.2014.05.005.

Ma, Y., M. Somayazulu, G. Shen, H.-k. Mao, J. Shu, and R. J. Hemley (2004), In situ X-ray diffraction studies of iron to Earth-core conditions, *Phys. Earth Planet. Inter.*, *143–144*, 455–467, doi:10.1016/j.pepi.2003.06.005.

Morard, G., D. Andrault, N. Guignot, C. Sanloup, M. Mezouar, S. Petitgirard, and G. Fiquet (2008), *In situ* determination of Fe–Fe$_3$S phase diagram and liquid structural properties up to 65 GPa, *Earth Planet. Sci. Lett.*, *272*, 620–626, doi:10.1016/j.epsl.2008.05.028.

Morard, G., D. Andrault, N. Guignot, J. Siebert, G. Garbarino, and D. Antonangeli (2011), Melting of Fe–Ni–Si and Fe–Ni–S alloys at megabar pressures: Implications for the core–mantle boundary temperature, *Phys. Chem. Minerals*, *38*, 767–776, doi:10.1007/s00269-011-0449-9.

Nguyen, J. H., and N. C. Holmes (2004), Melting of iron at the physical conditions of the Earth's core, *Nature*, *427*, 339–342, doi:10.1038/nature02248.

Nomura, R., K. Hirose, K. Uesugi, Y. Ohishi, A. Tsuchiyama, A. Miyake, and Y. Ueno (2014), Low core-mantle boundary temperature inferred from the solidus of pyrolite, *Science*, *343*, 522–525, doi:10.1126/science.1248186.

Ozawa, H., F. Takahashi, K. Hirose, Y. Ohishi, and N. Hirao (2011), Phase transition of FeO and stratification in Earth's outer core, *Science*, *334*, 792–794, doi:10.1126/science.1208265.

Ozawa, H., K. Hirose, T. Suzuki, Y. Ohishi, and N. Hirao (2013), Decomposition of Fe$_3$S above 250 GPa, *Geophys. Res. Lett.*, *40*, 4845–4849, doi:10.1002/grl.50946.

Poirier, J.-P. (1994), Light elements in the Earth's outer core: A critical review, *Phys. Earth Planet. Inter.*, *85*, 319–337, doi:10.1016/0031-9201(94)90120-1.

Pozzo, M., C. Davies, D. Gubbins, and D. Alfè (2012), Thermal and electrical conductivity of iron at Earth's core conditions, *Nature*, *485*, 355–358, doi:10.1038/nature11031.

Pozzo, M., C. Davies, D. Gubbins, and D. Alfè (2014), Thermal and electrical conductivity of solid iron and iron–silicon mixtures at Earth's core conditions, *Earth Planet. Sci. Lett.*, *393*, 159–164, doi:10.1016/j.epsl.2014.02.047.

Rohrbach, A., S. Ghosh, M. W. Schmidt, C. H. Wijbrans, and S. Klemme (2014), The stability of Fe–Ni carbides in the Earth's mantle: Evidence for a low Fe–Ni–C melt fraction in the deep mantle, *Earth Planet. Sci. Lett.*, *388*, 211–221, doi:10.1016/j.epsl.2013.12.007.

Rost, S., E. J. Garnero, Q. Williams, and M. Manga (2005), Seismological constraints on a possible plume root at the core–mantle boundary, *Nature*, *435*, 666–669, doi:10.1038/nature03620.

Sakamaki, K., E. Takahashi, Y. Nakajima, Y. Nishihara, K. Funakoshi, T. Suzuki, and Y. Fukai (2009), Melting phase relation of FeHx up to 20 GPa: Implication for the temperature of the Earth's core, *Phys. Earth Planet. Inter.*, *174*, 192–201, doi:10.1016/j.pepi.2008.05.017.

Salamat, A., R. A. Fischer, R. Briggs, M. I. McMahon, and S. Petitgirard (2014), *In situ* synchrotron X-ray diffraction in the laser-heated diamond anvil cell: Melting phenomena and synthesis of new materials, Coord. *Chem. Rev.*, *277–278*, 15–30, doi:10.1016/j.ccr.2014.01.034.

Saxena, S. K., G. Shen, and P. Lazor (1994), Temperatures in Earth's core based on melting and phase transformation experiments on iron, *Science*, *264*, 405–407, doi:10.1126/science.264.5157.405.

Seagle, C. T., D. L. Heinz, A. J. Campbell, V. B. Prakapenka, and S. T. Wanless (2008), Melting and thermal expansion in the Fe–FeO system at high pressure, *Earth Planet. Sci. Lett.*, *265*, 655–665, doi:10.1016/j.epsl.2007.11.004.

Seagle, C. T., E. Cottrell, Y. Fei, D. R. Hummer, and V. B. Prakapenka (2013), Electrical and thermal transport properties of iron and iron–silicon alloy at high pressure, *Geophys. Res. Lett.*, *40*, 5377–5381, doi:10.1002/2013GL057930.

Shen, G., V. B. Prakapenka, M. L. Rivers, and S. R. Sutton (2004), Structure of liquid iron at pressures up to 58 GPa, *Phys. Rev. Lett.*, *92*, 185701, doi:10.1103/PhysRevLett.92.185701.

Sola, E., and D. Alfè (2009), Melting of iron under Earth's core conditions from diffusion Monte Carlo free energy calculations, *Phys. Rev. Lett.*, *103*, 078501, doi:10.1103/PhysRevLett.103.078501.

Stewart, A. J., M. W. Schmidt, W. van Westrenen, and C. Liebske (2007), Mars: A new core-crystallization regime, *Science*, *316*, 1323–1325, doi:10.1126/science.1140549.

Terasaki, H., S. Kamada, T. Sakai, E. Ohtani, N. Hirao, and Y. Ohishi (2011), Liquidus and solidus temperatures of a Fe–O–S alloy up to the pressures of the outer core: Implication for the thermal structure of the Earth's core, *Earth Planet. Sci. Lett.*, *304*, 559–564, doi:10.1016/j.epsl.2011.02.041.

Urakawa, S., M. Kato, and M. Kumazawa (1987), Experimental study on the phase relations in the system Fe–Ni–O–S up to 15 GPa, in *High-Pressure Research in Mineral Physics*, edited by M. H. Manghnani and Y. Syono, pp. 95–111, Terra Scientific Pub., Tokyo, and AGU, Washington, D.C.

Williams, Q., R. Jeanloz, J. Bass, B. Svedsen, and T. J. Ahrens (1987), The melting curve of iron to 250 gigapascals: A constraint on the temperature at Earth's center, *Science*, *236*, 181–182, doi:10.1126/science.236.4798.181.

Yamazaki, D., E. Ito, T. Yoshino, A. Yoneda, X. Guo, B. Zhang, W. Sun, A. Shimojuku, N. Tsujino, T. Kunimoto, Y. Higo, and K.-i. Funakoshi (2012), P-V-T equation of state for ε-iron up to 80 GPa and 1900 K using the Kawai-type high pressure apparatus equipped with sintered diamond anvils, *Geophys. Res. Lett.*, *39*, L20308, doi:10.1029/2012GL053540.

Yoo, C. S., N. C. Holmes, M. Ross, D. J. Webb, and C. Pike (1993), Shock temperatures and melting of iron at Earth core conditions, *Phys. Rev. Lett.*, *70*, 3931–3934, doi:10.1103/PhysRevLett.70.3931.

Zhang, L., and Y. Fei (2008), Effect of Ni on Fe–FeS phase relations at high pressure and high temperature, *Earth Planet. Sci. Lett.*, *268*, 212–218, doi:10.1016/j.epsl.2008.01.028.

Zhang, Y., T. Sekine, H. He, Y. Yu, F. Liu, and M. Zhang (2014), Shock compression on Fe–Ni–Si system to 280 GPa: Implications for the composition of the Earth's outer core, *Geophys. Res. Lett.*, *41*, 4554–4559, doi:10.1002/2014GL060670.

2

Temperature of the Lower Mantle and Core Based on Ab Initio Mineral Physics Data

Taku Tsuchiya,[1,2] Kenji Kawai,[3] Xianlong Wang,[1,2] Hiroki Ichikawa,[1,2] and Haruhiko Dekura[1]

ABSTRACT

The thermal structure of Earth's interior plays a crucial role in controlling the dynamics and evolution of our planet. We have developed comprehensive models of geotherm from the lower mantle to the outer core and thermal conductivity of the lower mantle using the high-pressure and high-temperature physical properties of major constituent minerals calculated based on the ab inito computation techniques. Based on the models, the thermal property of the core-mantle boundary (CMB) region, including CMB heat flow is discussed.

2.1. INTRODUCTION

Temperature is one of the most important physical quantities controlling the thermal and dynamical evolution of planets. Many studies have considered Earth's thermal structure [*Poirier*, 2000], where seismic discontinuities in the mantle and core were used to infer the temperature (T) conditions at some fixed pressures (P) on Earth, as they can be associated with major phase transitions in the component minerals, including the olivine-spinel and post-spinel transitions in $(Mg, Fe)_2SiO_4$, perovskite (Pv) to post-perovskite (PPv) transition in $(Mg, Fe)SiO_3$, and melting transition in Fe alloys [*Alfè et al.*, 2004; *Boehler*, 1992; *Katsura et al.*, 2010; *Murakami et al.*, 2004; *Tsuchiya et al.*, 2004b]. Conditions are then adiabatically extrapolated from those fixed points

[*Anderson*, 1982; *Brown and Shanlkand*, 1981; *Katsura et al.*, 2010], which suggests that the temperature near the core-mantle boundary (CMB) is strongly depth dependent and forms a thermal boundary layer (TBL). The CMB heat flow, which is determined by the temperature gradient and conductivity in the TBL [*Turcotte and Schubert*, 2001], controls the deep Earth dynamics, such as the lower mantle basal heating, core cooling, and dynamo activity. Although empirical assumptions or large extrapolations long needed to be applied to estimate the thermal states of the deep Earth, recent advancements of the state-of-the-art density functional ab initio computation techniques now allow us to directly access the physical properties of mantle and core materials with complicated compositions under the lower mantle (LM) and core P, T conditions, including thermodynamics [*Metsue and Tsuchiya*, 2012; *Tsuchiya and Wang*, 2013], elasticity [*Badro et al.*, 2014; *Ichikawa et al.*, 2014; *Wang et al.*, 2015], and thermal conductivity [*Dekura et al.*, 2013]. Those cover not only high-P, T thermodynamics but also elasticity and transport property, leading to more quantitative modeling of deep Earth thermal states, which will be briefly summarized in this chapter.

[1]Geodynamics Research Center, Ehime University, Matsuyama, Ehime Prefecture, Japan

[2]Earth-Life Science Institute, Tokyo Institute of Technology, Meguro, Tokyo, Japan

[3]Department of Earth Science and Astronomy, Graduate School of Arts and Sciences, University of Tokyo, Meguro, Japan

Deep Earth: Physics and Chemistry of the Lower Mantle and Core, Geophysical Monograph 217, First Edition.
Edited by Hidenori Terasaki and Rebecca A. Fischer.

2.2. AB INITIO METHODS FOR MINERAL PHYSICS

2.2.1. Density Functional Theory

Ab initio approaches are theoretical techniques that solve the fundamental equations of quantum mechanics with a bare minimum of approximations. The density functional theory (DFT) is, in principle, an exact theory for the ground state and allows us to reduce the interacting many-electron problem to a single-electron problem (the nuclei being treated as an adiabatic background) [*Hohenberg and Kohn*, 1964; *Kohn and Sham*, 1965]. Using the variational principle, the one-electron Schrödinger equation, known as the Kohn-Sham equation, can be derived from the all-electron Schrödinger equation. The local density approximation (LDA) [*Kohn and Sham*, 1965; *Ceperley and Alder*, 1980; *Perdew and Zunger*, 1981] replaces the exchange correlation potential, which contains all the quantum many-body effects, at each point by that of a homogeneous electron gas with a density equal to the local density at the point. The LDA works remarkably well for a wide variety of materials, especially in the calculations of the equation of state (EoS), elastic constants, and thermodynamics of silicates. Cell parameters and bulk moduli obtained from well-converged calculations often agree with the experimental data within a few percent.

The generalized gradient approximation (GGA [*Perdew et al.*, 1992; 1996]) was reported to substantially improve the LDA for certain transition metals [*Bagno et al.*, 1989] and hydrogen-bonded systems [*Hamann*, 1997; *Tsuchiya et al.*, 2002, 2005a]. There is however some evidence that GGA underbinds the chemical bonding of silicates and oxides, leading to overestimating the volumes and underestimating the elasticity compared to experimental measures [*Hamann*, 1996; *Demuth et al.*, 1999]. If one includes the thermal effect with zero-point motion, the LDA provides more reasonable structural and elastic quantities (typically within a few percent). On the other hand, a discrepancy of about 10–15 GPa is usually seen in transition pressures calculated with the LDA and GGA [e.g., *Hamann*, 1996; *Tsuchiya et al.*, 2004b; *Dekura et al.*, 2011], which provide lower and upper bounds, respectively. Experimental transition pressures usually fall between the values obtained within the LDA and GGA, or closer to the GGA values [e.g., *Hamann*, 1996; *Tsuchiya et al.*, 2004a,b].

Apart from these arguments, the LDA and GGA both are known to fail to describe strongly correlated iron oxide bonding correctly. Both methods cannot reproduce the insulating ground states of FeO due to crystal field splitting of Fe d states, for example. The internally consistent LSDA+U method, where the Hubbard correction (U) is applied to describe the screened onsite Coulomb interaction between the d electrons more accurately, is a practical one to overcome this issue [*Cococcioni and de Gironcoli*, 2005; *Tsuchiya et al.*, 2006a,b] and allows to simulate the structure and elasticity of iron-bearing silicates within the same reproducibility level of the LDA for iron-free systems. Within the internally consistent scheme, the correction parameter U can be determined non empirically based on the variational principle [*Cococcioni and de Gironcoli*, 2005], depending on chemical composition, spin state, and oxidation state. Calculations reported in this study were therefore conducted based on the LDA, internally consistent LSDA+U, and GGA for iron-free and iron-bearing silicates and metallic iron, respectively.

2.2.2. Phonon and Crystal Thermodynamics

The potential of nuclei is treated as a part of the static external field. Compared to electrons, nuclei are much heavier, so that electrons can usually be relaxed to a ground state within nuclei dynamics [*Born and Oppenheimer*, 1927]. When quantum effects are negligible for the nuclei dynamics, the classical Newton equation of motion (EoM) can be applicable to simulate the time evolution of a many-atom system on the Born-Oppenheimer energy surface. This technique is referred to as the ab initio molecular dynamics (AIMD) method. If using the plane wave basis set, which becomes utilizable by means of the pseudopotential, forces acting on nuclei can be efficiently and accurately calculated based on the Hellmann-Feynman theorem [*Feynman*, 1939] from the total energy. Constant-temperature or constant-pressure conditions can be achieved by the extended Lagrangian methods [*Hoover*, 1985; *Nose*, 1984; *Parrinello and Rahman*, 1980; *Rahman and Parrinello*, 1981] or the velocity (cell) scaling algorithms. The AIMD technique is also used for structure relaxation combined with the damped dynamics algorithm.

The Born-Oppenheimer energy (often simply called the total energy), in which the nuclei are moving, can be expanded in a Taylor series around the equilibrium geometry in three dimensions, leading to

$$
\begin{aligned}
&E_{\mathrm{tot}}\left(\mathbf{R}_1, \mathbf{R}_2, \cdots, \mathbf{R}_N\right) \\
&= E_{\mathrm{tot}}\left(\{\mathbf{R}_i^0\}\right) + \sum_i^N \sum_\alpha^3 \left[\frac{\partial}{\partial R_{i,\alpha}} E_{\mathrm{tot}}\left(\{\mathbf{R}_i\}\right)\right]_{\{\mathbf{R}_i^0\}} dR_{i,\alpha} \\
&\quad + \frac{1}{2} \sum_{i,j}^N \sum_{\alpha,\beta}^3 \left[\frac{\partial^2}{\partial R_{i,\alpha}\, \partial R_{j,\beta}} E_{\mathrm{tot}}\left(\{\mathbf{R}_i\}\right)\right]_{\{\mathbf{R}_i^0\}} dR_{i,\alpha}\, dR_{j,\beta} \\
&\quad + \frac{1}{3!} \sum_{i,j,k}^N \sum_{\alpha,\beta,\gamma}^3 \left[\frac{\partial^3}{\partial R_{i,\alpha}\, \partial R_{j,\beta}\, \partial R_{k,\gamma}} E_{\mathrm{tot}}\left(\{\mathbf{R}_i\}\right)\right]_{\{\mathbf{R}_i^0\}} dR_{i,\alpha}\, dR_{j,\beta}\, dR_{k,\gamma}
\end{aligned}
$$

$$(2.1)$$

with the displacement $d\mathbf{R} = \mathbf{R} - \mathbf{R}_0$. The calculation of the equilibrium geometry and the harmonic-level vibrational (phonon) properties of a system thus amounts to computing the first and second derivatives of its total energy surface. The equilibrium geometry of the system is given by the condition that the first term of the right-hand side of Equation (2.1), which corresponds to the forces acting on individual ions, vanishes. On the other hand, the second-order term, the Hessian of the total energy, is usually called the harmonic force constant matrix. The vibrational frequencies ω are determined by the eigenvalues of this matrix, scaled by the nuclear masses as

$$\det\left|\frac{1}{\sqrt{M_i M_j}}\frac{\partial^2 E_{\text{tot}}\left(\{\mathbf{R}_i\}\right)}{\partial \mathbf{R}_i\, \partial \mathbf{R}_j} - \omega^2 \delta_{ij}\right| = 0 \qquad (2.2)$$

within the harmonic lattice dynamics (LD) theory [*Srivastava*, 1990]. An efficient procedure to obtain the linear response of the electron charge density to a distortion of the nuclear geometry within DFT is the density functional perturbation theory (DFPT [*Zein*, 1984; *Baroni et al.*, 1987; *Gonze*, 1995]), which has been successfully applied to investigate the phonon property of complicated silicate minerals [e.g., *Karki et al.*, 2000; *Tsuchiya et al.*, 2005b; *Yu et al.*, 2007]. However, the efficiency of DFPT decreases with increasing system size, so that this technique is less useful when calculating supercells required to study effects of impurity doping. Calculations of the thermodynamic properties of iron-bearing $MgSiO_3$ Pv (PPv) and MgO periclase were therefore investigated using the direct (supercell) method combined with the internally consistent LSDA+U formalism [*Fukui et al.*, 2012; *Metsue and Tsuchiya*, 2011, 2012; *Tsuchiya and Wang*, 2013].

Sufficiently below the melting point, the phonon energy U_{ph} and phonon free energy F_{ph} can be conveniently calculated within the quasi-harmonic approximation (QHA) [*Wallace*, 1972]. This consists in calculating U_{ph} and F_{ph} in the harmonic approximation, retaining only the implicit volume dependence through the frequencies as

$$U_{\text{ph}}(T,V) = \sum_{i,k}\hbar\omega_{ik}(V)\left\{\frac{1}{2} + \frac{1}{\exp\left[\dfrac{\hbar\omega_{ik}(V)}{k_b T}\right] - 1}\right\} \qquad (2.3)$$

and

$$F_{\text{ph}}(T,V) = -k_B T \sum_{i,k}\ln\left\{2\sinh\left[\frac{\hbar\omega_{ik}(V)}{2k_B T}\right]\right\}, \qquad (2.3')$$

respectively. Once internal energy and free energy are available, other thermodynamic properties are readily derivable based on the standard thermodynamic definitions. For example, vibrational entropy is derived simply from the fundamental relationship

$$S_{\text{ph}}(T,V) = \frac{U_{\text{ph}}(T,V) - F_{\text{ph}}(T,V)}{T} \qquad (2.4)$$

or from the more statistical physical representation

$$S_{\text{ph}}(T,V) = k_B \sum_{i,k}\left\{-\ln\left[1 - \exp\left[-\frac{\hbar\omega_{ik}(V)}{k_B T}\right]\right] + \frac{1}{T}\frac{\hbar\omega_{ik}(V)}{\exp\left[\dfrac{\hbar\omega_{ik}(V)}{k_B T}\right] - 1}\right\}. \qquad (2.4')$$

Although the QHA accounts only partially for the effects of anharmonicity through the volume dependence of the phonon spectra, it is widely accepted that it works very well at the high pressures and high temperatures corresponding to the lower mantle condition. The properties that can be calculated within the QHA include finite-temperature equation-of-state and elastic constants, specific heat, and thermal expansion coefficients [e.g., *Tsuchiya*, 2003; *Tsuchiya et al.*, 2005b; *Wentzcovitch et al.*, 2004; 2006]. A free energy balance between different phases yields their thermodynamic stability as a function of pressure and temperature [e.g., *Tsuchiya et al.*, 2004a,b].

2.2.3. EoS from Molecular Dynamics

Finite-temperature thermodynamics, including the P-V-T EoS of liquids and anharmonic solids, are dealt with by the AIMD technique, not by the LD. It is usually carried out within the canonical (NVT) ensemble, where Newton's EoM is numerically integrated. The sufficiently large size of the calculation cell and long-time integration for the EoM have to be taken into account to obtain convergence in statistical samplings. Calculated results at discrete P,T are often fitted using an EoS model based on the Mie-Grüneisen equation

$$P(V,T) = P_{T0}(V) + \Delta P_{\text{th}}(V,T), \qquad (2.5)$$

where total pressure $P(V, T)$ is expressed by the sum of the pressure at a reference temperature (the first term) and thermal pressure (the second). For the isothermal part at a reference temperature T_0, model EoS functions such as the Vinet (Morse-Rydberg) equation [e.g., *Stacey and Davis*, 2004] or the third-order Birch-Murnaghan equation are typically used. Here K_{T0} and K'_{T0} are the isothermal bulk modulus and its pressure derivative at zero pressure at T_0. Since the Newton EoM is for the classical dynamics, crystal dynamics with quantum effects remarkable below the Debye temperature cannot be represented by the Molecular dynamics (MD) method, unlike the LD. Nevertheless, the quasi-harmonic Debye model has been widely used to represent the thermal pressure of oxide and silicate solids [*Kawai and Tsuchiya*, 2014], where the Debye function is used to represent the phonon energy (atomic contribution), which approaches $3nRT$ at $T >> \Theta_D$ (the Debye temperature) (the Dulong-Petit limit). The QHA is however clearly inadequate for metallic liquids. For metals, the electronic excitation effect is also not negligible, particularly in high-temperature conditions corresponding, for example, to Earth's core [e.g., *Boness et al.*, 1986; *Tsuchiya and Kawamura*, 2002]. A new function for the internal thermal energy suite for metallic liquids in a broad P, T range has therefore been proposed [*Ichikawa et al.*, 2014], which is simply represented by a second-order polynomial of temperature with a volume-dependent second-order coefficient. Combining the calculated internal thermal energy and thermal pressure, the total Grüneisen parameter (γ) is parameterized from the relation,

$$\Delta P_{th}(V,T) = \frac{\gamma(V)}{V} \Delta E_{th}(V,T). \quad (2.6)$$

In this analysis, γ is assumed to depend only on volume and its temperature dependence is still neglected. This is often acceptable. In that case, one can employ a practical functional form of γ to express the volume dependence [*Tange et al.*, 2009]. The P-V-T EoS model constructed in this procedure is widely applicable from insulator solids to metallic liquids. The EoS parameters are determined by least squares analyses on the E-P-V-T data set obtained from the AIMD calculations. To obtain physically reasonable regression, the parameters related to γ should be optimized separately from the other EoS parameters.

Within the MD approach, it is generally difficult to determine entropy (and then free energy), compared to internal energy, because of bypassing the calculation of number of states. However, its variation can be estimated easily using a standard thermodynamic equation as

$$dS(T,V) = \frac{C_V(T,V)}{T} dT + \left(\frac{\partial P}{\partial T}\right)_V dV. \quad (2.7)$$

Isentropes can be determined using this relation from any P, T once the thermal EoS is modeled.

2.2.4. Thermoelasticity

The elastic constant tensor for a single crystal can be calculated with high precision by a method described briefly below. Combined with the ab initio plane wave pseudopotential method, pioneering works by *Karki et al.* [1997a,b,c] applied the technique to determine the elasticity of major mantle minerals of MgO, SiO_2, $MgSiO_3$ Pv, and $CaSiO_3$ Pv as a function of pressure. Although some studies performed the modeling of the seismic velocity structure extrapolating experimental results of elastic properties at ambient temperature [e.g., *Cammarano et al.*, 2005], the reliability and uniqueness of the extrapolations of thermoelasticity are always heavily controversial. Direct comparison between the ab initio and seismologically inferred elastic properties of Earth's interior therefore now spawns a new line of geophysical research.

Simulation proceeds as follows: (1) at a given pressure (or volume) the crystal structure is first fully optimized using the damped MD or similar algorithms; (2) the lattice is slightly deformed by applying a small strain (ε) and the values of the elastic constants. In the determination of the elastic constants by means of the DFT, the ionic positions are first reoptimized in the strained lattice in order to incorporate any couplings between strains (ε) and vibrational modes in the crystal [*Nastar and Willaime*, 1995], and the stress (σ) in the strained configuration is calculated. The elastic constants are computed in the linear regime, which is appropriate for the geophysical condition, following the stress-strain relation

$$\sigma_{ij} = c_{ijkl}\varepsilon_{kl}. \quad (2.8)$$

After obtaining the single-crystal elastic constant tensor, polycrystalline isotropic elasticity, adiabatic bulk (K), and shear (μ) moduli can be calculated based on relevant averaging schemes. For the seismic wave frequency range, K should be the adiabatic value K_S, which is the same as the isothermal one, K_T, at the static temperature (0 K) but becomes larger with increasing temperature. Although the most appropriate method for averaging is still unclear for the geophysical condition, the most often applied one is the Voigt-Reess-Hill average [*Hill*, 1952] or Hashin-Shtrikman average [*Hashin and Shtrikman*, 1962]. The isotropic averaged compressional (**P**), shear (**S**), and bulk (**Φ**) wave velocities can then be calculated from the equations:

$$V_P = \sqrt{\frac{K_S + 4/3\mu}{\rho}}, \quad V_S = \sqrt{\frac{\mu}{\rho}}, \quad V_\Phi = \sqrt{\frac{K_S}{\rho}}, \quad (2.9)$$

respectively.

Computations of finite-temperature elastic constants are considerably heavier than that of static elasticity. There are two different techniques to include thermal contributions: one is the LD method and another is the MD method. In the former approach, one first calculates finite-temperature free energy for strained conditions by applying the QHA, determines isothermal stiffness $c_{ij}^T(P,T)$ based on the energy-strain relation (not on the stress-strain) as

$$c_{ij}^T(P,T) = \frac{1}{V}\left(\frac{\partial^2 G(P,T)}{\partial \varepsilon_i\, \partial \varepsilon_i}\right)_T, \qquad (2.10)$$

where G is the Gibbs free energy, and then converts them to adiabatic quantities as

$$c_{ij}^S(P,T) = c_{ij}^T(P,T) + \frac{VT\lambda_i\lambda_j}{C_V}, \qquad (2.11)$$

where

$$\lambda_i(P,T) = \left(\frac{\partial S(P,T)}{\partial \varepsilon_i}\right)_T \qquad (2.12)$$

[*Karki et al.*, 1999; *Wentzcovitch et al.*, 2004, 2006]. LD calculations must be performed for each strained cell at each static pressure, leading to huge data sets to be managed. In contrast, in the latter approach, $c_{ij}^T(P,T)$ are determined based on the usual stress-strain relation as the static calculations of elasticity, but sufficiently long simulations with large supercells are required to get accurate averages of the stress tensor, leading to significant computation time.

2.2.5. Lattice Thermal Conductivity

The intrinsic bulk thermal conduction of insulators is caused by anharmonic phonon-phonon interactions, and thus evaluation of the anharmonic coupling strength is a key to calculating lattice thermal conductivity. Within the single-mode relaxation time approximation for the phonon Boltzmann's transport equation [*Ziman*, 1960], k_{lat} is given by

$$k_{\text{lat}} = \frac{1}{3}\sum_s\int |\mathbf{v}_{\mathbf{q},s}|^2 c_{\mathbf{q},s}\, \tau_{\mathbf{q},s}\; d\mathbf{q} \qquad (2.13)$$

where $\mathbf{v}_{\mathbf{q},s}$, $c_{\mathbf{q},s}$, and $\tau_{\mathbf{q},s}$ are the phonon group velocity, the mode heat capacity, and the phonon lifetime at $\mathbf{q}$ for the branch s, respectively, with $\tau_{\mathbf{q},s}$ related to the phonon damping function $\Gamma_{\mathbf{q},s} = 1/2\tau_{\mathbf{q},s}$ that measures the lattice anharmonicity attributable to phonon-phonon interactions. When considering up to the three-phonon process, which is sufficient for ionic and covalent crystals with no soft

phonon modes [*Ecsedy and Klemens*, 1977] such as diamond [*Ward et al.*, 2009], silicon [*Ward and Broido*, 2010], MgO [*Tang and Dong*, 2010], and MgSiO$_3$ [*Dekura et al.*, 2013], the frequency-dependent dumping function for a phonon at $\mathbf{q}$ is represented by

$$\Gamma(\omega_{\mathbf{q},s}) = \frac{\pi}{2}\sum_{q's's''}|V_3(\mathbf{q}s,\mathbf{q}'s',\mathbf{q}''s'')|^2$$
$$\left[\{1+n_{\mathbf{q}',s'}\}\delta(\omega_{\mathbf{q}',s'}+\omega_{\mathbf{q}',s'}-\omega)\right.$$
$$\left.+2\{n_{\mathbf{q}',s'}-n_{\mathbf{q}'s'}\}\delta(\omega_{\mathbf{q}',s'}-\omega_{\mathbf{q},s'}-\omega)\right], \qquad (2.14)$$

where $n_{\mathbf{q},s} = 1/\left(e^{\hbar\omega_{\mathbf{q}s}/k_BT}-1\right)$ is the Bose-Einstein occupation number for a phonon with energy $\hbar\omega_{\mathbf{q},s}$, $V_3(\mathbf{q}s,\mathbf{q}'s',\mathbf{q}''s'')$ is the anharmonic three-phonon coupling coefficient for the creation and annihilation process described in square brackets, and $N_{\mathbf{q}'}$ is the number of $\mathbf{q}$ points sampled in the integration [*Deinzer et al.*, 2003]. Due to the umklapp quasi-momentum conservation, $\mathbf{q}'' = -\mathbf{q}-\mathbf{q}'+\mathbf{G}$. The first and second terms in the square brackets represent the so-called summation process ($D^{\uparrow}$), decay of a single phonon into two phonons with lower frequencies and the so-called difference process ($D^{\downarrow}$) up conversion of two phonons into a single phonon with higher energy, respectively. The term $V_3(\mathbf{q}s,\mathbf{q}'s',\mathbf{q}''s'')$ is further represented as

$$V_3(\mathbf{q}s,\mathbf{q}'s',\mathbf{q}''s'')$$
$$= \sqrt{\frac{\hbar}{8\omega_{\mathbf{q},s}\omega_{\mathbf{q}',s'}\omega_{\mathbf{q}',s''}}}\sum_{\kappa\alpha,\kappa'\alpha'\kappa''\alpha''}\psi_{\alpha\alpha'\alpha''}^{\kappa\kappa'\kappa''}(\mathbf{q},\mathbf{q}',\mathbf{q}'') \qquad (2.15)$$
$$\times\frac{e_\alpha^\kappa(\mathbf{q},s)e_{\alpha'}^{\kappa'}(\mathbf{q}',s')e_{\alpha''}^{\kappa''}(\mathbf{q}'',s'')}{\sqrt{M_\kappa M_{\kappa'} M_{\kappa''}}}e^{i(\mathbf{q}\cdot\mathbf{R}+\mathbf{q}'\cdot\mathbf{R}'+\mathbf{q}''\cdot\mathbf{R}'')},$$

where the summation is taken over indices of atom κ and component α. is $\psi_{\alpha\alpha'\alpha''}^{\kappa\kappa'\kappa''}(\mathbf{q},\mathbf{q}',\mathbf{q}'')$ The third-order anharmonic dynamical tensor and can be calculated based on the density functional perturbation theory combined with the $2n+1$ theorem [*Debernardi*, 1998; *Deinzer et al.*, 2003; *Gonze and Vigneron*, 1989] and also a more standard supercell method. The damping function in Equation. (2.14) is related to the temperature-dependent two-phonon density of states (TDoS) that measures the number of damping channels for the $D^{\uparrow}$ and $D^{\downarrow}$ processes.

2.3. ADIABATIC TEMPERATURE PROFILES

2.3.1. Mantle Adiabat

A nearly adiabatic temperature variation is likely in a convecting system [*Tozer*, 1972; *O'Connell*, 1977; *Davies*, 1977]. The geotherm follows an adiabat in the homogeneous regions where the so-called Bullen parameter is close to 1.

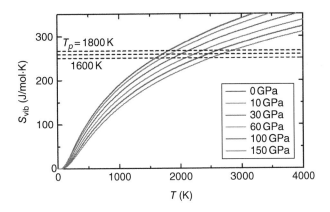

Figure 2.1 Vibrational entropy of pure MgPv calculated as a function of temperature at several pressures from 0 to 150 GPa (modified after *Tsuchiya et al.* [2005b]). Three adiabatic conditions with potential temperatures of 1600, 1700, and 1800 K are shown by dashed lines from bottom to top.

The lower mantle and outer core (OC) satisfy this well. Adiabatic profiles for Earth's interior therefore provide a helpful simplification to calculate further complicated situations. Those adiabats are usually anchored at the depths of seismic discontinuities identified with phase transitions whose P, T boundaries are well clarified. The principal discontinuities often considered are (1) the inner core boundary (ICB), identified with the freezing of the liquid core iron alloy, and (2) the 660 km discontinuity, identified with the post-spinel transition.

The entropy (S) of crystalline and liquid phases can be calculated by the LD and MD methods, respectively [Equations. (2.4), (2.4′), and (7)]. For metallic crystals, electronic contribution to entropy should also be considered [*Tsuchiya and Kawamura*, 2002]. Vibrational (phonon) entropy calculated for MgPv [*Tsuchiya et al.*, 2005b] is shown in Figure 2.1 as a function of temperature at several pressures from 0 to 150 GPa. A temperature profile that produces constant S can be determined from this. Figure 2.1 demonstrates adiabats of MgPv calculated for potential temperatures (T_p) of 1600, 1700, and 1800 K with previous models for the lower mantle geotherm including a standard adiabat model by *Brown and Shankland* [1981], which was made with an anchoring temperature of 1873 K at the 670 km depth. The most recent experiment reported that this condition coincides perfectly with the post-spinel phase boundary in the pyrolitic chemistry [*Ishii et al.*, 2011]. Our adiabat for pure MgPv with a potential temperature of 1600 K agrees well with the Brown-Shankland model. An adiabatic profile for 6.25% FeSiO$_3$-bearing MgPv is also shown in Figure 2.2 (open circles), which clearly indicates no significant effects of the iron incorporation on the adiabat of MgPv. The presently calculated three adiabats for pure MgPv are also tabulated in Table 2.1.

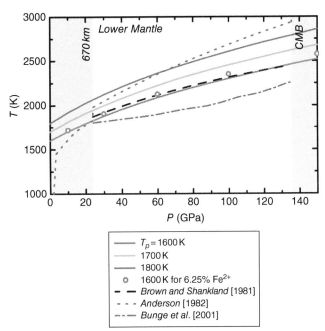

Figure 2.2 A diabatic temperature profiles for MgPv with potential temperatures of 1600, 1700, and 1800 K along with some previous models for geotherm [*Brown and Shanklnad*, 1981; *Anderson*, 1982; *Bunge et al.*, 2001]. Shaded areas are out of the LM pressure range.

Table 2.1 Adiabatic temperature profiles calculated for MgPv in the LM pressure range with three potential temperatures of 1600, 1700, and 1800 K.

P (GPa)	T(K)		
	$T_p = 1600$ K	1700 K	1800 K
23.8	1829	1946	2065
30	1880	1998	2124
40	1956	2078	2210
50	2025	2153	2288
60	2088	2223	2360
70	2146	2288	2428
80	2201	2347	2491
90	2252	2402	2552
100	2300	2454	2609
110	2348	2502	2663
120	2393	2550	2714
130	2438	2597	2763
135.8	2463	2623	2790

Superadiabatic [*Anderson*, 1982] and subadiabatic [*Bunge et al.*, 2001] models have also been proposed, that produce 300 K higher and lower than the adiabat with a potential temperature of 1, 600 K at 100 GPa, respectively. However, since a fact the velocities and density of the pyrolitic aggregate calculated along the adiabat with a potential temperature of 1630 K can reproduce the

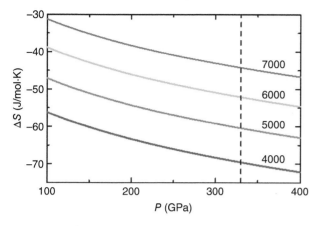

Figure 2.3 Relative entropy variation of pure liquid Fe from a value at reference condition of $T_0 = 8000$ K and $V_0 = 19.4685$ Å³/atom which was applied in the EoS analysis [*Ichikawa et al.*, 2014].

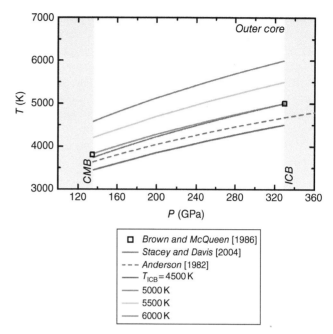

Figure 2.4 Adiabatic temperature gradients for liquid Fe with anchoring ICB temperatures of 4500, 5000, 5500, and 6000 K calculated using results of *Ichikawa et al.* [2014] along with some previous models for geotherm [*Brown and McQueen*, 1986; *Stacey and Davis*, 2004; *Anderson*, 1982]. Shaded areas are out of the OC pressure range.

seismological LM models quite well [*Wang et al.*, 2015] and also tomographic images, which suggest a global mantle circulation based on thermal convection [*Zhao*, 2015], those nonadiabatic temperature profiles might be less likely.

2.3.2. Core adiabat

The thermodynamic property of liquid iron, the major constituent of the core, has been investigated using the AIMD in detail [*Vocadlo et al.*, 2003; *Ichikawa et al.*, 2014]. The calculated entropy variation of liquid iron is shown in Figure 2.3 as a function of pressure from 100 to 400 GPa at different temperatures.

The adiabatic temperature profile of the outer core is determined from this entropy. The P, T condition of the ICB is used as an anchoring condition in general, which corresponds to the melting (solidus) point of the inner core material. Although the melting temperature of pure iron at the ICB pressure of 329 GPa was determined to be 6350 K [*Alfè et al.*, 2002b], T_{ICB} is in general inferred to be much lower due to light element incorporation [e.g., *Poirier*, 2000]. For four anchoring temperatures of 4500, 5000, 5500, and 6000 K, our EoS model yields the CMB temperatures (T_{CMB}) of 3452, 3829, 4206, and 4,581 K, respectively (Figure 2.4 and Table 2.2), with the average adiabatic temperature gradient of ~0.55±0.1 K/km [*Ichikawa et al.*, 2014]. A melting temperature of FeO was reported to be 3670 K at the CMB pressure [*Boehler*, 1992]. Considering this and that O is expected to be a major light element candidate, 3452 K might be lower than the lower bound of the T_{CMB}. Meanwhile, 4581 K, which is derived from the T_{ICB} of 6000 K near the melting temperature of pure iron at 329 GPa, would be close to its upper bound. Within this kind of analysis, there still

Table 2.2 Adiabatic temperature profiles of pure liquid Fe with four anchoring ICB temperatures in the outer core pressure range.

P (GPa)	T(K)			
	T_{ICB} = 4500 K	5000 K	5500 K	6000 K
136	3452	3829	4206	4581
150	3547	3936	4323	4710
175	3705	4113	4519	4926
200	3854	4279	4703	5127
225	3993	4435	4875	5316
250	4124	4581	5037	5493
275	4249	4720	5191	5661
300	4369	4853	5338	5976
325	4483	4981	5479	5976
329	4500	5000	5500	6000

remains ~1000 K uncertainty in T_{CMB}, which might be reduced by analyzing seismological properties in the lowermost LM, as described in the next section.

2.4. THERMAL STRUCTURE IN D″ LAYER

Analyses of the temperature profiles in the LM and OC indicate quite different CMB temperatures (T_{CMB}) at the mantle and core sides. The former and the latter are estimated to be ~2500 K and ~3500–4500 K, respectively. There are some conditions to constrain the core side temperature: (1) It should be higher than the melting temperatures of iron–light element alloys to accommodate the liquid OC but (2) lower than melting temperatures of silicates due to there beings no observation suggesting any global melt layer at the base of the mantle except for much more local ultra low-velocity zones (ULVZs). The core side T_{CMB} is also expected to be almost perfectly uniform, because any lateral temperature variation would be relaxed promptly in the liquid iron. These considerations suggest that the temperature near the CMB is strongly depth dependent and forms a large TBL in the lowermost LM with an ~1000 K variation. Such a large TBL could reasonably be expected to exist in a boundary region of a solid convection system based on the boundary layer theory [*Turcotte and Schubert*, 2001], where thermal energy is transported dominantly by a thermal conduction process rather than convection accompanied with advection.

Seismological information facilitates further constraints on the thermal structure in the CMB region. Tomographic images elucidated substantial lateral velocity variations in the deepest a few 100 km region being different from most other parts of the LM [e.g., *Masters et al.*, 2000]. The S wave travels faster beneath circum-Pacific than beneath central Pacific and Africa by ~3%. Also a small (~1.5−2.0 %) but distinct velocity jump, the so-called D" discontinuity, can often be observed at a depth of 200–300 km above the CMB in the S-wave velocity in the faster velocity regions [*Lay et al.*, 1998], while such discontinuous change is not detected or unclear in the low-velocity regions. Lateral temperature variations can produce lateral velocity variations that are colder in faster regions and hotter in slower regions [*Wentzcovitch et al.*, 2006], and are often discussed associated with down welling and up welling flows in the LM, respectively. Velocity jumps observed at the top of D" in high-velocity regions are thought to coincide with the high-pressure Pv-to-PPv phase change in Fe-bearing MgSiO₃, which is pushed up to near the CMB in high-temperature regions due to a large Clapeyron slope about 10 MPa/K [*Tsuchiya et al.*, 2004b]. Recently it has been further reported that the phase transition loop remains relatively sharp even in multicomponent systems and a discontinuity would thus still be seismologically detectable [*Tsuchiya and Tsuchiya*, 2008; *Metsue and Tsuchiya*, 2012]. This transition may therefore be a good indicator of the temperature in the deep mantle. The T_{CMB} at the core side should be homogeneous, so that the temperature heterogeneity in the D" region should be attributed to a lateral variation of the TBL thickness. Since the thermal energy is transported by conduction in the TBL, temperature profiles there can be calculated approximately by solving a one-

dimensional (1D) thermal conduction equation within the semi-infinite half space

$$\frac{\partial \theta}{\partial t} = \kappa \frac{\partial^2 \theta}{\partial z^2,} \qquad (2.16)$$

where $\kappa = k/\rho C$ is thermal diffusivity (k is thermal conductivity and C is heat capacity) and the position of CMB is set to $z = 0$. Also $\theta = (T - T_1)/(T_0 - T_1)$ is the dimensionless temperature ratio with $T_0 = T(z = 0,t) = T_{CMB}$ and $T_1 = T(z,t = 0) = T_{adiabat.}$ The solution of this equation is generally given by

$$\theta = 1 - \mathrm{erf}\left(\frac{z}{2\sqrt{\kappa t}}\right). \qquad (2.17)$$

Therefore, superadiabatic temperature deviation can be represented as

$$T(z,t) - T_1 = (T_0 - T_1)\left[1 - \mathrm{erf}\left(\frac{z}{2\sqrt{\kappa t}}\right)\right], \qquad (2.18)$$

where $2\sqrt{kt}$ is considered to be the characteristic thermal diffusion distance [*Turcotte and Schubert*, 2001], which can be related to the effective TBL thickness $\Delta Z_{TBL.}$

We evaluate temperature profiles with two different T_{CMB} values of 3800 and 4500 K to examine the corresponding 1D S-wave velocity structures. The PPv phase boundaries in pure and Fe- and Al-bearing MgSiO₃ were investigated carefully by *Tsuchiya et al.* [2004b], *Metsue and Tsuchiya* [2012], and *Tsuchiya and Tsuchiya* [2008] respectively, reporting that iron incorporation with a geophysically relevant amount has just a minor effect on the phase boundary. According to these studies, at a relatively lower T_{CMB} of 3800 K, the Pv phase remains stable at the CMB condition, while at the relatively higher T_{CMB} of 4500 K, which might be close or even higher than the solidus temperature of the pyrolitic aggregate, the PPv phase is stable at the CMB condition. In the latter case, the reverse transition from PPv to Pv occurs just below the CMB pressure [*Hernlund et al.*, 2005]. So these two temperature profiles are expected to produce very different S-wave velocity structures. Temperature profiles calculated with four $\sqrt{kt}$ values of 100, 200, 300, and 400 km for two T_{CMB} are shown in the upper panels of Figure 2.5. The PPv transition occurs along each profile when T_{CMB} = 3800 K, while only $\Delta Z_{TBL} = 2\sqrt{kt} <~ 190$ km when T_{CMB} = 4500 K, accompanied with the reverse transition.

Corresponding S-wave velocity profiles calculated using the high-P, T elasticity (B_S and G) of MgPv [*Wentzcovitch et al.*, 2006] are also shown in Figure 5.2 (lower panels). Here, the effects of iron incorporation on the velocity [*Tsuchiya and Tsuchiya*, 2006] were taken

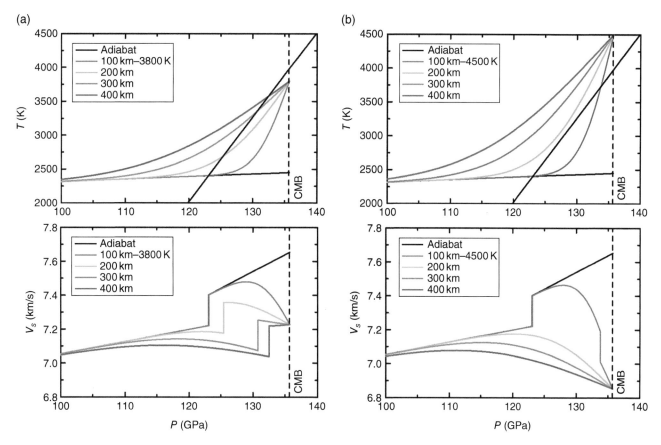

Figure 2.5 Temperature (upper) and S-wave velocity (lower) profiles in the TBL around the CMB calculated for two different T_{CMB} of (a) 3800 K and (b) 4500 K (modified after *Kawai and Tsuchiya* [2009]). Black thin lines in the temperature profiles represent the PPv phase boundary [*Tsuchiya et al.*, 2004b].

into account. The calculated profiles with T_{CMB} = 4500 K (Figure 2.5b) clearly show that (1) a large superadiabatic temperature increase (ΔT_{ad}) of ~2000 K leads to too slow S-wave velocity below ~6.9 km/s at the CMB (c.f., typical values of ~7.25 km/s [*Dziewonski and Anderson*, 1981]); (2) the PPv phase change can occur only in the coldest regions; (3) the velocity structure accompanied with the reverse transition, even in the colder regions, seems to being consistent with the velocity structure observed in the high-velocity regions [*Kawai et al.*, 2007a,b]; and (4) a characteristic S-shape structure observed in the low-velocity regions [*Kawai and Geller*, 2010] cannot be reproduced along hotter geotherms, while rather similar to a velocity structure suggesting the ULVZ, which is thought to be a local feature [*Avants et al.*, 2006]. In contrast, the S-wave velocity profiles calculated with T_{CMB} = 3800 K (Figure 2.5a) are (1) quite similar to observed profiles for both high- and low-velocity regions and (2) not too slow at the CMB because it is not too high and gives no reverse transition. Besides, (3) the velocities along the hot and cold geotherms show ~3–4% contrast at ~125–130 GPa, which is consistent with tomographic S-wave velocity heterogeneity. In addition to these analyses based on the

velocity structure modeling, T_{CMB} = 4500 K is also considered less likely, since it is higher or very close to the solidus temperature of peridotite [*Fiquet et al.*, 2010], which causes a universal melt layer just above the CMB, being incompatible with the locality of the ULVZ. After all, 3800 ± 200 K was proposed as the best estimated range of T_{CMB} [*Kawai and Tsuchiya*, 2009]. In this case, ΔT_{ad} is estimated to be ~1300 K.

According to the present analyses, two extensive low-velocity anomalies beneath central Pacific and Africa, often called large low S-wave velocity provinces (LLSVP) and discussed associated with large-scale upwelling, can be interpreted reasonably by a combination of thermal variation with several hundred Kelvin and the PPv transition. However, not only thermal but also compositional variations due to, e.g., deeply subducted materials such as subducted oceanic lithosphere or solidified basal magma ocean related to geochemically primordial reservoirs [*Kellogg et al.*, 1999] could be a source of lateral velocity heterogeneity. Density and elasticity of the midocean ridge basalt (MORB) composition at the LM pressures have been investigated in this context [*Tsuchiya*, 2011]. However results strongly suggested that the subducted

MORB piles do not match well the observed features of the LLSVP. In particular, its density is not high enough to be gravitationally stabilized at the bottom of the mantle for geological time. A similar conclusion has also been drawn by fluid dynamics modeling [*Li and McNamara*, 2013]. Perturbation from the compositional heterogeneity therefore seems to be higher-order phenomena. Nevertheless, some compositional variations are still expected in certain places even partially. Thin melt layers suggested in some limited areas (ULVZ) and sharp side boundaries observed at the edges of the LLSVP [e.g., *Ni et al.*, 2002, 2005] are possibly hard to explain by temperature heterogeneity alone. Those could be related to reaction products of the mantle and core [e.g., *Knittle and Jeanloz*, 1991] or subducted crustal materials [e.g., *Kawai et al.*, 2009].

An adiabatic extension of 3800 K of T_{CMB} leads to ~5000 K at the ICB, as shown in Figure 2.4 and Table 2.2. Since the melting temperature of pure iron at the ICB pressure of 329 GPa is reported to be 6350 K [*Alfè*, 2009], a reduction of melting temperature more than 1000 K is required by incorporation of light elements. This temperature drop seems relatively large for ~10 wt % light element concentration but might be possible if the melting is eutectic. End-member alloys with lower melting temperature than pure iron might be more acceptable for this. In addition, the OC composition should be more iron rich than the eutectic composition to preferentially partition light elements to the liquid phase.

2.5. CMB HEAT FLOW

2.5.1. Thermal Conductivity of LM Phases

The conductive energy flux expected across the CMB (q_{CMB}) can be represented based on Fourier's heat law as $q_{CMB} = k_{LM} \Delta T_{ad} / \Delta Z_{TBL}$, where $\Delta T_{ad}/\Delta Z_{TBL}$ is the temperature gradient in the TBL calculated using the superadiabatic temperature increase (ΔT_{ad}) and the effective TBL thickness (ΔZ_{TBL}), approximately estimated to be ~6 K/km and ~12 K/km for high-temperature (low-velocity) and low-temperature (high-velocity) regions, respectively (Figure 2.5), and k_{LM} is the thermal conductivity of the LM. Since $\Delta T_{ad}/\Delta Z_{TBL}$ is already constrained fairly well, calculating k_{LM} is a key to evaluating q_{CMB}.

Lattice thermal conductivity (k_{lat}) is not easy to measure under high pressure both experimentally and theoretically. A major reason of this is related to the fact that thermal conductivity is a nonequilibrium property. Experimental data of high-pressure thermal conductivity therefore have larger uncertainty than those of equilibrium properties, so that theoretical estimations, in particular based on nonempirical ab initio approaches, are meaningful. As described in Section 2.2.5, a key to calculating k_{lat} is the evaluation of phonon lifetime (τ). The theoretical background for calculating τ is based on the anharmonic (LD) theory, which is however more complicated than the process for equilibrium properties such as the thermodynamic property that can be evaluated within harmonic theory. The major difficulty in the anharmonic LD calculations is related to accurate evaluation of an enormous number of higher-order force constants, in particular in silicates. These studies have recently been tackled by several groups.

2.5.2. MgO

Ab initio theory has also been extended recently to evaluate the k_{lat} of MgO. That of MgO was calculated based on anharmonic LD [*Tang and Dong*, 2010], equilibrium MD [*de Koker*, 2009, 2010], and nonequilibrium MD (NEMD) [*Stackhouse et al.*, 2010] methods. Results are summarized in Figure 2.6, indicating that the MD studies give similar k_{lat} and reproduced low-P experimental values [*Kanamori et al.*, 1968; *Katsura*, 1997] fairly well. Pressure dependences of calculated k_{lat} are however considerably scattered. In particular, *Tang and Dong* [2010] reported very different k_{lat} at high pressures. Although the two MD studies appear consistent, these real-space approaches often have difficulties for sufficient sampling of phonons with long wavelengths comparable to their mean-free paths unless adopting an impractically large supercell. *De Koker* [2009, 2010] determined phonon lifetime from phonon peak widths calculated based on the equilibrium AIMD simulations with 2×2×2 and 3×3×3 supercells. This approach is simple to be implemented. However, these supercells are both likely insufficient to reach convergence, since the phonon mean-free path is a few nanometers even at the mantle temperatures. On the other hand, *Stackhouse et al.* [2010] evaluated k_{lat} of MgO based on the NEMD method. They applied elongate supercells up to 2×2×16 (512 atoms) to obtain convergence along the conducting direction and parameterized k_{MgO} as $k(\rho, T) = 5.9(\rho/3.3)^{4.6}(2000/T)$ ((W/m·K). In this approach, heat flux is imposed along the elongated direction and k_{lat} is computed as the ratio of an imposed heat flux to the resulting temperature gradient (dT/dx). Accurate evaluation of the temperature gradient is therefore a key. However, this is often not so easy, since the temperature gradient is nonlinear near the boundaries. A significant cell width is therefore required to achieve steady-state flow with linear gradient.

In contrast, *Tang and Dong* [2010] conducted anharmonic LD calculations rather than MD, where **q**-points were sampled on a 16×16×16 grid of the primitive Brillouin zone of MgO and derived a formulation $k(\rho, T) = \exp(9.31 + 35.9/\rho - 118.6/\rho^2)/T$ (W/m·K).

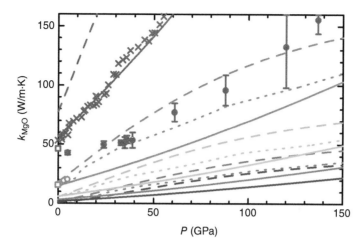

Figure 2.6 Lattice thermal conductivity of MgO as a function of temperature at 300 K (pink), 1000 K (orange), 2000 K (green), 3000 K (blue), and 4000 K (purple) proposed by *Stackhouse et al.* [2010] (solid lines), *de Koker* [2010] (dotted lines), and *Tang and Dong* [2010] (dashed lines). Experimental data at 500 and 1000 K by *Kanamori et al.* [1968] (open squares), 973 K by *Katsura* [1997] (open circles), and 300 K by *Dalton et al.* [2013] (crosses) and *Imada et al.* [2014] (filled circles).

This **q**-point sampling is equivalent to calculating a 16×16×16 supercell in real space, which includes 8192 atoms. Therefore, k_{lat} obtained in this way might be more reliable than others, but the *Stackhouse et al.* results are in better agreement with zero-pressure experiments than those of Tang and Dong which show systematic overestimations unless some corrections (which the authors, call isotope effects) are applied. Discrepancy between *Stackhouse et al.* and *Tang and Dong* k_{lat} is more significant under higher pressure and reaches as much as 50% at 135 GPa and 2000 K (Figure 2.6). *Tang and Dong* [2010] insisted that this considerable overestimation can be compensated for only by the isotopic effect. It should however be marginal at high temperature $(T > \Theta_D)$ because the umklapp scattering becomes stronger with increasing temperature and consequently drives the thermal conduction [*Ward et al.*, 2009]. The cause of overestimation seen in *Tang and Dong* [2010] is unclear, but one possibility is a lack of the LO-TO splitting at the Γ point in the direct LD calculation, which might be important when calculating k_{lat} of small unit cell crystals such as MgO.

High-P k_{lat} of MgO has also been measured experimentally very recently [*Dalton et al.*, 2013; *Imada et al.*, 2014], though those are still limited at 300 K (Figure 2.6). Both results are acceptably consistent with the earlier data near 0 GPa and a surprising agreement between computations [*de Koker*, 2010; *Stackhouse et al.*, 2012] and experiments [*Dalton et al.*, 2013] can be seen. However, the discrepancy between the two experiments becomes considerable under high pressure. This would be serious even though one is for a single crystal [*Dalton et al.*, 2013] and the other is for a polycrystal [*Imada et al.*, 2014], suggesting

that high-P measurements of k_{lat} might still remain challenging and clearly need further tests.

2.5.3. MgPv

In contrast to the studies on MgO, phonon lifetimes of Si and Ge were computed by a more sophisticated approach, where the third-order dynamical tensor was computed at an arbitrary wave vector (**q**) with a primitive cell based on the perturbation approach [*Debernardi*, 1998; *Deinzer et al.*, 2003; *Ward and Broido*, 2010]. Predicted k_{lat} values agree excellently with experiments, suggesting substantial prospects for other compounds. This technique, similar to the usual density functional perturbation theory [DFPT] for the harmonic dynamical matrix (Section 2.2.2), requires a single unit cell with no actual displacement. Therefore, it is considered more efficient and accurate than the other approaches described above. Its unique deficiency might be a difficulty of implementation. However, a similar technique was recently applied to k_{lat} of MgPv, the most dominant phase of the LM, nonempirically in a wide P, T range covering the entire LM conditions [*Dekura et al.*, 2013]. Calculated results are shown in Figure 2.7 and agree well with experimental data up to ~100 GPa, though those are available only at 300 K [*Ohta et al.*, 2012]. Results were also parameterized as a function of V and T as $k(V,T) = 6.7(V/24.71)^{-13.4+10.1V/24.71}(1-e^{-1.1.300/T})/(1-e^{-1.1})$ (W/m K). Conductivity k_{MgPv} increases with increasing pressure but decreases with temperature as expected. It varies from 2.0 W/m K at the shallow LM of 30 GPa and 2000 K to 4.7 W/m-K in the deep LM condition of 100 GPa and 2500 K, indicating that the pressure effect overcomes the temperature effect in most of the LM. However,

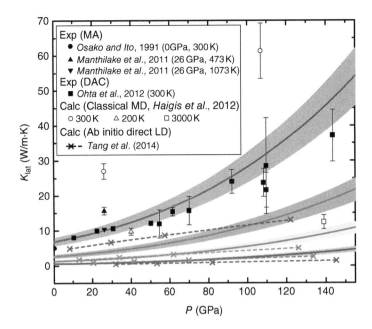

Figure 2.7 Calculated lattice thermal conductivity of MgPv as a function pressure at 300 K (red), 1000 K (orange), 2000 K (green), and 4000 K (blue) with computational uncertainties (shaded bands) [*Dekura et al.*, 2013]. Results of other studies are also shown for comparison [*Manthilake et al.*, 2011; *Osako and Ito*, 1991; *Ohta et al.*, 2012; *Tang et al.*, 2014; *Haigis et al.*, 2012].

it becomes 3.7 W/m-K at 135 GPa and 4000 K, the bottom of the LM, where k_{MgPv} distinctly decreases due to the high-temperature effects. According to *Stackhouse et al.* [2010] and *Dekura et al.* [2013], k_{MgPv} is roughly only 15% of k_{MgO} at the same P, T, which means MgPv is much more thermally resistive than MgO. This large difference in the k_{lat} of the two phases is caused primarily by significant differences in the phonon damping channels, which are determined by the energy conservation condition and the quasi-momentum conservation condition. For MgO with much fewer (1/10) phonon modes it is clearly harder to satisfy these two requisites than for MgPv so that phonons are long-lived in MgO.

It was pointed out based on direct lattice dynamics calculations that *Dekura et al.* [2013] potentially overestimated k_{lat} of MgPv (Figure 2.7) because a simple relationship between Γ and ω represented as $\Gamma_{q,s} \propto T\omega_{q,s}^2$ was applied for acoustic modes [*Tang et al.*, 2014]. This relation, called the Klemens model, is widely accepted for long-wavelength phonons at low temperature, where umklapp scattering is typically weak [*Klemens*, 1951]. Within this model, Γ of the acoustic phonons becomes zero at the zone center, so that τ is overestimated. However, the contribution of acoustic modes, dominant for the different processes ($D^{\downarrow}$), is not major for the phonon damping process in MgPv due to a larger number of optic modes, dominant for the summation process ($D^{\uparrow}$), as seen in TDS (Figure 2.8). There might be other reasons, maybe related to numerical accuracy, for the smaller

k_{lat} values proposed by *Tang et al.* [2014]. (Note that *Tang et al.* [2014] did not consider the effects of isotopic scattering for MgPv, in contrast to MgO.) In particular, the convergence of TDoS satisfying energy and quasi-momentum conservations generally requires careful treatment with quite dense **q**-point grids. In Figure 2.7, k_{MgPv} calculated by classical MD [*Haigis et al.*, 2012] is also shown and is considerably larger than our values. Reliability of this kind of simulation in general relies on the interatomic model potentials used, suggesting less quantitative accuracy compared to ab initio calculations, even though they can deal with a much larger number of atoms. The same arguments are applicable to other classical MD studies on lattice thermal conductivity [*Ammann et al.*, 2014].

2.5.4. Effects of Fe Incorporation and Radiative Contribution on k

To infer the heat flow across Earth's CMB, the total thermal conductivity of the actual LM (k_{LM}) is required. This was modeled by *Dekura et al.* [2013] as follows:

1. The LM can be approximated as an MgPv+MgO two-phase mixture with a pyrolitic ratio (8:2 in volume) [*Ringwood*, 1962]. A composite average was taken based on the Hashin-Shtrikman equation [*Hashin and Shtrikman*, 1962] using the k_{lat} of MgO [*Stackhouse et al.*, 2010].

2. At mantle conditions, iron is known to dissolve into both Pv and MgO to form solid solutions $(Mg,Fe)SiO_3$

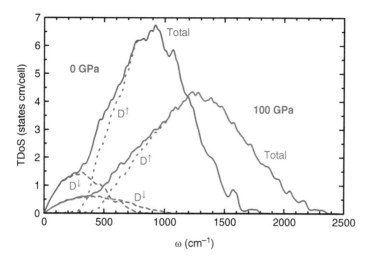

Figure 2.8 Calculated TDoS of Mg-Pv at 0 GPa (blue) and 100 GPa (red) [*Dekura et al.*, 2013]: total (solid lines), summation process ($D^{\uparrow}$) (dotted lines), and difference process ($D^{\downarrow}$) (dashed lines).

Table 2.3 Parameters optimized for the modeling of the effect of iron on k_{lat} of Mg Pv and MgO.

Phase	A	b_1	b_2	b_3	b_4	b_5
MgPv	-0.23^a	-1.776	5.576	-6.022	2.956	-0.551
MgO	-0.52^a	-1.215	2.681	-2.001	0.684	-0.089

^{a} From *Manthilake et al.* [2011].

and (Mg,Fe)O. Substantial reductions on k_{lat} by approximately half Pv and by one-third in MgO were recently measured for the phases with 3 and 20 mol % iron, respectively [*Manthilake et al.*, 2011]. Therefore the effect of iron incorporation cannot be neglected when modeling k_{LM}. Based on the measurements, the following simplified analytical form was proposed to represent this effect [*Dekura et al.*, 2013]:

$$\frac{k_{lat}^{Febearing}}{k_{lat}^{Pristine}} = \left(\frac{T_0}{T}\right)^{\alpha} \sum_{n=1}^{5} b_n \left(\frac{\rho}{\rho_0}\right)^{n-1} \qquad (2.19)$$

with two fitting parameters (a and b_n), which are listed in Table 2.3. Assume T_0 is set to 300 K and ρ_0 is 4.1 and 3.6 g/cm³ for pure Pv and MgO, respectively.

3. Other scatterers, such as grain boundary and point defect including isotope are not taken into account in this study, since their effects are likely expected to be marginal at LM temperatures [*Gilbert et al.*, 2003].

4. In contrast, the radiative energy transportation rad, also not calculated in this study, is generally thought to increase at high temperature. Although k_{rad} of MgPv and MgO is currently only roughly constrained at deep mantle conditions [*Goncharov et al.*, 2008; *Hofmeister*, 1999, 2008; *Keppler et al.*, 2008], $k_{rad}(T) = \chi T^3$ with

$\chi = 8.5 \times 10^{-11}$ W/m·K proposed for dense silicates and oxides [*Hofmeister*, 1999] is here employed.

Estimated total k of the LM (k_{LM}) [*Dekura et al.*, 2013] indicates the strong positive pressure dependence at low temperatures as k_{lat} of MgPv. However, k_{LM} also increases with temperature due to the increase in k_{rad} at high temperature. This contribution is not negligible at the deepest mantle conditions. The k_{LM} is consequently found to vary from 1.39 to 4.46 W/m·K along the LM adiabat (Section 2.4) from the top of the LM at $z = 660$ km depth ($P = 23.5$ GPa, $T = 1875$ K) to the base at $z = 2890$ km depth ($P = 136$ GPa, $T = 2450$ K) where X_{Fe} is 5 and 20 mol % for Pv and MgO, respectively. The k_{LM} further increases to 7.1 W/m K at the CMB temperature ($T = 3800$ K) due to radiative conduction.

In addition to ab initio approaches, some classical MD simulations using empirical interatomic potentials are also performed to calculate thermal transport in MgO and MgSiO₃ [*Haigis et al.*, 2012; *Ammann et al.*, 2014]. However, results of those simulations differ considerably, more than the discrepancy seen in the ab initio studies summarized above, suggesting that this kind of property is highly sensitive to the quality of interatomic model potentials. Highest-level anharmonic lattice dynamics computations are therefore expected to estimate high-P, T conductivity of minerals other than MgPv and impurity/isotope effects.

2.5.5. Heat Flow across CMB

Based on the current prediction of k_{LM}, the heat flow J across Earth's CMB can be inferred. The conductive energy flux across the CMB (q_{CMB}) can be represented based on Fourier's heat law as $q = k \Delta T / \Delta Z$. Here, we apply the mean thermal conductivity $\langle k \rangle = 5.3$ W/m·K and the thermal gradient $\Delta T/\Delta Z$ of 4 K/km (hotter gegions) to 7K/km (colder regions) in the deepest mantle TBL [*Kawai and Tsuchiya*, 2009]. This leads to $q_{CMB} = 0.02 - 0.04$/W/m², then $J_{CMB} = 4\pi r_{CMB}^2 \, q_{CMB} = 3, \cdots, 6$ TW 3 with Earth's core radius $r_{CMB} = 3480$ km. This heat flow, corresponding to 10% of the surface heat flow (46.3 TW) [*Lay et al.*, 2008], is larger than that required to sustain the geodynamo at the current magnitude (~2–3.5 TW) [*Buffett*, 2002]. However, since a high-pressure phase transition from MgPv to the PPv phase is expected near the CMB [*Tsuchiya et al.*, 2004b], k_{lat} of PPv, which is suggested to be larger than that of Pv at room temperature [*Ohta et al.*, 2012], must clearly be applied for a more accurate estimation of J_{CMB}. The effect of iron on the thermal conductivity applied in this modeling [*Manthilake et al.*, 2011] is considerable, but its applicability, as the well as the reliability of radiative conductivity, to the deepest mantle condition still remains unclear. If excluding the effect of iron and phase change, J_{CMB} reaches ~5–10 TW.

The anisotropy of k is another intriguing topic, since the strong shear wave splitting is generally observed in the D″ layer [*Lay et al.*, 1998]. Classical MD studies on k_{lat} of PPv report considerable (~40%) and rather small anisotropy in k_{lat} of this phase depending on the crystallographic direction of heat flow [*Haigis et al.*, 2012; *Ammann et al.*, 2014]. This effect should be included in the modeling of J_{CMB} after reliable high-P, T conductivity of PPv is determined.

In addition, J_{CMB} can be estimated from the OC properties [*de Koker et al.*, 2012; *Pozzo et al.*, 2012]. Proposed large thermal conductivity of liquid iron alloy suggests a significant conductive heat flow across the OC (~15 TW), even from the adiabatic contribution alone. The significant discrepancy between J_{CMB} estimated from the mantle side and the core side is a serious issue to be solved in future.

2.6. CONCLUSIONS

Studies on the key thermal properties of Earth's deep interior have been advanced by the latest ab initio computation methods. Major findings of the studies are as follows: (1) the adiabats of MgPv and liquid Fe estimated from ab initio thermodynamics agree well with previous geotherm models, (2) the CMB temperature is estimated to be 3800 ± 200 K by reproducing seismic S-wave velocity structures in the D″ region, and (3) the CMB heat flow was estimated to be 3–6 TW from the temperature structure and thermal conductivity in the D″ region, which is substantially smaller than that estimated from the OC side. Effects of iron incorporation, which were considered indirectly in the modeling, should be clarified in the future through direct calculations under pressure with reasonable treatments.

ACKNOWLEDGMENTS

We thank C. Shiraishi and S. Whitaker for helping to prepare the manuscript. This research was completed with the support of KAKENHI 26287137, JSPS Grants-in-Aid for JSPS Fellow No. 25 03023, and the X-ray Free Electron Laser Priority Strategy Program (MEXT).

REFERENCES

Alfè, D. (2009), Temperature of the inner-core boundary of the Earth: Melting of iron at high pressure from first-principles coexistence simulations, *Phys. Rev. B*, *79*(6), 060, 101, doi:10.1103/PhysRevB.79.060101.

Alfè, D., G. Price, and M. Gillan (2002), Iron under Earth's core conditions: Liquid-state thermodynamics and high-pressure melting curve from ab initio calculations, *Phys. Rev. B*, *65*(16), 165, 118, doi:10.1103/PhysRevB.65.165118.

Alfè, D., M. J. Gillan, and G. D. Price (2004), Composition and temperature of the Earth's core constrained by combining ab intio calculations and seismic data, *Earth Planet. Sci. Lett.*, *195*(1–2), 91–98, doi:10.1016/S0012-821X(01)00568-4.

Ammann, M. W., A. M. Walker, S. Stackhouse, J. Wookey, A. M. Forte, J. P. Brodholt, and D. P. Dobson (2014), Variation of thermal conductivity and heat flux at the Earth's core mantle boundary, *Earth Planet. Sci. Lett.*, *390*(15), 175–185, doi:10.1016/j.epsl.2014.01.009.

Anderson, O. L. (1982), The Earth's core and the phase diagram of iron, *Philos. Trans. R. Soc. Lond. A*, *306*(1492), 21–35, doi:10.1098/rsta.1982.0063.

Avants, M., T. Lay, S. A. Russell, and E. J. Garnero (2006), Shear velocity variation within the D″ region beneath the central Pacific, *J. Geophys. Res.*, *111*, B05, 305, doi:10.1029/2004JB003270.

Badro, J., A. S. Cote, and J. P. Brodholt (2014), A seismologically consistent compositional model of Earth's core, *Proc. Natl. Acad. Sci. USA*, *111*, 7542–7545, doi:10.1073/pnas.1316708111.

Bagno, P., O. Jepsen, and O. Gunnarsson (1989), Ground state properties of third-row elements with nonlocal density functionals, *Phys. Rev. B*, *40*(3), 1997(R)–2000, doi10.1103/PhysRevB.40.1997.

Baroni, S., P. Giannozzi, and A. Testa (1987), Green's-function approach to linear response in solids, *Phys. Rev. Lett.*, *58*(18), 1861–1864, doi:10.1103/PhysRevLett.58.1861.

Boehler, R. (1992), Melting of the Fe-FeO and the Fe-FeS systems at high-pressure: Constraints on core temperatures. *Earth Planet. Sci. Lett.*, *111*(2–4), 217–227, doi:10.1016/0012-821X(92)90180-4.

Boness, D. A., J. M. Brown, and A. K. McMahan (1986), The electronic thermodynamics of iron under Earth core conditions, *Phys. Earth Planet. Inter.*, *42*(4), 227–240, doi:10.1016/0031-9201(86)90025-7.

Born, M., and J. R. Oppenheimer (1927), On the quantum theory of molecules, *Ann. Phys. (Leipzig)*, *84*, 457–484.

Brown, J. M., and T. J. Shankland (1981), Thermodynamic parameters in the Earth as determined from seismic profiles,

Geophys. J. R. Astron. Soc., *66*(3), 579–596, doi:10.1111/j.1365-246X.1981.tb04891.x.

Brown, J. M., and R. G. McQueen (1986), Phase transitions, Grüneisen parameter, and elasticity for shocked iron between 77 GPa and 400 GPa, *J. Geophys. Res.*, *91*(B7), 7485–7494, doi:10.1029/JB091iB07p07485.

Buffett, B. A. (2002), Estimates of heat flow in the deep mantle based on the power requirements for the geodynamo, *Geophys. Res. Lett.*, *29*(12), 7-1-7-4, doi:10.1029/2001GL014649.

Bunge, H-P., Y. Ricard, and J. Matas (2001), Non-adiabaticity in mantle convection, *Geophys. Res. Lett.*, *28*(5), 879–882, doi:10.1029/2000GL011864.

Cammarano, F., S. Goes, A. Deuss, and D. Giardini (2005), Is a pyrolitic adiabatic mantle compatible with seismic data? *Earth Planet. Sci. Lett.*, *232*, 227–243, doi:10.1016/j.epsl.2005.01.031.

Ceperley, D. M., and B. J. Alder (1980), Ground state of the electron gas by a stochastic method, *Phys. Rev. Lett.*, *45*(7), 566–569, doi:10.1103/PhysRevLett.45.566.

Cococcioni, M., and S. de Gironcoli (2005), Linear response approach to the calculation of the effective interaction parameters in the LDA+U method, *Phys. Rev. B*, *71*, 035, 105, doi:10.1103/PhysRevB.71.035105.

Dalton, D. A., W.-P. Hsieh, G. T. Hohensee, D. G. Cahill, and A. F. Goncharov (2013), Effect of mass disorder on the lattice thermal conductivity of MgO periclase under pressure, *Sci. Rep.*, *3*, 2400, doi:10.1038/srep02400.

Davies, G. F. (1977), Whole-mantle convection and plate tectonics, *Geophys. J. Roy. Astron. Soc.*, *49*(2), 459–486, doi:10.1111/j.1365-246X.1977.tb03717.x.

Debernardi, A. (1998), Phonon linewidth in III-V semiconductors from density-functional perturbation theory, *Phys. Rev. B*, *57*(20), 12, 847–12, 858, doi:10.1103/PhysRevB.57.12847.

Deinzer, G., G. Birner, and D. Strauch (2003), Ab initio calculation of the linewidth of various phonon modes in germanium and silicon, *Phys. Rev. B*, *67*, 144, 304, doi:10.1103/PhysRevB.67.144304.

De Koker, N. (2009), Thermal conductivity of MgO periclase from equilibrium first principles molecular dynamics, *Phys. Rev. Lett.*, *103*, 125, 902, doi:10.1103/PhysRevLett.103.125902.

De Koker, N. (2010), Thermal conductivity of MgO periclase at high pressure: Implications for the D″ region, *Earth Planet. Sci. Lett.*, *292*(3–4), 392–398, doi:10.1016/j.epsl.2010.02.011.

De Koker, N., G. S. Neumann, and V. Vlček (2012), Electrical resistivity and thermal conductivity of liquid Fe alloys at high P and T, and heat flux in Earth's core, *Proc. Natl. Acad. Sci. USA*, *109*(11), 4070–4073, doi:10.1073/pnas.1111841109.

Dekura, H., T. Tsuchiya, Y. Kuwayama, and J. Tsuchiya (2011), Theoretical and experimental evidence for a new post-cotunnite phase of titanium dioxide with significant optical absorption, *Phys. Rev. Lett.*, *107*, 045, 701, doi:10.1103/PhysRevLett.107.045701.

Dekura, H., T. Tsuchiya, and J. Tsuchiya (2013), Ab initio lattice thermal conductivity of MgSiO₃ perovskite as found in Earth's lower mantle, *Phys. Rev. Lett.*, *110*, 025, 904, doi:10.1103/PhysRevLett.110.025904.

Demuth, T., Y. Jeanvoine, J. Hafner, and J. G. Ángyán (1999), Polymorphism in silica studied in the local density

and generalized-gradient approximations, *J. Phys. Cond. Mater.*, *11*(19), 3833–3874, doi:10.1088/0953-8984/11/19/306.

Dziewonski, A. M., and D. L. Anderson (1981), Preliminary reference Earth model, *Phys. Earth Planet. Inter.*, *25*(4), 297–356.

Ecsedy, D. J., and P. G. Klemens (1977), Thermal resistivity of dielectric crystals due to four-phonon processes and optical modes, *Phys. Rev. B*, *15*(12), 5957–5962, doi:10.1103/PhysRev B.15.5957.

Feynman, R. P. (1939), Forces in molecules, *Phys. Rev.*, *56*, 340 doi:http://dx.doi.org/10.1103/PhysRev.56.340

Fiquet, G., A. L. Auzende, J. Siebert, A. Corgne, H. Bureau, H. Ozawa, and G. Garbarino (2010), Melting of peridotite to 140 Gigapascals, *Science*, *329*(5998), 1516–1518, doi:10.1126/science.1192448.

Fukui, H., T. Tsuchiya, and A. Q. R. Baron (2012), Lattice dynamics calculations for ferropericlase with internally consistent LDA+U method, *J. Geophys. Res.*, *117*(B12), B12, 202, doi:10.1029/2012JB009591.

Gibert, B., F. R. Schilling, A. Tommasi, and D. Mainprice (2003), Thermal diffusivity of olivine single-crystals and polycrystalline aggregates at ambient conditions—A comparison, *Geophys. Res. Lett*, *30*(22), 1046, doi:10.1029/2003GL018459.

Goncharov, A. F., B. D. Haugen, V. V. Struzhkin, P. Beck, and S. D. Jacobsen (2008), Radiative conductivity in the Earth's lower mantle, *Nature*, *456*, 231–234, doi:10.1038/nature07412.

Gonze, X. (1995), Adiabatic density-functional perturbation theory, *Phys. Rev. A*, *52*(2), 1096–1114, doi:10.1103/PhysRev A.52.1096.

Gonze, X., and J. P. Vigneron (1989), Density-functional approach to nonlinear-response coefficients of solids, *Phys. Rev. B*, *39*(18), 13, 120–13, 128, doi:10.1103/PhysRevB.39.13120.

Haigis, V., M. Salanne, and S. Jahn (2012), Thermal conductivity of MgO, MgSiO₃ perovskite and post-perovskite in the Earth's deep mantle, *Earth Planet. Sci. Lett.*, *355-356*, 102–108, doi:10.1016/j.epsl.2012.09.002.

Hamann, D. R. (1996), Generalized gradient theory for silica phase transitions, *Phys. Rev. Lett.*, *76*(4), 660–663, doi:10.1103/PhysRevLett.76.660.

Hamann, D. R. (1997), H₂O hydrogen bonding in density functional theory, *Phys. Rev. B*, *55*(16), R10, 157–R10, 160, doi:10.1103/PhysRevB.55.R10157.

Hashin, Z., and S. Shtrikman (1962), A variational approach to the theory of the elastic behaviour of polycrystals, *J. Mech. Phys. Solids*, *10*, 343–352, doi:10.1016/0022-5096 (62)90005-4.

Hernlund, J. W., C. Thomas, and P. J. Tackley (2005), A doubling of the post-perovskite phase boundary and structure of the Earth's lowermost mantle, *Nature*, *434*, 882–886, doi:10.1038/nature03472.

Hill, R. (1952), The elastic behaviour of a crystalline aggregate, *Proc. Phys. Soc. A*, *65*(5), 349–355, doi:10.1088/0370-1298/65/5/307.

Hofmeister, A. M. (1999), Mantle values of thermal conductivity and the geotherm from phonon lifetimes, *Science*, *283*(5408), 1699–1706, doi:10.1126/science.283.5408.1699.

Hofmeister, A. M. (2008), Inference of high thermal transport in the lower mantle from laser-flash experiments and the damped harmonic oscillator model, *Phys. Earth Planet. Int.*, *170*(3–4), 201–206, doi:10.1016/j.pepi.2008.06.034.

Hohenberg, P., and W. Kohn (1964), Inhomogeneous electron gas, *Phys. Rev.*, *136*(3B), B864–B871, doi:10.1103/PhysRev.136.B864.

Hoover, W. G. (1985), Canonical dynamics: Equilibrium phase-space distributions, *Phys. Rev. A*, *31*(3), 1695–1697, doi:10.1103/PhysRevA.31.1695.

Ichikawa, H., T. Tsuchiya, and Y. Tange (2014), The P-V-T equation of state and thermodynamic properties of liquid iron, *J. Geophys. Res. Solid Earth*, *119*, 240–252, doi:10.1002/2013JB010732.

Imada, S., K. Ohta, T. Yagi, K. Hirose, H. Yoshida, and H. Nagahara (2014), Measurements of lattice thermal conductivity of MgO to core-mantle boundary pressures, *Geophys. Res. Lett.*, *41*, 4542–4547, doi:10.1002/2014GL060423.

Ishii, T., H. Kojitani, and M. Akaogi (2011), Post-spinel transitions in pyrolite and Mg_2SiO_4 and akimotoite–perovskite transition in $MgSiO_3$: Precise comparison by high-pressure high-temperature experiments with multi-sample cell technique, *Earth Planet. Sci. Lett.*, *309*, 185–197, doi:10.1016/j.epsl.2011.06.023.

Kanamori, H., N. Fujii, and H. Mizutani (1968), Thermal diffusivity measurement of rock-forming minerals from 300° to 1100°K, *J. Geophys. Res.*, *73*(2), 595–605, doi:10.1029/JB073i002p00595.

Karki, B. B., L. Stixrude, S. J. Clark, M. C. Warren, G. J. Ackland, and J. Crain (1997a), Structure and elasticity of MgO at high pressure, *Am. Mineral.*, *82*(1–2), 51–60.

Karki, B. B., L. Stixrude, S. J. Clark, M. C. Warren, G. J. Ackland, and J. Crain (1997b), Elastic properties of orthorhombic $MgSiO_3$ perovskite at lower mantle pressures, *Am. Mineral.*, *82*(5–6), 635–638.

Karki, B. B., L. Stixrude, and J. Crain (1997c), Ab initio elasticity of three high-pressure polymorphs of silica, *Geophys. Res. Lett.*, *24*(24), 3269–3272, doi:10.1029/97GL53196.

Karki, B. B., R. M. Wentzcovitch, S. de Gironcoli, and S. Baroni (1999), First-principles determination of elastic anisotropy and wave velocities of MgO at lower mantle conditions, *Science*, *286*(5445), 1705–1707, doi:10.1126/science.286.5445.1705.

Karki, B. B., R. M. Wentzcovitch, S. de Gironcoli, and S. Baroni (2000), Ab initio lattice dynamics of $MgSiO_3$ perovskite at high pressure, *Phys. Rev. B*, *62*(22), 14750–14756, doi:10.1103/PhysRevB.62.14750.

Katsura, T. (1997), Thermal diffusivity of periclase at high temperatures and high pressures, *Phys. Earth Planet. Inter.*, *101*(1–2), 73–77, doi:10.1016/S0031-9201(96)03223-2.

Katsura, T., A. Yoneda, D. Yamazaki, T. Yoshino, and E. Ito (2010), Adiabatic temperature profile in the mantle, *Phys. Earth Planet. Inter.*, *183*, 212–218.

Kawai, K., and R. J. Geller (2010), Waveform inversion for localized seismic structure and an application to D" structure beneath the Pacific, *J. Geophys. Res.*, *115*, B01, 305, doi:10.1029/2009JB006503.

Kawai, K., and T. Tsuchiya (2009), Temperature profile in the lowermost mantle from seismological and mineral physics joint modeling, *Proc. Natl. Acad. Sci. USA*, *106*(52), 22, 119–22, 123, doi:10.1073/pnas.0905920106.

Kawai, K., and T. Tsuchiya (2014), P-V-T equation of state of cubic $CaSiO_3$ perovskite from first-principles computation, *J. Geophys. Res. Solid Earth*, *119*, 2801–2809, doi:10.1002/2013JB010905.

Kawai, K., N. Takcuchi, R. J. Geller, and N. Fuji (2007a), Possible evidence for a double crossing phase transition in D" beneath Central America from inversion of seismic waveforms, *Geophys. Res. Lett.*, *34*, L09, 314, doi:10.1029/2007GL029642.

Kawai, K., R. J. Geller, and N. Fuji (2007b), D" beneath the Arctic from inversion of shear waveforms, *Geophys. Res. Lett.*, *34*, L21, 305, doi:10.1029/2007GL031517.

Kawai, K., T. Tsuchiya, J. Tsuchiya, and S. Maruyama (2009), Lost primordial continents, *Gondwana Res.*, *16*, 581–586.

Kellogg, L. H., B. H. Hager, and R. D. van der Hilst (1999), Compositional stratification in the deep mantle, *Science*, *283*(5409), 1881–1884, doi:10.1126/science.283.5409.1881.

Keppler, H., L. Dubrovinsky, O. Narygina, and I. Kantor (2008), Optical absorption and radiative thermal conductivity of silicate perovskite to 125 Gigapascals, *Science*, *322*(5907), 1529–1532, doi:10.1126/science.1164609.

Klemens, P. G. (1951), The thermal conductivity of dielectric solids at low temperatures (Theoretical), *Proc. R. Soc. Lond., Ser. A*, *208*(1952), 108, doi:10.1098/rspa.1951.0147.

Knittle, E., and R. Jeanloz (1991), Earth's core-mantle boundary: Results of experiments at high pressures and temperatures, *Science*, *251*, 1438–1443.

Kohn, W., and L. J. Sham (1965), Self-consistent equations including exchange and correlation effects, *Phys. Rev.*, *140*(4A), A1133–A1138, doi:10.1103/PhysRev.140.A 1133.

Lay, T., Q. Williams, and E. J. Garnero (1998), The core–mantle boundary layer and deep Earth dynamics, *Nature*, *392*, 461–468, doi:10.1038/33083.

Lay, T., J. Hernlund, and B. A. Buffett (2008), Core–mantle boundary heat flow, *Nat. Geosci.*, *1*, 25–32, doi:10.1038/ngeo.2007.44.

Li, M., and A. K. McNamara (2013), The difficulty for subducted oceanic crust to accumulate at the Earth's core-mantle boundary, *J. Geophys. Res. Solid Earth*, *118*(4), 1807–1816, doi:10.1002/jgrb.50156.

Manthilake, G. M., N. de Koker, D. J. Frost, and C. A. McCammon (2011), Lattice thermal conductivity of lower mantle minerals and heat flux from Earth's core, *Proc. Natl. Acad. Sci. U.S.A.*, *108*(44), 17, 901–17, 904, doi:10.1073/pnas.1110594108.

Masters, G., G. Laske, H. Bolton, and A. Dziewonski (2000), *Earth's Deep Interior: From Mineral Physics and Tomography from Atomic to the Global Scale, Geophysi. Monogr. Ser.*, vol. 117, edited by S., Karato, A. M., Forte, R.C., Liebermann, G. Masters, and L. Stixrude, pp. 63–87, (AGU, Washington, DC)

Metsue, A., and T. Tsuchiya (2011), Lattice dynamics and thermodynamic properties of $(Mg, Fe^{2+})SiO_3$ postperovskite, *J. Geophys. Res.*, *116*(B8), B08,207, doi:10.1029/2010JB008018.

Metsue, A., and T. Tsuchiya (2012), Thermodynamic properties of $(Mg, Fe^{2+})SiO_3$ perovskite at the lower-mantle pressures and temperatures: An internally consistent LSDA+U study, *Geophys. J. Int.*, *190*(1), 310–322, doi:10.1111/j.1365-246X.2012.05511.x.

Murakami, M., K. Hirose, K. Kawamura, N. Sata, and Y. Ohishi (2004), Post-perovskite phase transition in MgSiO$_3$, *Science, 304*, 855–858, doi:10.1126/science.1095932.

Nastar, M., and F. Willaime (1995), Tight-binding calculation of the elastic constants of fcc and hcp transition metals, *Phys. Rev. B, 51*, 6896–6907.

Ni, S., T. Tan, M. Gurnis, and D. Helmberger (2002), Sharp sides to the African superplume, *Science, 296*(5574), 1850–1852, doi:10.1126/science.1070698.

Ni, S., D. V. Helmberger, and J. Tromp (2005), Three-dimensional structure of the African superplume from waveform modelling, *Geophys. J. Int., 161*(2), 283–294, doi:10.1111/j.1365-246X.2005.02508.x.

Nose, S. (1984), A unified formulation of the constant temperature molecular-dynamics methods, *J. Chem. Phys., 81*(1), 511–519, doi:10.1063/1.447334.

O'Connell, R. J. (1977), On the scale of mantle convection, *Tectonophyics, 38*(1–2), 119–136, doi:10.1016/0040-1951(77)90203-7.

Ohta, K., T. Yagi, N. Taketoshi, K. Hirose, T. Komabayashi, T. Baba, Y. Ohishi, and J. Hernlund (2012), Lattice thermal conductivity of MgSiO$_3$ perovskite and post-perovskite at the core–mantle boundary, *Earth Planet. Sci. Lett., 349–350*, 109–115, doi:10.1016/j.epsl.2012.06.043.

Osako, M., and E. Ito (1991), Thermal diffusivity of MgSiO$_3$ perovskite, *Geophys. Res. Lett., 18*(2), 239–242, doi:10.1029/91GL00212.

Parrinello, M., and A. Rahman (1980), Crystal structure and pair potentials: A molecular-dynamics study, *Phys. Rev. Lett., 45*, 1196, doi:10.1103/PhysRevLett.45.1196.

Perdew, J. P., J. A. Chevary, S. H. Vosko, K. A. Jackson, M. R. Pederson, D. J. Singh, and C. Fiolhais (1992), Atoms, molecules, solids, and surfaces: Applications of the generalized gradient approximation for exchange and correlation, *Phys. Rev. B, 46*(11), 6671–6687, doi:10.1103/PhysRevB.46.6671.

Perdew, J. P., and A. Zunger (1981), Self-interaction correction to density-functional approximations for many-electron systems, *Phys. Rev. B, 23*, 5048–5079, doi:10.1103/PhysRevB.23.5048.

Perdew, J. P., K. Burke, and M. Ernzerhof (1996), Generalized gradient approximation made simple, *Phys. Rev. Lett., 77*(18), 3865–3868, doi:10.1103/PhysRevLett.77.3865.

Poirier, J-P (2000), *Introduction to the Physics of the Earth's Interior*, Cambridge Univ. Press, Cambridge.

Pozzo, M., C. Davies, D. Gubbins, and D. Alfé (2012), Thermal and electrical conductivity of iron at Earth's core conditions. *Nature, 485*, 355–385, doi:10.1038/nature11031.

Rahman, A., and N. Parrinello (1981), Polymorphic transitions in single crystals: A new molecular dynamics method, *J. Appl. Phys., 52*(12), 7182–7190, doi:10.1063/1.32869.

Ringwood, A. E. (1962), Model for upper mantle, *J. Geophys. Res. 67*(2), 857–867, doi:10.1029/JZ067i002p00857.

Srivastava, G. P. (1990), *The Physics of Phonons, Taylor & Francis*, New York.

Stacey, F. D., and P. M. Davis (2004), High pressure equations of state with applications to the lower mantle and core, *Phys. Earth Planet. Inter., 142*(3–4), 137–184, doi:10.1016/j.pepi.2004.02.003.

Stackhouse, S., L. Stixrude, and B. B. Karki (2010), Thermal conductivity of periclase (MgO) from first principles, *Phys. Rev. Lett., 104*, 208,501, doi:10.1103/PhysRevLett.104.208501.

Tang, X., and J. Dong (2010), Lattice thermal conductivity of MgO at conditions of Earth's interior, *Proc. Natl. Acad. Sci. USA, 107*(10), 4539–43, doi: 10.1073/pnas.0907194107.

Tang, X., M. C. Ntam, J. Dong, E. S. G. Rainey, and A. Kavner (2014), The thermal conductivity of Earth's lower mantle, *Geophys. Res. Lett., 41*, 2746–2752, doi:10.1002/2014GL059385.

Tange, Y., Y. Nishihara, and T. Tsuchiya (2009), Unified analyses for P-V-T equation of state of MgO: A solution for pressure-scale problems in high P-T experiments, *J. Geophys. Res., 114*(B3), B03, 208, doi:10.1029/2008JB005813.

Tozer, D. C. (1972), The present thermal state of the terrestrial planets, *Phys. Earth Planet. Inter., 6*(1–3), 182–197, doi:10.1016/0031-9201(72)90052-0.

Tsuchiya, T. (2003), First-principles prediction of the P-V-T equation of state of gold and the 660-km discontinuity in Earth's mantle, *J. Geophys. Res., 108*(B10), 2462, doi:10.1029/2003JB002446.

Tsuchiya, T. (2011), Elasticity of subducted basaltic crust at the lower mantle pressures: Insights on the nature of deep mantle heterogeneity, *Phys. Earth Planet. Inter., 188*(3–4), 142–149, doi:10.1016/j.pepi.2011.06.018.

Tsuchiya, T., and K. Kawamura (2002), First-principles electronic thermal pressure of metal Au and Pt, *Phys. Rev. B, 66*(9), 94, 115, doi:10.1103/PhysRevB.66.094115.

Tsuchiya, T., and J. Tsuchiya (2006), Effect of impurity on the elasticity of perovskite and postperovskite: Velocity contrast across the postperovskite transition in (Mg, Fe, Al)(Si, Al)O$_3$, *Geophys. Res. Lett., 33*(12), L12, S04, doi:10.1029/2006GL025706.

Tsuchiya, J., and T. Tsuchiya (2008), Post-perovskite phase equilibria in the MgSiO$_3$-Al$_2$O$_3$ system. *Proc. Natl. Acad. Sci. U.S.A., 105*, 19160–19164, doi:10.1073/pnas.0805660105.

Tsuchiya, T., and X. Wang (2013), Ab initio investigation on the high-temperature thermodynamic properties of Fe^{3+}-bearing MgSiO$_3$ perovskite, *J. Geophys. Res. Solid Earth, 118*(1), 83–91, doi:10.1029/2012JB009696.

Tsuchiya, J., T. Tsuchiya, S. Tsuneyuki, and T. Yamanaka (2002), First principles calculation of a high pressure hydrous phase, δ-AlOOH, *Geophys. Res. Lett., 29*(19), 15-1–15-4, doi:10.1029/2002GL015417.

Tsuchiya, T., R. Caracas, and J. Tsuchiya (2004a), First principles determination of the phase boundaries of high-pressure polymorphs of silica, *Geophys. Res. Lett., 31*(11), L11, 610, doi:10.1029/2004GL019649.

Tsuchiya, T., J. Tsuchiya, K. Umemoto, and R. M. Wentzcovitch (2004b), Phase transition in MgSiO$_3$ perovskite in the earth's lower mantle, *Earth Planet. Sci. Lett., 224*, 241–248, doi:10.1016/j.epsl.2004.05.017.

Tsuchiya, J., T. Tsuchiya, and S. Tsuneyuki (2005a), First principles study of hydrogen bond symmetrization of phase D under high pressure, *Am. Mineral., 90*(1), 44–49, doi:10.2138/am.2005.1628.

Tsuchiya, J., T. Tsuchiya, and R. M. Wentzcovitch (2005b), Vibrational and thermodynamic properties of MgSiO$_3$ post-perovskite, *J. Geophys. Res., 110*(B2), B02, 204, doi:10.1029/2004JB003409.

Tsuchiya, T., R. M. Wentzcovitch, C. R. S. da Silva, and S. de Gironcoli (2006a), Spin transition in magnesiowüstite in Earth's lower mantle, *Phys. Rev. Lett.*, *96*, 198501, doi: 10.1103/PhysRevLett.96.198501.

Tsuchiya, T., R. M. Wentzcovitch, C. R. S. da Silva, S. de Gironcoli, and J. Tsuchiya (2006b), Pressure induced high spin to low spin transition in magnesiowüstite, *Phys. Stat. Sol. B*, *243*, 2111–2116.

Turcotte, D. L., and G. Schubert (2001), *Geodynamics*, Cambridge Univ. Press, Cambridge.

Vocadlo, L., D. Alfè, M. J. Gillan, and G. D. Price (2003), The properties of iron under core conditions from first principles calculations, *Phys. Earth Planet. Inter.*, *140*(1–3), 101–125, doi:10.1016/j.pepi.2003.08.001.

Wallace, D. C. (1972), *Thermodynamics of Crystals*, Wiley, New York.

Wang, X., T. Tsuchiya, and A. Hase (2015), Computational support for a pyrolitic lower mantle containing ferric iron, *Nat. Geosci.*, *8*, 556–558, doi:10.1038/ngeo2458.

Ward, A., and D. A. Broido (2010), Intrinsic phonon relaxation times from first-principles studies of the thermal conductivities of Si and Ge, *Phys. Rev. B*, *81*, 085,205, doi:10.1103/PhysRevB.81.085205.

Ward, A., D. A. Broido, D. A. Stewart, and G. Deinzer (2009), Ab initio theory of the lattice thermal conductivity in diamond, *Phys. Rev. B*, *80*, 125, 203, doi:10.1103/PhysRevB.80.125203.

Wentzcovitch, R. M., B. B. Karki, M. Cococcioni, and S. de Gironcoli (2004), Thermoelastic properties of $MgSiO_3$-perovskite: Insights on the nature of the Earth's lower mantle, *Phys. Rev. Lett.*, *92*(1), 018, 501, doi:10.1103/PhysRevLett.92.018501.

Wentzcovitch, R. M., T. Tsuchiya, and J. Tsuchiya (2006), $MgSiO_3$ post-perovskite at D conditions, *Proc. Natl. Acad. Sci. USA*, *103*(3), 543–546, doi:10.1073/pnas.0506879103.

Yu, Y. G., R. M. Wentzcovitch, T. Tsuchiya, K. Umemoto, and D. J. Weidner (2007), First principles investigation of the postspinel transition in Mg_2SiO_4, *Geophys. Res. Lett.*, *34*(10), L10, 306, doi:10.1029/2007GL029462.

Zein, N. E. (1984), Density functional calculations of elastic moduli and phonon spectra of crystals, *Sov. Phys. Solid State*, *26*, 1825–1828 (1984).

Zhao, D. (2015), *Multiscale Seismic Tomography*, Springer Japan, Tokyo.

Ziman, J. M. (1960), *Electrons and Phonons*, Oxford Univ. Press, London.

3

Heat Transfer in the Core and Mantle

Abby Kavner[1] and Emma S. G. Rainey[1,2]

ABSTRACT

We review the importance of thermal conductivity in Earth interior processes and assess the current state of our physical understanding of thermal conductivity material properties of the core and mantle. The electron-based physical processes governing thermal conductivity in metals are reviewed, with an emphasis on their pressure and temperature dependencies relevant for iron alloy in Earth's core. Estimated ranges for the thermal conductivity of Earth's core are summarized from recent theoretical and experimental constraints. Current theoretical and experimental estimates for phonon-based thermal conductivity for Earth's mantle are summarized, with both theory and experiment providing a wide range of thermal conductivity models for the lower mantle. We reconcile two high-pressure optical absorption data sets with a consistent approach to place an upper bound on the contribution of radiative heat flow to the total lower mantle thermal conductivity. Finally, we examine core-mantle boundary thermal conductivity and its implications for the entire heat budget of Earth.

3.1. INTRODUCTION

Earth is a chemically active planet defined by large-scale transfer of its internal primordial heat to the surface. The details of this trajectory determine the thermochemical evolution of the entire Earth, including the core geodynamo, mantle convection, and ultimately the surface expression of plate tectonics. These processes are powered by Earth's thermal budget and geographic distribution of heat sources and are governed by the flow laws that describe thermal transport by radiation, conduction, and convection.

In this chapter, we focus on the mineral properties governing thermal conductivity, i.e., diffusive heat transfer.

The thermal conductivity of core and mantle materials plays important roles both in boundary layers and in large-scale convection zones. Thermal conductivity governs present-day heat flow in mechanical boundary layers where convection cannot take place and thus constrains the current thermal budget and heat flow history of Earth. In addition, the thermal conductivity and its variations with temperature and pressure can have important effects on the mode and style of convection in the core and mantle as well as the overall heat transfer.

In the next two sections, core and mantle thermal conductivities are considered separately because electrons govern thermal conductivity in metals such as the iron alloy of the core and phonons (quantized lattice waves) and photons are mostly responsible for heat conduction in oxides and silicates. In each section, we review the importance of thermal conductivity in geophysical processes and then discuss the key physics governing the material property of thermal conductivity. We review

[1]Earth, Planetary, and Space Sciences Department, University of California, Los Angeles, California, USA

[2]The Johns Hopkins University Applied Physics Lab, Laurel, Maryland, USA

Deep Earth: Physics and Chemistry of the Lower Mantle and Core, Geophysical Monograph 217, First Edition.
Edited by Hidenori Terasaki and Rebecca A. Fischer.
© 2016 American Geophysical Union. Published 2016 by John Wiley & Sons, Inc.

theoretical and experimental constraints on the thermal conductivity with an emphasis on recent estimates. In the final section, constraints on thermal conductivity in the lowermost mantle are summarized and estimates of the total heat crossing the core-mantle boundary are provided.

3.2. THERMAL CONDUCTIVITY OF EARTH'S CORE

The thermal conductivity of Earth's core is one of the parameters governing the amount of heat available to drive the geodynamo generation of Earth's magnetic field, the field decay time, and the influence of the inner core on the magnetic field. Earth's magnetic field has persisted for ~3.5 Gyr, and this constraint has important implications for the current thermal state of Earth and its time trajectory from the uncertain initial state to the current state [*Buffett*, 2002], with the particulars depending in part on the thermal conductivity of the core. Generally, higher values of electrical and thermal conductivity imply larger power requirements for generating and maintaining the geodynamo but imply a late-forming inner core, high internal temperatures in the early Earth's deep interior, and a relatively rapidly changing heat budget throughout geological time. Lower values of core thermal conductivity relax some of the constraints on powering the geodynamo and ease the requirements for heat flux across the core-mantle boundary. Uncertainties in the thermal and electrical conductivity of the core contribute the major source of uncertainty to the power accounting at the core-mantle boundary and the history of the thermal evolution of the core, including the timing for inner core growth.

For context in the following discussion, ambient-pressure thermal conductivity values for iron and its alloys vary widely, with pure iron having a thermal conductivity of ~80 W/m·K at room temperature [*Fulkerson et al.*, 1966], dropping to ~40 W/m·K at its melting point. Alloying tends to lower the thermal conductivity; for example, cast irons have lower conductivities by ~30%, and Ni-bearing steels such as Invar and stainless steels have much lower thermal conductivities, ~10 W/m·K (Engineering Toolbox.com).

In metals, thermal conductivity is primarily governed by electron-electron scattering and is therefore related to electrical resistivity via the Wiedemann-Franz law, which states that the ratio of thermal conductivity (κ_{cond}) to electrical conductivity (σ_{cond}) is linearly proportional to temperature (T):

$$\kappa_{cond}/\sigma_{cond} = \mathbf{L}T. \tag{3.1}$$

Stacey and Anderson [2001] extrapolated the then-existing iron conductivity measurements at high pressures

[*Matassov*, 1977] to conditions at Earth's core. They obtained values of 46 W/m·K at the core-mantle boundary, rising to values of 79 W/m·K at the center of the inner core. A series of additional theoretical assumptions were necessary for this extrapolation, including (a) constant resistivity along the melting curve of a pure metal, (b) resistivity proportional to concentration of impurity with proportionality constant independent of impurity type, and (c) inclusion of a temperature-dependent phonon contribution to the conductivity. A follow-up paper [*Stacey and Loper*, 2007] motivated by newer shock wave measurements [*Bi et al.*, 2002] generated a significant downward revision of the thermal conductivity to constant value of ~30 W/m·K throughout the core.

This picture of relatively low core thermal conductivity was recently upended by a series of theoretical calculations on the electrical and thermal conductivity of iron and alloys at the high-pressure (~300 GPa) and high-temperature (~5000 K) conditions of the core [*de Koker et al.*, 2012; *Pozzo et al.*, 2012, 2013, 2014]. These papers provide first-principles calculations of electrical and thermal conductivity of iron and alloys at high pressures and temperatures corresponding to the core. They show larger thermal conductivities than implied by the newer shock wave estimates. An independent theoretical treatment of the core conductivity problem by *Zhang et al.* [2015] includes additional scattering effects arising from electron-phonon interactions that were not included in the Pozzo et al. series of papers. The resulting calculations suggest lower values for thermal conductivity in the core (Figure 3.1b).

The calculation of thermal conductivity as a function of depth through the core is dependent on the assumption of the temperature profile, which in turn depends on the thermoelastic equation of state of iron. Figure 3.1a presents four temperature profiles encompassing bounds generated by uncertainty in absolute temperature (4050 ± 500K) and the Grüneisen parameter (1.3–1.8), which controls the temperature gradient down an adiabat [*Stacey and Davis*, 2004]. The 3500–4500 K bounds encompass the uncertainties as determined by the measurements by *Anzellini et al.* [2013]. Figure 3.1b shows two different sets of thermal conductivity profiles of pure iron as a function of depth in the core corresponding to the two different theoretical estimates and calculated for the two different temperature profiles shown in Figure 3.1a corresponding to the Grüneisen parameter of 1.3. Figure 3.1b demonstrates the very different temperature dependencies of the two theoretical models, with the Pozzo et al. model strongly temperature dependent and the Zhang et al. model not very temperature dependent.

For simplicity and ease of comparison, these calculations represent the behavior of pure iron. As shown in a series of

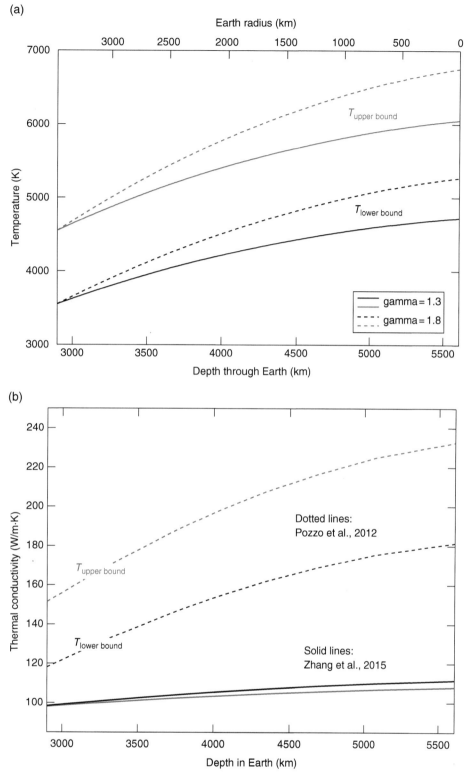

Figure 3.1 Temperature-dependent thermal conductivity estimates for Earth's core. (a) (Top) Approximate upper and lower bounds of current estimates of temperature gradient through core. Constraints on the melting temperature of iron are represented by the "Cool" and "Hot" bounds and come from *Anzellini et al.* [2013]. High and low temperature gradients are based on two bounds on the Grüneisen parameters for iron in Earth's core [*Stacey and Davis*, 2004]. (b) (Bottom) Calculations of thermal conductivity for pure solid iron throughout Earth's core as function of depth (pressure and temperature) based on two different computational models describing iron electrical conductivity. The bounds correspond to core temperatures shown in Figure 3.1a with a Grüneisen parameter of 1.3. To calculate thermal conductivity relevant for Earth's outer core, downward corrections for molten iron ranging from 30% [*Pozzo et al.*, 2012] to 50% [*Stacey and Anderson*, 2001] and for the presence of impurities/alloying elements (~10%) [*Stacey and Anderson*, 2001] need to be taken into account.

theoretical papers by *Pozzo et al.* [2012, 2013, 2014], the thermal conductivity decreases with alloying (due to increased scatter) and the conductivity decrease is proportional to the alloy content. The identity of the alloying component is of secondary importance in reducing the overall conductivity values.

Besides the shock wave studies cited above, a series of diamond anvil cell experiments have sought to measure electrical and thermal conductivity of iron and alloys as a function of pressure, temperature, and alloy content. Experiments by *Konopkova et al.* [2011], *Seagle et al.* [2013], *Deng et al.* [2013], *Gomi et al.* [2013], and *Gomi and Hirose* [2015] show roughly similar pressure and temperature dependencies and alloy effects, showing agreement to first order of experimental results. However, these measurements tend to be much closer to ambient pressures and temperatures than to the pressures and temperatures of the core. In order to extrapolate measured values to the core, physical models are required for electrical and thermal conductivity behaviors over wide pressure and temperature ranges.

Experimentally, electrical conductivity tends to be more easily measured than thermal conductivity, so experimental constraints on core conductivity are mostly determined via electrical conductivity measurements. However, some measurement techniques in development that rely on ultrafast transient temperature detection in the diamond anvil cell coupled with models of heat flow during laser heating are showing promise in directly measuring thermal conductivity at simultaneous high-temperature and high-pressure conditions of the core [*Beck et al.*, 2007; *Rainey and Kavner*, 2014; *McWilliams et al.*, 2015].

The divergent estimates of core conductivity arising from theory and extrapolation of data tend to arise from two classes of assumptions about physical behaviors of electron scatterers in metals at extreme conditions: To what extent do additional scattering processes such as electron-phonon scattering begin to dominate? To what extent do possible saturation effects arising from limits on scattering length scales accessed at extremely high pressures play a role in changing the pressure and/or temperature dependence across wide ranges? Experimental and theoretical input assessing these questions will help narrow the bounds on core thermal conductivity.

3.3. MANTLE

The thermal conductivity properties of mantle materials and their pressure and temperature dependence help govern the mode and style of convection throughout geological history and therefore the thermo chemical evolution of Earth's surface and mantle. In this section, the lattice thermal conductivity of mantle materials is considered, along with possible contributions to overall thermal conductivity from radiative effects. In the following section, the specific issue of thermal conductivity in the lowermost mantle adjacent to the core, a primary constraint on whole-Earth thermal evolution, is discussed.

The mantle is comprised principally of crystalline oxides and silicates at high pressures and temperatures. The Boltzmann transport equation provides a good starting model for considering the microphysics underlying lattice thermal conductivity in the insulating materials of the mantle and examining the dependencies on temperature and pressure. The Boltzmann transport equation describing lattice thermal conductivity κ_{latt} is given by

$$\kappa_{\text{latt}} = \left(\frac{1}{V_{\text{BZ}}}\right) \int_{\text{BZ}} \sum_i \left\{ \left[c_v\left(\vec{k},i\right) \cdot v_g^2\left(\vec{k},i\right) \cdot \tau\left(\vec{k},i\right) \right] / 3 \right\} dk, \quad (3.2)$$

where V_{BZ} is the unit cell volume, c_v is the phonon mode heat capacity, v_g is the phonon mode group velocity, and τ is the phonon mode lifetime. The product of these three values is summed over all phonon modes (i) and over all directions in reciprocal space (k). The pressure and temperature dependences of the thermal conductivity are determined by the pressure and temperature dependences of each term.

The coupled pressure and temperature dependence of the lattice thermal conductivity is often given by the empirical relationship

$$\kappa = \kappa_{\text{ref}} \left(\frac{T_{\text{ref}}}{T}\right)^a \left(\frac{\rho}{\rho_{\text{ref}}}\right)^g \quad (3.3)$$

where $a = 1$ for the normal inverse temperature relationship and g is material and model dependent and is related to the Grüneisen parameter $\gamma = \gamma_0 \left(\frac{V}{V_0}\right)^q$; the exponent q is obtained from the Mie-Grüneisen-Debye model of the thermoelastic equation of state for the material.

Similar to electrical conductivity, thermal conductivity values are determined by scattering energy propagation—the more scatterers, the lower the conductivity values. Quasi-harmonic crystals, such as diamond, can propagate kinetic energy via phonons relatively unhindered by scattering, with the phonon lifetimes on the order of nanoseconds and the group velocities dominated by the high values of the acoustic phonon modes at the Brillouin zone center. Anharmonicity, multiple optic modes in lower-symmetry materials with larger numbers of atoms per unit cell, high temperatures, and impurities all promote phonon scattering, which lowers the overall lattice thermal conductivity. For example, approximate values for thermal conductivities at ambient conditions of diamond, MgO, olivine, and granite are 100–1000, 50, 10, and 1–2 W/m·K, respectively.

The mantle consists of a mineral assemblage, likely dominated by silicate perovskite (bridgmanite) with ~20% Fe-bearing MgO. With its relatively simple structure and bonding, MgO provides a testbed for theoretical calculations and experimental measurements of thermal conductivity. For MgO, different theoretical estimates of the thermal conductivity and its pressure and temperature dependence are generally consistent with each other [*Tang and Dong*, 2009, 2010; *Stackhouse et al.*, 2010] and with experiments [*Katsura*, 1997; *Dalton et al.*, 2013; *Imada et al.*, 2014; *Hofmeister*, 2014; *Rainey and Kavner*, 2014]. Pressure and temperature dependence of thermal conductivity of MgO follows the relationship in equation (3.3), with $a = 1$ and $g = 3\gamma + 2q - 1/3$ [*de Koker*, 2010].

In contrast to the consensus for MgO, theoretical and experimental assessments of thermal conductivity values for bridgmanite, the dominant silicate perovskite of the lower mantle, vary widely. Molecular dynamics studies [*Haigis et al.*, 2012; *Stackhouse et al.*, 2015] which track the evolution of kinetic energy of an assemblage subjected to a temperature gradient and thus calculate the thermal conductivity and ab initio methods based on directly solving the Boltzmann transport equation [*Dekura et al.*, 2013; *Tang et al.*, 2014] provide different values for thermal conductivity and different pressure and temperature dependencies. For example, both the Dekura and Tang papers predict a $1/T$ temperature dependence while a recent molecular dynamics study [*Stackhouse et al.*, 2015] shows a weaker temperature dependence in the lower mantle, consistent with a thermal saturation of conductivity [*Roufosse and Klemens*, 1973]. On the other hand, the Dekura and Stackhouse studies show a similar linear pressure dependence, and the study of Tang et al. results in a linear pressure dependence with a much lower slope. The differences in pressure and temperature behavior of these three models result in thermal conductivity estimates for bridgmanite that vary by a factor of 6, with Tang's results lowest (at ~1–2 W/m·K), Stackhouse et al. largest (~6 W/m·K), and Dekura's intermediate (~3 W/m·K) at conditions of the lowermost mantle. While some of the differences arise from different approximations used—such as possible truncation of anharmonic terms in the necessarily approximate solutions to the Boltzmann transport equation or with incomplete convergence with the molecular dynamics simulations—a significant portion of the variation in results arises from different assumptions about the pressure and temperature dependence on the populations of scatterers and questions of how to handle possible saturation effects caused when the phonon mean free path approaches the nearest-neighbor distance between adjacent scatterers.

As with the case for core materials, there are few experimental constraints and none at the elevated temperature and pressure conditions of the mantle. While the experimental values have some variation, the greater contribution to uncertainty in the lower mantle is the assumed physical models for extrapolation in pressure and temperature. Experimental results on the thermal conductivity of bridgmanite without extrapolation are summarized in Figure 3.2.

Determination of lower mantle thermal conductivity values requires assessing additional contributions to the thermal conductivity, such as radiative effects, and combining the material thermal conductivities in a composite model. Due to the high temperatures in the lower mantle, the question of radiative contributions to thermal conductivity has been raised multiple times. In addition to a potential increase in the effective magnitude of the thermal conductivity, radiative transport may induce strong nonlinearities in the thermal behavior due to temperature and/or compositional variation in lower mantle optical absorption. In the next section, we provide an assessment of contributions of the radiative heat flow in the lower mantle based on a consistent analysis of two existing optical absorption data sets.

3.3.1. Photons: Radiative Heat Flow in Mantle

It has long been hypothesized that radiation could contribute to heat transfer in the high-temperature lower mantle [e.g., *Clark*, 1957]. Recent estimates of the radiative contribution to bulk thermal conductivity, k_{rad}, in the lowermost mantle range from ~0.5 W/m·K [*Goncharov et al.*, 2008] up to ~10 W/m·K [*Keppler et al.*, 2008] and include intermediate values of ~2–5 W/m·K [*Hofmeister*, 2005; *Hofmeister and Yuen*, 2007], broader than the range of estimated lattice thermal conductivities of major lower mantle minerals. Therefore, the role of radiation relative to lattice conduction in lower mantle heat transfer remains unclear.

Due to the difficulties in measuring optical properties of minerals at high pressure and temperature, k_{rad} is typically calculated using optical properties measured at ambient temperature and/or atmospheric pressure and extrapolated to lower mantle conditions. Estimates of k_{rad} for high-pressure minerals may differ due to differences in composition or mineral structure and the corresponding optical properties of the particular samples used (e.g., olivine versus perovskite or low versus high iron content). Differences in measurement techniques or baseline corrections may also be a factor [*Hofmeister*, 2014] as well as particular assumptions made in the radiative thermal conductivity models. To address this final source of

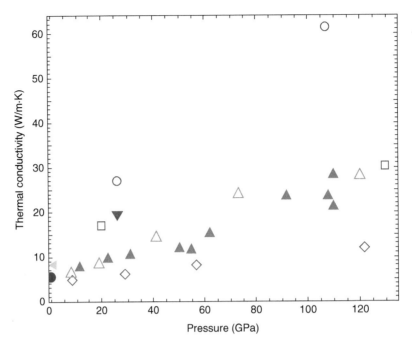

Figure 3.2 Literature values of thermal conductivity of perovskite-structured MgSiO$_3$ (bridgmanite) at ambient temperature as function of pressure. Experimentally determined values: red downward triangle [*Manthilake et al.*, 2011; Ångstrom method], dark green upward triangles [*Ohta et al.*, 2012, DAC thermoreflectance], light green leftward triangle [*Ohta et al.*, 2014; microspot angstrom method], purple filled circle [*Osako and Ito*, 1991; angstrom method]. First-principles calculations: dark red open circles [*Haigis et al.*, 2012; molecular dynamics], red open squares [*Ammann et al.*, 2014; molecular dynamics], orange open triangles [*Dekura et al.*, 2013; lattice dynamics], blue open diamonds [*Tang et al.*, 2014; lattice dynamics]. All data points shown in the figure were measured or calculated at 300 K except for those of *Manthilake et al.* [2011] and *Ammann et al.* [2014], which were extrapolated to 300 K from values reported at 473 and 1000 K, respectively, using the thermal conductivity temperature fits determined in each study.

uncertainty, in this section we use a consistent approach to reassess the radiative thermal conductivity profile of the lower mantle using two previously published data sets on high-pressure optical properties of lower mantle minerals.

Earth's mantle is optically thick, with relatively shallow temperature gradients such that individual grains are essentially isothermal (e.g., even for a very steep thermal boundary layer temperature gradient of 1000 K over 10 km, the temperature difference across a 1 mm grain is only 0.0001 K). For an optically thick medium with isotropic scattering and slowly varying material properties, radiative heat transfer can be treated as a diffusion process, and heat transfer can be modeled using a total thermal conductivity defined as the sum of lattice and radiative contributions. The radiative contribution to thermal conductivity can be calculated using the Rosseland mean approximation [*Rosseland*, 1924; *Siegel and Howell*, 2002]:

$$k_{rad}(T) = \frac{16n^2\sigma T^3}{3\beta_R} \qquad (3.4)$$

with

$$\frac{n^2}{\beta_R} = \frac{\pi}{4\sigma T^3} \int_0^\infty \frac{n_\lambda^2}{\beta_\lambda} \frac{dI_{b,\lambda}}{dT} \, d\lambda. \qquad (3.5)$$

where n is the index of refraction, σ is the Stefan-Boltzmann constant, T is temperature, β_R is the Rosseland mean extinction coefficient, n_λ is the index of refraction at wavelength λ, β_λ is the spectral extinction coefficient, and $I_{b,\lambda}$ is the Planck blackbody intensity function. Note that although β_R includes contributions from all wavelengths, generally the integral in equation (3.5) is calculated only over wavelengths for which absorbance data are available.

Note also that, in addition to the explicit T^3 dependence shown in equation (3.4), radiative thermal conductivity has a temperature dependence through the mean extinction coefficient $\beta_R(T)$ in equation (3.5). The mean extinction is calculated using the temperature derivative of the Planck distribution as a weighting function. Therefore, β_R is temperature dependent to the extent that the blackbody spectrum shifts with respect to the absorption spectrum with temperature. Since peaks in the

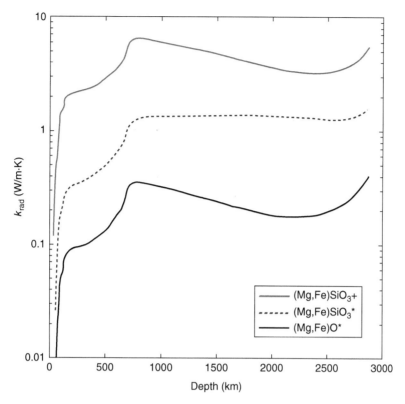

Figure 3.3 Estimated radiative thermal conductivity k_{rad} of (Mg,Fe)SiO$_3$ perovskite and (Mg,Fe)O ferropericlase as function of depth in mantle. The profiles denoted with * were calculated using absorption coefficients from *Goncharov et al.* [2009] for ferropericlase with 15 mol % Fe and *Goncharov et al.* [2008] for perovskite with 10 mol % Fe. The profile denoted with + was calculated using the absorption coefficient for perovskite with 10 mol % Fe from *Keppler et al.* [2008]. The assumed mantle adiabat appropriate for a convecting mantle with no midmantle thermal boundary layer is adopted from the model *of Jeanloz and Morris* [1986].

absorption spectra of mantle minerals also shift in wavelength with pressure [*Keppler et al.*, 2008], the radiative thermal conductivity of the mantle is expected to also depend on pressure. If significant windows or edges exist in the absorption spectrum, then k_{rad} may have complicated dependences on temperature [*Hofmeister and Yuen*, 2007].

Generally, the extinction coefficient β_λ is the sum of the absorption and scattering coefficients. Based on measured absorption coefficients, the photon mean free path in perovskite at high pressure is on the order of 10 μm [*Keppler et al.*, 2008; *Goncharov et al.*, 2008]. The scattering coefficient is on the order of the reciprocal of the grain size [*Shankland et al.*, 1979; *Hofmeister*, 2005], which is likely to be at least 0.1–1 mm in the lower mantle [*Solomatov et al.*, 2002]. Therefore, the photon mean free path for absorption in the lower mantle is much shorter than the expected scattering length. In the calculation that follows, we consider absorption only. However, if grains are small or the absorption coefficient very low, scattering must be accounted for, and the resulting effect on the radiative

thermal conductivity may be complex and nonlinear [*Hofmeister*, 2005].

Figure 3.3 shows recalculated profiles of k_{rad} for perovskite with 10 mol % iron and ferropericlase with 15 mol % iron constructed using absorption coefficients measured by *Keppler et al.* [2008] and *Goncharov et al.* [2006, 2008]. To create the profiles, k_{rad} was first calculated as a function of temperature at several pressures using equations (3.4) and (3.5). Values were then interpolated to calculate k_{rad} as a function of depth and temperature using a Preliminary Reference Earth Model (PREM) pressure profile [*Dziewonski and Anderson*, 1981] and the hot geotherm reported by *Jeanloz and Morris* [1986]. In addition, we used a model for depth-dependent index of refraction following a linear relationship between index of refraction and pressure [*Anderson and Schreiber*, 1965] and depth-dependent mantle densities from PREM [*Dziewonski and Anderson*, 1981]. This result shows that when treated consistently, the absorption data sets from *Keppler et al.* [2008] and *Goncharov et al.* [2008] differ by far less than the order-of-magnitude variation arising from the original presentation.

Note that in this study we neglected any temperature effect on the absorption coefficient when constructing the k_{rad} profiles. It is expected that high temperature might cause the absorption coefficient to increase, as has been measured for olivine at low pressure [*Shankland et al.*, 1979; *Hofmeister*, 2005]. The use of low-temperature absorption spectra, the choice of the higher-temperature geotherm, the assumption of limited grain boundary scattering, and the neglect of minor impurities all bolster the upper bound nature of this calculation.

Pressure and temperature have opposing effects on k_{rad}. Increasing temperature and density or refractive index increases k_{rad}; however, increasing pressure also tends to decrease k_{rad} due to increased optical absorption. As a result, the estimated k_{rad} for perovskite and ferropericlase shown in Figure 3.3 is constant or shallowly decreasing with depth throughout most of the lower mantle. Just above the core-mantle boundary, k_{rad} increases with depth in the conductive thermal boundary layer, where the temperature gradient is relatively steep.

Since the radiative thermal conductivities of perovskite and ferropericlase differ significantly, the total radiative thermal conductivity of the mantle is sensitive to how the total conductivity for different phases are combined, which depends on the proportion and geometry of the phases present and how the radiative conductivities combine in an effective media approximation. For example, a combination of 80 vol % perovskite and 20 vol % ferropericlase yields a range of radiative thermal conductivity of ~0.7–5 W/m·K throughout most of the lower mantle, with constant values through the convecting lower mantle and depth-dependent values through the core-mantle thermal boundary layer. This range of values encompasses the Reuss and Voigt bounds as well as the range of uncertainty implied by the difference in measured values of the perovskite absorption coefficient from *Goncharov et al.* [2008] and *Keppler et al.* [2008]. It should be noted, however, that this analysis assumes that the bulk radiative thermal conductivity can be calculated independently of the bulk lattice thermal conductivity.

At all depths in the lower mantle the radiative thermal conductivity of perovskite is larger by roughly a factor of 3 compared with that of ferropericlase (Figure 3.3). Generally, increasing transition metal impurities, especially iron, decreases optical absorption. There may be additional effects on optical absorption arising from depth-dependent electronic spin transitions in iron [*Badro et al.*, 2004; *Goncharov et al.*, 2006, 2010], with suggestions that the high-pressure low-spin state may have decreased absorption and thus enhanced radiative conductivity [*Badro et al.*, 2004]. Note that for ferropericlase the k_{rad} profile was calculated based on data for samples with higher iron content than that of the perovskite samples. However, if iron is preferentially partitioned into

ferropericlase in the lower mantle [*Badro et al.*, 2003; *Irifune et al.*, 2010], using higher-iron ferropericlase and lower-iron perovskite to construct profiles of k_{rad} for the lower mantle is at least qualitatively correct.

As temperature increases, the electron population in the conduction band increases, especially for the transition metal rich materials of the mantle. This increases the probability of electron scatter participation in the thermal conductivity, following the physics of the behavior of thermal conductivity in semiconductors. This may provide an additional contribution to the thermal conductivity that increases exponentially with temperature. When combined with the $1/T$ phonon temperature dependence, this serves to dampen the temperature-dependent decrease of thermal conductivity. As populations of charge carriers in the conduction band increase, this also serves to increase the optical absorption of minerals—they become more opaque. Therefore, semiconductor-like contributions and radiative contributions are likely mutually exclusive.

3.4. HEAT TRANSPORT ACROSS CORE-MANTLE BOUNDARY

The thermal conductivity in Earth's core-mantle boundary region governs the heat extracted from the core to the mantle and therefore ties together the thermal histories of these two major regions [*Lay et al.*, 2008]. In the thermal boundary layer of the lowermost mantle adjacent to the outer core, possible phase transformations, including the post-perovskite structure [*Murakami et al.*, 2004] and/or the presence of melt [*Williams and Garnero*, 1996], may change the conductivity compared with the bulk mantle. Measurements and theory suggest that the post-perovskite structure may have higher thermal conductivity [*Hunt et al.*, 2012; *Stackhouse et al.*, 2015]. The presence of partial melt lowers the thermal conductivity due to both an intrinsic effect arising from long-range lattice disorder and a likely decrease in any radiative transport [*Murakami et al.*, 2014]. Therefore, localized areas of partial melt and post-perovskite have opposite effects on the core-mantle boundary heat flux, both reinforcing increased heat flux in cooler regions.

Dynamical perturbations of the thermal boundary layer from above may have an additional important influence on the local thermal behavior by introducing lateral variations in thickness, influencing both the localized temperature gradient and the stable phase assemblage. Since the bridgmanite/post-perovskite phase boundary is likely to be favored at lower temperatures at core-mantle boundary pressures, the post-perovskite phase may be dynamically stabilized from mantle downwelling. Even if the thermal conductivity of the post-perovskite phase is identical to bridgmanite, the dynamical depression of

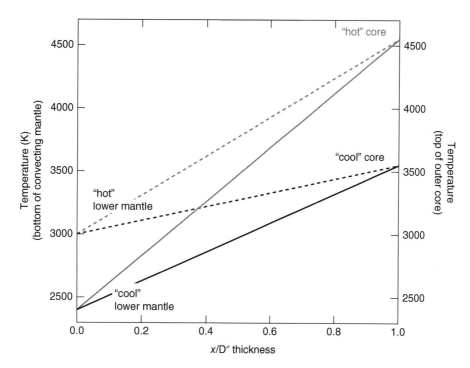

Figure 3.4 Range of estimated temperatures and temperature gradients in D″ region. The plot shows four different temperature profiles across the thermal boundary layer in the lowermost mantle constructed assuming constant thermal gradients between the temperature at the bottom of the convecting mantle and the temperature at the top of the outer core. The left axis shows two estimates for the temperature at the bottom of the convecting mantle. The lower temperature corresponds to a whole-mantle convecting system with little internal heating. The high temperature corresponds to a mantle with a thermal boundary layer and/or significant internal heating. The right axis shows the two bounds for best estimates of temperatures at the top of the outer core.

the thermal boundary layer will locally increase flux, reinforcing lateral variations in core-mantle boundary heat flow.

As is the case for the core, the estimates for thermal conductivity across the core-mantle boundary are strongly dependent on the assumption of the temperature profile across the core-mantle boundary. Figure 3.4 presents a range of estimates for the temperature profile from the bottom of the convecting mantle to the top of the outer core. For the range of temperatures at the bottom of D″ (top of the outer core), we use the same bounds presented in Figure 3.1a. For the top of D″ (bottom of the convecting mantle) we show two possible estimates, representing a low potential mantle temperature (if the whole mantle is convecting adiabatically) and a second, higher, potential temperature, which may be relevant either in the presence of an additional thermal boundary in the lower mantle or if there are significant radioactive sources within the D″ region. The result, are four different possible temperature profiles for the D″ layer.

The four corresponding estimates for the total average heat flux across the core-mantle boundary as a function of boundary thickness, temperature gradient, and bounds

on lower mantle thermal conductivity (4–8 W/m·K) are shown in Figure 3.5. Estimates for high average thermal conductivity of the core and low thermal conductivity for the lower mantle assemblage appear to provide a narrow solution space for satisfying the core-mantle boundary heat flux requirements—which must be large enough to sustain the core geodynamo and generate convection at that base of the mantle [e.g., *Roberts and Glatzmeier*, 2000; *Labrosse and Macouin*, 2003; *Hirose et al.*, 2013], yet not so large that it is inconsistent with the best estimates for internal heating of the mantle from radioactivity [e.g., *KamLAND*, 2011]. Given the uncertainty in the average thickness of the thermal boundary layer in the lowermost mantle and the strong relationship with total heat flux, invocation of a spatially variable heat flux at the core-mantle boundary, analogous to the situation at Earth's surface, is a practical and likely solution. Geophysical observations of the seismic structure at the core-mantle boundary suggest a strongly heterogeneous region, with patches of ultralow velocities, and areas showing larger-than-average seismic anisotropy [e.g., *Garnero and McNamara*, 2008]. This suggests that, in analogy with Earth's surface, the core-mantle boundary

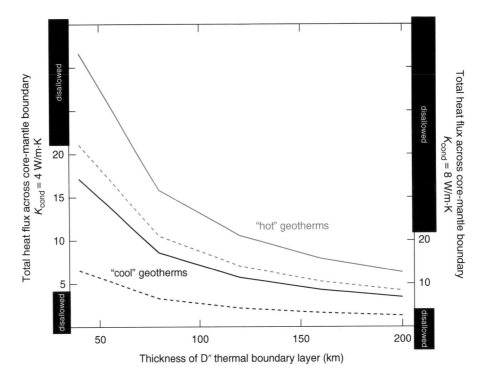

Figure 3.5 Estimates for total heat flux crossing core-mantle boundary as function of average thickness of thermal boundary layer. The left and right axes show calculations of the total heat flux crossing the core-mantle boundary for bounds on lower mantle thermal conductivities of 4 W/m·K (left axis) and for 8 W/m·K (right axis). Vertical black bars on the right and left axes block out regimes that are highly unlikely because a total heat flux of ~4 TW or more is required to power mantle convection and/or the geodynamo and a total heat flux greater than ~20 TW is not easily permitted by geochemical constraints [e.g., *KamLAND*, 2011]. The estimates for the amount of heat required to power the dynamo range from ~5 to 10 TW. Note that this plot represents an average thickness and does not account for the (likely) variations in the thickness of the thermal boundary layer in the lowermost mantle.

is likely a surface of heterogeneous heat flux, with small areas creating disproportionate contributions to the total heat flow, due to local variations in the thermal boundary layer. For example, localized high heat flux through the core-mantle boundary might be generated by dense, downgoing slabs which generate a locally higher temperature difference while also mechanically depressing the local boundary layer thickness. This would create localized areas of high heat flux at the core-mantle boundary, mapping to areas where downgoing slabs may have penetrated through the lower mantle.

In summary, while some models of core and mantle thermal conductivity can be excluded, uncertainties in the physical parameters of thermal conductivity at extreme conditions as well as the geophysical nature of the lowermost mantle still permit a wide range of models for whole-Earth thermal behavior. Both theory and experiments must work together in dual fashion to provide estimates of thermal conductivity values for the minerals comprising Earth's mantle and core and to investigate the underlying physical processes governing temperature, pressure, and compositional dependence of thermal conductivities to aid in extrapolating values to the extreme conditions of planetary interiors.

REFERENCES

Ammann, M. W., A. M. Walker, S. Stackhouse, J. Wookey, A. M. Forte, J. P. Brodholt, and D. P. Dobson, (2014), Variation of thermal conductivity and heat flux at the Earth's core mantle boundary, *Earth Planet. Sci. Lett.*, *390*, 175–185.

Anderson, O. L., and E. Schreiber (1965), The relation between refractive index and density of minerals related to the Earth's mantle, *J. Geophys. Res.*, *70*(6), 1463–1471, doi:10.1029/JZ070i006p01463.

Anzellini, S., A. Dewaele, M. Mezouar, P. Loubeyre, and G. Morard, (2013), Melting of Iron at Earth's inner core boundary based on fast X-ray diffraction, *Science*, *340*, 464–466.

Badro, J., G. Fiquet, F. Guyot, J.-P. Rueff, V. V. Struzhkin, G. Vanko, and G. Monaco (2003), Iron partitioning in Earth's mantle: Toward a deep lower mantle discontinuity, *Science*, *300*, 789–791, doi:10.1126/science.1081311.

Badro, J., J. P. Rueff, G. Vanko, G. Monaco, G. Fiquet, and F. Guyot (2004), Electronic transitions in perovskite: Possible nonconvecting layers in the lower mantle, *Science*, *305*(5682), 383–386.

Beck, P., A. F. Goncharov, V. V. Struzhkin, B. Militzer, H. K. Mao, and R. J. Hemley (2007), Measurement of thermal diffusivity at high pressure using a transient heating technique, *Appl. Phys. Lett.*, *91*(18), 181914.

Bi, Y., H. Tan, and F. Jing (2002), Electrical conductivity of iron under shock compression up to 200 GPa. *J. Phys. Condens. Matter.*, 14, 10,849–10,854; also Chinese Physi. Lett., *19*(2), 243–245.

Buffett, B. A. (2002), Estimates of heat flow in the deep mantle based on the power requirements for the geodynamo, *Geophys. Res. Lett.*, *29*(12), 7-1.

Clark, S. P. (1957), Radiative transfer in the Earth's mantle, *Trans. AGU.*, *38*, 931–938.

Dalton, D. A., W. P. Hsieh, G. T. Hohensee, D. G. Cahill, and A. F. Goncharov (2013), Effect of mass disorder on the lattice thermal conductivity of MgO periclase under pressure, *Sci. Reps.*, *3*, 2400, doi:10.1038/srep02400.

De Koker, N. (2010), Thermal conductivity of MgO periclase at high pressure: Implications for the D″ region, *Earth Planet. Sci. Lett.*, *292*(3), 392–398.

De Koker, N., G. Steinle-Neumann, and V. Vlček (2012), Electrical resistivity and thermal conductivity of liquid Fe alloys at high P and T, and heat flux in Earth's core, *Proc. Natl. Acad. Sci.*, *109*(11), 4070–4073.

Dekura, H., T. Tsuchiya, and J. Tsuchiya (2013), Ab initio lattice thermal conductivity of MgSiO$_3$ perovskite as found in Earth's lower mantle, *Phys. Rev. Lett.*, *110*, 025,904.

Deng, L., C. Seagle, Y. Fei, and A. Shahar (2013), High pressure and temperature electrical resistivity of iron and implications for planetary cores, *Geophys. Res. Lett.*, *40*, 33–37.

Dziewonski, A. M., and D. L. Anderson (1981), Preliminary reference Earth model, *Phys. Earth Planet. Inter.*, *25*, 297–356.

Fulkerson, W., J. P. Moore, and D. L. McElroy (1966), Comparison of the thermal conductivity, electrical resistivity, and Seebeck coefficient of a high purity iron and an Armco iron to 1000°C, *J. Appl. Phys.*, *37*, 2639, doi:10.1063/1.1782098.

Garnero, E. J., and A. K. McNamara (2008), Structure and dynamics of Earth's lower mantle, *Science*, *320*(5876), 626–628.

Gomi, H., and K. Hirose (2015), Electrical resistivity and thermal conductivity of hcp Fe–Ni alloys under high pressure: Implications for thermal convection in the Earth's core, *Phys. Earth Planet. Inter.*, *247*, 2–10, doi:10.1016/j.pepi.2015.04.003.

Gomi, H., K. Ohta, K. Hirose, S. Labrosse, R. Caracas, M. J. Verstraete, and J. W. Hernlund (2013), The high conductivity of iron and thermal evolution of the Earth's core, *Phys. Earth Planet. Inter.*, *224*, 88–103.

Goncharov, A. F., V. V. Struzhkin, and S. D. Jacobsen (2006), Reduced radiative conductivity of low-spin (Mg,Fe)O in the lower mantle, *Science*, *312*(5777), 1205–1208.

Goncharov, A. F., B. D. Haugen, V. V. Struzhkin, P. Beck, and S. D. Jacobsen (2008), Radiative conductivity in the Earth's lower mantle, *Nature*, *456*(7219), 231–234.

Goncharov, A. F., P. Beck, V. V. Struzhkin, B. D. Haugen, and S. D. Jacobsen (2009), Thermal conductivity of lower-mantle minerals, *Phys. Earth Planet. Inter.*, *174*, 24–32.

Goncharov, A. F., V. V. Struzhkin, J. A. Montoya, S. Kharlamova, R. Kundargi, J. Siebert, and W. Mao (2010), Effect of composition, structure, and spin state on the thermal conductivity of the Earth's lower mantle. *Phys. Earth Plane. Inter.*, *180*(3), 148–153.

Haigis, V., M. Salanne, and S. Jahn (2012), Thermal conductivity of MgO, MgSiO$_3$ perovskite and post-perovsite in the Earth's deep mantle, *Earth Planet. Sci. Lett.*, *355–356*, 102–108.

Hirose, K., S. Labrosse, and J. Hernlund (2013), Composition and state of the core, *Annu. Rev. Earth Planet Sci.*, *41*, 657–691.

Hofmeister, A. M. (2005), Dependence of diffusive radiative transfer on grain-size, temperature, and Fe-content: Implications for mantle processes, *J. Geodynam.*, *40*(1), 51–72.

Hofmeister, A. M. (2014), Thermal diffusivity and thermal conductivity of single-crystal MgO and Al$_2$O$_3$ and related compounds as a function of temperature. *Phys. Chem. Minerals*, *41*(5), 361–371.

Hofmeister, A. M., and D. A. Yuen (2007), Critical phenomena in thermal conductivity: Implications for lower mantle dynamics, *J. Geodynam.*, *44*(3–5), 186–199, doi:10.1016/j.jog.2007.03.002.

Hunt, S. A., D. R. Davies, A. M. Walker, R. J. McCormack, A. S. Wills, D. P. Dobson, and L. Li (2012), On the increase in thermal diffusivity caused by the perovskite to post-perovskite phase transition and its implications for mantle dynamics, *Earth Planet. Sci. Lett.*, *319*, 96–103.

Imada, S., K. Ohta, T. Yagi, K. Hirose, H. Yoshida, and H. Nagahara (2014), Measurements of lattice thermal conductivity of MgO to core-mantle boundary pressures, *Geophys. Res. Lett.*, *41*(13), 4542–4547.

Irifune, T., T. Shinmei, C. A. McCammon, N. Miyajima, D. C. Rubie, and D. J. Frost (2010), Iron partitioning and density changes of pyrolite in Earth's lower mantle, *Science*, *327*, 193–195, doi:10.1126/science.1181443.

Jeanloz, R., and S. Morris (1986), Temperature distribution in the crust and mantle, *Annu. Rev. Earth Planet. Sci.*, *14*, 377–415.

KamLAND Collaboration (2011), Partial radiogenic heat model for Earth revealed by geoneutrino measurements, *Nat. Geosci.*, *4*, 647–651.

Katsura, T. (1997), Thermal diffusivity of periclase at high temperatures and high pressures, *Phys. Earth Planet. Int.*, *101*, 73–77.

Keppler, H., L. S. Dubrovinsky, O. Narygina, and I. Kantor (2008), Optical absorption and radiative thermal conductivity of silicate perovskite to 125 gigapascals, *Science*, *322*(5907), 1529–1532.

Konopkova, Z., P. Lazor, A. F. Goncharov, and V. V. Struzhkin (2011), Thermal conductivity of hcp iron at high pressure and temperature, *High Press. Res.*, *31*(1), 228–236.

Labrosse, S., and M. Macouin (2003), The inner core and the geodynamo, *Comptes Rendus Geosci.*, *335*(1), 37–50.

Lay, T., J. Hernlund, and B. Buffett (2008), Core–mantle boundary heat flow, *Nature Geosci.*, *1*, 25–32.

Manthilake, G. M., N. de Koker, D. J. Frost, and C. A. McCammon (2011), Lattice thermal conductivity of lower mantle minerals and heat flux from Earth's core, *Proc. Natl. Acad. Sci. USA*, *108*, 17,901–17,904.

Matassov, G. (1977), *The electrical conductivity of iron-silicon alloys at high pressures and the Earth's core, Ph.D. thesis, Lawrence Livermore Laboratory Rep. UCRL-52322, 1977.*

McWilliams, R. S., Z. Konôpková, and A. F. Goncharov (2015), A flash heating method for measuring thermal conductivity at high pressure and temperature: Application to Pt, *Phys. Earth Planet. Inter.*, *247*, 17–26.

Murakami, M., K. Hirose, K. Kawamura, N. Sata, and Y. Ohishi (2004), Post-perovskite phase transition in MgSiO$_3$, *Science*, *304*(5672), 855–858.

Murakami, M., A. F. Goncharov, N. Hirao, R. Masuda, T. Mitsui, S. M. Thomas, and C. R. Bina (2014), High-pressure radiative conductivity of dense silicate glasses with potential implications for dark magmas. *Nature Communi.*, *5*, 5428, doi:10.1038/ncomms642.

Ohta, K., T. Yagi, N. Taketoshi, K. Hirose, T. Komabayashi, T. Baba, Y. Ohishi, and J. Hernlund (2012), Lattice thermal conductivity of MgSiO$_3$ perovskite and post-perovskite at the core-mantle boundary, *Earth Planet. Sci. Lett.*, 349–350, 109.

Ohta, K., T. Yagi, and K. Hirose (2014), Thermal diffusivities of MgSiO$_3$ and Al-bearing MgSiO$_3$ perovskites, *American Mineralogist*, *99*(1), 94–97.

Osako, M. and E. Ito (1991), Thermal diffusivity of MgSiO$_3$ perovskite, *Geophys. Res. Lett.*, *18*, 239–242.

Pozzo, M., C. Davies, D. Gubbins, and D. Alfè (2012), Thermal and electrical conductivity of iron at Earth's core conditions, *Nature*, *485*, 355–358.

Pozzo, M., C. Davies, D. Gubbins, and D. Alfè (2013), Transport properties for liquid silicon–oxygen–iron mixtures at Earth's core conditions, *Phys. Rev. B*, *87*, 014,110.

Pozzo, M., C. Davies, D. Gubbins, and D. Alfé (2014), Thermal and electrical conductivity of solid iron and iron–silicon mixtures at Earth's core conditions, *Earth Planet. Sci. Lett.*, *393*, 159–164.

Rainey, E. S. G., and A. Kavner (2014), Peak scaling method to measure temperatures in the laser-heated diamond anvil cell and application to the thermal conductivity of MgO, *J. Geophys. Res. Solid Earth*, *119*, 1–17, doi:10.1002/2014JB011267.

Roberts, P. H., and G. A. Glatzmaier, (2000), Geodynamo theory and simulations, *Rev. Modern Phys.*, *72*(4), 1081.

Rosseland, S. (1924), Electrical state of a star, *Monthly Notice R. Astron. Soc.*, *84*, 720.

Roufosse, M., and P. G. Klemens (1973), Thermal conductivity of complex dielectric crystals, *Phys. Rev. B*, *7*(12), 5379.

Seagle, C. T., E. Cottrell, Y. Fei, D. R. Hummer, and V. B. Prakapenka (2013), Electrical and thermal transport properties of iron and iron-silicon alloy at high pressure, *Geophys. Res. Lett.*, *40*, 20.

Shankland, T. J., U. Nitsan, and A. G. Duba (1979), Optical-absorption and radiative heat-transport in olivine at high-temperature, *J. Geophys. Res.*, *84*(NB4), 1603–1610.

Siegel, R., and J. R. Howell, (2002), *Thermal Radiation Heat Transfer*, Taylor & Francis, New York.

Solomatov, V. S., R. El-Khozondar, and V. Tikare (2002), Grain size in the lower mantle: constraints from numerical modeling of grain growth in two-phase systems, *Phys. Earth Planet. Inter.*, *129*, 265–282.

Stacey, F. D., and O. L. Anderson (2001), Electrical and thermal conductivities of Fe–Ni–Si alloy under core conditions, *Phys. Earth Planet. Inter.*, *124*(3), 153–162.

Stacey, F. D., and P. M. Davis (2004), High pressure equations of state with applications to the lower mantle and core, *Phys. Earth Planet. Inter.*, *142*(3), 137–184.

Stacey, F. D., and D. E. Loper (2007), A revised estimate of the conductivity of iron alloy at high pressure and implications for the core energy balance, *Phys. Earth Planet. Inter.*, *161*(1), 13–18.

Stackhouse, S., L. Stixrude, and B. B. Karki (2010), Thermal conductivity of periclase (MgO) from first principles, *Phys. Rev. Lett.*, *104*, 208,501–208,504.

Stackhouse, S., L. Stixrude, and B. B. Karki (2015), First-principles calculations of the lattice thermal conductivity of the lower mantle, *Earth Planet. Sci. Lett.*, *427*, 11–17.

Tang, X., and J. J. Dong (2009), Pressure dependence of harmonic and anharmonic lattice dynamics in MgO: A first-principles calculation and implications for lattice thermal conductivity, *Phys. Earth Planet. Inter.*, *174*, 33–38.

Tang, X., and J. J. Dong (2010), Lattice thermal conductivity of MgO at conditions of Earth's interior, *Proc. Nat. Acad. Sci.*, *107*(10), 4539–4543.

Tang, X., M. C. Ntam, J. Dong, E. S. G. Rainey, and A. Kavner (2014), The thermal conductivity of Earth's lower mantle, *Geophys. Res. Lett.*, *41*, 2746–2752, doi:10.1002/2014GL059385.

Williams, Q., and E. J. Garnero (1996), Seismic evidence for partial melt at the base of Earth's mantle, *Science*, *273*(5281), 1528–1530.

Zhang, P., R. E. Cohen, and K. Haule (2015), Effects of electron correlations on transport properties of iron at Earth's core conditions, *Nature*, *517*, 605–607.

4

Thermal State and Evolution of the Earth Core and Deep Mantle

Stéphane Labrosse

ABSTRACT

The thermal evolution of the Earth is generally studied in the context of lithospheric dynamics and subduction. However, the increasingly complex images of the deep Earth suggest that it can hold a key to understanding the early Earth and its evolution since then. After reviewing the way the thermal evolution of the Earth is classically adressed, recent findings on the evolution of the core are presented. It is shown that including a large core heat flow as a heat source for the mantle could help to solve the long-standing problem of the thermal evolution of the Earth. Implications for the evolution of the lower mantle with a basal magma ocean are then discussed.

4.1. INTRODUCTION

The thermal evolution of the Earth has been a debated subject for centuries and still is today. The reason for the importance of this question can be traced back to the first page of Sadi Carnot's book *Reflections on the Motive Power of Fire* (1824, p. 1) where it is stated that "to heat also are due the vast movements which take place on the earth. It causes the agitations of the atmosphere, the ascension of clouds, the fall of rain and of meteors, the currents of water which channel the surface of the globe, and of which man has thus far employed but a small portion. Even earthquakes and volcanic eruptions are the result of heat." Indeed, understanding heat transfer on Earth and the associated evolution is key to understanding its dynamics. The question has switched from that of the age of Earth to the understanding of plate tectonics, but it is still puzzling.

Whereas the problem of the thermal evolution of the Earth is usually adressed in terms of the dynamics of the lithosphere and its recycling, the role played by the deep Earth in the global evolution has not been adressed as

Univ Lyon, Ens de Lyon, Université Lyon 1, CNRS, UMR 5276 LGL-TPE, F-69364, Lyon, France

systematically. Several papers considered coupled models for the evolution of the mantle and core [e.g., *Stevenson et al.*, 1983; *Mollett*, 1984; *Yukutake*, 2000; *Grigné* and *Labrosse*, 2001; *Butler et al.*, 2005; *Nakagawa and Tackley*, 2005, 2010], but the question deserves to be reviewed in light of some recent findings. In particular, observations on the structure of the lower mantle suggest the existence of compositional variations and the presence of partially molten regions [see *Hernlund and McNamara*, 2015; *Labrosse et al.*, 2015, for recent reviews]. These structures result from the integrated evolution of the deep Earth which may hold clues of the early Earth [*Boyet and Carlson*, 2005]. These questions need to be integrated in a thermal evolution perspective [*Labrosse et al.*, 2007].

In addition, recent developments on the transport properties of the core have led to a reevaluation of the heat flow necessary at the core-mantle boundary (CMB) to sustain the geomagnetic field. That again links heat transfer to dynamics and needs to be integrated in our understanding of the thermal evolution of Earth.

This chapter will start by discussing the thermal evolution of the Earth as it is usually addressed (Section 4.2). Then, recent work on the evolution of the core will be reviewed, in particular the implications of the large

Deep Earth: Physics and Chemistry of the Lower Mantle and Core, Geophysical Monograph 217, First Edition.
Edited by Hidenori Terasaki and Rebecca A. Fischer.

thermal conductivity found in several independent studies (Section 4.3). Section 4.4 will discuss how these recent developments could change our view of the global evolution of the Earth as well as their implications for melting and crystallization processes in the deep mantle.

4.2. PRIMER ON THERMAL EVOLUTION OF EARTH

The question of the thermal evolution of the Earth, initially linked to the age of the Earth, has occupied researchers for several centuries, starting with count de Buffon, Joseph Fourier, and Lord Kelvin. (For an historical perspective on Kelvin's implication, see *Burchfield* [1975].) The discovery of radioactivity provided direct means of dating rocks, and the question has shifted and could be stated as follows: How can we reconcile thermal evolution models with the known age of the Earth and the constraints on the present rates of heat loss and radiogenic heat production? With the advent of plate tectonics, this question is related to the question of heat transfer by mantle convection and is still not solved today.

The thermal evolution of the Earth is classically adressed [see *Jaupart et al.*, 2015, for a complete review] by solving a time evolution equation expressing the conservation of the energy for the whole Earth,

$$MC\frac{dT}{dt} = -Q_S(T) + H(t), \qquad (4.1)$$

with M the mass of the Earth, C its specific heat, T its potential temperature, i.e., its temperature isentropically extrapolated to the surface, Q_S its heat loss, which is assumed to depend on T, and H the radiogenic heat production that depends explicitly on time t. In this equation C is an effective specific heat that contains all the effects of depth dependence of physical parameters, isentropic temperature variation, and so on. The present state of Earth is assumed to be known and can be constrained from present-day observations. In particular, the present potential temperature $T_0 = 1600\,K$, the total heat flow $Q_0 = 46\,TW$, and the radiogenic heat production rate $H_0 = 20\,TW$ have been obtained with reasonable confidence [*Jaupart et al.*, 2015]. Equation (4.1) can then be solved backward in time to get the evolution of the Earth provided the function $Q_S(T)$ is known.

The surface heat loss is controlled by the rate at which mantle convection transports heat toward the surface and can be shown to scale with the Rayleigh number Ra as [e.g., *Turcotte and Oxburgh*, 1967; *McKenzie et al.*, 1974; *Christensen*, 1984; *Sotin and Labrosse*, 1999; *Schubert et al.*, 2001]

$$Q_S = AS\frac{k\,\Delta T}{d}Ra^\beta\left(\frac{T}{\Delta T}\right)^{1+\beta}, \qquad (4.2)$$

with A a dimensionless coefficient, S the surface of the Earth, k the thermal conductivity of the mantle, ΔT the superisentropic temperature difference across the mantle, and

$$Ra = \frac{\alpha g\,\Delta T d^3}{\kappa v}, \qquad (4.3)$$

α being the coefficient of thermal expansion, g the acceleration of gravity, d the thickness of the mantle, κ the thermal diffusivity, and v the kinematic viscosity. Typically, experiments and numerical models show that the exponent $\beta \simeq 1/3$. This can be understood as expressing the fact that convection at a high Rayleigh number is controlled by processes happening in a thin boundary layer, the lithosphere in the case of the mantle, and therefore the heat flow should be independent of the thickness of the whole mantle, d, which is the behavior obtained for $\beta = 1/3$.

Assuming equation (4.2) applies to mantle convection, all the physical parameters need not be obtained independently since the present values T_0 and Q_0 also satisfy the scaling law. However, the viscosity needs special care since it depends strongly on temperature. Denoting $v_0 = v(T_0)$, the total surface heat flow can be written as

$$Q_S = Q_0\left(\frac{v(T)}{v_0}\right)^{-\beta}\left(\frac{T}{T_0}\right)^{1+\beta}. \qquad (4.4)$$

Including this expression in the energy equation (4.1) allows to compute the evolution of the temperature, and therefore the heat flow, forward or backward in time, from any known state. The only time for which constraints exist is the present time, and it is therefore more logical to start from that and compute the time-backward evolution.

Christensen [1985] followed the procedure explained above and explored systematically the effect of varying the exponent β and the present amount of radiogenic heating. More precisely, he introduced the dimensionless Urey number defined as the fraction of the present-day surface heat flow that can be attributed to present-day radiogenic heating:

$$Ur = \frac{H_0}{Q_0}. \qquad (4.5)$$

Figure 4.1 shows the results obtained. For $\beta = 0.3$, which is typically the value obtained in experiments and numerical models of convection, the range of Urey numbers leading to acceptable evolution is very narrow, too narrow to be written on the figure. As explained by *Christensen* [1985], the timescale for adjustment of the mantle to changes of conditions, and in particular radiogenic heating rate, can be computed by linearizing equation

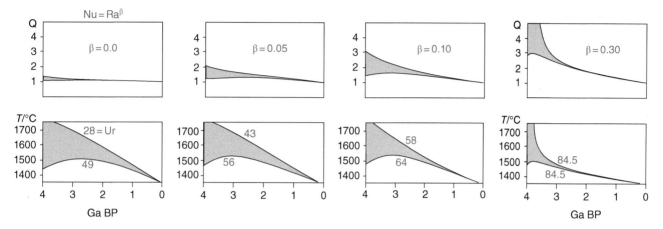

Figure 4.1 Evolution of potential temperature (bottom row) and surface heat flow (relative to its present value, top row) in time-backward calculation of thermal history of Earth as function of the present-day Urey number (whose value is given multiplied by 100) and the value of the β exponent entering the heat flow scaling law. For each value of β, the maximum and minimum values of the Urey number are given in the lower panel and provide the minimum and maximum acceptable values of the temperature in the grey shaded range, respectively. Adapted from *Christensen* [1985].

(4.1) around the present state and is around 800 Myr for $\beta = 0.3$ and reasonable rheological parameters. This time is much shorter than the age of the Earth and this implies that the thermal evolution is insensitive to initial conditions: after a few times the adjustment timescale, the Urey number reaches a constant value. However, a simple ratio of the present heat content of the Earth to its present heat loss gives a timescale of about 10 Gyr [*Labrosse and Jaupart*, 2007], which shows that Earth has not yet reached a quasi-stationary state. *Albarède* [2005] proposed an analogy between this timescale and the residence time of any chemical element.

In order to solve that problem, *Christensen* [1985] proposed to reduce the value of β, which effectively increases the adjustment timescale [*Labrosse and Jaupart*, 2007], and indeed, as shown in Figure 4.1, it allows to enlarge the range of acceptable present Urey numbers and decrease its value toward the geochemically constrained one. As a justification for such a low value of β, *Christensen* [1985, p. 3000] states: "The driving forces of plate motion are balanced by local resisting forces in the subduction region. The cause for this resistance may be (1) shear on the thrust fault toward the overriding plate, (2) resistance against the bending of the downgoing plate, (3) resistance against penetration into a high viscosity layer at greater depth, (4) resistance against bending or viscous deformation at the boundary between upper and lower mantle, or (5) resistance against penetration through an endothermic phase boundary. In cases 1, 2, and 5 the resisting force would be entirely independent from the asthenospheric temperature and viscosity, possibly also in case 4." Since that original proposal, many studies have followed that idea and argued for a small value of β, mainly assuming that resistance against plate bending controls the rate of subduction [e.g., *Conrad and Hager*, 1999a, b; *Sleep*,

2000; *Korenaga*, 2003, 2006] or that the maximum age of subducting plates is controlled by geometrical factors [*Labrosse and Jaupart*, 2007].

However, the assumptions behind the scaling laws proposed to have a small value of β, i.e., the fact that the buoyancy of plates is balanced by resistance to bending at subduction zones with a given average radius of curvature, need to be validated using fully self-consistent mantle convection models or experiments in which these assumptions are not imposed a priori. To date, it has not been the case. *Grigné et al.* [2005] used a model of convection with temperature-dependent viscosity and pseudo-plastic yielding in a plate-tectonic-like regime and showed that indeed it followed the classical scaling law with $\beta = 1/3$. *Korenaga* [2010] performed a more systematic study with a similar model and also obtained $\beta \sim 1/3$. Therefore, it seems that our understanding of heat transport by mantle convection cannot be reconciled with the constraints on the thermal evolution of Earth, and this problem has often received the qualification of paradox.

However, equation (4.1) is parameterized by one temperature only, the potential temperature of the mantle. Implicitly, this assumes that the temperatures of the core and the mantle evolve at the same rate. It is therefore necessary to question the validity of this assumption and see if an explicit contribution of the core could help solve the problem of the thermal evolution of the Earth.

4.3. THERMAL EVOLUTION OF CORE

The thermodynamics of the thermal evolution of the core has been worked out for some time for the case of a fully convective core [*Gubbins et al.*, 1979; *Lister and Buffett*, 1995; *Braginsky and Roberts*, 1995; *Labrosse et al.*,

1997; *Labrosse*, 2003; *Gubbins et al.*, 2004; *Nimmo et al.*, 2004; *Nimmo*, 2015; *Labrosse*, 2015] and for the case of a thermal stratification at the top of the core [*Labrosse et al.*, 1997; *Lister and Buffett*, 1998, *Gomi et al.*, 2013], and the requirements for the maintenance of dynamo action by convection are well established. The theory relies on the development of each variable as the sum of time- and depth-dependent reference profiles and the fluctuations around those profiles that average to zero when computing global quantities. The energy and entropy balances of the core then depend only on the reference profiles and their evolution on the large timescale relevant to the secular evolution of the core. The full theory need not be developed here and only its results will be presented below, focusing mostly on the effect of the thermal conductivity. For a recent account and detailed model, the readers are referred to *Labrosse* [2015].

In the classical view of the geodynamo driven by thermocompositional convection, the outer core is assumed to be, on average, well mixed and isentropic. Using, as state variables, pressure P, specific entropy s, and concentrations in light elements ξ_i, all quantities in the average state only vary with pressure. Using classical thermodynamics, the derivatives of these quantities with pressure can be expressed and provide differential equations that can be integrated. For example, the temperature derivative can be written as

$$\left(\frac{\partial T}{\partial P}\right)_{s,\xi_i} = \frac{\alpha T}{\rho C_p}, \qquad (4.6)$$

with ρ the density, α the coefficient of thermal expansion, and C_p the specific heat at constant pressure. This defines the isentropic temperature gradient. It is often termed "adiabatic" but it is quite improper since (1) heat is transported by conduction along it and its qualification of adiabatic has an oxymoron character and (2) if it is to be understood as the temperature change experienced by a fluid parcel moving vertically without exchanging heat (which is the classical view), it should include the effect of heat production from radioactivity and dissipation [*Jaupart et al.*, 2015]. Even though all these terms produce entropy, when using equation (4.6) we assume that convection is efficient enough to mix the fluid back to a uniform specific entropy, at least when averaged on a convective timescale.

Other derivatives can also be expressed in similar ways, and this permits the calculation of reference profiles for these quantities as a function of pressure or radius by using the hydrostatic balance that is also assumed to hold on average provided the right-hand-side terms can be expressed as a function of pressure or radius and an anchor point (boundary condition) can be defined. The first condition can usually be met, possibly using some thermodynamic identities. The boundary condition is best defined by a place where some thermodynamic equilibrium can be assumed. The inner core boundary is such a place since the equilibrium between the solid and the liquid implies that the temperature is equal to the liquidus of the outer core material and the chemical potentials must be continuous. The core liquidus depends on pressure and composition and is not perfectly known, but the discrepancies among different teams and methods of investigation have reduced in the past years [see *Hirose et al.*, 2013, for a review]. This dependence of the liquidus implies a dependence of the core reference state on the radius of the inner core, both from the variation of pressure with radius and from the change of composition due to inner core growth.

At the inner core boundary, freezing of the outer core alloy leads to fractionation of light elements. This phenomenon has profound implications for the dynamics and evolution of the core. The presence of light elements in the outer core and their unequal partitioning between the solid and liquid phases lead to a decrease of the liquidus compared to the melting temperature of pure iron [e.g., *Alfè et al.*, 2007], and the progressive growth of the inner core makes the concentration of the outer core in light elements increase with time. One implication is that the liquidus decreases more than from the pressure effect only. In addition, the release of light elements at the inner core boundary makes the fluid surrounding it lighter than the bulk of the outer core, which drives compositional convection. This motion stems from the gradient of the chemical potential in the outer core and the associated compositional energy depends on the profile of the chemical potential.

The inner core (IC) plays an important role at present in defining the average profiles in the outer core (OC). Since the different energy terms in the balance come from time derivatives of these profiles, they can be parameterized by the inner core radius r_{IC}. The energy balance equation can then be written as

$$Q_{CMB} = F(r_{IC})\frac{dr_{IC}}{dt} + H_{core}(t) \qquad (4.7)$$

with H_{core} the time-dependent radiogenic heating rate and F a function of the inner core radius that sums contributions from secular cooling, latent heat, and compositional energy. Detailed expressions for these terms can be found in *Labrosse* [2015].

Similarly, the entropy balance of the inner core can be expressed as a function of the inner core radius and its growth rate, with an efficiency factor η for conversion of each energy source into dissipation that can be expressed as a function of temperature in the core and

$$\Phi + T_\Phi \Sigma = \eta_{IC} F(r_{IC})\frac{dr_{IC}}{dt} + \eta_H H_{core}(t), \qquad (4.8)$$

with Φ the total "useful" dissipation, i.e., related to flow and dynamo action (viscous and Ohmic dissipation), T_Φ the temperature at which this dissipation effectively occurs, η_{IC} the combined efficiency for all sources related to inner core growth and η_H that for radiogenic heating, and

$$\Sigma = \int_{V_{OC}} k\left(\frac{\nabla T}{T}\right)^2 dV \qquad (4.9)$$

is the entropy production from conduction, mostly along the isentropic temperature gradient, with k the thermal conductivity. This term represents a toll on the energy sources that needs to be paid for the isentropic gradient to be maintained. Because of the large conductivity of the core, this term turns out to be quite large, as discussed further below.

These two energy equations are only valid when an inner core is present. However, this has not always been the case [e.g., *Stevenson et al.*, 1983; *Buffett et al.*, 1992; *Labrosse et al.*, 2001], and before the onset of the inner core crystallization, the average state of the core can be parameterized by the temperature at the center, T_c, or at the CMB. Using the former, the energy and entropy equations can be written as

$$Q_{CMB} = G\left(T_c\right)\frac{dT_c}{dt} + H_{core}\left(t\right) \qquad (4.10)$$

and

$$\Phi + T_\Phi \Sigma = \eta_G G\left(T_c\right)\frac{dT_c}{dt} + \eta_H H_{core}\left(t\right). \qquad (4.11)$$

These equations make clear that, using the temperature at the center as the control parameter for the thermal structure, the energy and entropy terms associated with the cooling of the core are proportional to the evolution rate of that temperature, with the function $G(T_c)$ a factor in the energy balance equation and η_G the efficiency factor for the conversion to the entropy balance.

The energy and efficiency equations can be used to constrain the thermal evolution of the Earth. The total dissipation in the core is poorly constrained [e.g., *Roberts et al.*, 2003; *christensen*, 2010] but must be positive for convection to occur. Figure 4.2 shows how the core cooling rate just before the inner core started to crystallize (Figure 4.2) and the present inner core growth rate (Figure 4.2) depend on the assumed total dissipation. Calculation of the inner core growth with time [e.g., *Buffett et al.*, 1992; *Labrosse et al.*, 1997; *Labrosse* 2015] shows that it typically follows a power law function, $r_{IC} = r_{IC}(0)\left(1+t/a_{IC}\right)^\delta$, with a_{IC} the age of the inner core, t time (negative in the past, zero at present), and δ an exponent that is around 0.4, the value used here to compute the age of the inner core from its present growth rate. Figure 4.2 shows that the age of the inner core is less than 1.5 Gyr, even including 200 ppm of

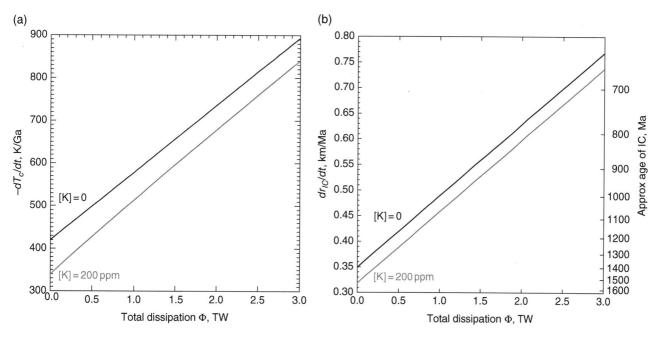

Figure 4.2 Core cooling rate just before onset of inner core crystallization (a) and present inner core growth rate (b) as function of total dissipation. Also given on the right axis of (b) is the approximate age of the inner core in Myr. Black lines are without radioactivity whereas red lines assume a 200 ppm concentration of potassium in the present core. Adapted from *Labrosse* [2015].

potassium in the core, which is a large upper bound [*Rama Murthy et al.*, 2003; *McDonough*, 2003; *Hirose et al.*, 2013]. Additionally, the cooling rate of the core before the existence of the inner core is found to be larger than 400 K/Gyr.

This figure and the associated calculations assume a large thermal conductivity [*Labrosse*, 2015], as suggested by several recent independent studies [*Sha and Cohen*, 2011; *de Koker et al.*, 2012; *Pozzo et al.*, 2012; *Gomi et al.*, 2013]. However, these values are still disputed. *Seagle et al.* [2013], performed ambient temperature–high pressure measurements of electrical conductivity, from which the thermal conductivity is computed using the Wiedemann-Franz law [*Poirier*, 2000], and obtained results quite similar to that of *Gomi et al.* [2013]. The two studies differ in their extrapolation to core temperatures. The classical theory uses the Bloch-Grüneisen formula, which in the range of interest here gives a linear increase

with temperature of the electrical resistivity. The combination of this linear increase with the Wiedemann-Franz law makes the thermal conductivity independent of temperature [*Poirier*, 2000]. *Gomi et al.* [2013], on the other hand, include a saturation effect at high temperature which makes the electrical resistivity level off to a saturation value. This effect is well documented in the metallurgy literature [*Gunnarsson et al.*, 2003] and has only been experimentally confirmed recently for iron at high pressure by *Ohta et al.* [2014].

Another question has been raised by *Zhang et al.* [2015], who used ab initio techniques to compute electron-electron scattering and the effect it has on the electrical resistivity of iron at high pressure and high temperature. They found the effect significant, contrary to the assumption made in previous ab initio calculations [*Sha and Cohen*, 2011; *de Koker et al.*, 2012; *Pozzo et al.*, 2012]. Of course, this question is important but does not affect

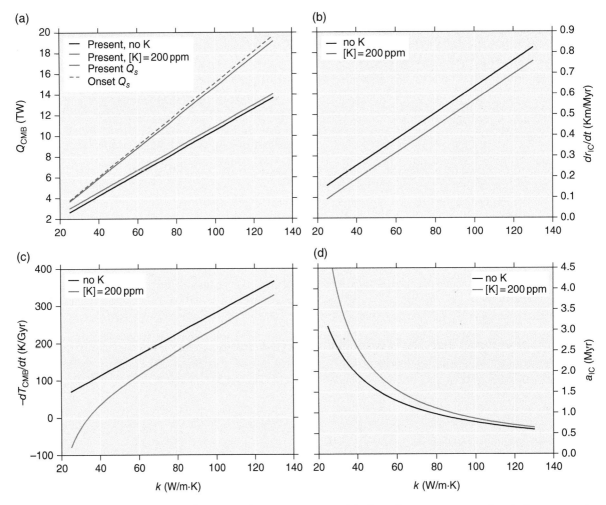

Figure 4.3 Effect of thermal conductivity value on minimum CMB heat flow at present or just before onset of inner core crystallization (a), minimum present growth rate of the inner core (b), minimum cooling rate of core before onset of inner core (c), and maximum age of inner core (d). The thermal conductivity is assumed constant in this calculation, which minimizes the demands on the CMB heat flow.

experimental results, which of course contain all the physical ingredients at no extra cost. In that respect, the results by *Ohta et al.* [2014] support the saturation effect and therefore the low value of the electrical resistivity and argue against the importance of electron-electron interactions.

The value of the core thermal conductivity being still debated, it is worth considering the effect this uncertainty has on the thermal evolution of the core. Figure 4.3 shows the results of a calculation of the minimum requirements to run a dynamo at present and just before the onset of the inner core as a function of the thermal conductivity in the core. The requirements are minimized by considering a negligible "useful" dissipation and by assuming the thermal conductivity does not increase with depth in the core, contrary to what has been found by recent studies on the topic. The range of conductivities covers the full range of values proposed in the literature. For the period just before the onset of inner core crystallization, the minimum CMB heat flow is considered to be the isentropic value, Q_S.

Figure 4.3 shows that the recent high values of the thermal conductivity ($k \geq 90\,\mathrm{W/m\cdot K}$) imply that the present CMB heat flow must be larger than about 10 TW and larger than about 14 TW before the onset of the inner core crystallization. Including 200 ppm of potassium makes little difference [*Labrosse*, 2015] and in fact requires a larger CMB heat flow because it reduces the rate of inner core growth and decreases the amount of associated compositional energy, which has a higher efficiency than radiogenic heating. For the same large values of thermal conductivity, the age of the inner core is found to be less than about 1 Gyr (Figure 4.3), which means that the cooling rate shown in Figure 4.3 applies to a large part of the history of the core. Of course, these minimum requirements are reduced by considering a low value of the thermal conductivity. For example, taking $k = 50\,\mathrm{W/m\cdot K}^1$, which is in the middle of the range advocated by *Seagle et al.* [2013], gives an inner core age around 1.5 Gyr and cooling rate prior to that of about 150 K/Gyr. The CMB heat flow must still exceed 8 TW to run a dynamo by thermal convection prior to the existence of the inner core. And one should remember that these estimates are extreme lower bounds that leave no room for Ohmic and viscous dissipation.

4.4. IMPLICATIONS FOR THERMAL EVOLUTION OF EARTH

4.4.1. Modified Urey Number

As discussed in the previous section, the recent upward revision of the core thermal conductivity [even using the lower values proposed by *Seagle et al.*, 2013] pushes

upward the demand on the heat flow across the CMB to maintain the geodynamo for all its known history. This means that the thermal evolution of the Earth cannot be modeled with only one energy balance equation (4.1) and one temperature. Without writing a detailed equation for both the mantle and the core, let us consider them as parameterized by two independent temperatures, T_m for the mantle and T_c for the core. Two energy conservation equations are then written to describe the evolution of these two temperatures:

$$M_m C_m \frac{dT_m}{dt} = -Q_S(T_m) + H(t) + Q_{\mathrm{CMB}}(T_m, T_c), \quad (4.12)$$

$$M_c C_c \frac{dT_c}{dt} = -Q_{\mathrm{CMB}}(T_m, T_c), \quad (4.13)$$

with Q_{CMB} the heat flow across the CMB and indices c and m for core and mantle quantities, respectively. Note that, according to equations (4.2) and (4.3) for $\beta = 1/3$, the surface heat flow depends only on the temperature of the mantle and not on ΔT. On the other hand, the heat flow across the CMB a priori depends on both mantle and core temperatures. In writing equation (4.13), the possibility of core radioactivity and the effects of compositional and latent heat have been neglected for the sake of simplicity.

Equation (4.12) makes clear that, for the mantle, the cooling of the core represents a heat source that supplements radiogenic heating. Therefore, we can define a modified mantle Urey number as the total heat sources to the mantle divided by the total heat loss,

$$\mathrm{Ur}^\star = \frac{H + Q_{\mathrm{CMB}}}{Q_S}. \quad (4.14)$$

Using the constraints on the CMB heat flow from the requirements of maintaining the geodynamo for at least 3.5 Gyr, a minimum of Q_{CMB} is around 16 TW [*Labrosse*, 2015]. Taking $H = 20$ TW and $Q_S = 46$ TW [*Jaupart et al.*, 2015] gives $\mathrm{Ur}^\star \geq 0.78$, a value much more likely to allow a reasonable thermal evolution than the standard one. Such a large value of the CMB heat flow is above the classicaly considered range, 5–15 TW [e.g., *Lay et al.*, 2008], and, on the core side, critically depends on the value of the thermal conductivity, as discussed above and displayed in Figure 4.3. The lower end values recently proposed [*Seagle et al.*, 2013; *Zhang et al.*, 2015] would give a minimum CMB heat flow around 8 TW, which would clearly be in the classical range but it would be of less help to solve the thermal evolution problem since it would give $\mathrm{Ur}^\star \geq 0.6$. Note that the value $H = 20$ TW for the current radiogenic heat production derives from classical "chondritic" Earth models. Alternative models based on enstatite chondrites [*Javoy*, 1999; *Javoy et al.*, 2010,] lead to a somewhat lower value of radiogenic heating and so do models assuming collisional erosion of the primordial crust [*O'Neill and Palme*, 2008; *Caro et al.*, 2008; *Jackson and Jellinek*, 2013; *Bonsor et al.*, 2015]. In that case, a larger value of the CMB heat flow is necessary to avoid the thermal catastrophe in thermal evolution models.

If a lower thermal conductivity is assumed for the core, the minimum requirement on the CMB heat flow is reduced, as explained in Section 4.3. However, it does not mean that the CMB heat flow cannot be much larger than this minimum requirement. In fact, fluctuations with time of the CMB heat flow, which is inherent to mantle convection [e.g., *Nakagawa and Tackley*, 2010], would lead to periods of extinctions of the dynamo if the long-term average is too close to the minimum required. A value at least 30% larger is necessary to avoid these extinction periods [*Nakagawa and Tackley*, 2010].

In order to clarify the difference with what has been assumed in classical models of Earth thermal evolution, we can estimate the CMB heat flow corresponding to a core cooling at a rate equal to that of the mantle, about 50 K/Gyr on average over the last 4 Gyr [*Abbott et al.*, 1994; *Jaupart et al.*, 2015]. Such a cooling rate corresponds to a CMB heat flow equal to 2.8 TW. This is not enough to solve the thermal catastrophe in the thermal evolution model, and this is clearly not enough to maintain the dynamo prior to the existence of the inner core, even if the thermal conductivity is taken as low as 50 W/m·K. In that case, about 200 ppm of potassium at present in the core would be just enough to maintain a thermal dynamo before the onset of the inner core (Figure 4.3), but this is a much larger value than can be accepted on geochemical and mineralogical grounds [*Hirose et al.*, 2013]. *Wohlers and Wood* [2015] have recently proposed inclusion of U and Th in the core, producing up to 2.4 TW of radiogenic heat. This would help in maintaining the geodynamo while reducing slightly the cooling rate of the core, but not down to 50 K/Gyr.

4.4.2. Importance of Melt in the Evolution of the Deep Mantle

Computation of the thermal evolution of the core with a high CMB heat flow required to maintain the geodynamo leads to large core temperatures in the past. The minimum core temperature drop over the last 3.5 Gyr, which is the period for which the existence of the magnetic field is documented [*Usui et al.*, 2009], is 2350 K in the absence of any radiogenic heating [*Labrosse* 2015]. Using the low estimate of core conductivity ($k = 50$ W/m·K) would give at least 450 K variation for the same period (Figure 4.3). According to the recent work on the melting temperature of silicates at high pressure

[*Stixrude et al.*, 2009; *de Koker and Stixrude*, 2009; *Fischer and Campbell*, 2010; *Andrault et al.*, 2011, 2014; *Fiquet et al.*, 2010; *Nomura et al.*, 2014], the present temperature of the CMB is close to the solidus value of complex compositions that can be encountered in the mantle. The observation of ultralow-velocity zones can easily be explained by the presence of partially molten patches at the CMB [e.g., *Williams and Garnero*, 1996; *Lay et al.*, 2004; *Rost et al.*, 2005; *Hernlund and Jellinek*, 2010; *Wimert and Hier-Majumder*, 2012; *Labrosse et al.*, 2015]. Alternatively, the seismically observed ultralow velocities could be explained by Fe-rich solids [*Dobson and Brodholt*, 2005; *Mao et al.*, 2006; *Wicks et al.*, 2010], but their melting temperature is likely to be lower or very close to the present temperature at the CMB. In any case, the amount of cooling the core must have experienced to explain the persistence of the magnetic field for at least 3.5 Gyr ensures that a large part of the lower mantle must have been molten in the past. This is the main argument behind the proposition of a basal magma ocean (BMO) [*Labrosse et al.*, 2007].

The existence of a BMO for a large part of Earth history changes many thing for the thermal evolution of Earth. *Labrosse et al.* [2007] showed how it can be included in a simple parameterized model. The timescale to crystallize completely the BMO is of the order of the age of Earth because of the foreseen influence of composition on the liquidus of the magma and of the large heat capacity of the core. However, the model for the evolution of the BMO and core has not been coupled yet to a fully consistent model for heat transfer in the solid mantle, and the possibility of solving the long-standing problem of the thermal evolution of Earth by including a large cooling rate of the core and the implied basal magma ocean is not proved. Several aspects of the problem need to be considered to this end. First, a proper treatment of fractional crystallization is necessary, with implications on the partitioning of heat-producing elements (U, Th, and K) between the BMO and solid mantle and on the evolution of the density of both the solid and the liquid. *Labrosse et al.*, [2007] proposed that the solid crystallizing from the BMO should get denser with progression of the crystallization and could eventually stabilize against entrainment by mantle convection, and it could explain some of the seismic observations of the deep mantle [*Labrosse et al.*, 2015]. This would also influence heat transfer in the solid mantle, and including that aspect of the problem in a mantle convection model is still a challenge.

Another important aspect to include in the theory regards the possibility of melting and freezing at the bottom of the mantle for a long period of Earth's evolution. Specifically, one can expect that, early on, downwelling currents coming from the surface, e.g., in the form of subducted plates, could melt when getting in contact with the hot magma. If melting happens fast enough, i.e., at the same rate as material is brought down, it does not need to slow down when getting close to the bottom surface, contrary to what happens in current models of mantle convection, where a nonpenetrable interface is usually considered at the bottom [*Deguen*, 2013]. A similar situation is encountered in the dynamics of inner core convection [*Aboussiére et al.*, 2010; *Deguen et al.*, 2013]. Such a process would enhance heat transfer by mantle convection and needs to be considered in the future.

4.5. CONCLUSIONS

The long-standing question of the thermal evolution of the Earth has generally focused on the efficiency of heat transfer by convection in the solid mantle, in particular on the dynamics of subduction, since it appears to be the main limiting factor. In doing so, most studies did not consider explicitly the deep Earth, the lower mantle and core, with a few exceptions [e.g., *Stevenson et al.*, 1983; *Mollett*, 1984; *Yukutake*, 2000; *Grigné and Labrosse*, 2001; *Butler et al.*, 2005; *Nakagawa and Tackley*, 2005, 2010], and implicitly assumed the core to cool at the same rate as the mantle. However, the thermal conductivity of the core has been recently evaluated by several teams using ab initio calculations and high-pressure experiments, and, despite some remaining discrepancies between the different studies, all show that the CMB heat flow must be larger than previously thought in order to maintain the geodynamo. As shown in Sections 4.3 and 4.4.1, an average core cooling rate at least twice that of the mantle is required even with the most conservative values of the thermal conductivity. Including this into a modified Urey number for the mantle may allows one to reconcile thermal evolution models of Earth with the classical power law scaling of heat transfer by mantle convection as a function of the Rayleigh number.

The large cooling rate of the core, implied by the thermodynamics of the geodynamo and considered as a potential solution to the problem of the thermal evolution of Earth, bears important consequences for the deep Earth. The present proximity of the CMB temperature to the solidus of potential mantle assemblages, as well as seismic evidences for the actual presence of partially molten regions at the bottom of the mantle [*Williams and Garnero*, 1996], indicate that a large part of the lowermost mantle must have been extensively molten in the past. Several conditions are necessary for this BMO model to be acceptable [*Labrosse et al.*, 2007]. For example, the density of the melt must be larger than that of the overlying solid mantle, which requires a Fe-rich composition.

In order to advance on this question, several steps are needed. On the mineral physics front, a lot of progress has been made on the melting relations at the relevant pressure [*Fiquet et al.*, 2010; *Andrault et al.*, 2011; *Thomas et al.*, 2012; *Mosenfelder et al.*, 2007; *Thomas et al.*, 2012; *Andrault et al.*, 2014; *de Koker et al.*, 2013; *Labrosse et al.*, 2015, for a review], and *Boukaré et al.* [2015] have proposed a phase diagram based on self-consistent thermodynamics using results from these recent studies. Inclusion of the effect of minor phases is still needed. Also, more work is needed to resolve the discrepancies remaining between the different studies on the core thermal conductivity, which impacts the constraints on the evolution of the core.

On the geodynamics front, a lot needs to be done in order to incorporate the possibility of melting and freezing at the bottom of the solid mantle. The challenges are great since it means dealing with a moving boundary, a rheological transition, fractional freezing and melting, and therefore a self-consistent treatment of composition in both the solid mantle and the basal magma ocean. Additionally, two-phase aspects are key to understanding the present-day observations in the deep Earth [*Hernlund and Jellinek*, 2010] and that brings a whole other dimension to the problem.

ACKNOWLEDGMENTS

I am thankful to the editors, Hidenori Terasaki and Rebecca Fisher, for soliciting this chapter and two anonymous reviewers for their constructive remarks.

REFERENCES

Abbott, D., L. Burgess, and J. Longhi (1994), An empirical thermal history of the Earth's upper mantle, *J. Geophys. Res.*, *99*, 13,835–13,850.

Albarède, F. (2005), The survival of mantle geochemical heterogeneities, in *Structure, Composition, and Evolution of Earth's Mantle*, Geophys. Monogr. Ser., Vol. 160, edited by R. van der Hilst, J. Bass, J. Matas, and J. Trampert, pp. 27–46, AGU, Washington, D.C.

Aboussière, T., R. Deguen, and M. Melzani (2010), Melting-induced stratification above the Earth's inner core due to convective translation, *Nature*, *466*(7307), 744–747.

Alfè, D., M. J. Gillan, and G. D. Price (2007), Temperature and composition of the Earth's core, *Contemp. Phys.*, *48*, 63–80, doi:10.1080/00107510701529653.

Andrault, D., N. Bolfan-Casanova, G. L. Nigro, M. A. Bouhifd, G. Garbarino, and M. Mezouar (2011), Solidus and liquidus profiles of chondritic mantle: Implication for melting of the Earth across its history, *Earth Planet. Sci. Lett.*, *304*, 251–259, doi:10.1016/j.epsl.2011.02.006.

Andrault, D., G. Pesce, M. A. Bouhifd, N. Bolfan-Casanova, J.-M. Hénot, and M. Mezouar (2014), Melting of subducted basalt at the core-mantle boundary, *Science*, *344*, 892–895.

Bonsor, A., Z. M. Leinhardt, P. J. Carter, T. Elliott, M. J. Walter, and S. T. Stewart (2015), A collisional origin to Earth's non-chondritic composition? *Icarus*, *247*, 291–300, doi:10.1016/j.icarus.2014.10.019.

Boukaré, C.-E., Y. Ricard, and G. Fiquet (2015), Thermo-dynamics of the MgO-FeO-SiO_2 system up to 140 GPa: Application to the crystallization of Earth's magma ocean, *J. Geophys. Res.*, *120*, 6085–6101, doi:10.1002/2015JB011929.

Boyet, M., and R. W. Carlson (2005), [142]Nd evidence for early (> 4.53 Ga) global differentiation of the silicate Earth, *Science*, *309*(5734), 576–581, doi:10.1126/science.1113634.

Braginsky, S. I., and P. H. Roberts (1995), Equations governing convection in Earth's core and the geodynamo, *Geophys. Astrophys. Fluid Dyn*, *79*, 1–97.

Buffett, B. A., H. E. Huppert, J. R. Lister, and A. W. Woods (1992), Analytical model for solidification of the Earth's core, *Nature*, *356*, 329–331.

Burchfield, J. D. (1975), *Kelvin and the Age of the Earth*, Unive. of Chicago Press, Chicago, Ill.

Butler, S. L., W. R. Peltier, and S. O. Costin (2005), Numerical models of the Earth's thermal history: Effects of inner-core solidification and core potassium, *Phys. Earth Planet. Inter.*, *152*(1–2), 22–42, doi:10.1016/j.pepi.2005.05.005.

Caro, G., B. Bourdon, A. N. Halliday, and G. Quitté (2008), Super-chondritic Sm/Nd ratios in Mars, the Earth and the Moon, *Nature*, *452*(7185), 336–339.

Carnot, S. (1824), Réflexions sur la puissance motrice du feu, Bachelier, Paris (freely available at http://gallica.bnf.fr/ark:/12148/btv1b86266609). English translation by R. H. Thurston available as Reflections on the motive power of heat (1897), John Wiley & Sons, New York. Citation on p. 37 of that edition (freely available at http://archive.org/stream/reflectionsonmot00carnrich)

Christensen, U. R. (1984), Heat transport by variable viscosity convection and implications for the Earth's thermal evolution, *Phys. Earth Planet. Inter.*, *35*, 264–282.

Christensen, U. R. (1985), Thermal evolution models for the Earth, *J. Geophys. Res.*, *90*, 2995–3007.

Christensen, U. R. (2010), Dynamo scaling laws and applications to the planets, *Space Sci. Rev.*, *152*, 565–590, doi:10.1007/s11214-009-9553-2.

Conrad, C. P., and B. H. Hager (1999a), The thermal evolution of an Earth with strong subduction zones, *Geophys. Res. Lett.*, *26*(19), 3041–3044.

Conrad, C. P., and B. H. Hager (1999b), Effects of plate bending and fault strength at subduction zones on plate dynamics, *J. Geophys. Res.*, *104*(B8), 17,551–17,571.

De Koker, N., and L. Stixrude (2009), Self-consistent thermo-dynamic description of silicate liquids, with application to shock melting of mgo periclase and mgsio3 perovskite, *Geophys. J. Int.*, *178*(1), 162–179.

Deguen, R. (2013), Thermal convection in a spherical shell with melting/freezing at either or both of its boundaries, *J. Earth Sci.*, *24*, 669–682.

Deguen, R., T. Alboussière, and P. Cardin (2013), Thermal convection in Earth's inner core with phase change at its boundary, *Geophys. J. Int.*, *194*, 1310–1334, doi:10.1093/gji/ggt202.

De Koker, N., G. Steinle-Neumann, and V. Vlček (2012), Electrical resistivity and thermal conductivity of liquid Fe alloys at high P and T, and heat flux in Earth's core, *Proc. Natl. Acad. Sci. USA*, *109*, 4070–4073.

De Koker, N., B. B. Karki, and L. P. Stixrude (2013), Thermodynamics of the MgO–SiO_2 liquid system in Earth's lowermost mantle from first principles, *Earth Planet. Sci. Lett.*, *361*(C), 58–63.

Dobson, D. P., and J. P. Brodholt (2005), Subducted banded iron formations as a source of ultralow-velocity zones at the core-mantle boundary, *Nature*, *434*(7031), 371–374.

Fiquet, G., A. L. Auzende, J. Siebert, A. Corgne, H. Bureau, H. Ozawa, and G. Garbarino (2010), Melting of peridotite to 140 GigaPascals, *Science*, *329*, 1516–1518, doi:10.1126/science.1192448.

Fischer, R. A., and A. J. Campbell (2010), High-pressure melting of wustite, *Am. Mineral.*, *95*(10), 1473–1477, doi:10.2138/am.2010.3463.

Gomi, H., K. Ohta, K. Hirose, S. Labrosse, R. Caracas, M. J. Verstraete, and J. W. Hernlund (2013), The high conductivity of iron and thermal evolution of the Earth's core, *Phys. Earth Planet. Inter.*, *224*, 88–103, doi:10.1016/j.pepi.2013.07.010.

Grigné, C., and S. Labrosse (2001), Effects of continents on Earth cooling: Thermal blanketing and depletion in radioactive elements, *Geophys. Res. Lett.*, *28*, 2707–2710.

Grigné, C., S. Labrosse, and P. J. Tackley (2005), Convective heat transfer as a function of wavelength: Implications for the cooling of the Earth, *J. Geophys. Res.*, *110*(B3), B03,409, doi:10.1029/2004JB003376.

Gubbins, D., T. G. Masters, and J. A. Jacobs (1979), Thermal evolution of the Earth's core, *Geophys. J. R. Astron. Soc.*, *59*, 57–99.

Gubbins, D., D. Alfè, G. Masters, G. D. Price, and M. J. Gillan (2004), Gross thermodynamics of 2-component core convection, *Geophys. J. Int.*, *157*, 1407–1414.

Gunnarsson, O., M. Calandra, and J. E. Han (2003), *Colloquium*: Saturation of electrical resistivity, *Rev. Mod. Phys.*, *75*, 1085–1099, doi:10.1103/RevModPhys.75.1085.

Hernlund, J., and A. McNamara (2015), 7.11 — The core–mantle boundary region, in *Treatise on Geophysics*, 2nd ed., edited by G. Schubert, pp. 461 – 519, Elsevier, Oxford, doi:10.1016/B978-0-444-53802-4.00136-6.

Hernlund, J. W., and A. M. Jellinek (2010), Dynamics and structure of a stirred partially molten ultralow-velocity zone, *Earth Planet. Sci. Lett.*, *296*(1–2), 1–8, doi:10.1016/j.epsl.2010.04.027.

Hirose, K., S. Labrosse, and J. W. Hernlund (2013), Composition and state of the core, *Annu. Rev. Earth Planet. Sci.*, *41*, 657–691, doi:10.1146/annurev-earth-050212-124007.

Jackson, M. G., and A. M. Jellinek (2013), Major and trace element composition of the high [3]He/[4]He mantle: Implications for the composition of a nonchonditic Earth, *Geochem, Geophys, Geosyst*, *14*(8), 2954–2976, doi:10.1002/ggge.20188.

Jaupart, C., S. Labrosse, F. Lucazeau, and J.-C. Mareschal (2015), 7.06 — Temperatures, heat, and energy in the mantle of the Earth, in *Treatise on Geophysics*, 2nd ed., edited by G. Schubert, pp. 223 – 270, Elsevier, Oxford, doi:10.1016/B978-0-444-53802-4.00126-3.

Javoy, M. (1999), Chemical Earth models, *C. R. Acad. Sci. Paris*, *329*, 537–555.

Javoy, M., E. Kaminski, F. Guyot, D. Andrault, C. Sanloup, M. Moreira, S. Labrosse, A. Jambon, P. Agrinier, A. Davaille, and C. Jaupart (2010), The chemical composition of the earth: Enstatite chondrite models, *Earth Planet. Sci. Lett.*, *293*(3–4), 259–268, doi:10.1016/j.epsl.2010.02.033.

Korenaga, J. (2003), Energetics of mantle convection and the fate of fossil heat, *Geophys. Res. Lett.*, *30*, 1437, doi:10.1029/2003GL016982.

Korenaga, J. (2006), Archean geodynamics and the thermal evolution of the Earth, in *Archean Geodynamics and Environments,*

Geophys. Monogr. Ser., vol. *164*, edited by K. Benn, J. Mareschal, and K. C. Condie, pp. 7–32, AGU, Washington, D.C.

Korenaga, J. (2010), Scaling of plate tectonic convection with pseudoplastic rheology, *J. Geophys. Res.*, *115*(B11), B11,405.

Labrosse, S. (2003), Thermal and magnetic evolution of the Earth's core, *Phys. Earth Planet. Inter.*, *140*, 127–143.

Labrosse, S. (2015), Thermal evolution of the core with a high thermal conductivity, *Phys. Earth Planet. Inter.*, *247*, 36–55, doi:10.1016/j.pepi.2015.02.002.

Labrosse, S., and C. Jaupart (2007), Thermal evolution of the Earth: Secular changes and fluctuations of plate characteristics, *Earth Planet. Sci. Lett.*, *260*, 465–481, doi:10.1016/j.epsl.2007.05.046.

Labrosse, S., J.-P. Poirier, and J.-L. Le Mouël (1997), On cooling of the Earth's core, *Phys. Earth Planet. Inter.*, *99*, 1–17.

Labrosse, S., J.-P. Poirier, and J.-L. Le Mouël (2001), The age of the inner core, *Earth Planet. Sci. Lett.*, *190*, 111–123.

Labrosse, S., J. W. Hernlund, and N. Coltice (2007), A crystallizing dense magma ocean at the base of the Earth's mantle, *Nature*, *450*, 866–869.

Labrosse, S., J. W. Hernlund, and K. Hirose (2015), Fractional melting and freezing in the deep mantle and implications for the formation of a basal magma ocean, in *The Early Earth: Accretion and Differentiation*, Geophys. Monogr. Ser., Vol. 212, edited by J. Badro and M. J. Walter, pp. 123–142, AGU, Washington, D.C.

Lay, T., E. J. Garnero, and Q. William (2004), Partial melting in a thermo-chemical boundary layer at the base of the mantle, *Phys. Earth Planet. Inter.*, *146*(3–4), 441–467.

Lay, T., J. Hernlund, and B. A. Buffett (2008), Core-mantle boundary heat flow, *Nature Geosci.*, *1*, 25–32.

Lister, J. R., and B. A. Buffett (1995), The strength and efficiency of the thermal and compositional convection in the geodynamo, *Phys. Earth Planet. Inter.*, *91*, 17–30.

Lister, J. R., and B. A. Buffett (1998), Stratification of the outer core at the core-mantle boundary, *Phys. Earth Planet. Inter.*, *105*, 5–19.

Mao, W. L., H.-k. Mao, W. Sturhahn, J. Zhao, V. B. Prakapenka, Y. Meng, J. Shu, Y. Fei, and R. J. Hemley (2006), Iron-rich post-perovskite and the origin of ultralow-velocity zones, *Science*, *312*(5773), 564–565.

McDonough, W. F. (2003), Compositional model for the Earth's core, *Treatise on Geochem.*, *2*, 547–568, doi:10.1016/B0-08-043751-6/02015-6.

McKenzie, D. P., J. M. Roberts, and N. O. Weiss (1974), Convection in the earth's mantle: Towards a numerical simulation, *J. Fluid Mech.*, *62*, 465–538.

Mollett, S. (1984), Thermal and magnetic constraints on the cooling of the Earth, *Geophys. J. R. Astron. Soc.*, *76*, 653–666.

Mosenfelder, J., P. Asimow, and T. Ahrens (2007), Thermodynamic properties of Mg_2SiO_4 liquid at ultra-high pressures for shock measurements to 200 GPa on forsterite and wadsleyite, *J. Geophys. Res.*, *112*, B06,208, doi:10.1029/2006JB004364.

Nakagawa, T., and P. J. Tackley (2005), Deep mantle heat flow and thermal evolution of the earth's core in thermochemical multiphase models of mantle convection, *Geochem. Geophys. Geosyst.*, *6*, Q08,003, doi:10.1029/2005GC000967.

Nakagawa, T., and P. J. Tackley (2010), Influence of initial CMB temperature and other parameters on the thermal evolution of Earth's core resulting from thermochemical spherical mantle convection, *Geochem. Geophyst. Geosys.*, *11*, Q06,001, doi:10.1029/2010GC003031.

Nimmo, F. (2015), 8.02 — Energetics of the core, in *Treatise on Geophysics*, 2nd ed., edited by G. Schubert, pp. 27–55, Elsevier, Oxford, doi:http://dx.doi.org/10.1016/B978-0-444-53802-4.00139-1.

Nimmo, F., G. D. Price, J. Brodholt, and D. Gubbins (2004), The influence of potassium on core and geodynamo evolution, *Geophys. J. Int.*, *156*, 263–376.

Nomura, R., K. Hirose, K. Uesugi, Y. Ohishi, A. Tsuchiyama, A. Miyake, and Y. Ueno (2014), Low core-mantle boundary temperature inferred from the solidus of pyrolite, *Science*, *343*(6170), 522–525, doi:10.1126/science.1248186.

Ohta, K., Y. Kuwayama, K. Shimizu, T. Yagi, K. Hirose, and Y. Ohishi (2014), Measurements of electrical and thermal conductivity of iron under earth's core conditions, *AGU Fall Meeting Abstracts*, AGU, Washington, D.C.

O'Neill, H. S. C., and H. Palme (2008), Collisional erosion and the non-chondritic composition of the terrestrial planets, *Philos. Trans. Ser. A Math., Phys. Eng. Sci.*, *366*(1883), 4205–4238, doi:10.1098/rsta.2008.0111.

Poirier, J.-P. (2000), *Introduction to the Physics of the Earth's Interior*, 2nd ed., Cambridge Univ. Press, Cambridge.

Pozzo, M., C. J. Davies, D. Gubbins, and D. Alfè (2012), Thermal and electrical conductivity of iron at Earth's core conditions, *Nature*, *485*, 355–358.

Rama Murthy, V., W. van Westrenen, and Y. Fei (2003), Radioactive heat sources in planetary cores: Experimental evidence for potassium, *Nature*, *423*, 163–165.

Roberts, P. H., C. A. Jones, and A. R. Calderwood (2003), Energy fluxes and ohmic dissipation in the Earth's core, in *Earth's Core and Lower Mantle*, edited by C. A. Jones, A. M. Soward, and K. Zhang, pp. 100–129, Taylor & Francis, London.

Rost, S., E. J. Garnero, Q. Williams, and M. Manga (2005), Seismological constraints on a possible plume root at the core-mantle boundary, *Nature*, *435*, 666–669, doi:10.1038/nature03620.

Schubert, G., D. L. Turcotte, and P. Olson (2001), *Mantle Convection in the Earth and Planets*, Cambridge Univ. Press, Cambridge.

Seagle, C. T., E. Cottrell, Y. Fei, D. R. Hummer, and V. B. Prakapenka (2013), Electrical and thermal transport properties of iron and iron-silicon alloy at high pressure, *Geophys. Res. Lett.*, *40*, 5377–5381, doi:10.1002/2013GL057930.

Sha, X., and R. E. Cohen (2011), First-principles studies of electrical resistivity of iron under pressure, *J. Phys. Condens. Matter*, *23*(7), 075,401.

Sleep, N. H. (2000), Evolution of the mode of convection within terrestrial planets, *J. Geophys. Res.*, *105*, 17,563–17,578.

Sotin, C., and S. Labrosse (1999), Three-dimensional thermal convection of an isoviscous, infinite-Prandtl-number fluid heated from within and from below: Applications to heat transfer in planetary mantles, *Phys. Earth Planet. Inter.*, *112*(3–4), 171–190.

Stevenson, D., T. Spohn, and G. Schubert (1983), Magnetism and thermal evolution of the terrestrial planets, *Icarus*, *54*, 466–489.

Stixrude, L., N. de Koker, N. Sun, M. Mookherjee, and B. B. Karki (2009), Thermodynamics of silicate liquids in the deep Earth, *Earth Planet. Sci. Lett.*, *278*, 226–232, doi:10.1016/j.epsl.2008.12.006.

Thomas, C. W., Q. Liu, C. B. Agee, P. D. Asimow, and R. A. Lange (2012), Multitechnique equation of state for Fe_2SiO_4 melt and the density of Fe-bearing silicate melts from 0 to 161 GPa, *J. Geophys. Res.*, *117*, B10,206.

Turcotte, D. L., and E. R. Oxburgh (1967), Finite amplitude convective cells and continental drift, *J. Fluid Mech.*, *28*, 29–42.

Usui, Y., J. A. Tarduno, M. Watkeys, A. Hofmann, and R. D. Cottrell (2009), Evidence for a 3.45-billion-year-old magnetic remanence: Hints of an ancient geodynamo from conglomerates of South Africa, *Geochem. Geophys. Geosyst.*, *10*, Q09Z07.

Wicks, J. K., J. M. Jackson, and W. Sturhahn (2010), Very low sound velocities in iron-rich (Mg,Fe)O: Implications for the core-mantle boundary region, *Geophys. Res. Lett.*, *37*(15), L15,304, doi:10.1029/2010GL043689.

Williams, Q., and E. J. Garnero (1996), Seismic evidence for partial melt at the base of the Earth's mantle, *Science*, *273*, 1528.

Wimert, J., and S. Hier-Majumder (2012), A three-dimensional microgeodynamic model of melt geometry in the Earth's deep interior, *J. Geophys. Res.*, *117*(B4), 2156–2202, doi:10.1029/2011 JB009012.

Wohlers, A., and B. J. Wood (2015), A Mercury-like component of early Earth yields uranium in the core and high mantle [142]Nd, *Nature*, *520*(7547), 337–340, doi:10.1038/nature14350.

Yukutake, T. (2000), The inner core and the surface heat flow as clues to estimating the initial temperature of the Earth's core, *Phys. Earth Planet. Inter.*, *121*, 103–137.

Zhang, P., R. E. Cohen, and K. Haule (2015), Effects of electron correlations on transport properties of iron at Earth's core conditions, *Nature*, *517*, 605–607.

Part II
Structures, Anisotropy, and Plasticity of Deep Earth Materials

5

Crystal Structures of Core Materials

Razvan Caracas

ABSTRACT

The Fe-based solid crystalline structure(s) present in Earth's core is(are) compact, whether hexagonal or cubic. Here we review the different possibilities as were measured in experiments or computed in ab initio simulations. Then we look at the effect of impurities: Ni, other Fe group elements, and the H, C, O, Si, P, and S light elements. We present the different stoichiometric compounds in the Fe-rich parts of the binary phase diagrams and their crystal structures under pressure. We finish with a few considerations on multicomponent systems.

5.1. INTRODUCTION

The extreme thermodynamic conditions and the remoteness make Earth's core one of the most enigmatic parts of our planet. The pressure at the top of the outer core is on the order of 135 GPa, and it increases up to 329 GPa at the top of the inner core and up to 364 GPa in the middle of the inner core. There is considerably larger incertitude on the temperature profile in the core, with most likely values in the 4500–7000 K range [*Gomi et al.*, 2013; *Tateno et al.*, 2010]. Until recently these thermodynamic conditions have been out of reach of experiments. Today developments in diamond-anvil cells coupled with new performant facilities allow us to measure *in situ* a growing range of properties. On the theoretical side, first-principles calculations, usually based on density functional theory (DFT), take full advantage of the advances in algorithms, implementations, and computing power of the last decades. Even if not easy, solving crystal structures

CNRS, Ecole Normale Supérieure de Lyon, Université Claude Bernard Lyon 1, Laboratoire de Géologie de Lyon, Lyon, France

and obtaining physical properties for core materials becomes almost a matter of routine.

Geochemical evidence [*McDonough and Sun*, 1995] suggests that the core must be formed mainly of alloyed Fe and Ni. But the physical properties of such alloys fail to some extent to explain the observed seismic and geophysical properties of both the outer and the inner core, like the density, the seismic wave velocities, and the seismic anisotropy [*Anderson*, 2003]. These observations can be explained by including light elements dissolved in the alloy [*Poirier*, 1994; *Wood*, 1993; *Stixrude et al.*, 1997; *Badro et al.*, 2007] and considering the effects of temperature [*Martorell et al.*, 2013a].

The nature of the light element (or elements) has been the subject of considerable speculation because of its bearing on the overall bulk composition of Earth, the conditions under which the core formed, the temperature regime in the core, and the continuing processes of core-mantle reactions [*Poirier*, 1994; *Anderson and Isaak*, 1999; *Badro et al.*, 2007; *Lin et al.*, 2002; *Wood*, 1993]. While any element lighter than iron could potentially compensate the density deficit if in the proper amount, geochemical availability and solubility in the Fe-Ni alloys are the main constraining factors for the resulting real alloy.

Deep Earth: Physics and Chemistry of the Lower Mantle and Core, Geophysical Monograph 217, First Edition.
Edited by Hidenori Terasaki and Rebecca A. Fischer.

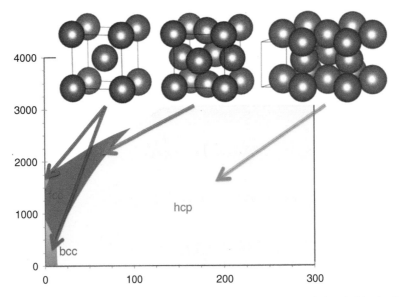

Figure 5.1 Compact structures of iron and their pressure and temperature stability fields: body-centered cubic (bcc), face-centered cubic (fcc), and hexagonal close packed (hcp). The hcp structure is stable at the inner core conditions, though thermal and chemical effects might turn bcc energetically competitive.

The comparisons with cosmic abundances and meteorite compositions suggest H, C, O, S, Si, and P as preferred light elements [*Anderson*, 2003]. These will be our major focus in the following discussion.

5.2. Fe AND Fe-GROUP ALLOYS

For such an important and simple material, it is striking that the crystal structure of iron is still a matter of debate for exactly the most relevant conditions: Earth's core. However, the phase diagram and the structure of iron at core conditions have been explored for a long time using both experimental and theoretical techniques.

Iron exhibits three crystal structures: body-centered cubic (bcc) stable at ambient conditions, face-centered cubic (fcc) stable at high temperatures, and hexagonal close packed (hcp) stable at high pressures (Figure 5.1). Several recent reviews provide nice detailed discussions about the sequence of structures at high pressure and temperatures [*Hirose et al.*, 2013; *Ohtani*, 2013; *Morard et al.*, 2014; *Saxena and Eriksson*, 2015]. The general consensus is that hcp is the stable structure at core pressures. But thermal conditions of the core fall exactly in the region where most uncertainties regarding the stability of the hcp phase exist. There are both experimental [*Belonoshko et al.*, 2003; *Dubrovinsky et al.*, 2007] and computational [*Vocadlo et al.*, 2003a] studies that suggest bcc structure to be stable at high pressures and high temperatures. Recent first-principles calculations show that bcc is mechanically unstable under inner core conditions [*Godwal et al.*, 2015].

This leaves hcp as the stable crystal structure for pure iron at core pressures and temperatures, as suggested in various earlier studies [*Stixrude and Cohen*, 1995; *Steinle-Neumann et al.*, 2001; *Vocadlo et al.*, 2003b, 2009; *Tateno et al.*, 2010; *Sha and Cohen*, 2010; *Antonangeli et al.*, 2012; *Martorell et al.*, 2013a; *Ohtani et al.*, 2013; *Gleason and Mao*, 2013; *Sakai et al.*, 2014]. Some amount of polytypism is possible [*Cottenier et al.*, 2011] and might explain observations of a double-hcp (dhcp) phase [*Saxena and Dubrovinsky*, 2000].

Up to about 10 at % Ni would preserve the structure of hcp iron [*Tateno et al.*, 2012]. In larger amounts, nickel, as well as probably other alloying elements, can affect the relative stability of the different phases of iron. With increasing amount of Ni, the bcc and then the fcc phases became stable at high pressures and high temperatures [*Kuwayama et al.*, 2008; *Côté et al.*, 2012; *Martorell et al.*, 2013b]. There are no data at high pressure for Fe-Co or Fe-Cr alloys, but one can safely assume that for levels in the core, they would simply dissolve into the hcp Fe without affecting its structure.

5.3. THE Fe-H SYSTEM

This system is one of the most challenging from the experimental point of view. Typical synthesis paths involve compressing iron after gas loading the cell with hydrogen in diamond anvil cell experiments [*Badding et al.*, 1992; *Antonov et al.*, 1998, 2002; *Mao*, 2004] or multianvil press experiments with LiAlH$_4$ as the H source [*Sakamaki et al.*, 2009]. Consequently, the stoichiometry of the resulting FeH_x alloy is not controlled, while the escape of hydrogen implies difficulties for a good chemical analysis of the recovered samples. Hydrogen would typically enter the hcp structure of iron interstitially, as an impurity, most likely with a high mobility. As such, it is assumed that it will expand the host Fe lattice linearly. Thus, this is the preferred measure for estimating the amount of H that was dissolved in Fe during the experiments. DFT simulations confirm this linear dependence [*Caracas*, 2015].

First-principles static calculations showed that hydrogen tends to stabilize the close-packed structures. The transition pressures between the Fe polymorphs are weakly affected by the presence of hydrogen [*Skorodumova*, 2004].

If present in large enough amounts, then FeH may form what has a dhcp structure [*Isaev et al.*, 2007]. The dhcp phase was observed in several experiments [*Sakamaki et al.*, 2009; *Terasaki et al.*, 2012; *Shibazaki et al.* 2012; *Mao*, 2004], with a collapse of the magnetic spin at around 22 GPa [*Mao*, 2004]. Its stability was proved beyond doubt in recent lattice dynamical calculations [*Isaev et al.*, 2007]. These simulations predict a sequence of phase transitions under pressure dhcp-hcp-fcc for the ordered FeH compound. A FeH_2 compound was synthesized at 67 GPa with tetragonal $I4/mmm$ symmetry, and a FeH_3 at 86 GPa with cubic $Pm3m$ symmetry in the same set of experiments [*Pepin et al.*, 2014]. For the structural determinations, the position of the Fe atoms was obtained directly from diffraction and the position of the H atoms from ab initio simulations.

The crystal structures of the stoichiometric FeH_x phases, with $x = 1, 2, 3$, are represented in Figure 5.2. A detailed discussion about the effect and the behavior of hydrogen in iron is presented later in this monograph [*Murphy*, 2016].

5.4. THE Fe-C SYSTEM

The Fe-C system in the Fe-rich part of the phase diagram is characterized by two phases with intermediate stoichiometry that have been shown to exist beyond doubt: Fe_3C and Fe_7C_3.

Fe_3C, the mineral cohenite, is present in meteorites but also as the major constituent of cast iron in metallurgy, where it is known as cementite. It has an orthorhombic structure with *Pnma* space group, archetype for a whole class of materials with cementite structure and Me_3X stoichiometry, where Me is a metal. The crystal structure is represented in Figure 5.3. The Fe atoms occupy $4c$ and $8d$ Wyckoff positions and the C atoms $4c$ Wyckoff positions. The structure can be seen as a distortion of a hexagonal lattice of iron in which the Fe plane makes zig-zag folds perpendicular to the a and b axis. The C atoms are placed in the void spaces, in the middle of Fe-based triangular

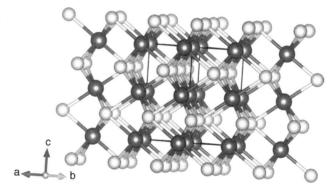

Figure 5.3 Representations of the Fe_3C cementite structure along two different directions. It is characterized by the presence of pairs of Fe_6C prisms that act like quasi-rigid bodies during compression. Fe and C atoms are, respectively, drawn with gold and black spheres.

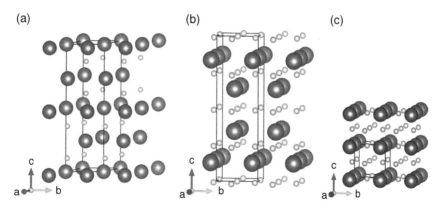

Figure 5.2 Crystal structures of several stoichiometric phases identified experimentally in the Fe-H system: (a) double hcp (dhcp) FeH, (b) tetragonal FeH_2, and (c) cubic FeH_3.

prisms. These carbon-iron coordination polyhedra form pairs that act as quasi-rigid bodies during compression. A theoretical study employing evolutionary algorithms [*Bazhanova et al.*, 2012] suggests possible monoclinic (*C2/m*) or tetragonal modifications (*I4*) as being more stable than the standard *Pnma* one at Earth's core pressures.

Under pressure, it has been extensively studied in experiments [*Lin et al.*, 2004; *Wood et al.*, 2004; *Duman et al.*, 2005; *Gao et al.*, 2008, 2011; *Rouquette et al.*, 2008; *Fiquet et al.*, 2009; *Sata et al.*, 2010; *Ono and Mibe*, 2010; *Prescher et al.*, 2012; *Litasov et al.*, 2013] and first-principles calculations [*Wang et al.*, 2005; *Vocadlo et al.*, 2002; *Mookherjee*, 2011]: Its equation of state, magnetic spin transition, and seismic wave velocities have been both measured and computed. It was shown that at low pressures it is ferromagnetic, with the magnetic spin decreasing with pressure. Early X-ray emission spectroscopy measurements [*Lin et al.*, 2004] place the disappearance of the magnetization around 25 GPa. Later inelastic X-ray scattering measurements [*Fiquet et al.*, 2009] place this transition above 68 GPa, in agreement to previous first-principles calculations [*Vocadlo et al.*, 2002]. A magnetic transition at low pressures, between 4.3 and 6.5 GPa, observed by nuclear resonant scattering is probably just an observation of the reduction in magnetism at the beginning of compression [*Gao et al.*, 2008]. At ambient pressure Fe_3C becomes paramagnetic at 483 K.

A first experimental study showed the formation of Fe_7C_3 above 7–8 GPa [*Lord et al.*, 2009], suggesting that the Fe-C system at core conditions should involve Fe_7C_3 rather than Fe_3C. Structures with hexagonal $P6_3mc$ or with orthorhombic *Pnma* symmetries have been proposed for Fe_7C_3. First-principles simulations suggest that the orthorhombic ferromagnetic modification is more stable [*Fang et al.*, 2009]. The latter study investigates in detail the role of the iron sublattice in the formation enthalpy and the magnetic state of the structure, including various magnetic configurations. The energy of Fe_7C_3 is comparable to that of Fe_3C in static, i.e., 0 K, simulations, at ambient pressure conditions. These two proposed crystal structures of Fe_7C_3 are represented in Figure 5.4.

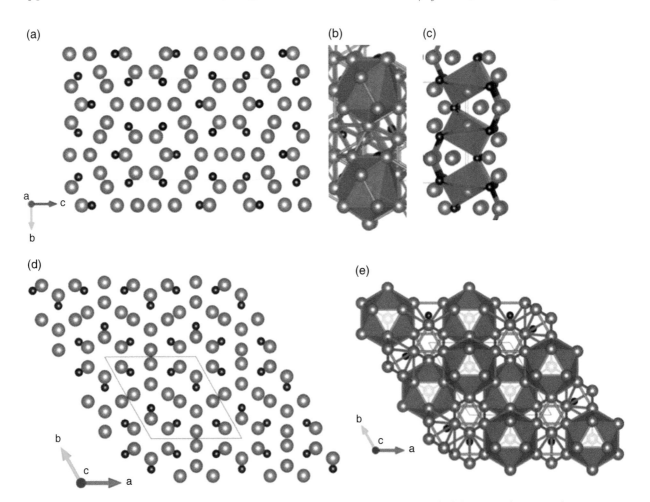

Figure 5.4 Two of the proposed structures of Fe_7C_3: (a) orthorhombic *Pnma* and (d) hexagonal $P6_3mc$. The *Pnma* structure is distorted hexagonal; it contains Fe_{10} decagons (b) and Fe_4C tetrahedra (c). In the hexagonal structure the decagons are more symmetric (e).

Recent measurements on single crystals of Fe_7C_3 suggest yet another orthorhombic modification with *Pbca* symmetry. The latter yields Poisson ratio values that are remarkably close to the measured ones for the inner core [*Prescher et al.*, 2015].

A series of other Fe-C phases, with various stoichiometries, have been investigated computationally [*Bazhanova et al.*, 2012; *Zhao et al.*, 2015]. In general their theoretical structures can be seen as various derivations of the Laves phases: hexagonal $P6_3/mmc$, orthorhombic *Pnma*, and cubic $Pm3m$ for Fe_2C, monoclinic *C2/c* Fe_5C_2, and monoclinic *P2/m*, tetragonal *I4/m* and cubic *P-43m* Fe_4C. Static calculations suggest that Fe_2C is particularly stable in this system.

5.5. THE Fe-O SYSTEM

The iron oxides exhibit strong correlation effects due to the partial occupancy of the *d* electrons and their hybridization with the O *p* electrons. Because of this, they are a major headache for computational scientists because DFT fails to properly describe the correlation term in the current exchange correlation part of the Kohn-Sham Hamiltonian. Some advanced techniques using, for example, the Hubbard U correction [*Anisimov et al.*, 1991] or dynamical mean-field theory [*Georges et al.*, 1996] are able to improve the treatment of this term for a nonnegligible computational or implementation cost.

There are several intermediate stoichiometric compositions known for a long time, due to their practical importance: FeO wüstite, Fe_2O_3 hematite, and Fe_3O_4 magnetite.

FeO is a Mott insulator at room temperature and pressure. The mineral wüstite has B1 structure with vacancies on the Fe site, the actual stoichiometry being $Fe_{1-x}O$, with x less than 0.1. Because of the iron vacancies, some of the iron is ferric [*Höfer et al.*, 2000; *Otsuka et al.*, 2010; *Longo et al.*, 2011]. The structure is then locally distorted with the iron atoms in various coordinations. The ordering of the defects influences the electronic, magnetic, and transport properties [*Hazen and Jeanloz*, 1984]. With the emergence of the multidimensional crystallography and the description of modulated structures, the defect ordering was described as a modulation wave with an incommensurate wave vector along the <111> cubic direction [*Yamamoto*, 1982]. Synthesis at high pressures remove the vacancies in the structures [*McCammon*, 1993].

Recently, it was predicted theoretically and confirmed experimentally that the electronic behavior of FeO under both temperature and pressure is much more complicated than previously assumed. Corrections of the correlation term in the standard DFT using dynamical mean-field theory [*Ohta et al.*, 2012] showed the presence of an insulator-metal transition under pressure in the B1 phase.

At core pressure conditions the stable structure is B8, NiAs type, which is a derivation of the hcp structure (Figure 5.5). At some pressures but high temperatures the B1 phase occurs before the melting is reached [*Fischer et al.*, 2011].

In general FeO has been studied at relatively moderate pressures because of its implications as a major mineral of Earth's lower mantle. However, for the Earth's core, even if FeO has the largest amount of oxygen from all the iron oxides, because of its reducing character it remains a very important Fe-O phase when assessing thermodynamically the oxygen amount in the core.

Fe_2O_3, the mineral hematite, is an oxidized iron oxide where all the Fe atoms are in the Fe^{3+} valence state. There are several modifications under pressure, starting with

(a)

(b)

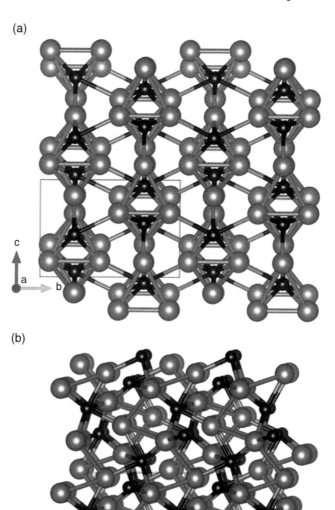

Figure 5.5 NiAs-type structure adopted by both FeO and FeS.

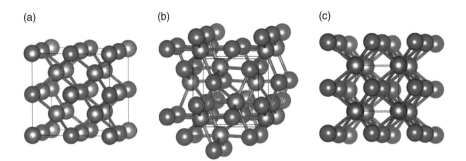

Figure 5.6 The (a) Fe_3Si structures belong to the DO_3 type, typical for M_3N alloys, like Au_3Al. (b) FeSi B20 exhibits unusual sevenfold coordination for both Fe and Si. (c) FeSi B2 is derived from body-centered cubic. Fe and Si atoms are represented by, respectively, gold and blue spheres.

rhombohedral $R\text{-}3c$ space group at ambient pressure conditions. Its high-pressure phases have been investigated up to the lower mantle conditions, up to which pressures, it undergoes at least one Mott electronic transition [*Pasternak et al.*, 1999] and structural changes into $Rh_2O_3(II)$-type structure around 70 GPa. At higher pressures transitions to perovskite, and post-perovskite have been recorded, but the transition pressures are still debatable [*Ito et al.*, 2009; *Shim et al.*, 2009; *Ono et al.*, 2004; *Badro et al.*, 2002; *Olsen et al.*, 1991].

Fe_3O_4 is the stoichiometric phase richest in iron in the whole Fe-O system. Magnetite has the inverse spinel structure at ambient conditions with the $Fd3m$ space group. Fe atoms populate two sublattices, with Fe^{2+} occupying the octahedral sites and Fe^{3+} occupying in equal amounts octahedral and tetrahedral sites. A phase transition was observed around 20 GPa toward monoclinic symmetry [*Morris and Williams*, 1997]. No data are known for magnetite or hematite at core conditions.

5.6. THE Fe-SI SYSTEM

At low pressures, up to 100 GPa, there are two stoichiometric phases in this system: Fe_3Si and FeSi. Both phases are metallic. Several other stoichiometries along the Fe-Si phase, such as Fe_2Si, Fe_5Si_3, $FeSi_2$, $FeSi_3$, and Fe_3Si_7, some proposed experimentally, turned out unstable against these two major phases [*Brosh et al.*, 2009; *Zhang and Oganov*, 2010].

Fe_3Si has the DO_3 crystal structure, with Au_3Al archetype. It is cubic, with the $Fm3m$ space group (Figure 5.6a). It consists of two interpenetrating face-centered-cubic (fcc) sublattices: one occupied entirely by Fe and one from each second Fe atom is replaced by Si.

The structure is ferromagnetic, with the magnetization decreasing with increasing pressure [*Rhee and Harmon*, 2004]. The Fe_3Si structure is unstable under compression and decomposes into a mixture of hcp Fe and B2 FeSi above about 1 Mbar [*Fischer et al.*, 2012]; there is an

immiscibility gap between the two phases. At high temperatures the melting of the mixture is reached before the closure of the gap.

The FeSi phase at ambient pressure conditions comes into a B20 structure type (Figure 5.6b). This is a cubic structure with $P2_13$ symmetry, where both Fe and Si atoms occupy $4a$ (x,x,x) Wyckoff positions. This exhibits an unusual sevenfold coordination for both the Fe and Si atoms. At high pressures the stable structure is the B2 structure type, which derives from bcc where the Fe and Si atoms orderly alternate along the <111> diagonal axes of the cube (Figure 5.6c). There is no recorded transition between the B20 and the B2 structures in experimental studies, rather the two phases have been synthesized at different conditions.

The transition pressure at static 0 K conditions was predicted in first-principles calculations based on DFT at 30 or 40 GPa [*Caracas and Wentzcovitch*, 2004] depending on the exchange correlation: respectively, in the local density approximation (LDA) and the generalized gradient approximation (GGA). It is worthwhile noting that both LDA and GGA underestimate the volume for both B20 and B2—typically LDA overestimates bond length and specific volume and GGA underestimates them [*Payne et al.*, 1992].

5.7. THE FE-P SYSTEM

This is a rich system with numerous stoichiometric compounds: Fe_4P, Fe_3P, Fe_2P, FeP, and FeP_4.

Fe_4P is known as the melliniite mineral, discovered in meteorites [*Pratesi*, 2006]. At ambient pressure conditions it is cubic with the $P2_13$ space group; it has the $AlAu_4$-type structure. The Fe atoms occupy $4a$ and $12b$ Wyckoff positions and the P atoms $4a$ positions. The structure is formed of a packing of distorted $Fe_{12}P$ icosahedra (Figure 5.7a).

At high pressure first-principles calculations show the disappearance of the magnetic moment above 80 GPa.

(a)

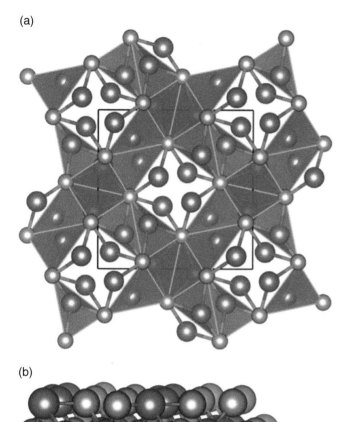

(b)

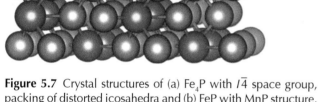

Figure 5.7 Crystal structures of (a) Fe_4P with $I\bar{4}$ space group, packing of distorted icosahedra and (b) FeP with MnP structure, distorted NiAs with orthorhombic *Pnma* space group. Fe and P atoms are, respectively, the golden and bluish spheres.

At static conditions the cubic modification appears more stable than a mixture of Fe_3P and hcp Fe at least up to 400 GPa [*Wu et al.*, 2011]. As always with static simulations, predictions need to be verified at high temperatures; if such a stoichiometric phase is indeed stable at core pressures and temperatures, the entire phase diagram of the Fe-P system needs to be revisited.

Fe_3P, schreibersite, is another abundant mineral in iron meteorites. At ambient pressures it has $I\bar{4}$ space group. Under compression it undergoes phase transitions at 17 and 30 GPa [*Scott et al.*, 2007]. A first theoretical study suggested that at high pressure Fe_3P has cementite structure with a *Pnma* space group [*Wu et al.*, 2011]. A later detailed study [*Gu et al.*, 2014] analyzing four different tetragonal structures, cementite, and DO_3 showed a possible phase transition above 70 GPa toward a $P4/mmc$ structure. It was confirmed by in situ diffraction measurements

performed in the same study. The transition is accompanied by a spin collapse. They showed that the cementite structure becomes stable above 200 GPa and stays as such up to inner core pressures.

Fe_2P, allabogdanite, is another rare mineral present in iron meteorites. At ambient conditions it has a hexagonal structure with a P-$62m$ space group. In the early 1980s a series of experimental works showed a strong decrease of the magnetization at the beginning of the compression [*Fujiwara et al.*, 1980]. At 8 GPa it undergoes a structural transition to a β-Co_2Si orthorhombic *Pnma* structure. The transition consists of a small crystallographic change, with the P atoms changing two possible neighboring sites separated by 1.73 Å [*Dera et al.*, 2008]. Magnetization decreases with pressure, but does not vanish at least up to 75 GPa, as shown in theoretical DFT simulations that used the Hubbard U parameter explicitly [*Wu and Qin*, 2010]. Because of correlation effects, the actual value of the magnetization and the local magnetic spin are highly dependent on the choice of U. The same calculations predict a phase transition to a P-$3m$ trigonal phase at 153 GPa. Interestingly, Fe_2P serves as a model for the prediction of high-pressure phases of silica, with such a transition to the hexagonal phase of Fe_2P predicted to occur in SiO_2 at 640 GPa [*Tsuchiya and Tsuchiya*, 2011].

Finally FeP is in the MnP-type structure at ambient conditions (distorted NiAs with a *Pnma* space group) and stays at least up to 16 GPa [*Gu et al.*, 2011]. Fe atoms are in distorted octahedral coordination that form a close-packed framework (Figure 5.7b). Due the structural and electronic similarity between the Fe-P and Fe-S phases, as well as using hints from the NiP phase diagram, most probably at core conditions FeP would take a NiAs structure, with nondistorted octahedra [*Dera et al.*, 2011, 2012].

The natural iron phosphides present a certain chemical variability, with large amounts of Ni substituting Fe. At high pressures, the similarity between the phase diagrams of Fe-S and Fe-P suggest possible extended solid solutions. Formation of new stoichiometric compounds in a ternary Fe(+Ni)-P-S or even distinct quaternary Fe-Ni-P-S system are plausible and should be taken into account when constructing the thermodynamics of the solid phases from Earth's inner core.

5.8. THE Fe-S SYSTEM

There are several stoichiometric compounds in the iron-rich half of the Fe-S system: Fe_3S, Fe_2S, Fe_3S_2, and FeS.

Fe_3S is isostructural with the Fe_3P tetragonal phase at ambient conditions with the $I\bar{4}$ space group [*Fei et al.*, 2000; *Li et al.*, 2001; *Morard et al.*, 2008]. It exhibits a ferromagnetic to non-magnetic transition around 21 GPa [*Lin*, 2005]. Upon further compression it remains in the same tetragonal structure up to at least 200 GPa [*Seagle et al.*, 2006; *Kamada et al.*, 2014]. The Fe-Fe$_3$S system has an eutectic whose position at 65 GPa is around 18 at % S, moving toward Fe with increasing pressure [*Morard et al.*, 2008].

Fe$_3$S$_2$ is monoclinic and was synthesized under pressure [*Fei et al.*, 1997]. A hexagonal $P6_3/mmc$ phase with Fe$_{3.2}$S stoichiometry, as well as hexagonal $P\bar{6}2m$ Fe$_2$S have been obtained in multianvil press experiments at 22 GPa [*Koch-Muller et al.*, 2002].

FeS is widely found in meteorites as the troillite mineral. It exhibits a rich phase diagram with several phase transitions under compression, which were extensively studied [*Sherman*, 1995; *Fei et al.*, 1995; *Kusaba et al.*, 1998; *Nelmes et al.*, 1999; *Kavner et al.*, 2001; *Urakawa et al.*, 2004; *Ono and Kikegawa*, 2006; *Ono et al.*, 2008; *Sata et al.*, 2010; *Cuda et al.*, 2011]. Troillite is hexagonal, NiAs type (Figure 5.5). Upon compression it transforms to hexagonal FeS IV, monoclinic FeS III. Due to its stability field, this latter phase is important in the context of Mars' core [*Fei et al.*, 1995; *Urakawa et al.*, 2004]. At higher pressures the FeS VI, nonmagnetic MnP-type orthorhombic phase with *Pnma* space group [*Ono and Kikegawa*, 2006; *Ono et al.*, 2008] was predicted to transform into another orthorhombic modification with *Pmmn* symmetry [*Ono et al.*, 2008].

5.9. OTHER LIGHT ELEMENTS

Because of chemical and geochemical considerations, most of the other light elements would be found in Earth's core in the level of parts per million or less [*McDonough*, 2003]. From all these possibilities the most likely to challenge this view might be magnesium. Mg-Fe alloys with hcp structure have been obtained in several high-pressure experiments [*Dubrovinskaia et al.*, 2005]. A series of first-principles calculations addressed the relative stability in various compact structures fcc [*Asker et al.*, 2010] and hcp [*Kádas et al.*, 2008]. They showed that small amounts of Mg can have a strong effect on the physical properties of iron. However, it is not certain if in real Earth conditions Mg can partition in Fe rather than in silicates.

5.10. CONCLUSION AND PERSPECTIVE

The core formation, whether a continuous process throughout the accretion [*Wade and Wood*, 2005; *Wood*, 2008; *Wood and Halliday*, 2005] or a multistage process with several large-size impacts [*O'Brien et al.*, 2006; *Rubie et al.*, 2007, 2011], would carry its signature on the content in light elements. All of the geochemically favored light elements possibly present in the core, H, C, O, Mg, Si, P, or S would be found in small amounts [*Poirier*, 1994; *Antonangeli et al.*, 2010; *Badro et al.*, 2007; *Lin et al.*, 2002; *Takafuji*, 2005; *Wood*, 2008; *Wood et al.*, 2008; *Rubie et al.*, 2011], i.e., typically less than 5 wt. %.

As we have seen above, there are no such stoichiometric compounds known to date. Hence these impurities should be dissolved in the high-pressure and high-temperature solid Fe-Ni alloy.

Interestingly enough, ternary alloys like Fe-Ni-Si or Fe-Ni-O are found in the hcp structure through synthesis [*Badro et al.*, 2007; *Antonangeli et al.*, 2012; *Morard et al.*, 2014]. However, realistic temperature conditions, actual element partitioning during core formation, or chemical reactions at the core-mantle boundary [*Dubrovinsky et al.*, 2004] might change this image, in stabilizing the bcc structure. This leaves important question marks that wait for answers in the years to come.

REFERENCES

Anderson, D. L. (2003), *Theory of the Earth*, 1989 ed., Blackwell Sci. Pub. Boston.

Anderson, O. L., and D. G. Isaak (1999), Elastic constants of mantle minerals at high temperature, in *Mineral Physics and Crystallography. A Handbook of Physical Constants*, edited by T. J. Ahrens, pp. 1–34, AGU, Washington, D.C.

Anisimov, V., J. Zaanen, and O. Andersen (1991), Band theory and Mott insulators—Hubbard-U instead of Stoner-I, *Phys. Rev. B*, *44*(3), 943–954.

Antonangeli, D., J. Siebert, J. Badro, D. Farber, G. Fiquet, G. Morard, and F. Ryerson (2010), Composition of the Earth's inner core from high-pressure sound velocity measurements in Fe–Ni–Si alloys, *Earth Planet. Sci. Lett*, *295*(1), 292–296.

Antonangeli, D., T. Komabayashi, F. Occelli, E. Borissenko, A. C. Walters, G. Fiquet, and Y. Fei (2012), Simultaneous sound velocity and density measurements of hcp iron up to 93GPa and 1100K: An experimental test of the Birch's law at high temperature, *Earth Planet. Sci. Lett.*, *331–332*, 210–214.

Antonov, V. E., K. Cornell, V. K. Fedotov, A. I. Kolesnikov, E. G. Ponyatovsky, V. I. Shiryaev, and H. Wipf (1998), Neutron diffraction investigation of the dhcp and hcp iron hydrides and deuterides, *J. Alloys Compounds*, *264*(1), 214–222.

Antonov, V. E., M. Baier, B. Dorner, V. K. Fedotov, G. Grosse, A. I. Kolesnikov, E. G. Ponyatovsky, G. Schneider, and F. E. Wagner (2002), High-pressure hydrides of iron and its alloys, *J. Phys. Condens. Matter*, *14*(25), 6427.

Asker, C., U. Kargén, L. Dubrovinsky, and I. A. Abrikosov (2010), Equation of state and elastic properties of face-centered-cubic Fe–Mg alloy at ultrahigh pressures from first-principles, *Earth Planet. Sci. Lett.*, *293*(1–2), 130–134.

Badding, J. V., H. K. Mao, and R. J. Hemley (1992), High-pressure crystal structure and equation of states of iron hydride, *Geophys. Monogr. Ser.*, *67*, 363–371.

Badro, J., G. Fiquet, V. V. Struzhkin, M. Somayazulu, H.-K. Mao, G. Shen, and T. Le Bihan (2002), Nature of the high-pressure transition in Fe$_2$O$_3$ hematite, *Phys. Rev. Lett.*, *89*(20), 205,504.

Badro, J., G. Fiquet, F. Guyot, E. Gregoryanz, F. Occelli, D. Antonangeli, and M. D'astuto (2007), Effect of light elements on the sound velocities in solid iron: Implications for the composition of Earth's core, *Earth Planet. Sci. Lett.*, *254*(1–2), 233–238.

Bazhanova, Z. G., A. R. Oganov, and O. Gianola (2012), Fe–C and Fe–H systems at pressures of the Earth's inner core, *Phys. Uspekhi*, *55*(5), 489–497.

Belonoshko, A. B., R. Ahuja, and B. Johansson (2003), Stability of the body-centred-cubic phase of iron in the Earth's inner core, *Nature*, *424*(6952), 1032–1034.

Brosh, E., G. Makov, and R. Shneck (2009), Thermodynamic analysis of high-pressure phase equilibria in Fe–Si alloys, implications for the inner-core, *Phys. Earth Planet. Inter.*, *172*(3–4), 289–298.

Caracas, R. (2015), The influence of hydrogen on the seismic properties of solid iron, *Geophys. Res. Lett.*, *42*(10), 3780–3785.

Caracas, R., and R. Wentzcovitch (2004), Equation of state and elasticity of FeSi, *Geophys. Res. Lett.*, *31*(20), doi: 10.1029/2004GL020601.

Côté, A. S., L. Vocadlo, and J. P. Brodholt (2012), Ab initio simulations of ironnickel alloys at Earths core conditions, *Earth Planet. Sci. Lett.*, *345–348*, 126–130.

Cottenier, S., M. I. J. Probert, T. Van Hoolst, V. Van Speybroeck, and M. Waroquier (2011), Crystal structure prediction for iron as inner core material in heavy terrestrial planets, *Earth Planet. Sci. Lett.*, *312*(1–2), 237–242.

Cuda, J., T. Kohout, J. Tucek, J. Haloda, J. Filip, R. Prucek, and R. Zboril (2011), Low-temperature magnetic transition in troilite: A simple marker for highly stoichiometric FeS systems, *J. Geophys. Res.*, *116*(B11), B11,205.

Dera, P., B. Lavina, L. A. Borkowski, V. B. Prakapenka, S. R. Sutton, M. L. Rivers, R. T. Downs, N. Z. Boctor, and C. T. Prewitt (2008), High-pressure polymorphism of Fe_2P and its implications for meteorites and Earth's core, *Geophys. Res. Lett.*, *35*(10), L10,301.

Dera, P., J. D. Lazarz, and B. Lavina (2011), Pressure-induced development of bonding in NiAs type compounds and polymorphism of NiP, *J. Solid State Chem.*, *184*(8), 1997–2003.

Dera, P., J. Nisar, R. Ahuja, S. Tkachev, and V. B. Prakapenka (2012), New type of possible high-pressure polymorphism in NiAs minerals in planetary cores, *Phys. Chem. Minerals*, *40*(2), 183–193.

Dubrovinskaia, N., et al. (2005), Beating the miscibility barrier between iron group elements and magnesium by high-pressure alloying, *Phys. Rev. Lett.*, *95*(24), 1–4.

Dubrovinsky, L., N. Dubrovinskaia, F. Langenhorst, D. Dobson, D. Rubie, C. Gessmann, T. L. Bihan, and W. A. Crichton (2004), Reaction of iron and silica at core–mantle boundary conditions, *Phys. Earth Planet. Inter.*, *146*(1–2), 243–247.

Dubrovinsky, L., *et al.* (2007), Body-centered cubic iron-nickel alloy in Earth's core, *Science*, *316*(5833), 1880–1883.

Duman, E., M. Acet, E. F. Wassermann, J. P. Itié, F. Baudelet, O. Mathon, and S. Pascarelli (2005), Magnetic instabilities in Fe_3C cementite particles observed with Fe *k*-edge x-ray circular dichroism under pressure, *Phys. Rev. Lett.*, *94*, 075502.

Fang, C. M., M. A. van Huis, and H. W. Zandbergen (2009), Structural, electronic, and magnetic properties of iron carbide Fe_7C_3 phases from first-principles theory, *Phys. Rev. B*, *80*(22), 224, 108.

Fei, Y., C. T. Prewitt, H.-K. Mao, and C. M. Bertka (1995), Structure and density of FeS at high pressure and high temperature and the internal structure of Mars, *Science*, *268*(5219), 1892–1894.

Fei, Y., C. M. Bertka, and L. W. Finger (1997), High-pressure iron-sulfur compound, Fe_3S2, and melting relations in the Fe-FeS system, *Science*, *275*(5306), 1621–1623.

Fei, Y., J. Li, C. M. Bertka, and C. T. Prewitt (2000), Structure type and bulk modulus of Fe_3S, a new iron-sulfur compound, *Am. Mineral.*, *85*(11–12), 1830–1833.

Fiquet, G., J. Badro, E. Gregoryanz, Y. Fei, and F. Occelli (2009), Sound velocity in iron carbide (Fe_3C) at high pressure: Implications for the carbon content of the Earth's inner core, *Phys. Earth Planet. Inter.*, *172*(1–2), 125–129.

Fischer, R. A., A. J. Campbell, G. A. Shofner, O. T. Lord, P. Dera, and V. B. Prakapenka (2011), Equation of state and phase diagram of FeO, *Earth Planet. Sci. Lett.*, *304*(3–4), 496–502.

Fischer, R. A., A. J. Campbell, R. Caracas, D. M. Reaman, P. Dera, and V. B. Prakapenka (2012), Equation of state and phase diagram of Fe–16Si alloy as a candidate component of Earth's core, *Earth Planet. Sci. Lett.*, *357–358*, 268–276.

Fujiwara, H., H. Kadomatsu, K. Tohma, H. Fujii, and T. Okamoto (1980), Pressure-induced magnetic transition in Fe 2 P, *J. Magnet. Magnetic Mater.*, *21*(3), 262–268.

Gao, L., et al. (2008), Pressure-induced magnetic transition and sound velocities of Fe_3C: Implications for carbon in the Earth's inner core, *Geophys. Res. Lett.*, *35*(17), L17,306.

Gao, L., B. Chen, J. Zhao, E. E. Alp, W. Sturhahn, and J. Li (2011), Effect of temperature on sound velocities of compressed Fe_3C, a candidate component of the Earth's inner core, *Earth Planet. Sci. Lett.*, *309*(3), 213–220.

Georges, A., G. Kotliar, W. Krauth, and M. Rozenberg (1996), Dynamical mean-field theory of strongly correlated fermion systems and the limit of infinite dimensions, *Rev. Modern Phys.*, *68*(1), 13–125.

Gleason, A. E., and W. L. Mao (2013), Strength of iron at core pressures and evidence for a weak Earth's inner core, *Nature Geosci.*, *6*(7), 571–574.

Godwal, B. K., F. González-Cataldo, A. K. Verma, L. Stixrude, and R. Jeanloz (2015), Stability of iron crystal structures at 0.3-1.5 TPa, *Earth Planet. Sci. Lett.*, *409*, 299–306.

Gomi, H., K. Ohta, K. Hirose, S. Labrosse, R. Caracas, M. J. Verstraete, and J. W. Hernlund (2013), The high conductivity of iron and thermal evolution of the Earth's core, *Phys. Earth Planet. Inter.*, *224*, 88–103.

Gu, T., X. Wu, S. Qin, and L. Dubrovinsky (2011), In situ high-pressure study of FeP: Implications for planetary cores, *Phys. Earth Planet. Inter.*, *184*(3–4), 154–159.

Gu, T., Y. Fei, X. Wu, and S. Qin (2014), High-pressure behavior of Fe_3P and the role of phosphorus in planetary cores, *Earth Planet. Sci. Lett.*, *390*, 296–303.

Hazen, R. M., and R. Jeanloz (1984), Wüstite (Fe1x O): A review of its defect structure and physical properties, *Rev. Geophys.*, *22*(1), 37–46.

Hirose, K., S. Labrosse, and J. Hernlund (2013), Composition and state of the core, *Annu. Rev. Earth Planet. Sci.*, *41*(1), 657–691.

Höfer, H. E., S. Weinbruch, and C. A. McCammon (2000), Comparison of two electron probe microanalysis techniques to determine ferric iron in synthetic wüstite samples, *Eur. J. Mineral.*, *12*(1), 63–71.

Isaev, E. I., N. V. Skorodumova, R. Ahuja, Y. K. Vekilov, and B. Johansson (2007), Dynamical stability of Fe-H in the Earth's mantle and core regions, *Proc. Nat. Acad. Sci.*, *104*(22), 9168–9171.

Ito, E., et al. (2009), Determination of high-pressure phase equilibria of Fe_2O_3 using the Kawai-type apparatus equipped with sintered diamond anvils, *Am. Mineral.*, *94*(2–3), 205–209.

Kádas, K., L. Vitos, and R. Ahuja (2008), Elastic properties of iron-rich hcp Fe–Mg alloys up to Earth's core pressures, *Earth Planet. Sci. Lett.*, *271*(1–4), 221–225.

Kamada, S., E. Ohtani, H. Terasaki, T. Sakai, S. Takahashi, N. Hirao, and Y. Ohishi (2014), Equation of state of Fe_3S at room temperature up to 2 megabars, *Phys. Earth Planet. Inter.*, *228*, 106–113.

Kavner, A., T. S. Duffy, and G. Shen (2001), Phase stability and density of FeS at high pressures and temperatures: Implications for the interior structure of Mars, *Earth Planet. Sci. Lett.*, *185*(1), 25–33.

Koch-Muller, M., Y. Fei, R. Wirth, and C. M. Bertka (2002), Characterization of high-pressure iron-fulfur compounds, *Lunar Planet. Sci.*, *33*, 1–2.

Kusaba, K., Y. Syono, T. Kikegawa, and O. Shimomura (1998), High pressure and temperature behavior of FeS, *J. Phys. Chem. Solids*, *59*(6), 945–950.

Kuwayama, Y., K. Hirose, N. Sata, and Y. Ohishi (2008), Phase relations of iron and iron–nickel alloys up to 300 GPa: Implications for composition and structure of the Earth's inner core, *Earth Planet. Sci. Lett.*, *273*(3–4), 379–385.

Li, J., Y. Fei, H. Mao, K. Hirose, and S. Shieh (2001), Sulfur in the Earth's inner core, *Earth Planet. Sci. Lett.*, *193*(3–4), 509–514.

Lin, J., D. Heinz, A. Campbell, J. Devine, and G. Shen (2002), Iron-silicon alloy in Earth's core? *Science*, *295*(5553), 313–315.

Lin, J.-F. (2005), Sound velocities of hot dense iron: Birch's law revisited, *Science*, *308*(5730), 1892–1894.

Lin, J.-F., V. V. Struzhkin, H.-K. Mao, R. J. Hemley, and J. Li (2004), Magnetic transition in compressed Fe_3C from x-ray emission spectroscopy, *Phys. Rev. B*, *70*(21), 212, 405.

Litasov, K., I. Sharygin, A. Shatskii, P. Gavryushkin, P. Dorogokupets, T. Sokolova, E. Ohtani, A. Dymshits, and T. Alifirova (2013), P-V-T equations of state for iron carbides Fe_3C and Fe_7C_3 and their relationships under the conditions of the earth's mantle and core, *Doklady Earth Sci.*, *453*(2), 1269–1273, doi:10.1134/S1028334X13120192.

Longo, M., C. A. McCammon, and S. D. Jacobsen (2011), Microanalysis of the iron oxidation state in (Mg,Fe)O and application to the study of microscale processes, *Contrib. Mineral. Petrol.*, *162*(6), 1249–1257.

Lord, O., M. Walter, R. Dasgupta, D. Walker, and S. Clark (2009), Melting in the Fe-C system to 70 GPa, *Earth Planet. Sci. Lett.*, *284*(1–2), 157–167.

Mao, W. L. (2004), Nuclear resonant x-ray scattering of iron hydride at high pressure, *Geophys. Res. Lett.*, *31*(15), L15, 618.

Martorell, B., L. Vočadlo, J. Brodholt, and I. G. Wood (2013a), Strong premelting effect in the elastic properties of hcp-Fe under inner-core conditions, *Science*, *342*(6157), 466–468.

Martorell, B., J. Brodholt, I. G. Wood, and L. Vočadlo (2013b), The effect of nickel on the properties of iron at the conditions of Earth's inner core: Ab initio calculations of seismic wave velocities of FeNi alloys, *Earth Planet. Sci. Lett.*, *365*, 143–151.

McCammon, C. (1993), Effect of pressure on the composition of the lower mantle end member Fe_xO, *Science*, *259*(5091), 66–68.

McDonough, W. F. (2003), Compositional model for the Earth's core, *Treatise Geochem.*, *2*, 547–568.

McDonough, W. F., and S.-S. Sun (1995), The composition of the Earth, *Chem. Geol.*, *120*(3), 223–253.

Mookherjee, M. (2011), Elasticity and anisotropy of Fe_3C at high pressures, *Am. Mineral.*, *96*(10), 1530.

Morard, G., D. Andrault, N. Guignot, C. Sanloup, M. Mezouar, S. Petitgirard, and G. Fiquet (2008), In situ determination of Fe–Fe_3S phase diagram and liquid structural properties up to 65 GPa, *Earth Planet. Sci. Lett.*, *272*(3–4), 620–626.

Morard, G., D. Andrault, D. Antonangeli, and J. Bouchet (2014), Comptes Rendus geoscience, *Comptes Rendus Geosci.*, *346*(5–6), 130–139.

Morris, E., and Q. Williams (1997), Electrical resistivity of Fe_3O_4 to 48 GPa: Compression-induced changes in electron hopping at mantle pressures, *J. Geophys. Res. Solid Earth*, *102*(B8), 18, 139–18, 148.

Murphy, C.A. (2016), Hydrogen in core: Review of structural, elastic, and thermodynamic properties of iron-hydrogen alloys, *Geophys. Monogr. Ser.*, *217*, 255–264.

Nelmes, R., M. McMahon, S. Belmonte, and J. Parise (1999), Structure of the high-pressure phase III of iron sulfide, *Phys. Rev. B*, *59*(14), 9048–9052.

O'Brien, D. P., A. Morbidelli, and H. F. Levison (2006), Terrestrial planet formation with strong dynamical friction, *Icarus*, *184*(1), 39–58.

Ohta, K., R. E. Cohen, K. Hirose, K. Haule, K. Shimizu, and Y. Ohishi (2012), Experimental and theoretical evidence for pressure-induced metallization in FeO with rocksalt-type structure, *Phys. Rev. Lett.*, *108*(2), 26, 403.

Ohtani, E. (2013), Chemical and physical properties and thermal state of the core, Phys. Chem. Deep Earth, 244–270.

Ohtani, E., et al. (2013), Sound velocity of hexagonal closepacked iron up to core pressures, *Geophys. Res. Lett.*, *40*(19), 5089–5094.

Olsen, J., C. Cousins, L. Gerward, H. Jhans, and B. Sheldon (1991), A study of the crystal-structure of Fe_2O_3 in the pressure range up to 65 GPa using synchrotron radiation, *Phys. Scripta*, *43*(3), 327–330.

Ono, S., and T. Kikegawa (2006), High-pressure study of FeS, between 20 and 120 GPa, using synchrotron X-ray powder diffraction, *Am. Mineral.*, *91*(11–12), 1941–1944.

Ono, S., and K. Mibe (2010), Magnetic transition of iron carbide at high pressures, *Phys. Earth Planet. Inter.*, *180*(1), 1–6.

Ono, S., T. Kikegawa, and Y. Ohishi (2004), High-pressure phase transition of hematite, Fe_2O_3, *J. Phys. Chem. Solids*, *65*(8–9), 1527–1530.

Ono, S., A. R. Oganov, J. P. Brodholt, L. Vocadlo, I. G. Wood, A. Lyakhov, C. W. Glass, A. S. Côté, and G. D. Price (2008), High-pressure phase transformations of FeS: Novel phases at conditions of planetary cores, *Earth Planet. Sci. Lett.*, *272*(1–2), 481–487.

Otsuka, K., C. A. McCammon, and S.-i. Karato (2010), Tetrahedral occupancy of ferric iron in (Mg,Fe)O: Implications for point defects in the Earth's lower mantle, *Phys. Earth Planet. Int.*, *180*(3–4), 179–188.

Pasternak, M., G. Rozenberg, G. Machavariani, O. Naaman, R. Taylor, and R. Jeanloz (1999), Breakdown of the Mott-Hubbard state in Fe_2O_3: A first-order insulator-metal transition with collapse of magnetism at 50 GPa, *Phys. Rev. Lett.*, *82*(23), 4663–4666.

Payne, M. C., M. P. Teter, D. C. Allan, T. A. Arias, and J. D. Joannopoulos (1992), Iterative minimization techniques for ab initio total energy calculations — Molecular dynamics and conjugate gradients, *Rev. Modern Phys.*, *64*(4), 1045–1097.

Pepin, C. M., A. Dewaele, G. Geneste, P. Loubeyre, and M. Mezouar (2014), New iron hydrides under high pressure, *Phys. Rev. Lett.*, *113*(26), 265504.

Poirier, J. P. (1994), Light elements in the Earth's outer core — A critical review, *Phys. Earth Planet. Inter.*, *85*(3–4), 319–337.

Pratesi, G. (2006), Icosahedral coordination of phosphorus in the crystal structure of melliniite, a new phosphide mineral from the Northwest Africa 1054 acapulcoite, *Am. Mineral.*, *91*(2–3), 451–454.

Prescher, C., L. Dubrovinsky, C. McCammon, K. Glazyrin, Y. Nakajima, A. Kantor, M. Merlini, and M. Hanfland (2012), Structurally hidden magnetic transitions in Fe_3C at high pressures, *Phys. Rev. B*, *85*, 140,402.

Prescher, C., et al. (2015), High Poisson's ratio of Earth's inner core explained by carbon alloying, *Nature Geosci.*, *8*(3), 220–223.

Rhee, J., and B. Harmon (2004), Metamagnetic behavior of Fe_3M (M = Al and Si) alloys at high pressure, *Phys. Rev. B*, *70*(9), 094, 411.

Rouquette, J., D. Dolejš, I. Kantor, C. McCammon, D. Frost, V. Prakapenka, and L. Dubrovinsky (2008), Iron-carbon interactions at high temperatures and pressures, *Appl. Phys. Lett.*, *92*, 121, 912.

Rubie, D., F. Nimmo, and H. Melosh (2007), Formation of Earth's core, in *Evolution of the Earth: Treatise on Geophysics*, edited by D. J. Stevenson, pp. 51–90, AGU, Washington, D.C.

Rubie, D. C., D. J. Frost, U. Mann, Y. Asahara, F. Nimmo, K. Tsuno, P. Kegler, A. Holzheid, and H. Palme (2011), Heterogeneous accretion, composition and core–mantle differentiation of the Earth, *Earth Planet. Sci. Lett.*, *301*(1–2), 31–42.

Sakai, T., S. Takahashi, N. Nishitani, I. Mashino, E. Ohtani, and N. Hirao (2014), Equation of state of pure iron and $Fe_{0.9}Ni_{0.1}$ alloy up to 3 Mbar, *Phys. Earth Planet. Inter.*, 1–13, 114–126.

Sakamaki, K., E. Takahashi, Y. Nakajima, Y. Nishihara, K. Funakoshi, T. Suzuki, and Y. Fukai (2009), Melting phase relation of FeH_x up to 20 GPa: Implication for the temperature of the Earth's core, *Phys. Earth Planet. Inter*, *174*(1–4), 192–201.

Sata, N., K. Hirose, G. Shen, Y. Nakajima, Y. Ohishi, and N. Hirao (2010), Compression of FeSi, Fe_3C, $Fe_{0.95}O$, and FeS under the core pressures and implication for light element in the Earth's core, *J. Geophys. Res.*, *115*, B09,204.

Saxena, S., and L. Dubrovinsky (2000), Iron phases at high pressures and temperatures: Phase transition and melting, *Am. Mineral.*, *85*(2), 372–375.

Saxena, S. K., and G. Eriksson (2015), *J. Phys. Chem. Solids*, 1–5.

Scott, H. P., S. Huggins, M. R. Frank, S. J. Maglio, C. D. Martin, Y. Meng, J. Santillán, and Q. Williams (2007), Equation of state and high-pressure stability of Fe_3P-schreibersite: Implications for phosphorus storage in planetary cores, *Geophys. Res. Lett.*, *34*(6), L06, 302.

Seagle, C. T., A. J. Campbell, D. L. Heinz, G. Shen, and V. B. Prakapenka (2006), Thermal equation of state of Fe_3S and implications for sulfur in Earth's core, *J. Geophys. Res.*, *111*(B6), B06,209.

Sha, X., and R. E. Cohen (2010), First-principles thermal equation of state and thermoelasticity of hcp Fe at high pressures, *Phys. Rev. B*, *81*(9), 094,105.

Sherman, D. (1995), Stability of possible Fe-FeS and Fe-FeO alloy phases at high-pressure and the composition of the Earth's core, *Earth Planet. Sci. Lett.*, *132*(1–4), 87–98.

Shibazaki, Y., et al. (2012), Sound velocity measurements in dhcp-FeH up to 70GPa with inelastic X-ray scattering: Implications for the composition of the Earth's core, *Earth Planet. Sci. Lett.*, *313–314*, 79–85.

Shim, S.-H., A. Bengtson, D. Morgan, W. Sturhahn, K. Catalli, J. Zhao, M. Lerche, and V. Prakapenka (2009), Electronic and magnetic structures of the postperovskite-type Fe_2O_3 and implications for planetary magnetic records and deep interiors, *Proc. Nat. Acad. Sci.*, *106*(14), 5508–5512.

Skorodumova, N. V. (2004), Influence of hydrogen on the stability of iron phases under pressure, *Geophys. Res. Lett.*, *31*(8), 1–3.

Steinle-Neumann, G., L. Stixrude, R. E. Cohen, and O. Gülseren (2001), Elasticity of iron at the temperature of the Earth's inner core, *Nature*, *413*(6851), 57–60.

Stixrude, L., and R. E. Cohen (1995), High-pressure elasticity of iron and anisotropy of Earth's inner core, *Science*, *267*(5206), 1972–1972.

Stixrude, L., E. Wasserman, and R. E. Cohen (1997), Composition and temperature of Earth's inner core, *J. Geophys. Res.*, *102*(B11), 24,729.

Takafuji, N. (2005), Solubilities of O and Si in liquid iron in equilibrium with $(Mg,Fe)SiO_3$ perovskite and the light elements in the core, *Geophys. Res. Lett.*, *32*(6), 1–4.

Tateno, S., K. Hirose, Y. Ohishi, and Y. Tatsumi (2010), The structure of iron in Earth's inner core, *Science*, *330*(6002), 359–361.

Tateno, S., K. Hirose, T. Komabayashi, H. Ozawa, and Y. Ohishi (2012), The structure of Fe-Ni alloy in Earth's inner core, *Geophys. Res. Lett.*, *39*(12), L12305, doi:10.1029/2012GL052103.

Terasaki, H., et al. (2012), Stability of Fe-Ni hydride after the reaction between Fe-Ni alloy and hydrous phase (delta-AlOOH) up to 1.2 Mbar: Possibility of H contribution to the core density deficit, *Phys. Earth Planet. Inter.*, *194*, 18–24.

Tsuchiya, T., and J. Tsuchiya (2011), Prediction of a hexagonal SiO_2 phase affecting stabilities of $MgSiO_3$ and $CaSiO_3$ at multimegabar pressures, *Proc. Nat. Acad. Sci. USA*, *108*(4), 1252–1255.

Urakawa, S., et al. (2004), Phase relationships and equations of state for FeS at high pressures and temperatures and implications for the internal structure of Mars, *Phys. Earth Planet. Inter.*, *143*, 469–479.

Vocadlo, L., J. Brodholt, D. Dobson, K. Knight, W. Marshall, G. Price, and I. Wood (2002), The effect of ferromagnetism on the equation of state of Fe_3C studied by first-principles calculations, *Earth Planet. Sci. Lett.*, *203*(1), 567–575.

Vocadlo, L., D. Alfe, M. J. Gillan, I. G. Wood, J. P. Brodholt, and G. D. Price (2003a), Possible thermal and chemical stabilization of body-centred-cubic iron in the Earth's core, *Nature*, *424*(6948), 536–539.

Vocadlo, L., D. Alfe, M. Gillan, and G. Price (2003b), The properties of iron under core conditions from first principles calculations, *Phys. Earth Planet. Interi.*, *140*(1–3), 101–125.

Vocadlo, L., D. P. Dobson, and I. G. Wood (2009), Ab initio calculations of the elasticity of hcp-Fe as a function of temperature at inner-core pressure, *Earth Planet. Sci. Lett.*, *288*(3–4), 534–538.

Wade, J., and B. J. Wood (2005), Core formation and the oxidation state of the Earth, *Earth Planet. Sci. Lett.*, *236*(1–2), 78–95.

Wang, X., S. Scandolo, and R. Car (2005), Carbon phase diagram from ab initio molecular dynamics, *Phys. Rev. Lett.*, *95*(18), 185,701.

Wood, B. (1993), Carbon in the core, *Earth Planet. Sci. Lett.*, *117*(3–4), 593–607.

Wood, B. J. (2008), Accretion and core formation: Constraints from metal-silicate partitioning, *Philos. Trans. R. Soc. A Math. Phys. Eng. Sci.*, *366*(1883), 4339–4355.

Wood, B. J., and A. N. Halliday (2005), Cooling of the Earth and core formation after the giant impact, *Nature*, *437*(7063), 1345–1348.

Wood, B. J., J. Wade, and M. R. Kilburn (2008), Core formation and the oxidation state of the Earth: Additional constraints from Nb, V and Cr partitioning, *Geochim. Cosmochim. Acta*, *72*(5), 1415–1426.

Wood, I., L. Vocadlo, K. Knight, D. Dobson, W. Marshall, G. Price, and J. Brodholt (2004), Thermal expansion and crystal structure of cementite, Fe_3C, between 4 and 600 K determined by time-of-flight neutron powder diffraction, *J. App. Crystallogr.*, *37*, 82–90.

Wu, X., and S. Qin (2010), First-principles calculations of the structural stability of Fe_2P, *J. Phys. Conf. Ser.*, *215*, 012,110.

Wu, X., M. Mookherjee, T. Gu, and S. Qin (2011), Elasticity and anisotropy of iron-nickel phosphides at high pressures, *Geophys. Res. Lett.*, *38*(20).

Yamamoto, A. (1982), Modulated structure of wustite ($Fe_{1-x}O$) (three-dimensional modulation), *Acta Crystallogr. Sect. B Struct. Crystallogr. Crystal Chem.*, *38*(5), 1451–1456.

Zhang, F., and A. R. Oganov (2010), Iron silicides at pressures of the Earth's inner core, *Geophys. Res. Lett.*, *37*, L02305, doi:10.1029/2009GL041224.

Zhao, S., X.-W. Liu, C.-F. Huo, Y.-W. Li, J. Wang, and H. Jiao (2015), Determining surface structure and stability of ε-Fe_2C, χ-Fe_5C_2, θ-Fe_3C and Fe_4C phases under carburization environment from combined DFT and atomistic thermodynamic studies, *Catal. Struct. Reactivity*, *1*(1), 44–60.

6

Crystal Structures of Minerals in the Lower Mantle

June K. Wicks and Thomas S. Duffy

ABSTRACT

The crystal structures of lower mantle minerals are vital components for interpreting geophysical observations of Earth's deep interior and in understanding the history and composition of this complex and remote region. The expected minerals in the lower mantle have been inferred from high pressure-temperature experiments on mantle-relevant compositions augmented by theoretical studies and observations of inclusions in natural diamonds of deep origin. While bridgmanite, ferropericlase, and $CaSiO_3$ perovskite are expected to make up the bulk of the mineralogy in most of the lower mantle, other phases such as SiO_2 polymorphs or hydrous silicates and oxides may play an important subsidiary role or may be regionally important. Here we describe the crystal structure of the key minerals expected to be found in the deep mantle and discuss some examples of the relationship between structure and chemical and physical properties of these phases.

6.1. INTRODUCTION

Earth's lower mantle, which spans from 660 km depth to the core-mantle boundary (CMB), encompasses nearly three quarters of the mass of the bulk silicate Earth (crust and mantle). Our understanding of the mineralogy and associated crystal structures of this vast region has greatly expanded over the course of the past two decades with the development of new capabilities for reproducing in the laboratory the extreme pressures and temperatures expected to be found in the lower mantle (24–135 GPa and 1800–4000 K) [*Ricolleau et al.*, 2010; *Irifune and Tsuchiya*, 2015]. Most structural studies of lower mantle minerals have been carried out using the laser-heated diamond anvil cell with X-ray diffraction techniques, most commonly powder diffraction, as the primary diagnostic. Several recent review articles have summarized the major experimental methods for exploring the structures and

properties of lower mantle minerals [*Duffy*, 2005; *Mao and Mao*, 2007; *Shen and Wang*, 2014; *Ito*, 2015].

The crystal structure is the most fundamental property of a mineral and is intimately related to its major physical and chemical characteristics, including compressibility, density, and sound velocities. As an example, observations of seismic anisotropy in the lowermost mantle are connected to the elastic anisotropy of the associated minerals, and this anisotropy is ultimately dictated by the mineral's chemical composition and the details of its crystal structure [*Karki et al.*, 1997b; *Marquardt et al.*, 2009; *Dobson et al.*, 2013]. Crystal structure also strongly influences the partitioning of elements between different phases and so plays an important role in the chemical state of the lower mantle as well.

The expected phase assemblages and the corresponding crystal structures are largely a function of the chemical makeup of the deep mantle. Figure 6.1 illustrates the expected mineralogies of the two major expected lithologies of the lower mantle. The pyrolite model, the most widely accepted model for the bulk lower mantle, is based

Department of Geosciences, Princeton University, Princeton, New Jersey, USA

Deep Earth: Physics and Chemistry of the Lower Mantle and Core, Geophysical Monograph 217, First Edition.
Edited by Hidenori Terasaki and Rebecca A. Fischer.
© 2016 American Geophysical Union. Published 2016 by John Wiley & Sons, Inc.

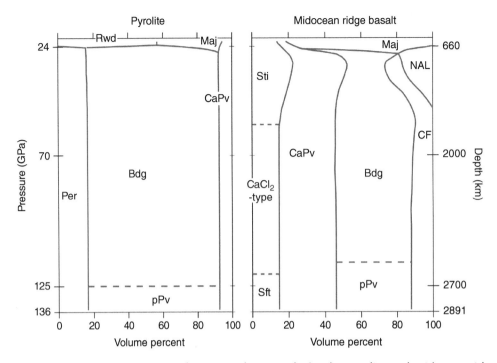

Figure 6.1 Expected volume fractions of lower mantle minerals for the pyrolite and midocean ridge basalt compositional models. Dashed lines indicate isochemical phase transitions, the exact locations of which depend on the mantle temperature profile. Mineral abbreviations are: Bdg, bridgmanite; Per, (ferro)periclase; Maj, majorite; Rwd, ringwoodite; Sti, stishovite; Sft, seifertite. Other abbreviations: CaPv, calcium-silicate perovskite; CF, calcium ferrite-type phase; NAL, new aluminous phase; pPv, post-perovskite. Adapted from *Hirose et al.* [2005]; *Irifune et al.* [2010], and *Ricolleau et al.* [2010].

on experimental and petrological studies of the peridotites and basalts of the uppermost mantle and assumes that the lower mantle composition is broadly similar to that of the upper mantle [*Ringwood*, 1975]. In pyrolite, bridgmanite, periclase, and perovskite-structured $CaSiO_3$ are the major minerals found in experiments across most of the lower mantle pressure range [*Kesson et al.*, 1998; *Irifune et al.*, 2010; *Irifune and Tsuchiya*, 2015]. There is geophysical evidence that subducting oceanic lithosphere penetrates the 660 km discontinuity, thereby transporting oceanic crust in the deep mantle, perhaps reaching as far as the core-mantle boundary. Subducted basaltic crust is expected to produce a different mineralogy, with increased Ca-perovskite together with free silica and aluminum-bearing phases (Figure 6.1) [*Irifune and Ringwood*, 1993; *Ono et al.*, 2001; *Hirose et al.*, 2005; *Ricolleau et al.*, 2010].

6.2. BRIDGMANITE, (Mg,Fe)SiO₃

The major mineral phase in the lower mantle and the most abundant mineral in Earth is $(Mg,Fe)SiO_3$ in the perovskite structure, now known as bridgmanite [*Tschauner et al.*, 2014]. The perovskite structure has

the general formula ABX_3 and is adopted by materials encompassing a wide range of compositions. The A site is occupied by a large-radius cation, the B site contains a smaller cation, and the X site is an anion, typically oxygen or fluorine. The structure can be described as a framework of BX_6 octahedra forming a corner-sharing network with A cations nestled in the framework cavities (Figure 6.2). In an ideal cubic perovskite (space group $Pm\overline{3}m$), the large A cations are in 12-fold dodecahedral coordination with the anion, but when a smaller cation (such as Mg^{2+}) occupies the A site, the octahedral framework collapses around the cation such that the coordination decreases to eightfold [*Horiuchi et al.*, 1987].

Considerable structural diversity in perovskites can be introduced by rotation and tilting of octahedra, as well as through the effects of cation offsets, cation ordering, and nonstoichiometry [*Wang and Angel*, 2011]. Octahedral tilting alone can lead to a number of variants with tetragonal, orthorhombic, or monoclinic symmetry. Perovskite-structured $MgSiO_3$, first synthesized by *Liu* [1975], is orthorhombic with the $GdFeO_3$-type structure (space group *Pbnm* or *Pnma*, Table 6.1) and is isostructural with the mineral perovskite, $CaTiO_3$. The space group *Pnma* is equivalent to *Pbnm*, with the following

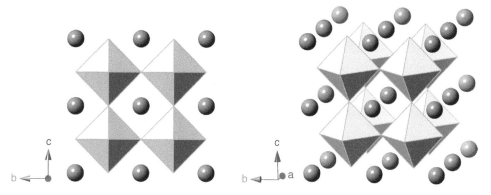

Figure 6.2 Cubic perovskite structure (space group $Pm\bar{3}m$, Table 6.1) of $CaSiO_3$ at conditions of the lower mantle. Spheres are Ca^{2+} cations each surrounded by 12 O^{2-} anions. Si^{4+} cations are centered among 6 O^{2-} anions in an octahedral arrangement, corner shared with adjacent octahedra.

Table 6.1 Crystallographic parameters for major lower mantle structures.

	x/a	y/b	z/c
Bridgmanite, $(Mg,Fe)SiO_3$[a]			
SG = $Pnma$[b], a = 5.02(3) Å, b = 6.90(3) Å, c = 4.81(2) Å			
Mg/Fe (4c)	0.557(2)	1/4	0.513(3)
Si/Fe (4b)	0	0	1/2
O (4c)	0.931(1)	1/4	0.381(3)
O (8d)	0.176(3)	0.575(1)	0.160(2)
Ferropericlase, $(Mg,Fe)O$[c]			
SG = $Fm\bar{3}m$, a = 4.211(1) Å			
Mg/Fe (4a)	0	0	0
O (4b)	1/2	1/2	1/2
Ca-perovskite, $CaSiO_3$[d]			
SG = $Pm\bar{3}m$, a = 3.546 Å			
Ca (1b)	1/2	1/2	1/2
Si (1a)	0	0	0
O (3d)	1/2	0	0
CaIrO$_3$-type post-perovskite, $(Mg,Fe)SiO_3$[e]			
SG = $Cmcm$, a = 2.466(1) Å, b = 8.130(6) Å, c = 6.108(10) Å			
Mg (4c)	0	0.256(2)	1/4
Si (4a)	0	0	0
O1 (4c)	0	0.929(5)	1/4
O2 (8f)	0	0.639(4)	0.437(5)

[a] $(Mg_{0.75}Fe_{0.20}Na_{0.03}Ca_{0.02}Mn_{0.01})Si_{1.00}O_3$, ambient pressure [Tschauner et al., 2014].
[b] The space group $Pnma$ is equivalent to $Pbnm$.
[c] MgO, ambient pressure [Hazen, 1976].
[d] Theoretical, ambient pressure [Caracas and Wentzcovitch, 2006].
[e] $Mg_{0.93}Fe_{0.07}SiO_3$, 121 GPa [Zhang et al., 2013].

conversion: $(a,b,c)_{Pnma} \rightarrow (b,c,a)_{Pbnm}$. Deviation from the ideal cubic arrangement is achieved through both rotation (11.2° about the c axis at room pressure in $Pbnm$) and tilting (16.7° with respect to the c axis) of SiO_6

octahedra and offset in position of the central Mg atom [Horiuchi et al., 1987] (Figure 6.3). These structural distortion increase with increasing pressure [Fiquet et al., 2000].

While bridgmanite can be quenched to ambient conditions in the laboratory, direct evidence for the natural existence of this phase proved elusive for many years. Inclusions in rare diamonds of deep origin were interpreted as breakdown products of bridgmanite [Harte and Harris, 1994; Stachel et al., 2000]. Recently, the first natural occurrence of this phase was definitively identified in a chondritic meteorite [Tschauner et al., 2014], allowing assignment of the name bridgmanite. The natural sample had the composition $(Mg_{0.75}Fe_{0.20}Na_{0.03}Ca_{0.02}Mn_{0.01})Si_{1.00}O_3$ and was found in association with akimotoite, a magnesium silicate with the ilmenite-type structure. The formation conditions for the bridgmanite sample were estimated to be 23–25 GPa and 2200–2400 K.

A wide range of cations can occupy the A and B sites in perovskites. Bridgmanite itself can accommodate increasing amounts of Fe and Al under compression [Mao et al., 1997; Ito et al., 1998]. Substitution of Fe^{2+} for Mg^{2+} in the A site of perovskite expands the structure and decreases the degree of distortion [Kudoh et al., 1990]. Substitution of Fe^{3+} similarly expands the structure but can increase the degree of distortion as a result of a coupled substitution mechanism involving both the A and B sites [Catalli et al., 2010]. Incorporation of aluminum into bridgmanite can occur by either a Tschermak-like coupled stoichiometric substitution:

$$Mg^{2+} + Si^{4+} = 2Al^{3+}, \tag{6.1}$$

or through nonstoichiometric substitution involving an oxygen vacancy, $V_O^{\cdot\cdot}$:

$$2Si^{4+} = 2Al^{3+} + V_O^{\cdot\cdot}. \tag{6.2}$$

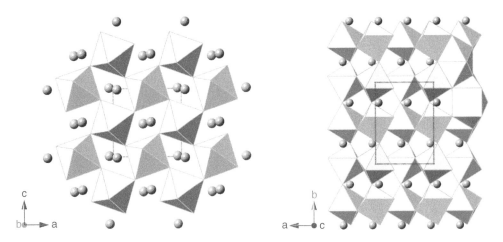

Figure 6.3 View along the *b* (left) and *c* axis (right) of bridgmanite (*Pnma*, Table 6.1). Gray dashed line shows the outline of the unit cell. Tilting of SiO₆ octahedra (blue) results in shortening of crystallographic axes and produces a displacement of Mg cations (yellow spheres and representative MgO₆ polyhedra) reducing the coordination from 12 to 8.

At pressures of the lower mantle, the stoichiometric substitution (6.1) is energetically favorable [*Brodholt*, 2000; *Yamamoto et al.*, 2003; *Akber-Knutson and Bukowinski*, 2004], yet there is experimental evidence for the operation of both substitution mechanisms in laboratory-synthesized samples. Nuclear magnetic resonance results are consistent with two distinct Al sites, indicating coupled substitution [*Stebbins et al.*, 2001], while the enhanced compressibility observed in some equation-of-state studies is more consistent with vacancy substitution [*Andrault et al.*, 2007]. Al³⁺ substitution increases both the unit cell volume and its degree of distortion, possibly as a result of less efficient packing due to substitution in both the A and B sites, leading to an (Mg,Al)(Si,Al)O₃ composition [*Weng et al.*, 1982]. This is consistent with the results of a structure refinement study comparing perovskites synthesized with both substitution mechanisms: "stoichiometric" Al-containing MgSiO₃ perovskites are more distorted than their "non-stoichiometric" counterparts [*Kojitani et al.*, 2007a]. The perovskite structure can accommodate up to 25 mol.% aluminum, as both pyrope and almandine garnets have been shown to adopt the perovskite structure at sufficiently high pressures [*Ito et al.*, 1998; *Kesson et al.*, 1995; *Dorfman et al.*, 2012].

Coupled substitution of Fe³⁺ and Al³⁺ allows both cations to be preferentially incorporated into the bridgmanite phase, stabilizing Fe³⁺ even under the more reducing conditions of the lower mantle [*Frost and McCammon*, 2008]. This capacity to host Fe with such high oxidation state allows for a lower mantle enriched in Fe³⁺ and one that potentially contains metallic iron as part of the equilibrium assemblage. A disproportiona-tion of Fe²⁺ to Fe³⁺ + Fe⁰ has been proposed on the basis of closed-system experiments observing the latter [*Frost et al.*, 2004; *Auzende et al.*, 2008].

The nature of the spin-pairing crossover of Fe in bridgmanite has been the subject of considerable debate from both experimental and theoretical viewpoints [e.g., *Lin et al.*, 2013]. The behavior of Fe in Mg-perovskite is inherently complex due to iron's potential to adopt multiple structural sites (A or B), valence states (2+, 3+), and electronic configurations (high, low, and potentially intermediate spin). It appears that Fe²⁺ in bridgmanite does not undergo a spin crossover at lower mantle pressures but does exhibit a large change in quadrupole splitting as a function of pressure, consistent with a marked increase in lattice distortion of the A site [*Jackson et al.*, 2005; *Hsu et al.*, 2010]. For ferric iron-containing perovskite, the evidence indicates that Fe³⁺ in the smaller B undergoes a transitions to a low-spin state by 60 GPa, whereas Fe³⁺ in the larger A site remains in the high-spin configuration [*Catalli et al.*, 2010].

Over the pressure range of the lower mantle, the maximum solubility of Fe²⁺ in bridgmanite increases. In Al-free samples, a maximum solubility of Fe/(Mg+Fe) ranges from 0.16 at 25 GPa, 1500°C [*Tange et al.*, 2009] to at least 0.74 at 80 GPa [*Dorfman et al.*, 2013]. Recent laser-heated diamond anvil experiments reported unexpectedly that at pressures of 95 GPa (~2100 km depth) and temperatures above 2200 K, Fe-bearing bridgmanite disproportionates into an iron-poor bridgmanite phase and a previously unknown iron-rich silicate [*Zhang et al.*, 2014]. If confirmed, this finding could have major relevance for understanding the deep mantle.

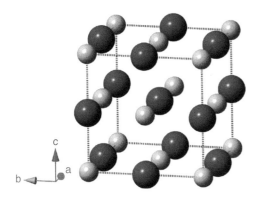

Figure 6.4 Rocksalt structure of ferropericlase (Mg,Fe)O ($Fm\bar{3}m$, Table 6.1). Interpenetrating Mg/Fe (yellow spheres) and O (red spheres) lattices result in octahedral coordination of both cations and anions.

6.3. FERROPERICLASE, (Mg,Fe)O

Ferropericlase, (Mg,Fe)O, is the second most abundant phase in pyrolite compositions in the lower mantle (Figure 6.1). This phase adopts the simple rocksalt (B1) structure (space group $Fm\bar{3}m$) consisting of interpenetrating face-centered-cubic lattices producing alternating Mg^{2+}/Fe^{2+} cations and O^{2-} anions (Figure 6.4). The structure can also be described as an edge-share array of $(Mg,Fe)O_6$ octahedra in which all edges are shared with neighboring octahedra. High pressure-temperature (*P-T*) partitioning studies between ferropericlase and bridgmanite indicate that iron partitions preferentially into ferropericlase such that this mineral is expected to have compositions around Mg/(Mg+Fe) = 0.8 under mantle conditions [*Auzende et al.*, 2008; *Sakai et al.*, 2009; *Tange et al.*, 2009] and that iron in (Mg,Fe)O is predominately in the ferrous state [*McCammon et al.*, 1998]. Ferropericlase also undergoes a spin-pairing transition at lower mantle conditions, and the effects of this transition on density, elastic properties, partitioning behavior, and transport properties have been the subjects of intensive recent study [*Badro et al.*, 2003; *Sturhahn et al.*, 2005; *Bower et al.*, 2009; *Lin et al.*, 2013; *Badro*, 2014].

While the Mg end-member periclase is expected to be stable throughout the lower mantle pressure range [*Duffy et al.*, 1995], the Fe end-member, wüstite, displays more complex behavior. At room temperature, FeO first undergoes a distortion to a rhombohedral structure (space group $R\bar{3}m$) [*Mao et al.*, 1996; *Shu et al.*, 1998] followed by transition to the hexagonal B8 phase (space group $P6_3mc$) [*Fei and Mao*, 1994; *Murakami et al.*, 2004a] at ~120 GPa. At the high *P-T* conditions expected along a mantle geotherm, FeO is expected remain in the B1 structure, yet experiments suggest it will undergo an insulator-metal transition near 70 GPa and 1900 K [*Fischer et al.*, 2011]. Such a phase transition in the end-member composition

suggests that the Mg^{2+}-Fe^{2+} solid solution, complete at low pressures, would no longer be continuous above the transition pressure. At present, experimental confirmation of this is inconclusive as evidence for and against dissociation has been reported [*Dubrovinsky et al.*, 2000; *Lin et al.*, 2003; *Kondo et al.*, 2004; *Ohta et al.*, 2014].

FeO may exist at the core-mantle boundary as a result of reaction with the liquid outer core metallic alloy [*Manga and Jeanloz*, 1996; *Buffett and Seagle*, 2010]. Such a metallic component in the lowermost mantle could have implications for phase relations, electrical conductivity, and coupling between the core and mantle [*Seagle et al.*, 2008]. Iron-rich ferropericlase may also provide an explanation for ultra-low seismic velocity zones observed near the CMB [*Wicks et al.*, 2010].

6.4. CaSiO$_3$ PEROVSKITE

Calcium silicate perovskite ($CaSiO_3$, Ca-Pv) is considered to be the third most abundant phase in the lower mantle, comprising ~10% by volume of a pyrolite composition and up to ~25 vol % of the abundance in basaltic regions [*Irifune et al.*, 2010; *Ricolleau et al.*, 2010] (Figure 6.1). This phase was first synthesized at high pressure and temperature by *Liu and Ringwood* [1975]. It is unquenchable and transforms to glass at ambient conditions [*Tamai and Yagi*, 1989]. The crystal structure at 300 K was originally reported to be cubic perovskite ($Pm\bar{3}m$) [*Liu and Ringwood*, 1975; *Mao et al.*, 1989; *Wang and Weidner*, 1994] (Figure 6.2). Later theoretical calculations indicated that the cubic form was unstable at low temperatures as a result of slight rotations of the SiO_6 octahedra [*Stixrude et al.*, 1996]. A structural distortion was subsequently confirmed by experiment [*Shim et al.*, 2002]. The nature of this low-temperature noncubic distortion has been the subject of extensive theoretical investigations with both tetragonal [*Stixrude et al.*, 1996; *Caracas et al.*, 2005] and orthorhombic [*Jung and Oganov*, 2005; *Adams and Oganov*, 2006; *Li et al.*, 2006a] forms being proposed.

At elevated temperatures, Ca-Pv transforms into the cubic structure. While one study has found that the noncubic form remains stable up to 18 GPa and 1600 K [*Uchida et al.*, 2009], most studies find that Ca-Pv becomes cubic at relatively low temperatures (<600 K) [*Kurashina et al.*, 2004; *Ono et al.*, 2004; *Komabayashi et al.*, 2007; *Noguchi et al.*, 2012; *Sun et al.*, 2014]. Thus, the cubic form is expected to be the relevant one for Earth's mantle. Theoretical calculations predict a low shear modulus that could potentially be relevant to low shear velocity anomalies in the deep mantle [*Kawai and Tsuchiya*, 2015].

Experiments exploring the mutual solubility of the two perovskites (Mg into Ca-Pv and Ca into bridgmanite) find substitution to be very limited in the uppermost lower mantle, but that up to 10% Mg can be substituted into Ca-Pv by 55 GPa [*Armstrong et al.*, 2012]. This effect is enhanced with pressure, temperature, and the substitution of Ti^{4+} for Si^{4+}, with Mg-Ca forming a complete solution at 97 GPa for modest amounts of Ti substitution (Ti/(Ti+Si) = 0.05).

6.5. CaIrO₃-TYPE POST-PEROVSKITE PHASE

While $MgSiO_3$ bridgmanite is stable throughout most of the lower mantle, at pressures near 125 GPa and ~2500 K (corresponding to the D″ region near the core-mantle boundary), it transforms to a high-pressure polymorph known as "post-perovskite" [*Murakami et al.*, 2004b; *Oganov and Ono*, 2004]. This phase adopts the orthorhombic $CaIrO_3$-type structure (space group *Cmcm*). In post-perovskite, Si^{4+} and Mg^{2+} have the same coordination environment as in perovskite, but there are profound differences in their structural arrangement. In contrast to the corner-sharing network in perovskite, the SiO_6 octahedra in post-perovskite share edges along the *a* axis and corners along *c*. The structure is thus sheet-like with layers of SiO_6 octahedra alternating with MgO_8 layers along the *b* axis (Figure 6.5). The Mg sites in post-perovskite are smaller and less distorted than those in Pv, resulting in a volume reduction of ~1–1.5% across the transition.

Experimental studies of the structure of the post-perovskite phase have mainly been carried out using powder X-ray diffraction techniques [*Murakami et al.*, 2004b; *Shim et al.*, 2008; *Hirose et al.*, 2015]. First-principles studies of the structure and its pressure evolution are generally consistent with experiment results [*Oganov and Ono*, 2004; *Lin et al.*, 2014]. Some evidence for slight structural modifications of the post-perovskite structure have been reported on the basis of polycrystalline X-ray studies in aluminum-rich [*Tschauner et al.*, 2008] and iron-rich compositions [*Yamanaka et al.*, 2012], but these have not yet been confirmed. Recently, a single-crystal structure refinement of (Mg,Fe)SiO₃ post-perovskite was carried out to very high pressures of 120 GPa by isolating individual crystals for study from a coarse multigrain aggregate that was synthesized in the diamond anvil cell [*Zhang et al.*, 2013].

The anisotropic nature of the post-perovskite structure likely has a number of geophysical implications. The compressibility in the *b* direction is substantially higher than in *a* or *c* due to the presence of the relatively soft MgO_8 layer [*Iitaka et al.*, 2004]. Theoretical calculations indicate that post-perovskite also has larger elastic shear wave anisotropy than perovskite. The mode of deformation is being studied extensively in both silicate and analog compositions to attempt to explain the observed seismic anisotropy in the deep mantle in terms of the deformation behavior and elastic anisotropy of post-perovskite, with no definitive answer as of yet [*Oganov et al.*, 2005; *Yamazaki et al.*, 2006; *Merkel et al.*, 2007; *Miyagi et al.*, 2010; *Dobson et al.*, 2013]. Theoretical studies have also shown that, in contrast to bridgmanite, diffusion in post-perovskite is highly anisotropic [*Ammann et al.*, 2010] and that thermal conductivity of post-perovskite is both more anisotropic and larger than in perovskite [*Ammann et al.*, 2014]. These factors may be important for understanding the viscosity and heat flow at the base of the mantle.

6.6. SiO₂ POLYMORPHS

Mid-ocean ridge basalt compositions are rich in SiO_2 relative to pyrolite, and the resulting mineral assemblage is expected to contain ~10–20% free silica at lower mantle conditions (Figure 6.1) [*Irifune and Ringwood*, 1993;

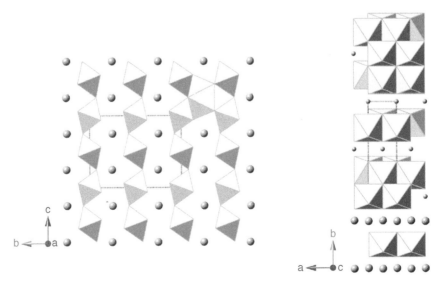

Figure 6.5 Crystal structure of CaIrO₃-type post-perovskite (space group *Cmcm*, Table 6.1) viewed along *a* (left) and *c* axis (right). Gray dashed line shows the outline of unit cell. The structure consists of "sheets" of edge-shared SiO₆ octahedra (gray) that are intercalated by Mg ions (yellow spheres and polyhedra).

(a)   (b)   (c)

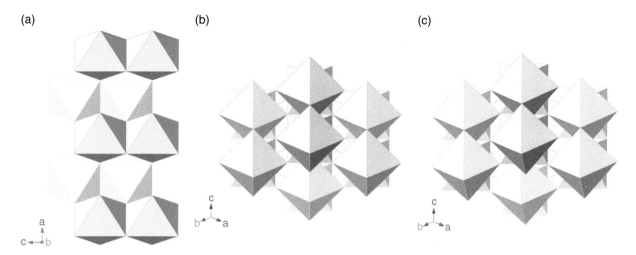

Figure 6.6 (a) Structure of rutile-type stishovite (space group $P4_2/mnm$, Table 6.2) along the b axis. SiO_6 octahedra (blue) are organized into edge-sharing chains that extend along the c direction and connect to neighboring chains by corner sharing. The figure is depth shaded to illustrate offset in the octahedra. Slight rotation difference between the octahedra of (b) stishovite and (c) $CaCl_2$-type (space group $Pnnm$, Table 6.2) structures can be seen when viewed 45° from the a, b, and c axes. Octahedral tilting reduces the symmetry from tetragonal to orthorhombic and the amount of tilt increases with pressure [*Andrault et al.*, 1998].

Table 6.2 Crystallographic parameters of SiO_2 polymorphs in the lower mantle.

	x/a	y/b	z/c
Stishovite, SiO_2[a]			
SG = $P4_2/mnm$, a = 4.1812(1) Å, c = 2.6662(3) Å			
Si (2a)	0	0	0
O (4f)	0.3063(1)	0.3063(1)	0
$CaCl_2$-type SiO_2[b]			
SG = $Pnnm$, a = 3.7201(9) Å, b = 3.9422(10) Å, c = 2.4913(3) Å			
Si (2a)	0	0	0
O (4g)	0.276	0.313	0
Seifertite, SiO_2[c]			
SG = $Pbcn$, a = 4.097(1) Å, b = 5.0462(9) Å, c = 4.4946(8) Å			
Si (4c)	0	0.1522(9)	¼
O (8d)	0.7336(16)	0.6245(12)	0.9186(29)

[a] Ambient pressure [*Yamanaka et al.*, 2002].
[b] 120 GPa [*Andrault et al.*, 1998].
[c] Ambient pressure [*Dera et al.*, 2002].

Ricolleau et al., 2010]. Stishovite, which becomes the stable form of SiO_2 at pressures above ~7 GPa, crystallizes in the tetragonal rutile-type structure (space group $P4_2/mnm$) [*Stishov and Belov*, 1962]. This structure consists of slightly distorted SiO_6 octahedra that share edges to form chains running parallel to the c axis (Figure 6.6a). Each octahedron is corner linked to four neighboring chains. The structure can also be described as a distorted

hexagonal close-packed array of O^{2-} anions with half of the octahedral sites occupied. A number of structural studies of stishovite have been reported at both ambient [*Sinclair and Ringwood*, 1978; *Hill et al.*, 1983] and high pressures [*Sugiyama et al.*, 1987; *Ross et al.*, 1990]. Stishovite has been observed naturally as an inclusion phase in diamonds of deep origin [*Joswig et al.*, 1999] and as a product of transient high pressure-temperature meteorite impact events [*Chao et al.*, 1962].

At ~50 GPa, stishovite undergoes a displacive phase transition to the orthorhombic $CaCl_2$-type structure (orthorhombic, space group $Pnnm$) [*Cohen*, 1987; *Tsuchida and Yagi*, 1989; *Kingma et al.*, 1995]. This structure differs from stishovite only by a small rotation of the octahedral chains (Figure 6.6c). The transition is driven by the softening of a zone-center optic mode that couples with acoustic modes to produce a marked softening of the shear elastic constants [*Cohen*, 1992; *Karki et al.*, 1997a; *Jiang et al.*, 2009; *Asahara et al.*, 2013]. As a result, even a small fraction of free silica may produce a detectable seismic signal in the mid-lower mantle [*Karki et al.*, 1997a]. Several studies have attempted to associate this transition with seismic reflectors observed at greater than 800 km depth [*Kawakatsu and Niu*, 1994; *Vinnik et al.*, 2001, 2010]. The *P-T* phase boundary for the post-stishovite transition has been examined a number of times [*Kingma et al.*, 1995; *Ono et al.*, 2002; *Tsuchiya et al.*, 2004; *Nomura et al.*, 2010]. *Nomura et al.* [2010] reported a positive Clapeyron slope for the transition and predicted a transition pressure of ~70 GPa along a typical mantle geotherm and 56 GPa for conditions appropriate to a subducting slab. These correspond

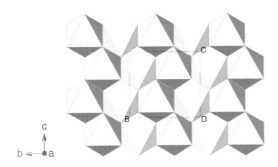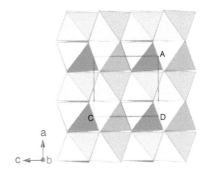

Figure 6.7 Structure of α-PbO$_2$-type SiO$_2$, seifertite (space group, *Pbcn*, Table 6.2), viewed along the *a* (left) and *b* axis (right). The octahedra are depth shaded to aid in distinguishing top layers of SiO$_6$ octahedra, dark blue, from underlying layers, lighter blue. Unlike the stishovite and the CaCl$_2$-type structures (Figure 6.6), the SiO$_6$ octahedra are organized into kinked chains that allow both for more efficient packing and less-distorted octahedra.

to depths of about 1400–1700 km in the mantle. Incorporation of aluminum and water into SiO$_2$, however, has been shown to markedly reduce the post-stishovite transition pressure [*Lakshtanov et al.*, 2007; *Bolfan-Casanova et al.*, 2009].

At yet higher pressures, SiO$_2$ undergoes a further transformation to seifertite, which has the scrutinyite (α-PbO$_2$-type, space group *Pbcn*) structure [*Murakami et al.*, 2003; *Dubrovinsky et al.*, 2001; *Belonoshko et al.*, 1996; *Grocholski et al.*, 2013]. In this phase, distorted SiO$_6$ octahedra are arranged into kinked chains extending along the *c* axis, resulting in a small density increase compared with stishovite and the CaCl$_2$-type phase (Figure 6.7). The reported transition pressure is close to that of Earth's CMB, but it is not yet clear if this phase would definitely be expected to exist in the D″ region just above the CMB [*Murakami et al.*, 2003; *Shieh et al.*, 2005; *Grocholski et al.*, 2013]. The presence of aluminum appears to have a modest effect in reducing the transition pressure, which may be enough to stabilize the phase in Earth's deep lower mantle [*Andrault et al.*, 2014].

Seifertite has also been found in a number of meteorites where it appears to have experienced shock pressures well below those of the expected stability field [*Sharp et al.*, 1999; *Dera et al.*, 2002; *Goresy et al.*, 2008]. Theoretical studies have shown that there exists a large family of closely related and energetically competitive phases of SiO$_2$ with closed-packed oxygen anions and different arrangements of Si in the octahedral sites [*Teter et al.*, 1998]. As a result, the metastable formation of silica structures may occur under various conditions, and this may explain the observation of seifertite in meteorites [*Kubo et al.*, 2015]. In laboratory studies, the observed phase may depend on starting material, stress conditions, or *P-T* path. There have been reports of the occurrence of seifertite or a related structure at much lower pressures (~80 GPa) in some high-temperature experiments

[*Dubrovinsky et al.*, 1997, 2001] or upon compression of cristobalite starting material at room temperature to ~50 GPa [*Dubrovinsky et al.*, 2001; *Shieh et al.*, 2005].

6.7. ALUMINOUS PHASES

Two aluminum-rich phases expected in basalt lithologies at high pressures are the "new aluminous phase" (NAL) and a phase with the Ca-ferrite structure (called the "CF phase") (Figure 6.1). Together, these two phases may account for 10–25 vol.% of subducting basaltic crust in the lower mantle [*Ono et al.*, 2001; *Ricolleau et al.*, 2010; *Irifune and Ringwood*, 1993]. The Ca-ferrite structure is a common high-pressure structure type adopted by several compositions including MgAl$_2$O$_4$ [*Irifune et al.*, 1991], CaAl$_2$O$_4$ [*Reid and Ringwood*, 1969], and NaAlSiO$_4$ [*Liu*, 1977]. It is one of a series of similar structures (CaFe$_2$O$_4$, CaTi$_2$O$_4$, CaMn$_2$O$_4$) that are common high-pressure polymorphs of spinel-structured phases [*Yamanaka et al.*, 2008]. Single-crystal diffraction experiments have recently shown that forsterite, Mg$_2$SiO$_4$, metastably adopts a related structure upon room temperature compression above 58 GPa [*Finkelstein et al.*, 2014].

The general formula for the Ca-ferrite structure is XY$_2$O$_4$ where the eight-fold coordinated X site can be occupied by the mono- and divalent cations K$^+$, Na$^+$, Ca^{2+}, and Mg^{2+}, and the six-coordinated Y sites are occupied by Al^{3+} and Si^{4+}. The structure is orthorhombic (space group *Pbnm*) (Figure 6.8), and consists of edge-sharing double chains of octahedra that form tunnels parallel to the *c* axis with the X cations occupying spaces between the double chains.

The NAL phase is hexagonal (space group *P6$_3$/m*) and has the general formula AX$_2$Y$_6$O$_{12}$ where A is a nine-fold coordinated channel site typically occupied by large mono or divalent cations such as Na$^+$ or K$^+$ or Ca^{2+} [*Gasparik et al.*, 2000; *Miura et al.*, 2000; *Miyajima*

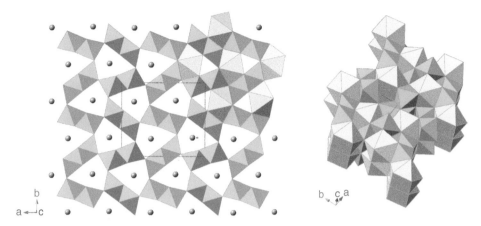

Figure 6.8 Calcium ferrite-type structure (space group *Pbnm*, Table 6.3) viewed along (left) and oblique (right) to the *c* axis. AlO_6 octahedra (light blue) edge share to form double chains extending along the *c* axis. Eight-fold coordinated cations (yellow spheres and polyhedra) are located in channels made up of four double chains–two edge-on and two face-on. This model is based on structural parameters for the $MgAl_2O_4$ end-member [*Kojitani et al.*, 2007b], whereas the Ca-ferrite-type phase in basaltic compositions of the mantle can exist within the $NaAlSiO_4$-$MgAl_2O_4$ system.

et al., 2001] (Figure, 6.9). The X site is six-coordinated but trigonal prismatic rather than octahedral and is typically occupied by Mg^{2+} or Fe^{2+}. The octahedrally coordinated Y site is occupied by Al^{3+} or Si^{4+}. Like the Ca-ferrite structure, the distorted Y octahedra form edge-sharing double chains extending along the *c* axis that are corner linked to form tunnels. Within the tunnels lie the larger A sites while the smaller X sites are surrounded by three double chains. With multiple sites that can each accept more than one cation, the NAL structure can accommodate a wide range of compositional variation.

The large cation sites in the NAL and CF structure allow them to be potential hosts for the alkali elements in subducted crust. Of the two structures, NAL more readily accommodates large ions due to its nine-fold site [*Miyajima et al.*, 2001]. Theoretical studies indicate that pressure tends to favor the stability of the CF phase with respect to NAL, but that NAL is stabilized by Mg^{2+} or alkali element enrichment [*Kawai and Tsuchiya*, 2012; *Mookherjee et al.*, 2012]. As a result, the relative abundance of these phases in the lower mantle will depend on composition. K-free NAL was shown to become unstable above 27 GPa and 1850 K [*Imada et al.*, 2011]. Previous studies of basaltic compositions showed that the NAL phase was only stable to ~40 GPa [*Ricolleau et al.*, 2010]. However, it has been shown that K-rich NAL [*Kojitani et al.*, 2011] is stable throughout the lower mantle pressure range [*Kato et al.*, 2013] and could exist in deeply subducted sediments or continental crust [*Kawai et al.*, 2009; *Komabayashi et al.*, 2009]. K-rich NAL may then be a host for K in the deep mantle and could contribute to radiogenic heating in the deep Earth [*Kato et al.*, 2013]. Recently, inclusion compositions in diamonds inferred to

originate from the lower mantle have provided evidence for the CF and NAL phases in natural samples for the first time [*Walter et al.*, 2011]. Structural differences between these two phases lead to distinct physical properties—density functional theory calculations show that the NAL phase has low seismic velocities compared to the CF phase, and NAL could contribute to low-velocity provinces in the deep mantle [*Mookherjee et al.*, 2012].

6.8. POSSIBLE HYDROUS PHASES

There is considerable interest in the potential role of hydrogen in deep-mantle crystal structures, as the presence of hydrogen, even in small quantities, can strongly affect physical and chemical properties such as melting, viscosity, phase transitions, and seismic velocities. In the upper mantle, a variety of crystal structures can accommodate hydrogen, especially under lower temperature conditions of subducting slabs. Nominally anhydrous minerals such as wadsleyite and ringwoodite are potential hosts for water in the transition zone [*Smyth*, 1987; *Kohlstedt et al.*, 1996; *Pearson et al.*, 2014]. The role of hydrous phases in the lower mantle is only beginning to be explored but has been a growing focus of study in recent years.

6.8.1. Phase D

Among the hydrous magnesium silicates that are potential hosts for water in the upper mantle and transition zone [*Prewitt and Parise*, 2000], phase D has the highest pressure stability (to ~44 GPa and 1400°C, corresponding to depths of ~1250 km) [*Shieh et al.*, 1998]. This phase

Table 6.3 Crystallographic parameters of aluminous and hydrous phases in the lower mantle.

	x/a	y/b	z/c
Calcium ferrite-type MgAl$_2$O$_4$[a]			
SG = *Pbnm*, a = 9.9498(6) Å, b = 8.6468(6) Å, c = 2.7901(2) Å			
Mg (4c)	0.3503(4)	0.7576(6)	1/4
Al1 (4c)	0.3854(4)	0.4388(5)	1/4
Al2 (4c)	0.8964(3)	0.4159(4)	1/4
O1 (4c)	0.8344(7)	0.2005(8)	1/4
O2 (4c)	0.5279(7)	0.1201(6)	1/4
O3 (4c)	0.2150(7)	0.5357(8)	1/4
O4 (4c)	0.5709(6)	0.4089(7)	1/4
New aluminous phase, NAL[b]			
SG = *P6$_3$/m*, a = 8.7225(4) Å, c = 2.7664(2) Å			
M1 (6h)	0.98946(8)	0.34353(9)	1/4
M2 (2d)	2/3	1/3	1/4
M3 (2a)	0	0	1/4
O1 (6h)	0.1283(2)	0.5989(2)	1/4
O2 (6h)	0.3124(2)	0.2024(2)	1/4
Phase D, MgSi$_2$H$_2$O$_6$[c]			
SG = *P$\bar{3}$1m*, a = 4.7453(4) Å, c = 4.3450(5) Å			
Mg (1a)	0	0	0
Si (2d)	2/3	1/3	1/2
O (6k)	0.6327(2)	0	0.2716(2)
H^d (6k)	0.536(13)	0	0.091(10)
Phase H, MgSiH$_2$O$_4$[e]			
SG = *Pnnm*, a = 4.733(2) Å, b = 4.325(1) Å, c = 2.842(1) Å			
Mg/Si (2a)	0	0	0
O (4g)	0.347(1)	0.230(1)	0
H^f (4g)	0.475	0.042	0

[a] Ambient pressure [*Kojitani et al.*, 2007].
[b] Na$_{0.41}$[Na$_{0.125}$Mg$_{0.79}$Al$_{0.085}$]$_2$[Al$_{0.79}$Si$_{0.21}$]$_6$O$_{12}$, ambient pressure [*Pamato et al.*, 2014].
[c] Mg$_{1.11}$Si$_{1.89}$H$_{2.22}$O$_6$, ambient pressure [*Yang et al.*, 1997].
[d] Partial H occupancy of 0.37.
[e] Ambient pressure [*Bindi et al.*, 2014].
[f] Tentative H position, assumed occupancy of 0.50.

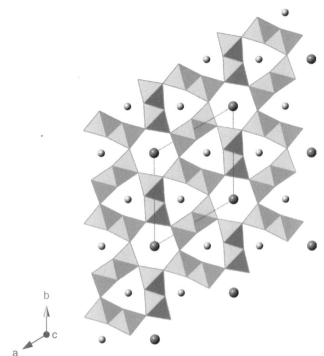

Figure 6.9 New aluminous phase, NAL (space group *P6$_3$/m*, Table 6.3) viewed along the c axis. Around each Mg^{2+} ion (small yellow spheres) are three double chains of AlO$_6$ octahedra (light blue) connected by corner-sharing oxygen ions. Large channels are formed by six SiO$_6$ octahedra in three double chains and are occupied typically by Na$^+$ or K$^+$ ions (large green spheres) at half occupancy.

can form upon breakdown of serpentine and could serve as a water carrier in cold lithosphere subducting into the lower mantle [*Liu*, 1986; *Frost and Fei*, 1998].

Phase D has the ideal formula MgSi$_2$O$_4$(OH)$_2$ but is typically nonstoichiometric. It crystallizes in the trigonal system (space group *P$\bar{3}$1m*) [*Kudoh et al.*, 1997; *Yang et al.*, 1997]. Oxygen anions form a hexagonal close-packed lattice with Si^{4+} and Mg^{2+} both occupying octahedral sites. The MgO$_6$ and SiO$_6$ octahedra are organized into alternating layers along the *c* direction. The SiO$_6$ octahedra form gibbsite-like layers in which each octahedron shares 3 edges, leaving one third of the octahedral

sites vacant. MgO$_6$ octahedra lie above and below vacant sites in the SiO$_6$ layer and are corner linked to the Si octahedra, and so two thirds of the Mg octahedral sites are empty (Figure 6.10).

The hydrogen positions are located in the MgO$_6$ layers, with O-H bonds facing away from SiO$_6$ octahedra. Typically, approximately one third of the proton positions are occupied. Phase D exhibits considerable disorder and compositional variability as a function of synthesis conditions, with reported Mg/Si ratios variable between 0.55 and 0.71 and H$_2$O content variable between 10 and 18 wt.% [*Frost and Fei*, 1998]. Phase D can incorporate both aluminum and iron: the presence of Al in phase D has been shown to extend its stability field, although the presence of Fe counteracts this expansion [*Ghosh and Schmidt*, 2014]. Very recently an aluminum-rich variant of phase D (Al$_2$SiO$_4$(OH)$_2$) has been synthesized, and its stability field was found to extend to over 2000°C at 26 GPa [*Pamato et al.*, 2015]. Disordering of Al and Si cations renders the two previously distinct octahedral sites equivalent, increasing the symmetry and enhancing the stability of this phase.

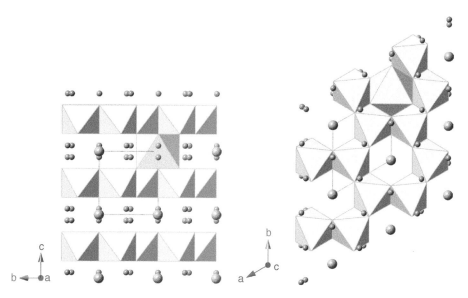

Figure 6.10 Dense hydrous magnesium silicate, phase D (space group $P\bar{3}1m$, Table 6.3), viewed along the a axis (left) and c axis (right). SiO_6 octahedra (blue) share edges with neighbors to form rings in layers perpendicular to the c axis. Mg cations (yellow spheres and representative MgO_6 polyhedron) partially occupy the spaces between the SiO_6 rings in alternating layers. Charge balance is achieved by hydrogen anions represented by small pink spheres.

Phase D exhibits anisotropic compression as a result of the layered nature of the structure, with strong SiO_6 layers alternating with weaker MgO_6 octahedra resulting in enhanced compressibility along the c axis [*Frost and Fei*, 1999]. Theoretical calculations of the elastic constants indicate that the anisotropy decreases with depth but significant anisotropy in seismic wave velocities is retained to the highest pressures [*Mainprice et al.*, 2007].

It has been suggested on theoretical grounds that hydrogen bonding in phase D will increasingly approach and finally reach a condition of symmetric bonding at ~40 GPa [*Tsuchiya et al.*, 2005]. A symmetric hydrogen bond is one in which the hydrogen atom is located at the midpoint between the two neighboring oxygen atoms, rather than the asymmetric O-H distances that characterize a conventional hydrogen bond. H bond symmetrization in phase D is predicted to affect the compression behavior, resulting in an ~20% increase in the bulk modulus [*Tsuchiya et al.*, 2005]. While a powder X-ray diffraction study has reported evidence for such an anomaly [*Hushur et al.*, 2011], a more recent high-resolution single-crystal X-ray study to 65 GPa found no evidence for a bulk modulus anomaly to at least this pressure [*Rosa et al.*, 2013]. No evidence for hydrogen bond symmetrization was found in an infrared spectroscopic study [*Shieh et al.*, 2009] or in other theoretical calculations [*Mainprice et al.*, 2007]. However, a spin-pairing transition in Fe^{3+}-bearing phase D is reported to produce a pronounced softening of the bulk modulus [*Chang et al.*, 2013].

6.8.2. δ-AlOOH and Phase H

Hydrous aluminum oxides and silicates, such as δ-AlOOH and phase H, are also candidate deep-mantle water carriers. δ-AlOOH is a high-pressure polymorph of diaspore that was initially synthesized at 21 GPa and 1300 K [*Suzuki et al.*, 2000]. Experiments have now shown that its stability extends over the entire range of conditions (to 134 GPa and 2300 K) expected in the lower mantle [*Sano et al.*, 2004, 2008]. At low pressures, δ-AlOOH crystallizes in a distorted rutile-type structure with the noncentrosymmetric orthorhombic space group $P2_1nm$ [*Suzuki et al.*, 2000; *Komatsu et al.*, 2006; *Vanpeteghem et al.*, 2007]. The structure contains both AlO_6 and HO_6 octahedra, and the oxygen anions form a slightly distorted hexagonal close-packed arrangement perpendicular to the b direction.

A recent single-crystal X-ray diffraction study indicates that δ-AlOOH transforms to the $CaCl_2$-type structure (space group $Pnnm$) at ~8 GPa [*Kuribayashi et al.*, 2013]. As described above, this structure consists of corner-sharing chains of edge-sharing octahedra (AlO_6, in this case). There is no discontinuity in unit cell volume across this displasive transition. The main difference between the two structures lies in the disordering of hydrogen atoms in the high-pressure structure. The substitution of small amounts of Mg^{2+} and Si^{4+} into δ-AlOOH also promotes the transformation from $P2_1nm$ to the $Pnnm$ $CaCl_2$-type structure [*Komatsu et al.*, 2011]. The nature of the hydrogen bonding in these phases has been

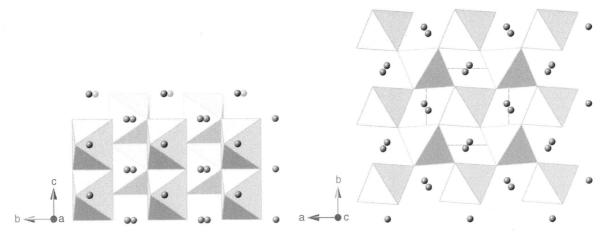

Figure 6.11 View of the phase H structure (space group *Pnnm*, Table 6.3) along the *a* (left) and *c* axis (right). Distorted Mg/Si octahedra (yellow) are charge compensated by H$^+$ ions (pink spheres), which may occupy one of two positions depicted between the octahedra.

extensively discussed, especially with regard to the formation of symmetric hydrogen bonds, but there is currently no consensus on this issue [*Tsuchiya et al.*, 2002; *Panero and Stixrude*, 2004; *Li et al.*, 2006b; *Xue et al.*, 2006; *Vanpeteghem et al.*, 2007; *Sano-Furukawa et al.*, 2008].

A related hydrous silicate, called phase H, $MgSiO_2(OH)_2$, was predicted theoretically and synthesized near 50 GPa [*Tsuchiya*, 2013; *Nishi et al.*, 2014]. The phase, which contains 15 wt.% water, can be quenched to ambient conditions [*Nishi et al.*, 2014]. A single-crystal X-ray diffraction study on a recovered sample of phase H shows that it adopts the orthorhombic $CaCl_2$-type structure (*Pnnm*), with Mg^{2+} and Si^{4+} disordered over the octahedral sites [*Bindi et al.*, 2014]. H$^-$ positions are expected to be disordered in this structure (Figure 6.11).

The stability field for end-member phase H appears to be relatively narrow and confined to conditions of the upper part of the lower mantle [*Tsuchiya*, 2013; *Ohtani et al.*, 2014]. However, based on structural similarity, it is expected that phase H and δ-AlOOH (whose formulas can be written, respectively, as $MgSiH_2O_4$ and $AlAlH_2O_4$) will exhibit extensive solid solution at high pressures [*Bindi et al.*, 2014]. Experiments have now shown that intermediate δ-AlOOH-$MgSiO_2(OH)_2$ compositions can be synthesized together with perovskite and post-perovskite under slab geotherm conditions, making them plausible candidate hydrous phases under lower mantle conditions [*Ohira et al.*, 2014].

6.9. SUMMARY

In this review, we have summarized the current status of our understanding of lower mantle minerals and their structures. Despite much activity over the last two decades, the lower mantle remains a region that is still poorly understood. Future progress is likely to come from technical advances in high-pressure crystallography. For example, single-crystal synchrotron X-ray diffraction techniques in the diamond anvil cell are now extending into the lower mantle pressure range [*Merlini and Hanfland*, 2013; *Zhang et al.*, 2013; *Duffy*, 2014], and capabilities for simultaneous high *P* and *T* single-crystal diffraction are being pioneered [*Dubrovinsky et al.*, 2010]. These advances will allow us to better understand the crystallographic consequences of cation substitution, water incorporation, and effects of temperature on minerals stable only at these high pressures.

The pressure range of multianvil press techniques can now reach well into the lower mantle range as a result of the development of sintered diamond anvil technology [*Yamazaki et al.*, 2014]. The combined larger sample size and more uniform heating at high pressures may resolve current discrepancies in phase stability and element partitioning between lower mantle phases. Concurrently, the capabilities of theoretical studies using density functional theory have continued to expand [*Wentzcovitch et al.*, 2010]. Each of these developments holds considerable future promise for further understanding of deep-mantle crystal structures in the coming years.

REFERENCES

Adams, D. J., and A. R. Oganov (2006), *Ab initio* molecular dynamics study of $CaSiO_3$ perovskite at P-T conditions of Earth's lower mantle, *Phys. Rev. B*, *73*(18), 184, 106, doi:10.1103/PhysRevB.73.184106.

Akber-Knutson, S., and M. S. T. Bukowinski (2004), The energetics of aluminum solubility into $MgSiO_3$ perovskite at lower mantle conditions, *Earth Planet. Sci. Lett.*, *220*(3–4), 317–330, doi:10.1016/S0012-821X(04)00065-2.

Ammann, M. W., J. P. Brodholt, J. Wookey, and D. P. Dobson (2010), First-principles constraints on diffusion in lower-mantle minerals and a weak D″ layer, *Nature*, *465*(7297), 462–465, doi:10.1038/nature09052.

Ammann, M. W., A. M. Walker, S. Stackhouse, J. Wookey, A. M. Forte, J. P. Brodholt, and D. P. Dobson (2014), Variation of thermal conductivity and heat flux at the Earth's core mantle boundary, *Earth Planet. Sci. Lett.*, *390*, 175–185, doi:10.1016/j.epsl.2014.01.009.

Andrault, D., G. Fiquet, F. Guyot, and M. Hanfland (1998), Pressure-induced Landau-type transition in stishovite, *Science*, *282*(5389), 720–724, doi:10.1126/science.282.5389.720.

Andrault, D., N. Bolfan-Casanova, M. A. Bouhifd, N. Guignot, and T. Kawamoto (2007), The role of Al-defects on the equation of state of Al–(Mg,Fe)SiO₃ perovskite, *Earth Planet. Sci. Lett.*, *263*(3–4), 167–179, doi:10.1016/j.epsl.2007.08.012.

Andrault, D., R. G. Trønnes, Z. Konôpková, W. Morgenroth, H. P. Liermann, G. Morard, and M. Mezouar (2014), Phase diagram and P-V-T equation of state of Al-bearing seifertite at lowermost mantle conditions, *Am. Mineral.*, *99*(10), 2035–2042, doi:10.2138/am-2014-4697.

Armstrong, L. S., M. J. Walter, J. R. Tuff, O. T. Lord, A. R. Lennie, A. K. Kleppe, and S. M. Clark (2012), Perovskite phase relations in the system CaO–MgO–TiO₂–SiO₂ and implications for deep mantle lithologies, *J. Petrol.*, *53*(3), 611–635, doi:10.1093/petrology/egr073.

Asahara, Y., K. Hirose, Y. Ohishi, N. Hirao, H. Ozawa, and M. Murakami (2013), Acoustic velocity measurements for stishovite across the post-stishovite phase transition under deviatoric stress: Implications for the seismic features of subducting slabs in the mid-mantle, *Am. Mineral.*, *98*(11–12), 2053–2062, doi:10.2138/am.2013.4145.

Auzende, A.-L., J. Badro, F. J. Ryerson, P. K. Weber, S. J. Fallon, A. Addad, J. Siebert, and G. Fiquet (2008), Element partitioning between magnesium silicate perovskite and ferropericlase: New insights into bulk lower-mantle geochemistry, *Earth Planet. Sci. Lett.*, *269*(1–2), 164–174, doi:10.1016/j.epsl.2008.02.001.

Badro, J. (2014), Spin transitions in mantle minerals, *Annu. Rev. Earth Planet. Sci.*, *42*(1), 231–248, doi:10.1146/annurev-earth-042711-105304.

Badro, J., G. Fiquet, F. Guyot, J.-P. Rueff, V. V. Struzhkin, G. Vankó, and G. Monaco (2003), Iron partitioning in Earth's mantle: Toward a deep lower mantle discontinuity, *Science*, *300*(5620), 789–791, doi:10.1126/science.1081311.

Belonoshko, A. B., L. S. Dubrovinsky, and N. A. Dubrovinsky (1996), A new high-pressure silica phase obtained by molecular dynamics, *Am. Mineral.*, *81*, 785–788, doi:10.2138/am-1996-5-632.

Bindi, L., M. Nishi, J. Tsuchiya, and T. Irifune (2014), Crystal chemistry of dense hydrous magnesium silicates: The structure of phase H, MgSiH₂O₄, synthesized at 45 GPa and 1000°C, *Am. Mineral.*, *99*(8–9), 1802–1805, doi:10.2138/am.2014.4994.

Bolfan-Casanova, N., D. Andrault, E. Amiguet, and N. Guignot (2009), Equation of state and post-stishovite transformation of Al-bearing silica up to 100 GPa and 3000 K, *Phys. Earth Planet. Inter.*, *174*(1–4), 70–77, doi:10.1016/j.pepi.2008.06.024.

Bower, D. J., M. Gurnis, J. M. Jackson, and W. Sturhahn (2009), Enhanced convection and fast plumes in the lower mantle

induced by the spin transition in ferropericlase, *Geophys. Res. Lett.*, *36*(10), L10,306, doi:10.1029/2009GL037706.

Brodholt, J. P. (2000), Pressure-induced changes in the compression mechanism of aluminous perovskite in the Earth's mantle, *Nature*, *407*(6804), 620–622, doi:10.1038/35036565.

Buffett, B. A., and C. T. Seagle (2010), Stratification of the top of the core due to chemical interactions with the mantle, *J. Geophys. Res. Solid Earth*, *115*(B4), B04,407, doi:10.1029/2009JB006751.

Caracas, R., and R. M. Wentzcovitch (2006), Theoretical determination of the structures of CaSiO₃ perovskites, *Acta Crystallogr.*, *B62*, 1025–1030, doi:10.1107/S0108768106035762.

Caracas, R., R. Wentzcovitch, G. D. Price, and J. Brodholt (2005), CaSiO₃ perovskite at lower mantle pressures, *Geophys. Res. Lett.*, *32*(6), L06,306, doi:10.1029/2004GL022144.

Catalli, K., S.-H. Shim, V. B. Prakapenka, J. Zhao, W. Sturhahn, P. Chow, Y. Xiao, H. Liu, H. Cynn, and W. J. Evans (2010), Spin state of ferric iron in MgSiO₃ perovskite and its effect on elastic properties, *Earth Planet. Sci. Lett.*, *289*(1–2), 68–75, doi:10.1016/j.epsl.2009.10.029.

Chang, Y.-Y., et al. (2013), Spin transition of Fe³⁺ in Al-bearing phase D: An alternative explanation for small-scale seismic scatterers in the mid-lower mantle, *Earth Planet. Sci. Lett.*, *382*, 1–9, doi:10.1016/j.epsl.2013.08.038.

Chao, E. C. T., J. J. Fahey, J. Littler, and D. J. Milton (1962), Stishovite, SiO₂, a very high pressure new mineral from Meteor Crater, Arizona, *J. Geophys. Res.*, *67*(1), 419–421, doi:10.1029/JZ067i001p00419.

Cohen, R. E. (1987), Calculation of elasticity and high pressure instabilities in corundum and stishovite with the Potential Induced Breathing Model, *Geophys. Res. Lett.*, *14*(1), 37–40, doi:10.1029/GL014i001p00037.

Cohen, R. E. (1992), First-principles predictions of elasticity and phase transitions in high pressure SiO₂ and geophysical implications, in *High-Pressure Research: Application to Earth and Planetary Sciences*, edited by Y. Syono and M. Manghnani, pp. 425–431, AGU, Washington, D.C.

Dera, P., C. T. Prewitt, N. Z. Boctor, and R. J. Hemley (2002), Characterization of a high-pressure phase of silica from the Martian meteorite Shergotty, *Am. Mineral.*, *87*(7), 1018–1023, doi: 10.2138/am-2002-0728.

Dobson, D. P., N. Miyajima, F. Nestola, M. Alvaro, N. Casati, C. Liebske, I. G. Wood, and A. M. Walker (2013), Strong inheritance of texture between perovskite and post-perovskite in the D″ layer, *Nat. Geosci.*, *6*(7), 575–578, doi:10.1038/ngeo1844.

Dorfman, S. M., S. R. Shieh, Y. Meng, V. B. Prakapenka, and T. S. Duffy (2012), Synthesis and equation of state of perovskites in the (Mg, Fe)₃Al₂Si₃O₁₂ system to 177 GPa, *Earth Planet. Sci. Lett.*, *357–358*, 194–202, doi:10.1016/j.epsl.2012.09.024.

Dorfman, S. M., Y. Meng, V. B. Prakapenka, and T. S. Duffy (2013), Effects of Fe-enrichment on the equation of state and stability of (Mg,Fe)SiO₃ perovskite, *Earth Planet. Sci. Lett.*, *361*, 249–257, doi:10.1016/j.epsl.2012.10.033.

Dubrovinsky, L., et al. (2010), Single-crystal X-ray diffraction at megabar pressures and temperatures of thousands of degrees, *High Press. Res.*, *30*(4), 620–633, doi:10.1080/08957959.2010.534092.

Dubrovinsky, L. S., S. K. Saxena, P. Lazor, R. Ahuja, O. Eriksson, J. M. Wills, and B. Johansson (1997), Experimental and theoretical identification of a new high-pressure phase of silica, *Nature*, *388*(6640), 362–365, doi:10.1038/41066.

Dubrovinsky, L. S., N. A. Dubrovinskaia, S. K. Saxena, H. Annersten, E. Hålenius, H. Harryson, F. Tutti, S. Rekhi, and T. L. Bihan (2000), Stability of ferropericlase in the lower mantle, *Science*, *289*(5478), 430–432, doi:10.1126/science.289.5478.430.

Dubrovinsky, L. S., N. A. Dubrovinskaia, S. K. Saxena, F. Tutti, S. Rekhi, T. Le Bihan, G. Shen, and J. Hu (2001), Pressure-induced transformations of cristobalite, *Chem. Phys. Lett.*, *333*(3–4), 264–270, doi:10.1016/S0009-2614(00)01147-7.

Duffy, T. (2014), Crystallography's journey to the deep Earth, *Nature*, *506*(7489), 427–429, doi:10.1038/506427a.

Duffy, T. S. (2005), Synchrotron facilities and the study of the Earth's deep interior, *Rep. Prog. Phys.*, *68*(8), 1811, doi:10.1088/0034-4885/68/8/R03.

Duffy, T. S., R. J. Hemley, and H. Mao (1995), Equation of state and shear strength at multimegabar pressures: Magnesium oxide to 227 GPa, *Phys. Rev. Lett.*, *74*(8), 1371–1374, doi:10.1103/PhysRevLett.74.1371.

Fei, Y., and H. Mao (1994), In situ determination of the NiAs phase of FeO at high pressure and temperature, *Science*, *266*(5191), 1678–1680, doi:10.1126/science.266.5191.1678.

Finkelstein, G., P. Dera, S. Jahn, A. R. Oganov, C. M. Holl, Y. Meng, and T. S. Duffy (2014), Phase transitions and equation of state of forsterite to 90 GPa from single-crystal X-ray diffraction and molecular modeling, *Am. Mineral.*, *99*(1), 35–43, doi:10.2138/am.2014.4526.

Fiquet, G., A. Dewaele, D. Andrault, M. Kunz, and T. Le Bihan (2000), Thermoelastic properties and crystal structure of $MgSiO_3$ perovskite at lower mantle pressure and temperature conditions, *Geophys. Res. Lett.*, *27*(1), 21–24, doi:10.1029/1999GL008397.

Fischer, R. A., A. J. Campbell, O. T. Lord, G. A. Shofner, P. Dera, and V. B. Prakapenka (2011), Phase transition and metallization of FeO at high pressures and temperatures, *Geophys. Res. Lett.*, *38*(24), L24,301, doi:10.1029/2011GL049800.

Frost, D. J., and Y. Fei (1998), Stability of phase D at high pressure and high temperature, *J. Geophys. Res. Solid Earth*, *103*(B4), 7463–7474, doi:10.1029/98JB00077.

Frost, D. J., and Y. Fei (1999), Static compression of the hydrous magnesium silicate phase D to 30 GPa at room temperature, *Phys. Chem. Minerals*, *26*(5), 415–418, doi:10.1007/s002690050202.

Frost, D. J., and C. A. McCammon (2008), The redox state of Earth's mantle, *Annu. Rev. Earth Planet. Sci.*, *36*(1), 389–420, doi:10.1146/annurev.earth.36.031207.124322.

Frost, D. J., C. Liebske, F. Langenhorst, C. A. McCammon, R. G. Trønnes, and D. C. Rubie (2004), Experimental evidence for the existence of iron-rich metal in the Earth's lower mantle, *Nature*, *428*(6981), 409–412, doi:10.1038/nature02413.

Gasparik, T., A. Tripathi, and J. B. Parise (2000), Structure of a new Al-rich, $[K,Na]_{0.9}[Mg,Fe]_2[Mg,Fe,Al,Si]_6O_{12}$, synthesized at 24 GPa, *Am. Mineral.*, *85*, 613–618, doi:10.2138/am-2000-0426.

Ghosh, S., and M. W. Schmidt (2014), Melting of phase D in the lower mantle and implications for recycling and storage of H_2O in the deep mantle, *Geochim. Cosmochim. Acta*, *145*, 72–88, doi:10.1016/j.gca.2014.06.025.

Goresy, A. E., P. Dera, T. G. Sharp, C. T. Prewitt, M. Chen, L. Dubrovinsky, B. Wopenka, N. Z. Boctor, and R. J. Hemley (2008), Seifertite, a dense orthorhombic polymorph of silica from the Martian meteorites Shergotty and Zagami, *Eur. J. Mineral.*, *20*(4), 523–528, doi:10.1127/0935-1221/2008/0020-1812.

Grocholski, B., S.-H. Shim, and V. B. Prakapenka (2013), Stability, metastability, and elastic properties of a dense silica polymorph, seifertite, *J. Geophys. Res. Solid Earth*, *118*(9), 4745–4757, doi:10.1002/jgrb.50360.

Harte, B., and J. W. Harris (1994), Lower mantle mineral associations preserved in diamonds, *Mineral. Mag.*, *58A*, 384–385.

Hazen, R. M. (1976), Effects of temperature and pressure on the cell dimension and X-ray temperature factors of periclase, *Am. Mineral.*, *61*, 266–271.

Hill, R. J., M. D. Newton, and G. V. Gibbs (1983), A crystal chemical study of stishovite, *J. Solid State Chem.*, *47*(2), 185–200, doi:10.1016/0022-4596(83)90007-5.

Hirose, K., N. Takafuji, N. Sata, and Y. Ohishi (2005), Phase transition and density of subducted MORB crust in the lower mantle, *Earth Planet. Sci. Lett.*, *237*(1–2), 239–251, doi:10.1016/j.epsl.2005.06.035.

Hirose, K., R. Wentzcovitch, D. A. Yuen, and T. Lay (2015), 2.05—Mineralogy of the deep mantle—The post-perovskite phase and its geophysical significance, in *Treatise on Geophysics*, 2nd ed., edited by G. Schubert, pp. 85–115, Elsevier, Oxford.

Horiuchi, H., E. Ito, and D. J. Weidner (1987), Perovskite-type $MgSiO_3$: Single-crystal X-ray diffraction, *Am. Mineral.*, *72*, 357–360.

Hsu, H., K. Umemoto, P. Blaha, and R. M. Wentzcovitch (2010), Spin states and hyperfine interactions of iron in $(Mg,Fe)SiO_3$ perovskite under pressure, *Earth Planet. Sci. Lett.*, *294*(1–2), 19–26, doi:10.1016/j.epsl.2010.02.031.

Hushur, A., M. H. Manghnani, J. R. Smyth, Q. Williams, E. Hellebrand, D. Lonappan, Y. Ye, P. Dera, and D. J. Frost (2011), Hydrogen bond symmetrization and equation of state of phase D, *J. Geophys. Res. Solid Earth*, *116*(B6), B06,203, doi:10.1029/2010JB008087.

Iitaka, T., K. Hirose, K. Kawamura, and M. Murakami (2004), The elasticity of the $MgSiO_3$ post-perovskite phase in the Earth's lowermost mantle, *Nature*, *430*(6998), 442–445, doi:10.1038/nature02702.

Imada, S., K. Hirose, and Y. Ohishi (2011), Stabilities of NAL and Ca-ferrite-type phases on the join $NaAlSiO_4$-$MgAl_2O_4$ at high pressure, *Phys. Chem. Minerals*, *38*(7), 557–560, doi:10.1007/s00269-011-0427-2.

Irifune, T., and A. E. Ringwood (1993), Phase transformations in subducted oceanic crust and buoyancy relationships at depths of 600–800 km in the mantle, *Earth Planet. Sci. Lett.*, *117*(1–2), 101–110, doi:10.1016/0012-821X(93)90120-X.

Irifune, T., and T. Tsuchiya (2015), 2.03—Mineralogy of the Earth—Phase transitions and mineralogy of the lower mantle, in *Treatise on Geophysics*, edited by G. Schubert, pp. 33–60, Elsevier, Oxford.

Irifune, T., K. Fujino, and E. Ohtani (1991), A new high-pressure form of $MgAl_2O_4$, *Nature*, *349*(6308), 409–411, doi:10.1038/349409a0.

Irifune, T., T. Shinmei, C. A. McCammon, N. Miyajima, D. C. Rubie, and D. J. Frost (2010), Iron partitioning and density changes of pyrolite in Earth's lower mantle, *Science*, *327*(5962), 193–195, doi:10.1126/science.1181443.

Ito, E. (2015), 2.10—Multi-anvil cells and high pressure experimental methods, in *Treatise on Geophysics*, 2nd ed., edited by G. Schubert, pp. 233–261, Elsevier, Oxford.

Ito, E., A. Kubo, T. Katsura, M. Akaogi, and T. Fujita (1998), High-pressure transformation of pyrope ($Mg_3Al_2Si_3O_{12}$) in a sintered diamond cubic anvil assembly, *Geophys. Res. Lett.*, *25*(6), 821–824, doi:10.1029/98GL00519.

Jackson, J. M., W. Sturhahn, G. Shen, J. Zhao, M. Y. Hu, D. Errandonea, J. D. Bass, and Y. Fei (2005), A synchrotron Mössbauer spectroscopy study of (Mg,Fe)SiO₃ perovskite up to 120 GPa, *Am. Mineral.*, *90*(1), 199–205, doi:10.2138/am.2005.1633.

Jiang, F., G. D. Gwanmesia, T. I. Dyuzheva, and T. S. Duffy (2009), Elasticity of stishovite and acoustic mode softening under high pressure by Brillouin scattering, *Phys. Earth Planet. Inter.*, *172*(3–4), 235–240, doi:10.1016/j.pepi.2008.09.017.

Joswig, W., T. Stachel, J. W. Harris, W. H. Baur, and G. P. Brey (1999), New Ca-silicate inclusions in diamonds—Tracers from the lower mantle, *Earth Planet. Sci. Lett.*, *173*(1–2), 1–6, doi:10.1016/S0012-821X(99)00210-1.

Jung, D. Y., and A. R. Oganov (2005), *Ab initio* study of the high-pressure behavior of CaSiO₃ perovskite, *Phys. Chem. Minerals*, *32*, 146–153, doi:10.1007/s00269-005-0453-z.

Karki, B. B., L. Stixrude, and J. Crain (1997a), Ab initio elasticity of three high-pressure polymorphs of silica, *Geophys. Res. Lett.*, *24*(24), 3269–3272, doi:10.1029/97GL53196.

Karki, B. B., L. Stixrude, S. J. Clark, M. C. Warren, G. J. Ackland, and J. Crain (1997b), Elastic properties of orthorhombic MgSiO₃ perovskite at lower mantle pressures, *Am. Mineral.*, *82*(5–6), 635–638.

Kato, C., K. Hirose, T. Komabayashi, H. Ozawa, and Y. Ohishi (2013), NAL phase in K-rich portions of the lower mantle, *Geophys. Res. Lett.*, *40*(19), 5085–5088, doi:10.1002/grl.50966.

Kawai, K., and T. Tsuchiya (2012), Phase stability and elastic properties of the NAL and CF phases in the $NaMg_2Al_5SiO_{12}$ system from first principles, *Am. Mineral.*, *97*(2–3), 305–314, doi:10.2138/am.2012.3915.

Kawai, K., and T. Tsuchiya (2015), Small shear modulus of cubic CaSiO₃ perovskite, *Geophys. Res. Lett.*, *42*, doi:10.1002/2015GL063446.

Kawai, K., T. Tsuchiya, J. Tsuchiya, and S. Maruyama (2009), Lost primordial continents, *Gondwana Res.*, *16*(3–4), 581–586, doi:10.1016/j.gr.2009.05.012.

Kawakatsu, H., and F. Niu (1994), Seismic evidence for a 920-km discontinuity in the mantle, *Nature*, *371*(6495), 301–305, doi:10.1038/371301a0.

Kesson, S. E., J. D. Fitz Gerald, J. M. G. Shelley, and R. L. Withers (1995), Phase relations, structure and crystal chemistry of some aluminous silicate perovskites, *Earth Planet. Sci. Lett.*, *134*(1–2), 187–201, doi:10.1016/0012-821X(95)00112-P.

Kesson, S. E., J. D. Fitz Gerald, and J. M. Shelley (1998), Mineralogy and dynamics of a pyrolite lower mantle, *Nature*, *393*(6682), 252–255, doi:10.1038/30466.

Kingma, K. J., R. E. Cohen, R. J. Hemley, and H.-k. Mao (1995), Transformation of stishovite to a denser phase at lower-mantle pressures, *Nature*, *374*(6519), 243–245, doi:10.1038/374243a0.

Kohlstedt, D. L., H. Keppler, and D. C. Rubie (1996), Solubility of water in the α, β and γ phases of (Mg,Fe)₂SiO₄, *Contrib. Mineral. Petrol.*, *123*(4), 345–357, doi:10.1007/s004100050161.

Kojitani, H., T. Katsura, and M. Akaogi (2007a), Aluminum substitution mechanisms in perovskite-type MgSiO₃: An investigation by Rietveld analysis, *Phys. Chem. Minerals*, *34*(4), 257–267, doi:10.1007/s00269-007-0144-z.

Kojitani, H., R. Hisatomi, and M. Akaogi (2007b), High-pressure phase relations and crystal chemistry of calcium ferrite-type solid solutions in the system $MgAl_2O_4$-Mg_2SiO_4, *Am. Mineral.*, *92*(7), 1112–1118, doi:10.2138/am.2007.2255.

Kojitani, H., T. Iwabuchi, M. Kobayashi, H. Miura, and M. Akaogi (2011), Structure refinement of high-pressure hexagonal aluminous phases $K_{1.00}Mg_{2.00}Al_{4.80}Si_{1.15}O_{12}$ and $Na_{1.04}Mg_{1.88}Al_{4.64}Si_{1.32}O_{12}$, *Am. Mineral.*, *96*(8–9), 1248–1253, doi:10.2138/am.2011.3638.

Komabayashi, T., K. Hirose, N. Sata, Y. Ohishi, and L. S. Dubrovinsky (2007), Phase transition in CaSiO₃ perovskite, *Earth Planet. Sci. Lett.*, *260*(3–4), 564–569, doi:10.1016/j.epsl.2007.06.015.

Komabayashi, T., S. Maruyama, and S. Rino (2009), A speculation on the structure of the D″ layer: The growth of anticrust at the core–mantle boundary through the subduction history of the Earth, *Gondwana Res.*, *15*(3–4), 342–353, doi:10.1016/j.gr.2008.11.006.

Komatsu, K., T. Kuribayashi, A. Sano, E. Ohtani, and Y. Kudoh (2006), Redetermination of the high-pressure modification of AlOOH from single-crystal synchrotron data, *Acta Crystallogr. Sect. E*, *62*(11), i216–i218, doi:10.1107/S160053680603916X.

Komatsu, K., A. Sano-Furukawa, and H. Kagi (2011), Effects of Mg and Si ions on the symmetry of δ-AlOOH, *Phys. Chem. Minerals*, *38*(9), 727–733, doi:10.1007/s00269-011-0445-0.

Kondo, T., E. Ohtani, N. Hirao, T. Yagi, and T. Kikegawa (2004), Phase transitions of (Mg,Fe)O at megabar pressures, *Phys. Earth Planet. Inter.*, *143–144*, 201–213, doi:10.1016/j.pepi.2003.10.008.

Kubo, T., T. Kato, Y. Higo, and K. Funakoshi (2015), Curious kinetic behavior in silica polymorphs solves seifertite puzzle in shocked meteorite, *Sci. Adv.*, *1*(4), e1500075, doi:10.1126/sciadv.1500075.

Kudoh, Y., C. T. Prewitt, L. W. Finger, A. Darovskikh, and E. Ito (1990), Effect of iron on the crystal structure of (Mg,Fe)SiO₃ perovskite, *Geophys. Res. Lett.*, *17*(10), 1481–1484, doi:10.1029/GL017i010p01481.

Kudoh, Y., T. Nagase, H. Mizohata, E. Ohtani, S. Sasaki, and M. Tanaka (1997), Structure and crystal chemistry of Phase G, a new hydrous magnesium silicate synthesized at 22 GPa and 1050°C, *Geophys. Res. Lett.*, *24*(9), 1051–1054, doi:10.1029/97GL00875.

Kurashina, T., K. Hirose, S. Ono, N. Sata, and Y. Ohishi (2004), Phase transition in Al-bearing CaSiO₃ perovskite: Implications for seismic discontinuities in the lower mantle, *Phys. Earth Planet. Inter.*, *145*(1–4), 67–74, doi:10.1016/j.pepi.2004.02.005.

Kuribayashi, T., A. Sano-Furukawa, and T. Nagase (2013), Observation of pressure-induced phase transition of δ-AlOOH by using single-crystal synchrotron X-ray diffraction method, *Phys. Chem. Minerals*, *41*(4), 303–312, doi:10.1007/s00269-013-0649-6.

Lakshtanov, D. L., et al. (2007), The post-stishovite phase transition in hydrous alumina-bearing SiO_2 in the lower mantle of the earth, *Proc. Natl. Acad. Sci.*, *104*(34), 13,588–13,590, doi:10.1073/pnas.0706113104.

Li, L., D. J. Weidner, J. Brodholt, D. Alfè, G. D. Price, R. Caracas, and R. Wentzcovitch (2006a), Phase stability of $CaSiO_3$ perovskite at high pressure and temperature: Insights from ab initio molecular dynamics, *Phys. Earth Planet. Inter.*, *155*(3–4), 260–268, doi:10.1016/j.pepi.2005.12.007.

Li, S., R. Ahuja, and B. Johansson (2006b), The elastic and optical properties of the high-pressure hydrous phase δ-AlOOH, *Solid State Commun.*, *137*(1–2), 101–106, doi:10.1016/j.ssc.2005.08.031.

Lin, J.-F., D. L. Heinz, H. Mao, R. J. Hemley, J. M. Devine, J. Li, and G. Shen (2003), Stability of magnesiowüstite in Earth's lower mantle, *Proc. Natl. Acad. Sci.*, *100*(8), 4405–4408, doi:10.1073/pnas.252782399.

Lin, J.-F., S. Speziale, Z. Mao, and H. Marquardt (2013), Effects of the electronic spin transitions of iron in lower mantle minerals: Implications for deep mantle geophysics and geochemistry, *Rev. Geophys.*, *51*(2), 244–275, doi:10.1002/rog.20010.

Lin, Y., R. E. Cohen, S. Stackhouse, K. P. Driver, B. Militzer, L. Shulenburger, and J. Kim (2014), Equations of state and stability of $MgSiO_3$ perovskite and post-perovskite phases from quantum Monte Carlo simulations, *Phys. Rev. B*, *90*(18), 184,103, doi:10.1103/PhysRevB.90.184103.

Liu, L. (1975), Post-oxide phases of forsterite and enstatite, *Geophys. Res. Lett.*, *2*(10), 417–419, doi:10.1029/GL002i010p00417.

Liu, L. (1977), High pressure $NaAlSiO_4$: The first silicate calcium ferrite isotype, *Geophys. Res. Lett.*, *4*(5), 183–186, doi:10.1029/GL004i005p00183.

Liu, L.-g. (1986), Phase transformations in serpentine at high pressures and temperatures and implications for subducting lithosphere, *Phys. Earth Planet. Inter.*, *42*(4), 255–262, doi:10.1016/0031-9201(86)90028-2.

Liu, L.-g., and A. E. Ringwood (1975), Synthesis of a perovskite-type polymorph of $CaSiO_3$, *Earth Planet. Sci. Lett.*, *28*(2), 209–211, doi:10.1016/0012-821X(75)90229-0.

Mainprice, D., Y. Le Page, J. Rodgers, and P. Jouanna (2007), Predicted elastic properties of the hydrous D phase at mantle pressures: Implications for the anisotropy of subducted slabs near 670-km discontinuity and in the lower mantle, *Earth Planet. Sci. Lett.*, *259*(3–4), 283–296, doi:10.1016/j.epsl.2007.04.053.

Manga, M., and R. Jeanloz (1996), Implications of a metal-bearing chemical boundary layer in D″ for mantle dynamics, *Geophys. Res. Lett.*, *23*(22), 3091–3094, doi:10.1029/96GL03021.

Mao, H.-k., J. Shu, Y. Fei, J. Hu, and R. J. Hemley (1996), The wüstite enigma, *Phys. Earth Planet. Inter.*, *96*(2–3), 135–145, doi:10.1016/0031-9201(96)03146-9.

Mao, H.-k., G. Shen, and R. J. Hemley (1997), Multivariable dependence of Fe-Mg partitioning in the lower mantle, *Science*, *278*(5346), 2098–2100, doi:10.1126/science.278.5346.2098.

Mao, H.-k., and W. L. Mao (2007), 2.09—Theory and practice—Diamond-anvil cells and probes for high P–T mineral physics studies, in *Treatise on Geophysics*, edited by G. Schubert, pp. 231–267, Elsevier, Amsterdam.

Mao, H.-k., L. C. Chen, R. J. Hemley, A. P. Jephcoat, Y. Wu, and W. A. Bassett (1989), Stability and equation of state of $CaSiO_3$-perovskite to 134 GPa, *J. Geophys. Res. Solid Earth*, *94*(B12), 17,889–17,894, doi:10.1029/JB094iB12p17889.

Marquardt, H., S. Speziale, H. J. Reichmann, D. J. Frost, F. R. Schilling, and E. J. Garnero (2009), Elastic shear anisotropy of ferropericlase in Earth's lower mantle, *Science*, *324*(5924), 224–226, doi:10.1126/science.1169365.

McCammon, C., J. Peyronneau, and J.-P. Poirier (1998), Low ferric iron content of (Mg,Fe)O at high pressures and temperatures, *Geophys. Res. Lett.*, *25*(10), 1589–1592.

Merkel, S., A. K. McNamara, A. Kubo, S. Speziale, L. Miyagi, Y. Meng, T. S. Duffy, and H.-R. Wenk (2007), Deformation of $(Mg,Fe)SiO_3$ post-perovskite and D″ anisotropy, *Science*, *316*(5832), 1729–1732, doi:10.1126/science.1140609.

Merlini, M., and M. Hanfland (2013), Single-crystal diffraction at megabar conditions by synchrotron radiation, *High Press. Res.*, *33*(3), 511–522, doi:10.1080/08957959.2013.831088.

Miura, H., Y. Hamada, T. Suzuki, M. Akaogi, N. Miyajima, and K. Fujino (2000), Crystal structure of $CaMg_2Al_6O_{12}$, a new Al-rich high pressure form, *Am. Mineral.*, *85*(11–12), 1799–1803.

Miyagi, L., W. Kanitpanyacharoen, P. Kaercher, K. K. M. Lee, and H.-R. Wenk (2010), Slip systems in $MgSiO_3$ post-perovskite: Implications for D″ anisotropy, *Science*, *329*(5999), 1639–1641, doi:10.1126/science.1192465.

Miyajima, N., T. Yagi, K. Hirose, T. Kondo, K. Fujino, and H. Miura (2001), Potential host phase of aluminum and potassium in the Earth's lower mantle, *Am. Mineral.*, *86*, 740–746.

Mookherjee, M., B. B. Karki, L. Stixrude, and C. Lithgow-Bertelloni (2012), Energetics, equation of state, and elasticity of NAL phase: Potential host for alkali and aluminum in the lower mantle, *Geophys. Res. Lett.*, *39*(19), L19,306, doi:10.1029/2012GL053682.

Murakami, M., K. Hirose, S. Ono, and Y. Ohishi (2003), Stability of $CaCl_2$-type and α-PbO_2-type SiO_2 at high pressure and temperature determined by in-situ X-ray measurements, *Geophys. Res. Lett.*, *30*(5), 1207, doi:10.1029/2002GL016722.

Murakami, M., K. Hirose, S. Ono, T. Tsuchiya, M. Isshiki, and T. Watanuki (2004a), High pressure and high temperature phase transitions of FeO, *Phys. Earth Planet. Inter.*, *146*(1–2), 273–282, doi:10.1016/j.pepi.2003.06.011.

Murakami, M., K. Hirose, K. Kawamura, N. Sata, and Y. Ohishi (2004b), Post-perovskite phase transition in $MgSiO_3$, *Science*, *304*(5672), 855–858, doi:10.1126/science.1095932.

Nishi, M., T. Irifune, J. Tsuchiya, Y. Tange, Y. Nishihara, K. Fujino, and Y. Higo (2014), Stability of hydrous silicate at high pressures and water transport to the deep lower mantle, *Nat. Geosci.*, *7*(3), 224–227, doi:10.1038/ngeo2074.

Noguchi, M., T. Komabayashi, K. Hirose, and Y. Ohishi (2012), High-temperature compression experiments of $CaSiO_3$ perovskite to lowermost mantle conditions and its thermal equation of state, *Phys. Chem. Minerals*, *40*(1), 81–91, doi:10.1007/s00269-012-0549-1.

Nomura, R., K. Hirose, N. Sata, and Y. Ohishi (2010), Precise determination of post-stishovite phase transition boundary and implications for seismic heterogeneities in the mid-lower mantle, *Phys. Earth Planet. Inter.*, *183*(1–2), 104–109, doi:10.1016/j.pepi.2010.08.004.

Oganov, A. R., and S. Ono (2004), Theoretical and experimental evidence for a post-perovskite phase of MgSiO$_3$ in Earth's D" layer, *Nature*, *430*(6998), 445–448, doi:10.1038/nature02701.

Oganov, A. R., R. Martoňák, A. Laio, P. Raiteri, and M. Parrinello (2005), Anisotropy of Earth's D" layer and stacking faults in the MgSiO$_3$ post-perovskite phase, *Nature*, *438*(7071), 1142–1144, doi:10.1038/nature04439.

Ohira, I., E. Ohtani, T. Sakai, M. Miyahara, N. Hirao, Y. Ohishi, and M. Nishijima (2014), Stability of a hydrous δ-phase, AlOOH–MgSiO$_2$(OH)$_2$, and a mechanism for water transport into the base of lower mantle, *Earth Planet. Sci. Lett.*, *401*, 12–17, doi:10.1016/j.epsl.2014.05.059.

Ohta, K., K. Fujino, Y. Kuwayama, T. Kondo, K. Shimizu, and Y. Ohishi (2014), Highly conductive iron-rich (Mg,Fe)O magnesiowüstite and its stability in the Earth's lower mantle, *J. Geophys. Res. Solid Earth*, *119*(6), 4656–4665, doi:10.1002/2014JB010972.

Ohtani, E., Y. Amaike, S. Kamada, T. Sakamaki, and N. Hirao (2014), Stability of hydrous phase H MgSiO$_4$H$_2$ under lower mantle conditions, *Geophys. Res. Lett.*, *41*(23), 8283–8287, doi:10.1002/2014GL061690.

Ono, S., E. Ito, and T. Katsura (2001), Mineralogy of subducted basaltic crust (MORB) from 25 to 37 GPa, and chemical heterogeneity of the lower mantle, *Earth Planet. Sci. Lett.*, *190*(1–2), 57–63, doi:10.1016/S0012-821X(01)00375-2.

Ono, S., K. Hirose, M. Murakami, and M. Isshiki (2002), Post-stishovite phase boundary in SiO$_2$ determined by in situ X-ray observations, *Earth Planet. Sci. Lett.*, *197*(3–4), 187–192, doi:10.1016/S0012-821X(02)00479-X.

Ono, S., Y. Ohishi, and K. Mibe (2004), Phase transition of Ca-perovskite and stability of Al-bearing Mg-perovskite in the lower mantle, *Am. Mineral.*, *89*(10), 1480–1485, doi:10.2138/am-2004-1016.

Pamato, M. G., A. Kurnosov, T. B. Ballaran, D. M. Trots, R. Caracas, and D. J. Frost (2014), Hexagonal Na$_{0.41}$[Na$_{0.125}$Mg$_{0.79}$Al$_{0.085}$]$_2$[Al$_{0.79}$Si$_{0.21}$]$_6$O$_{12}$ (NAL phase): Crystal structure refinement and elasticity, *Am. Mineral.*, *99*(8-9), 1562–1569, doi:10.2138/am.2014.4755.

Pamato, M. G., R. Myhill, T. Boffa Ballaran, D. J. Frost, F. Heidelbach, and N. Miyajima (2015), Lower-mantle water reservoir implied by the extreme stability of a hydrous aluminosilicate, *Nat. Geosci.*, *8*(1), 75–79, doi:10.1038/ngeo2306.

Panero, W. R., and L. P. Stixrude (2004), Hydrogen incorporation in stishovite at high pressure and symmetric hydrogen bonding in δ-AlOOH, *Earth Planet. Sci. Lett.*, *221*(1–4), 421–431, doi:10.1016/S0012-821X(04)00100-1.

Pearson, D. G., et al. (2014), Hydrous mantle transition zone indicated by ringwoodite included within diamond, *Nature*, *507*(7491), 221–224, doi:10.1038/nature13080.

Prewitt, C. T., and J. B. Parise (2000), Hydrous phases and hydrogen bonding at high pressure, *Rev. Mineral. Geochem.*, *41*(1), 309–333, doi:10.2138/rmg.2000.41.11.

Reid, A. F., and A. E. Ringwood (1969), Newly observed high pressure transformations in Mn$_3$O$_4$, CaAl$_2$O$_4$, and ZrSiO$_4$, *Earth Planet. Sci. Lett.*, *6*(3), 205–208, doi:10.1016/0012-821X(69)90091-0.

Ricolleau, A., J.-P. Perrillat, G. Fiquet, I. Daniel, J. Matas, A. Addad, N. Menguy, H. Cardon, M. Mezouar, and N. Guignot (2010), Phase relations and equation of state of a natural MORB: Implications for the density profile of subducted oceanic crust in the Earth's lower mantle, *J. Geophys. Res. Solid Earth*, *115*(B8), B08202, doi:10.1029/2009JB006709.

Ringwood, A. E. (1975), *Composition and Petrology of the Earth's Mantle*, McGraw-Hill, New York.

Rosa, A. D., M. Mezouar, G. Garbarino, P. Bouvier, S. Ghosh, A. Rohrbach, and C. Sanchez-Valle (2013), Single-crystal equation of state of phase D to lower mantle pressures and the effect of hydration on the buoyancy of deep subducted slabs, *J. Geophys. Res. Solid Earth*, *118*(12), 6124–6133, doi:10.1002/2013JB010060.

Ross, N. L., J. Shu, and R. M. Hazen (1990), High-pressure crystal chemistry of stishovite, *Am. Mineral.*, *75*(7–8), 739–747.

Sakai, T., E. Ohtani, H. Terasaki, N. Sawada, Y. Kobayashi, M. Miyahara, M. Nishijima, N. Hirao, Y. Ohishi, and T. Kikegawa (2009), Fe-Mg partitioning between perovskite and ferropericlase in the lower mantle, *Am. Mineral.*, *94*(7), 921–925, doi:10.2138/am.2009.3123.

Sano, A., E. Ohtani, T. Kubo, and K. Funakoshi (2004), In situ X-ray observation of decomposition of hydrous aluminum silicate AlSiO$_3$OH and aluminum oxide hydroxide d-AlOOH at high pressure and temperature, *J. Phys. Chem. Solids*, *65*(8–9), 1547–1554, doi:10.1016/j.jpcs.2003.12.015.

Sano, A., E. Ohtani, T. Kondo, N. Hirao, T. Sakai, N. Sata, Y. Ohishi, and T. Kikegawa (2008), Aluminous hydrous mineral δ-AlOOH as a carrier of hydrogen into the core-mantle boundary, *Geophys. Res. Lett.*, *35*(3), L03,303, doi:10.1029/2007GL031718.

Sano-Furukawa, A., K. Komatsu, C. B. Vanpeteghem, and E. Ohtani (2008), Neutron diffraction study of δ-AlOOD at high pressure and its implication for symmetrization of the hydrogen bond, *Am. Mineral.*, *93*(10), 1558–1567, doi:10.2138/am.2008.2849.

Seagle, C. T., D. L. Heinz, A. J. Campbell, V. B. Prakapenka, and S. T. Wanless (2008), Melting and thermal expansion in the Fe–FeO system at high pressure, *Earth Planet. Sci. Lett.*, *265*(3–4), 655–665, doi:10.1016/j.epsl.2007.11.004.

Sharp, T. G., A. E. Goresy, B. Wopenka, and M. Chen (1999), A post-stishovite SiO$_2$ polymorph in the meteorite Shergotty: Implications for impact events, *Science*, *284*(5419), 1511–1513, doi:10.1126/science.284.5419.1511.

Shen, G., and Y. Wang (2014), High-pressure apparatus integrated with synchrotron radiation, *Rev. Mineral. Geochem.*, *78*(1), 745–777, doi:10.2138/rmg.2014.78.18.

Shieh, S. R., H. Mao, R. J. Hemley, and L. C. Ming (1998), Decomposition of phase D in the lower mantle and the fate of dense hydrous silicates in subducting slabs, *Earth Planet. Sci. Lett.*, *159*(1–2), 13–23, doi:10.1016/S0012-821X(98)00062-4.

Shieh, S. R., T. S. Duffy, and G. Shen (2005), X-ray diffraction study of phase stability in SiO$_2$ at deep mantle conditions,

Earth Planet. Sci. Lett., *235*(1–2), 273–282, doi:10.1016/j. epsl.2005.04.004.

Shieh, S. R., T. S. Duffy, Z. Liu, and E. Ohtani (2009), High-pressure infrared spectroscopy of the dense hydrous magnesium silicates phase D and phase E, *Phys. Earth Planet. Inter.*, *175*(3–4), 106–114, doi:10.1016/j.pepi.2009.02.002.

Shim, S.-H., R. Jeanloz, and T. S. Duffy (2002), Tetragonal structure of CaSiO$_3$ perovskite above 20 GPa, *Geophys. Res. Lett.*, *29*(24), 2166, doi:10.1029/2002GL016148.

Shim, S.-H., K. Catalli, J. Hustoft, A. Kubo, V. B. Prakapenka, W. A. Caldwell, and M. Kunz (2008), Crystal structure and thermoelastic properties of (Mg$_{0.91}$Fe$_{0.09}$)SiO$_3$ postperovskite up to 135 GPa and 2,700 K, *Proc. Natl. Acad. Sci.*, *105*(21), 7382–7386, doi:10.1073/pnas.0711174105.

Shu, J., H. Mao, J. Hu, Y. Fei, and R. J. Hemley (1998), Single-crystal X-ray diffraction of wüstite to 30 GPa hydrostatic pressure, *Neues Jahrb. Für Mineral. Abh.*, *172*, 309–323.

Sinclair, W., and A. E. Ringwood (1978), Single crystal analysis of the structure of stishovite, *Nature*, *272*(5655), 714–715, doi:10.1038/272714a0.

Smyth, J. R. (1987), β-Mg$_2$SiO$_4$: A potential host for water in the mantle? *Am. Mineral.*, *72*, 1051–1055.

Stachel, T., J. W. Harris, G. P. Brey, and W. Joswig (2000), Kankan diamonds (Guinea) II: Lower mantle inclusion parageneses, *Contrib. Mineral. Petrol.*, *140*(1), 16–27, doi:10.1007/s004100000174.

Stebbins, J. F., S. Kroeker, and D. Andrault (2001), The mechanism of solution of aluminum oxide in MgSiO$_3$ perovskite, *Geophys. Res. Lett.*, *28*(4), 615–618, doi:10.1029/2000GL012279.

Stishov, S. M., and N. V. Belov (1962), Crystal structure of a new dense modification of silica SiO$_2$, *Dokl. Akad. NAUK SSSR*, *143*(4), 951.

Stixrude, L., R. E. Cohen, R. Yu, and H. Krakauer (1996), Prediction of phase transition in CaSiO$_3$ perovskite and implications for lower mantle structure, *Am. Mineral.*, *81*, 1293–1296, doi: 10.2138/am-1996-9-1030.

Sturhahn, W., J. M. Jackson, and J.-F. Lin (2005), The spin state of iron in minerals of Earth's lower mantle, *Geophys. Res. Lett.*, *32*(12), L12,307, doi:10.1029/2005GL022802.

Sugiyama, M., S. Endo, and K. Koto (1987), The crystal structure of stishovite under pressure up to 6 GPa, *Mineral. J.*, *13*(7), 455–466, doi:10.2465/minerj.13.455.

Sun, T., D.-B. Zhang, and R. M. Wentzcovitch (2014), Dynamic stabilization of cubic CaSiO$_3$ perovskite at high temperatures and pressures from *ab initio* molecular dynamics, *Phys. Rev. B*, *89*(9), 094,109, doi:10.1103/PhysRevB.89.094109.

Suzuki, A., E. Ohtani, and T. Kamada (2000), A new hydrous phase δ-AlOOH synthesized at 21 GPa and 1000°C, *Phys. Chem. Minerals*, *27*(10), 689–693, doi:10.1007/s002690000120.

Tamai, H., and T. Yagi (1989), High-pressure and high-temperature phase relations in CaSiO$_3$ and CaMgSi$_2$O$_6$ and elasticity of perovskite-type CaSiO$_3$, *Phys. Earth Planet. Inter.*, *54*(3–4), 370–377, doi:10.1016/0031-9201(89)90254-9.

Tange, Y., E. Takahashi, Y. Nishihara, K. Funakoshi, and N. Sata (2009), Phase relations in the system MgO-FeO-SiO$_2$ to 50 GPa and 2000°C: An application of experimental techniques using multianvil apparatus with sintered diamond

anvils, *J. Geophys. Res. Solid Earth*, *114*(B2), B02,214, doi:10.1029/2008JB005891.

Teter, D. M., R. J. Hemley, G. Kresse, and J. Hafner (1998), High pressure polymorphism in silica, *Phys. Rev. Lett.*, *80*(10), 2145–2148, doi:10.1103/PhysRevLett.80.2145.

Tschauner, O., B. Kiefer, H. Liu, S. Sinogeikin, M. Somayazulu, and S.-N. Luo (2008), Possible structural polymorphism in Al-bearing magnesiumsilicate post-perovskite, *Am. Mineral.*, *93*, 533–539, doi:10.2138/am.2008.2372.

Tschauner, O., C. Ma, J. R. Beckett, C. Prescher, V. B. Prakapenka, and G. R. Rossman (2014), Discovery of bridgmanite, the most abundant mineral in Earth, in a shocked meteorite, *Science*, *346*(6213), 1100–1102, doi:10.1126/science.1259369.

Tsuchida, Y., and T. Yagi (1989), A new, post-stishovite high-pressure polymorph of silica, *Nature*, *340*(6230), 217–220, doi:10.1038/340217a0.

Tsuchiya, J. (2013), First principles prediction of a new high-pressure phase of dense hydrous magnesium silicates in the lower mantle, *Geophys. Res. Lett.*, *40*(17), 4570–4573, doi:10.1002/grl.50875.

Tsuchiya, J., T. Tsuchiya, S. Tsuneyuki, and T. Yamanaka (2002), First principles calculation of a high-pressure hydrous phase, δ-AlOOH, *Geophys. Res. Lett.*, *29*(19), 1909, doi:10.1029/2002GL015417.

Tsuchiya, J., T. Tsuchiya, and S. Tsuneyuki (2005), First-principles study of hydrogen bond symmetrization of phase D under high pressure, *Am. Mineral.*, *90*(1), 44–49, doi:10.2138/am.2005.1628.

Tsuchiya, T., R. Caracas, and J. Tsuchiya (2004), First principles determination of the phase boundaries of high-pressure polymorphs of silica, *Geophys. Res. Lett.*, *31*(11), L11,610, doi:10.1029/2004GL019649.

Uchida, T., et al. (2009), Non-cubic crystal symmetry of CaSiO$_3$ perovskite up to 18 GPa and 1600 K, *Earth Planet. Sci. Lett.*, *282*(1–4), 268–274, doi:10.1016/j.epsl.2009.03.027.

Vanpeteghem, C. B., A. Sano, K. Komatsu, E. Ohtani, and A. Suzuki (2007), Neutron diffraction study of aluminous hydroxide δ-AlOOD, *Phys. Chem. Minerals*, *34*(9), 657–661, doi:10.1007/s00269-007-0180-8.

Vinnik, L., M. Kato, and H. Kawakatsu (2001), Search for seismic discontinuities in the lower mantle, *Geophys. J. Int.*, *147*(1), 41–56, doi:10.1046/j.1365-246X.2001.00516.x.

Vinnik, L. P., S. I. Oreshin, S. Speziale, and M. Weber (2010), Mid-mantle layering from SKS receiver functions, *Geophys. Res. Lett.*, *37*(24), L24,302, doi:10.1029/2010GL045323.

Walter, M. J., S. C. Kohn, D. Araujo, G. P. Bulanova, C. B. Smith, E. Gaillou, J. Wang, A. Steele, and S. B. Shirey (2011), Deep mantle cycling of oceanic crust: Evidence from diamonds and their mineral inclusions, *Science*, *334*(6052), 54–57, doi:10.1126/science.1209300.

Wang, D., and R. J. Angel (2011), Octahedral tilts, symmetry-adapted displacive modes and polyhedral volume ratios in perovskite structures, *Acta Crystallogr. B*, *67*(4), 302–314, doi:10.1107/S0108768111018313.

Wang, Y., and D. J. Weidner (1994), Thermoelasticity of CaSiO$_3$ perovskite and implications for the lower mantle, *Geophys. Res. Lett.*, *21*(10), 895–898, doi:10.1029/94GL00976.

Weng, K., H.-k. Mao, and P. Bell (1982), Lattice parameters of the perovskite phase in the system $MgSiO_3$–$CaSiO_3$–Al_2O_3, *Year b. Carnegie Instit. Wash, 81*, 273–277.

Wentzcovitch, R. M., Z. Wu, and P. Carrier (2010), First principles quasiharmonic thermoelasticity of mantle minerals, *Rev. Mineral. Geochem., 71*(1), 99–128, doi:10.2138/rmg.2010.71.5.

Wicks, J. K., J. M. Jackson, and W. Sturhahn (2010), Very low sound velocities in iron-rich (Mg,Fe)O: Implications for the core-mantle boundary region, *Geophys. Res. Lett., 37*(15), L15,304, doi:10.1029/2010GL043689.

Xue, X., M. Kanzaki, H. Fukui, E. Ito, and T. Hashimoto (2006), Cation order and hydrogen bonding of high-pressure phases in the Al_2O_3-SiO_2-H_2O system: An NMR and Raman study, *Am. Mineral., 91*(5-6), 850–861, doi:10.2138/am.2006.2064.

Yamamoto, T., D. A. Yuen, and T. Ebisuzaki (2003), Substitution mechanism of Al ions in $MgSiO_3$ perovskite under high pressure conditions from first-principles calculations, *Earth Planet. Sci. Lett., 206*(3–4), 617–625, doi:10.1016/S0012-821X(02)01099-3.

Yamanaka, T., T. Fukuda, and J. Mimaki (2002), Bonding character of SiO_2 stishovite under high pressures up to 30 Gpa, *Phys. Chem. Minerals, 29*(9), 633–641, doi:10.1007/s00269-002-0257-3.

Yamanaka, T., A. Uchida, and Y. Nakamoto (2008), Structural transition of post-spinel phases $CaMn_2O_4$, $CaFe_2O_4$, and $CaTi_2O_4$ under high pressures up to 80 GPa, *Am. Mineral., 93*(11–12), 1874–1881, doi:10.2138/am.2008.2934.

Yamanaka, T., K. Hirose, W. L. Mao, Y. Meng, P. Ganesh, L. Shulenburger, G. Shen, and R. J. Hemley (2012), Crystal structures of $(Mg_{1-x}Fe_x)SiO_3$ postperovskite at high pressures, *Proc. Natl. Acad. Sci., 109*(4), 1035–1040, doi:10.1073/pnas.1118076108.

Yamazaki, D., T. Yoshino, H. Ohfuji, J. Ando, and A. Yoneda (2006), Origin of seismic anisotropy in the D″ layer inferred from shear deformation experiments on post-perovskite phase, *Earth Planet. Sci. Lett., 252*(3–4), 372–378, doi:10.1016/j.epsl.2006.10.004.

Yamazaki, D., E. Ito, T. Yoshino, N. Tsujino, A. Yoneda, X. Guo, F. Xu, Y. Higo, and K. Funakoshi (2014), Over 1 Mbar generation in the Kawai-type multianvil apparatus and its application to compression of $(Mg_{0.92}Fe_{0.08})SiO_3$ perovskite and stishovite, *Phys. Earth Planet. Inter., 228*, 262–267, doi:10.1016/j.pepi.2014.01.013.

Yang, H., C. T. Prewitt, and D. J. Frost (1997), Crystal structure of the dense hydrous magnesium silicate, phase D, *Am. Mineral., 82*, 651–654.

Zhang, L., Y. Meng, P. Dera, W. Yang, W. L. Mao, and H.-k. Mao (2013), Single-crystal structure determination of (Mg,Fe)SiO_3 postperovskite, *Proc. Natl. Acad. Sci., 110*(16), 6292–6295, doi:10.1073/pnas.1304402110.

Zhang, L., et al. (2014), Disproportionation of (Mg,Fe)SiO_3 perovskite in Earth's deep lower mantle, *Science, 344*(6186), 877–882, doi:10.1126/science.1250274.

7

Deformation of Core and Lower Mantle Materials

Sébastien Merkel and Patrick Cordier

ABSTRACT

Large-scale processes such as mantle convection involve plastic deformation of rocks. Plastic deformation is also thought to be related to observations such as seismic anisotropy. Quantitative studies of plastic properties under deep Earth conditions are challenging and remain constrained by the development of new experimental and numerical methods. Here, we review the latest trends in experimental and numerical approaches to understand the plasticity of deep Earth materials. We then review the known plastic properties of lower mantle, D″, and inner core materials: ferropericlase, bridgmanite, $CaSiO_3$ in the perovskite structure, post-perovskite, and ε-Fe. These studies constrain the pressure, temperature, and, sometimes, strain rate dependence of deformation of high-pressure minerals, along with the slip systems controlling the evolution of microstructures with deformation. These results have important implications for deep Earth viscosity and the geodynamic interpretation of seismic anisotropy.

7.1. INTRODUCTION

Earth hosts large-scale dynamical processes such as mantle or core convection. The outer core consists of a low-viscosity fluid following the laws of fluid mechanics and will not be discussed here. Other layers of the deep Earth are composed of solid materials, and their dynamics involves the plastic deformation of rocks [*Karato and Wenk*, 2002]. For geodynamical applications, one would like to understand the rheology of deep Earth materials, i.e., the relationship between stress and strain rate, including the effect of various parameters such as pressure, temperature, strain rate, or composition. A major field of research also involves understanding the relationship between deformation, the induced microstructures, and observations such as seismic anisotropy.

Plastic deformation of rocks is inherently multiscale both in space and time and occurs through a variety of mechanisms. Moreover, quantitative studies of plastic properties under deep Earth conditions remain challenging, and major progress in this area is often associated with the development of new techniques. As such, deep Earth plasticity studies often focus on simplified issues, such as the identification of individual deformation mechanisms for a given phase or the study of typical microstructures induced by deformation in a simple geometry. Fewer studies have addressed the issue of stress and strain rate relationships under such conditions.

Plastic deformation can occur by a variety of mechanisms [*Poirier*, 1985]. All involve the motion of point defects, dislocations, and grain boundaries. These defects can be involved either in isolation or in combination, and for each deformation mechanism a particular type of flow law applies [*Frost and Ashby*, 1982]. Common mechanisms thought to be operating in deep Earth include

UMET, Unité Matériaux et Transformations, ENSCL, CNRS, Université de Lille, Lille, France

Deep Earth: Physics and Chemistry of the Lower Mantle and Core, Geophysical Monograph 217, First Edition.
Edited by Hidenori Terasaki and Rebecca A. Fischer.

(1) diffusion creep caused by stress-induced diffusion of atoms, (2) dislocation creep caused by the thermally activated motion of dislocations, and (3) grain boundary processes where deformation occurs predominantly at grain boundaries. Which of these mechanisms dominates deformation in the deep Earth remains unclear. Among these, dislocation motion is the most efficient at generating lattice preferred orientations (LPOs) in polycrystals. As such, observations of strong D″ [e.g., *Panning and Romanowicz*, 2004] and inner core [e.g., *Souriau*, 2007] anisotropy motivated studies of dislocation-related plasticity for deep Earth materials, such as the identification of dislocation types, slip systems, and typical induced microstructures. This does not preclude, however, the combined activation of other mechanisms, such as diffusion.

For the sake of clarity and following the advances in the recent literature, this chapter focuses mostly on dislocation-related plasticity. We first describe the experimental and numerical methods used for the study of plastic behavior of deep Earth phases. The following sections focus on the known properties of phases relevant for the lower mantle, D″, and the inner core.

7.2. METHODS

7.2.1. Experimental Deformation

Dominant deformation mechanisms can change with composition, impurities, water content [e.g., olivine; *Jung et al.*, 2006], and external parameters such as pressure, temperature, or strain rate [e.g., MgO; *Cordier et al.*, 2012]. In the deep Earth, pressure is an agent for phase transformations, producing denser crystal structures with different mechanical properties. Pressure also induces profound changes in the electronic structure of solids, which also affect mechanical properties. As such, plastic properties should be studied within the stability field of the phase and at relevant pressure and temperature conditions. This section reviews current methods, their capabilities, and limitations. Considering the topic of the current book, we will only describe devices operating at pressures above 5 GPa.

7.2.1.1. Large-Volume Presses (LVPs)

The deformation-DIA (D-DIA), rotational Drickamer apparatus (RDA), and the newly developed deformation T-Cup (DT-Cup) are the three main LVPs used for high-pressure plasticity studies. All allow the controlled deformation of millimeter-size samples.

The D-DIA [*Wang et al.*, 2003] is a modification of a device known as the DIA in which the cubic sample assembly is confined between six anvils. The D-DIA modification allows for the independent control of one pair of opposing anvils, thus allowing deformation at constant pressure, both compressional and extensional. Experiments in the D-DIA can be performed at strain rates between 10^{-3} and 10^{-7} s^{-1} and, currently, up to 18 GPa and 1900 K [*Kawazoe et al.*, 2013]. Similarly, the DT-Cup [*Hunt et al.*, 2014] is an extension of the T-Cup design in which the "top" and "bottom" anvils have been redesigned to allow for controlled deformation in axial geometry. To date, the DT-Cup is capable of deformation experiments up ~20 GPa at 300 K and ~10 GPa at ~1100 K with works currently under way for extending those ranges. In both cases, the sample stress state is axial, with cylindrical symmetry around the deformation direction. Axial strains are measured directly (see below) and lateral strains can be reconstructed from changes in sample pressure.

The Drickamer apparatus is an opposed-anvil device with a containment cylinder for aligning the anvils and preventing gasket flow. The RDA [*Yamazaki and Karato*, 2001] is an extension in which a rotational actuator is attached to one of the anvils, thus allowing a large rotational shear deformation of the sample. RDA experiments have reached pressures over 20 GPa at 1800 K [e.g., *Miyagi et al.*, 2014]. In the RDA, the stress boundary condition acting on the sample may be considered as a combination of simple shear and axial compression [*Xu et al.*, 2005].

7.2.1.2. Diamond Anvil Cell (DAC)

LVPs allow for controlled deformation experiments at constant pressure and temperature, but their P/T range is limited and does not yet reach deep lower mantle conditions. The DAC, on the other hand, in combination with heating methods, allows for experiments up to core conditions [e.g., *Tateno et al.*, 2010]. DACs may not only impose pressure but also a compressive stress that produces elastic and plastic deformation. A limitation of the DAC is that pressure and deformation cannot be decoupled. At 300 K, plastic deformation studies in the DAC were successfully performed up to 300 GPa [*Hemley et al.*, 1997]. Recent developments for plasticity studies include the design of resistive heating systems with current P/T ranges up to ~20 GPa at 1700 K and ~35 GPa at 1100 K [*Liermann et al.*, 2009; *Miyagi et al.*, 2013] or laser heating experiments up to ~100 GPa and 2000 K [*Hirose et al.*, 2010]. Stress in DAC experiments is typically assumed to be axial, with cylindrical symmetry around the compression direction and with low-pressure gradients across the sample. Note that this assumption relies on the use of small samples and gasket confinement, unlike older experiments that did not confine the sample within gaskets and, hence, induced sample flow in the radial directions [e.g., *Meade and Jeanloz*, 1988]. Sample strain along the diamond direction is compressive. Radial strain is dependent on sample, gasket, and sample loading.

7.2.2. Analytical Methods

Most lower mantle and core materials cannot be brought back to ambient pressure. Moreover, plastic properties depend on pressure and temperature. Hence, analytical methods that can be used in situ, as the sample is being deformed, are key for understanding such properties. Because of small sample sizes and geometrical constraints of the experimental devices, in situ analysis techniques rely on synchrotron radiation, either in the form of X-ray diffraction (XRD) or X-ray radiography.

7.2.2.1. Average Polycrystalline Properties

Typical experiments tend to investigate average properties inside a polycrystal, including strain and strain rate, average stress, and texture. In the D-DIA, RDA, and DT-Cup, sample strain and strain rates are measured using X-ray radiograph images of the sample [e.g., *Vaughan et al.*, 2000]. Similar measurements have been performed in the DAC in the radial diffraction geometry [e.g., *Merkel and Yagi*, 2005] with limited use, however, since sample deformation is not controlled. Typically, metal foils are placed above, below, or within the sample, and their displacement with time is used to evaluate the sample strain and strain rate.

Deviatoric stress is key for quantifying the rheological properties of a material. Stress itself provides information about strength, and in combination with strain rate it can yield flow laws for the sample. In situ measurements of stress rely on methods derived from XRD residual stress analysis [e.g., *Noyan and Cohen*, 1987]. The basic principle is that variations of elastic lattice strains with orientation relative to the deformation geometry can be used to invert the stress field within the polycrystalline sample. Such inversions are often performed using equations derived from theories based on linear elasticity [e.g., *Singh et al.*, 1998; *Xu et al.*, 2005]. Such equations do not account for plastic deformation, which, in fact, induces a local relaxation of stress in the deforming grains. This issue can be overcome by comparing the experimental lattice strains to results of self-consistent plasticity calculations that account for both elastic and plastic relaxation [e.g., *Li et al.*, 2004; *Merkel et al.*, 2009]. Such modeling, however, is complex and involves multiple parameters such as single-crystal elastic constants and slip systems. Hence, experimentalists often invert stresses based on the plain elastic theories that, for most cases, will provide the right order of magnitude [e.g., *Raterron et al.*, 2013].

LPO, or texture, is quantified with diffraction intensity variations along the Debye rings [e.g., *Merkel et al.*, 2002]. Assuming that the experimental LPOs arise from plastic deformation, they can be compared with polycrystal plasticity simulations to obtain information about the slip systems operating in the sample, with multiple applications to metals and minerals [*Wenk et al.*, 2006]. In addition, methods are currently being developed for tracking individual grains within a polycrystalline sample, and they have been applied in the DAC [*Nisr et al.*, 2014]. These allow for investigating microstructural parameters such as grain size distributions of LPOs grain by grain during dynamical processes such as deformation.

7.2.2.2. Individual Deformation Mechanisms

Transmission electron microscopy (TEM) is the method of choice for characterizing individual deformation mechanisms such as dislocation types and microstructures [a review in *Cordier*, 2002]. This applies only to materials that are stable or can be quenched to room pressure. Quenched high-pressure phases, moreover, often destabilize quite rapidly under the electron beam of the TEM leading to amorphization.

X-ray line profile analysis (XLPA) is an alternate method for identifying and characterizing defects in materials that cannot be looked at in the TEM [*Kerber et al.*, 2011]. In polycrystals, X-ray peak broadening increases with decreasing crystallite size and increasing density of lattice defects. Lattice defects induce heterogeneous local stresses and cause an asymmetric peak broadening, which lies at the core of XLPA. Recently, XLPA was successfully used for identifying dislocation types in deep Earth minerals quenched to ambient conditions [*Cordier et al.*, 2004] and in situ in a DAC [*Nisr et al.*, 2012].

7.2.3. Numerical Modeling

Novel methods relying on numerical modeling have emerged in the literature. They require limited input from experiments and allow exploring ranges of pressure, temperature, and strain rates, which are not easily accessible in the experiment. Numerical models, however, rely on theory, and the complexity of plastic flow has impeded the emergence of a general unified theory for such processes. Multiscale modeling of dislocation-induced plasticity was initiated near the end of the last century [*Bulatov and Kubin*, 1998] as a bottom-up approach, where properties calculated at one scale serve as an input for simulations at the scale above: atomic scale for the properties governed by the core structure of dislocations; mesoscopic scale for elastic interactions and reactions between dislocations in a complex microstructure; and macroscopic scale for constraints due to loading depending on boundary conditions, including grain-grain interactions in a polycrystal.

7.2.3.1. Numerical Modeling of Defect Properties

Dislocations induce a long-range displacement field proportional to $1/r$, where r is the distance to the dislocation line. Thus, direct atomistic modeling of dislocations requires a large number of atoms (e.g., 10,000) and an accurate description of the elastic fields to adjust the boundary conditions [*Walker et al.*, 2010], which is beyond the range of current ab initio calculations. Empirical potentials can be used, but, in practice, the validation of the potentials for describing the large inelastic displacements close to the dislocation core is not straightforward.

The Peierls-Nabarro (PN) model is a simplification, based on a continuous description of the dislocation that allows for direct atomistic calculations. The underlying assumption is that a dislocation line in a real crystal cannot be a singularity and that the inelastic displacements representing the core must have some extension in space. In the original model of *Peierls* [1940] and further refined by *Nabarro* [1947], it is considered that a dislocation gliding in a given plane must have its core spreading in that same plane. The amount of spreading is described by the balance between elastic forces in the bulk crystal, which tend to spread the core, and inelastic forces due to atomic mismatch, which tend to keep it narrow. The PN model attracted further attention when *Vitek* [1968] demonstrated that the inelastic forces could be obtained from the atomic misfit energy in crystallographic glide planes in two dimensions, the so-called γ surface, using ab initio methods based on density functional theory. For high-pressure phases, this allows for properly accounting for the influence of pressure on the electronic structure. The model has been demonstrated to describe dislocations with an excellent agreement with dislocation core observations by high-resolution TEM [*Carrez et al.*, 2007; *Ferré et al.*, 2008]. The PN model, however, assumes a planar core structure and only collinear dissociations can be considered.

The so-called Peierls-Nabarro-Galerkin (PNG) method is a generalization of the PN model, which can account for several potential glide planes simultaneously [*Denoual*, 2004]. It can also describe noncollinear dissociation in one or several planes. Besides describing potentially complex core structures (see, e.g., wadsleyite dislocation models of *Metsue et al.* [2010]), it is also very useful to predict slip system activity from the tendency of a screw dislocation to spread its core in a given plane [*Amodeo et al.*, 2012].

7.2.3.2. Numerical Modeling of Defect Interaction

Plastic deformation in a single crystal results from the collective behavior of a large number of dislocations that meet, react to form junctions or annihilate, and entangle or interact over large distances through their long-range elastic stresses. In dislocation dynamics [*Bulatov et al.*, 1998; *Devincre et al.*, 2001], the rheology of a crystal is obtained by simulating the dynamics and interactions of dislocation lines in an elastic continuum and accounting for external loading and internal stresses iteratively. These models provide an accurate description of the

three-dimensional (3D) dislocation interactions and multiplications that compare well with real crystals. They are, however, quite computer demanding. Less computer time consuming are the so-called 2.5D simulations where parallel linear dislocations only are considered [*Boioli et al.*, 2015]. Local rules are introduced to reproduce hardening, dislocation multiplication, the formation and destruction of junctions, etc. These rules are adjusted to reproduce the results of 3D models. It is then possible to reach larger strains at much lower computational costs. Several codes are available to perform dislocation dynamics simulations, e.g., ParaDiS (DD code developed at Lawrence Livermore National Laboratory) or microMegas (developed at the French Aerospace Laboratory ONERA).

7.2.3.3. From Single Crystals to Polycrystals

Rock-like lower mantle and core materials are polycrystals with many crystal orientations. The behavior of such materials can be modeled using finite-element models [e.g., *Castelnau et al.*, 2008] but, because of the enormous computational efforts and numerous parameters to adjust, there are only a few applications. Simplified models assume that each grain is homogeneous and deforms within an effective medium with the average properties of the polycrystal. Such models, labeled as "self-consistent," are very successful at accounting for the anisotropic response of plastically deforming aggregates. Typically, the interpretation of textures in high-pressure experiments and modeling of texture development in deep Earth processes are performed in the viscoplastic approximation, do not account for elastic deformation, and rely on the viscoplastic self-consistent (VPSC) implementation of *Lebensohn and Tomé* [1993] that of *Castelnau et al.* [2008], and its extensions. When the contribution of elastic deformation cannot be neglected, such as for the interpretation of elastic lattice strains measured in experiments, the elastoplastic self-consistent (EPSC) implementation of *Turner and Tomé* [1994] and its extensions are used. Codes for performing such calculations can be obtained upon request from their authors.

7.3. LOWER MANTLE

7.3.1. Ferropericlase

Ferropericlase (Mg,Fe)O crystallizes under the NaCl structure and is stable under ambient conditions. Hence, extensive low-pressure data are available regarding this material, and particularly MgO [e.g., *Copley and Pask*, 1965; *Weaver and Paterson*, 1969; *Paterson and Weaver*, 1970]. Dominant slip systems in such structures involve dislocations with $\frac{1}{2}\langle 110 \rangle$ Burgers vector gliding in {111}, {110}, or {100} (Figure 7.1).

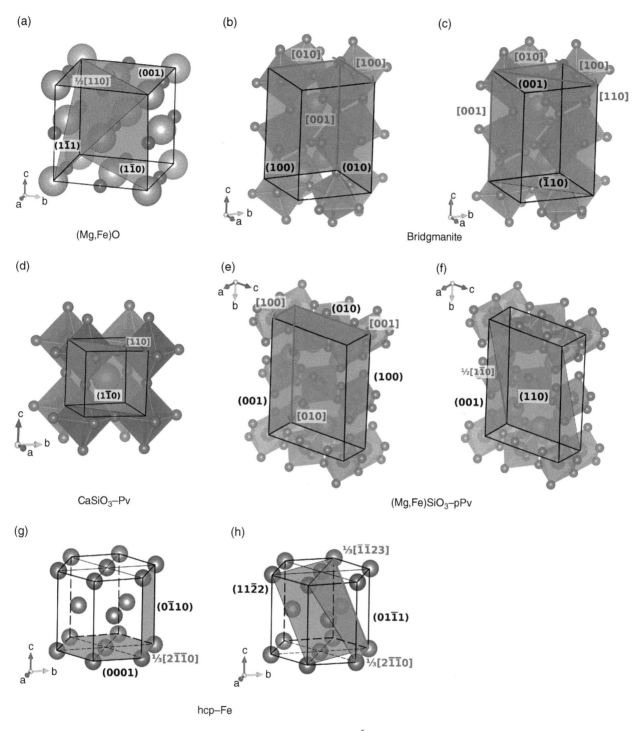

Figure 7.1 Potential slip systems for deep Earth materials. (a) $\frac{1}{2}\langle 110\rangle$ slip in either {100}, {110}, or {111} for (Mg,Fe)O. (b,c) Bridgmanite with (b) [010] slip in (100), [100] slip in (010), or [001] slip in either (100) or (010) and (c) [010], [100], and $\langle 110\rangle$ slip in (001) and [001] and $\langle 110\rangle$ slip in {$\bar{1}10$}. (d) $\langle 110\rangle$\{$1\bar{1}0$\} slip in Ca-Pv. (e,f) (Mg,Fe)SiO$_3$-pPv with (e) [100] slip in (001), [001] in (010), or [010] slip in either (001) or (010) and (f) $\frac{1}{2}\langle 1\bar{1}0\rangle$ slip in either (001) or (110). (g,h) hcp-Fe with (g) $\frac{1}{3}\langle 2\bar{1}\bar{1}0\rangle$ slip in (0001) (basal slip) or {$0\bar{1}10$} (prismatic slip) and (h) $\frac{1}{3}\langle 2\bar{1}\bar{1}0\rangle$ slip in {$01\bar{1}1$} (pyramidal $\langle a\rangle$ slip) and $\frac{1}{3}\langle 11\bar{2}\bar{3}\rangle$ slip in either {$01\bar{1}1$} or {$11\bar{2}2$} (pyramidal $\langle c+a\rangle$ slip).

At 300 K and low confining pressure, slip on {110} is an order of magnitude easier than that on {100}, and this difference decreases with increasing temperature [e.g., *Foitzik and Skrotzki*, 1989]. Under ambient pressure, the strength of both {110} and {100} slip decrease with temperature up to ~700 K and ~1300 K, respectively [e.g., *Foitzik and Skrotzki*, 1989]. Below those thresholds, the movement of dislocations is controlled by lattice friction. Above those temperatures, interactions between dislocations become the controlling factor.

The plastic behavior of polycrystalline MgO at 300 K is controlled by dominant slip on {110} [e.g., *Paterson and Weaver*, 1970], and this remains true up to 50 GPa [*Merkel et al.*, 2002], with little effect of the addition of Fe [*Tommaseo et al.*, 2006] nor of the spin transition in Fe in the 40–50 GPa range [*Lin et al.*, 2009].

According to single-crystal properties, both {100} and {110} slip should become active at higher temperatures. Deformation experiments on (Mg,Fe)O at 1200–1800 K have indeed shown that deformation textures at those temperatures do involve some contribution of {100} and, even, {111} slip [e.g., *Yamazaki and Karato*, 2002; *Heidelbach et al.*, 2003], but these results are somewhat contradictory and will need to be confirmed at lower mantle *P*/*T* range. Further experiments were performed in the D-DIA up to 10 GPa, leading to the first quantification of the flow stress of MgO, characterized by an apparent activation volume of $V^* \approx 2.4 \times 10^{-6}$ m^3/mol at $T = 1473$ K and for a strain rate of 3×10^{-5} s^{-1} [*Mei et al.*, 2008].

Based on numerical models, *Amodeo et al.* [2012] suggested that the activation of the {111} slip should always be negligible, that the {110} slip prevails at low pressure, and that a transition to the dominant {100} slip should occur between 30 and 60 GPa, except at high temperatures where both systems become active. These predictions were partially confirmed by single-crystal D-DIA experiments up to 9 GPa at 1370 K that demonstrated that slip on {110} hardens more significantly with pressure than slip on {100}, with an inversion of dominant slip system predicted at ~25 GPa [*Girard et al.*, 2012].

Finally, *Cordier et al.* [2012] used numerical models to investigate the effect of pressure, temperature, and strain rate on MgO plasticity and concluded that {110} and {100} slip are activated jointly in the whole lower mantle. Moreover, these simulations show that at a depth of less than ~2000 km, MgO deforms in an "athermal" regime in which stress is determined by the microstructure only and is independent of temperature or strain rate. Hence, the notion of viscosity cannot be defined. At depths below 2000 km, the viscosity of MgO should lie in the 10^{21}–10^{22} Pa.s range. Until now, these predictions have not been verified experimentally, although recent experiments indicate that the strength of (Mg,Fe)O increases by a factor of three at pressures between 20 and 65 GPa [*Marquardt and Miyagi*, 2015].

7.3.2. Bridgmanite

Bridgmanite, (Mg,Fe)SiO$_3$, is the main mineral of the lower mantle. It crystallizes as an orthorhombic perovskite (Pv, space group *Pbnm*) with $a = 4.7754$ Å, $b = 4.9292$ Å, and $c = 6.8969$ Å. Do note that a pseudocubic lattice (with $a_c \approx 3.4$ Å for bridgmanite) is often introduced for analyzing possible deformation mechanisms in all perovskites [see *Poirier et al.*, 1989; *Cordier*, 2002; or *Ferré et al.*, 2007]. However, multiple experiments and characterizations on analog Pv-structured materials have demonstrated that they do not constitute an isomechanical group [see *Cordier*, 2002, for a review]. They will, hence, not be discussed here, and slip systems will be described in the orthorhombic framework only.

Cordier et al. [2004] used XLPA to investigate dislocations in MgSiO$_3$ deformed during stress relaxation experiments at 25 GPa and 1400°C and identified [100](001) and [010](001) in their material. Later, *Miyajima et al.* [2009] used TEM on a sample recovered from synthesis at 26 GPa and 2023 K. Dislocations with Burgers vectors of ⟨110⟩ were observed, consistent with slip on ⟨110⟩(001). These works support the conclusion that slip on (001) is dominant at those pressures and that slip occurs in several slip directions.

Early deformation experiments on polycrystalline (Mg,Fe)SiO$_3$ at ambient temperature in DACs [*Meade et al.*, 1995; *Merkel et al.*, 2003; *Wenk et al.*, 2004] produced fairly weak and inconsistent textures, dependent on sample synthesis, from which the identification of a given slip system was difficult. *Wenk et al.* [2006] and *Miyagi* [2009] used an improved experimental method and identified that, at pressures below 55 GPa, slip on (001) appears the most active, with an ambiguous slip direction. *Miyagi* [2009] also identified that, at pressures above 55 GPa, the dominant slip plane changes from (001) to (100). DAC experiments also demonstrated that the observed textures are strongly "orthorhombic," requiring slip systems also to be orthorhombic and not pseudocubic, indicating, again, that cubic perovskites are poor analogs for understanding plasticity in the lower mantle.

Mainprice et al. [2008] investigated the possible slip systems in MgSiO$_3$ at mantle pressure using numerical models and concluded that, up to 100 GPa, [010] (001) glide is the easiest. They also relied on polycrystal plasticity simulations to demonstrate that, at low pressure, the most active system is not the lowest Peierls stress system[010](001), but ⟨110⟩(001) because of the two available ⟨110⟩ glide directions in the (001) plane. At higher pressures, 30 and 100 GPa, [010](100) becomes the most active and other systems in (001) and {110} have similar activities, consistent with the conclusions of *Miyagi* [2009]. Finally, numerical efforts are currently under way to characterize the dislocation structure in bridgmanite in more detail

[*Gouriet et al.*, 2014; *Hirel et al.*, 2014] in the hope of constraining the effect of temperature and strain rate on its plasticity.

7.3.3. CaSiO₃ Perovskite

$CaSiO_3$ in the lower mantle crystallizes in the Pv structure and was originally reported as cubic. A growing line of evidence suggests a lower symmetry. However, the deviation from cubic symmetry is small (0.7% or smaller deviation in the c/a ratio) and is difficult to discriminate either numerically or experimentally [e.g., *Shim et al.*, 2002; *Caracas et al.*, 2005]. We will hence be using the cubic reference frame.

Ambient temperature DAC deformation experiments between 25 and 50 GPa found compression textures characteristic of slip on $\langle 1\bar{1}0\rangle\{110\}$ [*Miyagi et al.*, 2009], confirmed by calculations indicating that, between 0 and 100 GPa, $\langle 1\bar{1}0\rangle\{110\}$ should be the easiest system [*Ferré et al.*, 2009]. *Ferré et al.* [2009] also showed that, unlike most other minerals, $CaSiO_3$ exhibits no or negligible lattice friction for this slip system due to a low-energy stacking fault allowing a wide spreading of the dislocation core. The plastic behavior of $CaSiO_3$ should hence be controlled by dislocation interactions rather than the interactions between defects and the lattice, whatever the temperature. Those conclusions as well as the effect of temperature or strain rate on $CaSiO_3$ plasticity remain to be investigated.

7.4. D″ LAYER

The D″ layer is associated with a transition in (Mg,Fe) SiO_3 from the Pv to the post-perovskite (pPv) structure. Other materials that could be present in the D″ layer include bridgmanite, ferropericlase, and $CaSiO_3$–Pv, and have been discussed above.

Based on the layered structure of pPv, it was suggested that it would slip along the layers, on the (010) plane [e.g., *Murakami et al.*, 2004]. This suggestion is supported by multiple analyses on the $CaIrO_3$ analog with results consistent with either (010) as a dominant slip plane or [100](010) as the dominant slip system [e.g., *Yamazaki et al.*, 2006; *Walte et al.*, 2007; *Niwa et al.*, 2007; *Miyagi et al.*, 2008a; *Miyajima and Walte*, 2009] in agreement with the most recent numerical calculations [*Goryaeva et al.*, 2015a, b]. Early DAC deformation experiments on $MgGeO_3$-pPv and $MgSiO_3$-pPv at 300 K in the 100–160 GPa range found textures consistent with slip on {110} [*Merkel et al.*, 2006, 2007], in line with first-principles calculations on energetics of stacking faults [*Oganov et al.*, 2005]. However, *Okada et al.* [2010] pointed out that textures in pPv are affected by sample synthesis. This was confirmed by *Miyagi et al.* [2010, 2011] who further demonstrated that, for both

$MgGeO_3$-pPv and $MgSiO_3$-pPv, high-pressure 300 K deformation textures are consistent with dominant slip in (001). Finally, *Hirose et al.* [2010] performed laser-heated DAC experiments at 2000 K in the 80–110 GPa range on the $MnGeO_3$ analog and also found results consistent with slip along (001). Experimental results hence indicate that, unlike materials such as $CaIrO_3$, germanates are good analogs for understanding $(Mg,Fe)SiO_3$ plasticity. Deformation textures of pPv are controlled by slip on (001) but are also strongly affected by synthesis and phase transformations, which could be of great significance for understanding processes in the D″ layer [*Dobson et al.*, 2013].

Nisr et al. [2012] performed XLPA on $MgGeO_3$-pPv at 300 K and 90 GPa in the DAC and found results most consistent with $\frac{1}{2}\langle 110\rangle$ dislocations gliding on $\{\bar{1}10\}$ or (001) and the [010](001) slip system. Earlier numerical calculations suggested [001](010) as the easy slip system in both $CaIrO_3$-pPv and $MgSiO_3$-pPv [*Carrez et al.*, 2007; *Metsue et al.*, 2009], while results for $MgGeO_3$-pPv lead to equivalent strengths for [100] (001), [001](010), and [010](001) [*Metsue et al.*, 2009]. Recent full atomistic models rather suggest easy slip along [100](010) in $MgSiO_3$-pPv. Obviously, those analyses on individual dislocation systems are not fully consistent with experimental textures for $MgGeO_3$-pPv and $MgSiO_3$-pPv and this issue remains to be resolved.

7.5. INNER CORE

Earth's inner core mostly consists of iron and, while other structures have been proposed, it is often assumed that iron in the inner core is hexagonal close packed (hcp) [e.g., *Tateno et al.*, 2010]. Observations of strong seismic anisotropy in the inner core [e.g., *Souriau*, 2007] suggest a deformation mechanism that effectively results in LPOs. Based on comparison with metals deformed at very low stresses, *Reaman et al.* [2011] concluded that most of the inner core should deform through dislocation creep. This conclusion is supported by *Gleason and Mao* [2013] who showed that iron at high pressure is rheologically weak and can indeed deform through dislocation creep in the inner core.

Proposed deformation systems for hcp-Fe include basal, prismatic, and pyramidal <c+a> (Figure 7.1), along with twinning (a low-temperature deformation mechanism). Based on stacking fault energy calculations, *Poirier and Price* [1999] suggested basal slip as a dominant deformation mechanism for hcp-Fe. This was confirmed by DAC compression experiments up to 220 GPa at 300 K leading to texture patterns consistent with dominant basal slip [*Wenk et al.*, 2000; *Merkel et al.*, 2004]. Do note, however, that the contribution of twinning on those low-temperature textures should not be overlooked

[*Kanitpanyacharoen et al.*, 2012]. In contrast, *Poirier and Langenhorst* [2002] used TEM to study an hcp Ni-Cr stainless steel analog compressed up to 13 GPa and 1073 K and observed dislocations consistent with pyramidal $\langle c + a \rangle$ slip. *Miyagi et al.* [2008b] attempted a compression experiment on hcp-Fe in a laser-heated DAC up to 30 GPa and a peak sample temperature of 1800 K. Despite large sample heterogeneities induced by temperature gradients, they concluded that, as temperature increases, the activity of pyramidal $\langle c + a \rangle$ slip increases and overruns that of basal slip. Finally, based on a comparison between results of D-DIA experiments up to 17.5 GPa and 600 K and EPSC models, *Merkel et al.* [2012] concluded that the plastic behavior of hcp-Fe is controlled by basal, prismatic, and pyramidal slip, along with twinning. Twinning activity decreases with increasing temperature, while the activity of pyramidal slip increases with increasing temperature. Basal slip, however, was dominant over all the investigated conditions, and some component of prismatic slip was necessary to obtain a good agreement between experiments and models.

Overall, experiments indicate that basal slip and twinning dominate the plastic behavior of hcp-Fe at low temperature. As temperature increases, the twinning disappears, and the activity of pyramidal $\langle c + a \rangle$ increases. This will have to be confirmed in future experiments.

7.6. CONCLUSIONS

Plasticity studies are essential for understanding the microscopic mechanisms underlying global-scale processes such as mantle convection. They, on the other hand, remain technically challenging and essentially driven by developments in high-pressure deformation experiments, characterization methods, and numerical modeling of multiscale processes.

Plasticity studies regarding deep mantle and core phases have long been limited to ambient temperature experiments or relied on the concept of analogs. Recent extensions of the P/T range of deformation experiments, along with the progress in modeling dislocation-related plasticity numerically now allow investigating the effects of pressure, temperature, and strain rate.

This chapter reviewed the current knowledge regarding the plastic behavior of lower mantle, D″, and inner core phases. The recent developments in experimental and numerical methods should allow a better understanding of the effect of parameters such as temperature and strain rates on such properties. They should also allow extending the range of studies to more complex systems such as multiphase materials or to conditions with competing deformation mechanisms, such as dislocation and grain boundary processes.

ACKNOWLEDGMENT

P.C. is supported by funding from the European Research Council under the Seventh Framework Programme (FP7), ERC grant N°290424 – RheoMan. S.M. is supported by the Institut Universitaire de France.

REFERENCES

Amodeo, J., P. Carrez, and P. Cordier (2012), Modelling the effect of pressure on the critical shear stress of MgO single crystals, *Philos. Mag.*, *92*, 1523–1541, doi:10.1080/14786435.2011.652689.

Boioli, F., Ph. Carrez, P. Cordier, B. Devincre, and M. Marquille (2015), Modeling the creep properties of olivine by 2.5-D dislocation dynamics simulations, *Physical Review B*, *92*(1), 014115, doi:10.1103/PhysRevB.92.014115.

Bulatov, V., F. F. Abraham, L. Kubin, B. Devincre, and S. Yip (1998), Connecting atomistic and mesoscale simulations of crystal plasticity, *Nature*, *391*, 669–672, doi:10.1038/35577.

Bulatov, V.V., and L. Kubin (1998), Dislocation modelling at atomistic and mesoscopic scales, *Curr. Opin. Solid State Mater. Sci.*, *3*, 558–561, doi:10.1016/S1359-0286(98)80025-9.

Carrez, P., D. Ferré, and P. Cordier (2007), Implications for plastic flow in the deep mantle from modelling dislocations in $MgSiO_3$ minerals, *Nature*, *446*, 68–70, doi:10.1038/nature05593.

Caracas, R., R. Wentzcovitch, G. D. Price, and J. Brodholt (2005), $CaSiO_3$ perovskite at lower mantle pressures, *Geophys. Res. Lett.*, *32*, L06306, doi:10.1029/2004GL022144.

Castelnau, O., D. K. Blackman, R. A. Lebensohn, and P. Ponte Castañeda (2008), Micromechanical modeling of the viscoplastic behavior of olivine, *J. Geophys. Res.*, *113*, B09202.

Copley, S. M., and J. A. Pask (1965), Plastic deformation of MgO single crystals up to 1600°C, *J. Am. Ceram. Soc.*, *48*, 139–146, doi:10.1111/j.1151-2916.1965.tb16050.x.

Cordier, P. (2002), Dislocations and slip systems of mantle minerals, in *Pastic Deformation of Minerals and Rocks*, edited by S. Karato and H. R. Wenk, pp. 137–179, Mineral. Soc. of Am., Washington, D.C.

Cordier, P., T. Ungár, L. Zsoldos, and G. Tichy (2004), Dislocation creep in $MgSiO_3$ perovskite at conditions of the earth's uppermost lower mantle, *Nature*, *428*, 837–840, doi:10.1038/nature02472.

Cordier, P., J. Amodeo, and P. Carrez (2012), Modelling the rheology of MgO under Earth's mantle pressure, temperature and strain rates, *Nature*, *481*, 177–180, doi:10.1038/nature10687.

Denoual, C. (2004), Dynamic dislocation modeling by combining Peierls Nabarro and Galerkin methods, *Phys. Rev. B*, *70*, 024106, doi:10.1103/PhysRevB.70.024106.

Devincre, B., L.P. Kubin, C. Lemarchand, and R. Madec (2001), Mesoscopic simulations of plastic deformation, *Mater. Sci. Eng. A*, *309–310*, 211–219, doi:10.1016/S0921-5093(00)01725-1.

Dobson, D. P., N. Miyajima, F. Nestola, M. Alvaro, N. Casati, C. Liebske, I. G. Wood, and A. M. Walker (2013), Strong inheritance of texture between perovskite and post-perovskite in the D″ layer, *Nat. Geosci.*, *6*, 575–578, doi:10.1038/ngeo1844.

Ferré, D., P. Carrez and P. Cordier (2008), Modeling dislocation cores in SrTiO$_3$ using the Peierls-Nabarro model, *Phys. Rev. B*, 77, 014106, doi:10.1103/PhysRevB.77.014106.

Ferré, D., P. Cordier, and P. Carrez (2009), Dislocation modeling in calcium silicate perovskite based on the Peierls-Nabarro model, *Am. Mineral.*, 94, 135–142, doi:10.2138/am.2009.3003.

Foitzik, A., and W. Skrotzki (1989), Correlation between microstructure, dislocation dissociation and plastic anisotropy in ionic crystals, *Mater. Sci. Eng. A*, 113, 399–407, doi:10.1016/0921-5093(89)90326-2.

Girard, J., J. Chen, and P. Raterron (2012), Deformation of periclase single crystals at high pressure and temperature: Quantification of the effect of pressure on slip-system activities, *J. Appl. Phys.*, 111, 112607, doi:10.1063/1.4726200.

Gleason, A. E., and W. L. Mao (2013), Strength of iron at core pressures and evidence for a weak Earth's inner core, *Nat. Geosci.*, 6, 571–574, doi:10.1038/ngeo1808.

Goryaeva, A. M., Ph. Carrez, and P. Cordier (2015a), Modeling defects and plasticity in MgSiO$_3$ post-perovskite: 2- screw and edge [100] dislocations. *Phys. and Chem. of Minerals.*, 42, 793–803, doi:10.1007/s00269-015-0763-8.

Goryaeva, A. M., Ph. Carrez, and P. Cordier (2015b), Modeling defects and plasticity in MgSiO$_3$ post-perovskite: 1- generalized stacking faults. *Phys. and Chem. of Minerals.*, 42, 781–792, doi:10.1007/s00269-015-0762-9.

Gouriet, K., P. Carrez, and P. Cordier (2014), Modelling [100] and [010] screw dislocations in MgSiO$_3$ perovskite based on the Peierls–Nabarro–Galerkin model, *Model. Simul. Mater. Sci. Eng.*, 22, 025020, doi:10.1088/0965-0393/22/2/025020.

Ferré, D., P. Carrez, and P. Cordier (2007), First principles determination of dislocations properties of MgSiO$_3$ perovskite at 30 GPa based on the Peierls–Nabarro model, *Phys. Earth Planet. Int.*, 163, 283–291.

Frost, H. J., and M. F. Ashby (1982), *Deformation Mechanisms Maps*, Pergamon, Oxford.

Heidelbach, F., I. Stretton, F. Langenhorst, and S. Mackwell (2003), Fabric evolution during high shear strain deformation of magnesiowüstite (Mg$_{0.8}$Fe$_{0.2}$O), *J. Geophys. Res.*, 108, 2154, doi:10.1029/2001JB001632.

Hemley, R. J., H. K. Mao, G. Shen, J. Badro, P. Gillet, M. Hanfland, and D. Häusermann (1997), X-ray imaging of stress and strain of diamond, iron, and tungsten at megabar pressures, *Science*, 276, 1242–1245, doi:10.1126/science.276.5316.1242.

Hirel, P., A. Kraych, P. Carrez, and P. Cordier (2014), Atomic core structure and mobility of [1 0 0](0 1 0) and [0 1 0](1 0 0] dislocations in MgSiO$_3$ perovskite, *Acta Mater.*, 79, 117–125, doi:10.1016/j.actamat.2014.07.001.

Hirose, K., Y. Nagaya, S. Merkel, and Y. Ohishi (2010), Deformation of MnGeO$_3$ post-perovskite at lower mantle pressure and temperature, *Geophys Res. Lett.*, 37, L20302, doi:10.1029/2010GL044977.

Hunt, S. A., D. J. Weidner, R. J. McCormack, M. L. Whitaker, E. Bailey, L. Li, M. T. Vaughan, and D. P. Dobson (2014), Deformation T-Cup: A new multi-anvil apparatus for controlled strain-rate deformation experiments at pressures above 18 GPa, *Rev. Sci. Instrum.*, 85, 085103, doi:10.1063/1.4891338.

Jung, H., I. Katayama, Z. Jiang, T. Hiraga, and S.I. Karato (2006), Effect of water and stress on the lattice-preferred orientation of olivine, *Tectonophysics*, 421, 1–22.

Kanitpanyacharoen, W., S. Merkel, L. Miyagi, P. Kaercher, C. N. Tomé, Y. Wang, and H. R. Wenk (2012), Significance of mechanical twinning in hexagonal metals at high pressure, *Acta Mater.*, 60, 430–442, doi:10.1016/j.actamat.2011.07.055.

Karato, S.-I., and H.-R. Wenk (2002), *Plastic Deformation of Minerals and Rocks*, Mineral. Soc. of Am., Washington, D.C.

Kawazoe, T., T. Ohuchi, Y. Nishihara, N. Nishiyama, K. Fujino, and T. Irifune (2013), Seismic anisotropy in the mantle transition zone induced by shear deformation of wadsleyite, *Phys. Earth Planet. Inter.*, 216, 91–98, doi:10.1016/j.pepi.2012.12.005.

Kerber, M. B., M. J. Zehetbauer, E. Schafler, F. C. Spieckermann, S. Bernstorff, and T. Ungar (2011), X-ray line profile analysis—An ideal tool to quantify structural parameters of nanomaterials, *JOM*, 63, 61–70, doi:10.1007/s11837-011-0115-1.

Lebensohn, R. A., and C. N. Tomé (1993), A selfconsistent anisotropic approach for the simulation of plastic deformation and texture development of polycrystals: Application to zirconium alloys, *Acta Metall. Mater.*, 41, 2611–2624.

Li, L., D. J. Weidner, J. Chen, M. T. Vaughan, M. Davis, and W. B. Durham (2004), X-ray strain analysis at high pressure: Effect of plastic deformation in MgO, *J. Appl. Phys.*, 95, 8357–8365, doi:10.1063/1.1738532.

Liermann, H.-P., S. Merkel, L. Miyagi, H.-R. Wenk, G. Shen, H. Cynn, and W. J. Evans (2009), New experimental method for in situ determination of material textures at simultaneous high-pressure and temperature by means of radial diffraction in the diamond anvil cell, *Rev. Sci. Instrum.*, 80, 104501, doi:10.1063/1.3236365.

Lin, J. F., H. R. Wenk, M. Voltolini, S. Speziale, J. Shu, and T. S. Duffy (2009), Deformation of lower-mantle ferroperi-clase (Mg,Fe)O across the electronic spin transition, *Phys. Chem. Minerals.*, 36, 585–592, doi:10.1007/s00269-009-0303-5.

Mainprice, D., A. Tommasi, D. Ferré, P. Carrez, and P. Cordier (2008), Predicted glide systems and crystal preferred orientations of polycrystalline silicate Mg-Perovskite at high pressure: Implications for the seismic anisotropy in the lower mantle, *Earth Planet. Sci. Lett.*, 271, 135–144.

Marquardt, H. and L. Miyagi (2015), Slab stagnation in the shallow lower mantle linked to an increase in mantle viscosity, *Nat. Geosci.*, 8, 311–314.

Meade, C., and R. Jeanloz (1988), Yield strength of MgO to 40 GPa, *J. Geophys. Res.*, 93, 3261–3269.

Meade, C., P. G. Silver, and S. Kanshima (1995), Laboratory and seismological observations of lower mantle isotropy, *Geophys. Res. Lett.*, 22, 1293–1296, doi:10.1029/95GL01091.

Mei, S., D. L. Kohlstedt, W. B. Durham, and L. Wang (2008), Experimental investigation of the creep behavior of MgO at high pressures, *Phys. Earth Planet. Inter.*, 170, 170–175, doi:10.1016/j.pepi.2008.06.030.

Merkel, S., H. R. Wenk, J. Shu, G. Shen, P. Gillet, H. K. Mao, and R. J. Hemley (2002), Deformation of polycrystalline MgO at pressures of the lower mantle, *J. Geophys. Res.*, 107, 2271, doi:10.129/2001JB000920.

Merkel, S., H. R. Wenk, J. Badro, G. Montagnac, P. Gillet, H. K. Mao, and R. J. Hemley (2003), Deformation of (Mg,Fe)

SiO$_3$ perovskite aggregates up to 32 GPa, *Earth Planet. Sci. Lett.*, *209*, 351–360, doi:10.1016/S0012-821X(03)00098-0.

Merkel, S., H. R. Wenk, P. Gillet, H. K. Mao, and R. J. Hemley (2004), Deformation of polycrystalline iron up to 30 GPa and 1000 K, *Phys. Earth Planet. Inter.*, *145*, 239–251.

Merkel, S., and T. Yagi (2005), X-ray transparent gasket for diamond anvil cell high pressure experiments, *Rev. Sci. Instrum.*, *76*, 046109, doi:10.1063/1.1884195.

Merkel, S., A. Kubo, L. Miyagi, S. Speziale, T. S. Duffy, H.-K. Mao, and H.-R. Wenk (2006), Plastic deformation of MgGeO$_3$ post-perovskite at lower mantle pressures, *Science*, *311*, 644–646, doi:10.1126/science.1121808.

Merkel, S., A. K. McNamara, A. Kubo, S. Speziale, L. Miyagi, Y. Meng, T. S. Duffy, and H.-R. Wenk (2007), Deformation of (Mg,Fe)SiO$_3$ post-perovskite and D″anisotropy, *Science*, *316*, 1729–1732, doi:10.1126/science.1140609.

Merkel, S., C. N. Tomé, and H.-R. Wenk (2009), A modeling analysis of the influence of plasticity on high pressure deformation of hcp-Co, *Phys. Rev. B*, *79*, 064110, doi:10.1103/PhysRevB.79.064110.

Merkel, S., M. Gruson, Y. Wang, N. Nishiyama, and C. N. Tomé (2012), Texture and elastic strains in hcp-iron plastically deformed up to 17.5 GPa and 600 K: Experiment and model, *Model. Simul. Mater. Sci. Eng.*, *20*, 024005, doi:10.1088/0965-0393/20/2/024005.

Metsue, A., P. Carrez, D. Mainprice, and P. Cordier (2009), Numerical modelling of dislocations and deformation mechanisms in CaIrO$_3$ and MgGeO$_3$ post-perovskites—Comparison with MgSiO$_3$ post-perovskite, *Phys. Earth Planet. Inter.*, *174*, 165–173, doi:10.1016/j.pepi.2008.04.003.

Metsue, A., Ph. Carrez, C. Denoual, D. Mainprice, and P. Cordier (2010), Plastic deformation of wadsleyite: IV dislocation core modelling based on the Peierls-Nabarro-Galerkin model, *Acta Mater.*, *58*, 1467–1478. doi:10.1016/j.actamat.2009.10.047.

Miyagi, L., N. Nishiyama, Y. Wang, A. Kubo, D. V. West, R. J. Cava, T. S. Duffy, and H. R. Wenk (2008a), Deformation and texture development in CaIrO$_3$ post-perovskite phase up to 6 GPa and 1300 K, *Earth Planet. Sci. Lett.*, *268*, 515–525, doi:10.1016/j.epsl.2008.02.005.

Miyagi, L., M. Kunz, J. Knight, J. Nasiatka, M. Voltolini, and H.-R. Wenk (2008b), In situ phase transformation and deformation of iron at high pressure and temperature, *J. Appl. Phys.*, *104*, 103510, doi:10.1063/1.3008035.

Miyagi, L., S. Merkel, T. Yagi, N. Sata, Y. Ohishi, and H.-R. Wenk (2009), Diamond anvil cell deformation of CaSiO$_3$ perovskite up to 49 GPa, *Phys. Earth Planet. Inter.*, *174*, 159–164, doi:10.1016/j.pepi.2008.05.018.

Miyagi, L., W. Kanitpanyacharoen, P. Kaercher, K. K. M. Lee, and H.-R. Wenk (2010) Slip systems in MgSiO$_3$ post-perovskite: Implications for D″ anisotropy, *Science*, *329*, 1639–1641, doi:10.1126/science.1192465.

Miyagi, L., W. Kanitpanyacharoen, S. Stackhouse, B. Militzer, and H.-R. Wenk (2011), The enigma of post-perovskite anisotropy: Deformation versus transformation textures, *Phys. Chem. Minerals.*, *38*, 665–678, doi:10.1007/s00269-011-0439-y.

Miyagi, L., W. Kanitpanyacharoen, S. V. Raju, P. Kaercher, J. Knight, A. MacDowell, H.-R. Wenk, Q. Williams, and E. Z. Alarcon (2013), Combined resistive and laser heating technique for in situ radial X-ray diffraction in the diamond anvil cell at high pressure and temperature, *Rev. Sci. Instrum.*, *84*, 025118, doi:10.1063/1.4793398.

Miyagi, L., G. Amulele, K. Otsuka, Z. Du, R. Farla, and S.-I. Karato (2014), Plastic anisotropy and slip systems in ringwoodite deformed to high shear strain in the Rotational Drickamer Apparatus, *Phys. Earth Planet. Int.*, *228*, 244–253, doi:10.1016/j.pepi.2013.09.012.

Miyagi, L. M. (2009) Deformation and texture development in deep Earth mineral phases: Implications for seismic anisotropy and dynamics, PhD thesis, Univ. of Calif., Berkeley.

Miyajima, N., T. Yagi, and M. Ichihara (2009), Dislocation microstructures of MgSiO$_3$ perovskite at a high pressure and temperature condition, *Phys. Earth Planet. Int.*, *174*, 153–158, doi:10.1016/j.pepi.2008.04.004.

Miyajima, N., and N. Walte (2009), Burgers vector determination in deformed perovskite and post-perovskite of CaIrO$_3$ using thickness fringes in weak-beam dark-field images, *Ultramicroscopy*, *109*, 683–692, doi:10.1016/j.ultramic.2009.01.010.

Murakami, M., K. Hirose, K. Kawamura, N. Sata, and Y. Ohishi (2004), Post-perovskite phase transition in MgSiO$_3$, *Science*, *304*, 855–858, doi:10.1126/science.1095932.

Nabarro, F. R. N. (1947), Dislocations in a simple cubic lattice, *Proc. Phys. Soc.*, *59*, 256–272, doi:10.1088/0959-5309/59/2/309.

Nisr, C., G. Ribárik, T. Ungár, G. B. M. Vaughan, P. Cordier, and S. Merkel (2012), High resolution three-dimensional X-ray diffraction study of dislocations in grains of MgGeO3 post-perovskite at 90 GPa, *J. Geophys. Res.*, *117*, B03201, doi:10.1029/2011JB008401.

Nisr, C., G. Ribárik, T. Ungár, G. B. Vaughan, and S. Merkel (2014), Three-dimensional X-ray diffraction in the diamond anvil cell: Application to stishovite, *High Press. Res.*, *34*, 158–166, doi:10.1080/08957959.2014.885021.

Niwa, K., T. Yagi, K. Ohgushi, S. Merkel, N. Miyajima, and T. Kikegawa (2007), Lattice preferred orientation in CaIrO$_3$ perovskite and post-perovskite formed by plastic deformation under pressure, *Phys. Chem. Minerals.*, *34*, 679–686, doi:10.1007/s00269-007-0182-6.

Noyan, I., and J. Cohen (1987), *Residual Stress: Measurements by Diffraction and Interpretation*, Springer-Verlag, New York.

Oganov, A. R., R. Martoňák, A. Laio, P. Raiteri, and M. Parrinello (2005), Anisotropy of Earth's D″ layer and stacking faults in the MgSiO$_3$, *Nature*, *438*, 1142–1144, doi:10.1038/nature04439.

Okada, T., T. Yagi, and K. N. T. Kikegawa (2010), Lattice-preferred orientations in post-perovskite-type MgGeO$_3$ formed by transformations from different pre-phases, *Phys. Earth Planet. Inter.*, *180*, 195–202, doi:10.1016/j.pepi.2009.08.002.

Panning, M., and B. Romanowicz (2004), Inferences on flow at the base of Earth's mantle based on seismic anisotropy, *Science*, *303*, 351–353, doi:10.1126/science.1091524.

Paterson, M. S., and C. W. Weaver (1970), Deformation of polycrystalline MgO under pressure, *J. Am. Ceram. Soc.*, *53*, 463–471.

Peierls, R. E. (1940), The size of a dislocation, *Proc. Phys. Soc.*, *52*, 34–37, doi:10.1088/0959-5309/52/1/305.

Poirier, J. P. (1985), *Creep of Crystals*, Cambridge Univ. Press, Cambridge.

Poirier, J. P., and F. Langenhorst (2002), TEM study of an analogue of the Earth's inner core ε-Fe, *Phys. Earth Planet. Inter.*, *129*, 347–358, doi:10.1016/S0031-9201(01)00300-4.

Poirier, J. P., and G. D. Price (1999), Primary slip system of ε-iron and anisotropy of the Earth's inner core, *Phys. Earth Planet. Inter.*, *110*, 147–156, doi:10.1016/S0031-9201(98)00131-9.

Poirier, J. P., S. Beauchesne, and F. Guyot (1989), Deformation mechanisms of crystals with perovskite structure, in *Perovskite: A Structure of Great Interest to Geophysics and Materials Science*, edited by A. Navrotsky and D. Weidner, pp. 119–123, AGU, Washington D.C.

Raterron, P., S. Merkel, and C. W. Holyoke III (2013), Axial temperature gradient and stress measurements in the deformation-DIA cell using alumina pistons, *Rev. Sci. Instrum.*, *84*, 043906, doi:10.1063/1.4801956.

Reaman, D. M., G. S. Daehn, and W. R. Panero (2011), Predictive mechanism for anisotropy development in the Earth's inner core, *Earth Planet. Sci. Lett.*, *312*, 437–442, doi:10.1016/j.epsl.2011.10.038.

Shim, S. H., R. Jeanloz, and T. S. Duffy (2002), Tetragonal structure of CaSiO$_3$ perovskite above 20 GPa, *J. Geophys. Res.*, *29*, 2166, doi:10.1029/2002GL016148.

Singh, A. K., C. Balasingh, H. K. Mao, R. J. Hemley, and J. Shu (1998), Analysis of lattice strains measured under non-hydrostatic pressure, *J. Appl. Phys.*, *83*, 7567–7575, doi:10.1063/1.367872.

Souriau, A. (2007), Deep earth structure — The Earth's cores, in *Treatise on Geophysics*, edited by G. Schubert, vol. 1, pp. 655–693, Elsevier, Amsterdam, doi:10.1016/B978-044452748-6.00023-7.

Tateno, S., K. Hirose, Y. Ohishi, and Y. Tatsumi (2010), The structure of iron in Earth's inner core, *Science*, *330*, 359–361, doi:10.1126/science.1194662.

Tommaseo, C. E., J. Devine, S. Merkel, S. Speziale, and H.-R. Wenk (2006), Texture development and elastic stresses in magnesiowüstite at high pressure, *Phys. Chem. Minerals.*, *33*, 84–97, doi:10.1007/s00269-005-0054-x.

Turner, P. A., and C. N. Tomé (1994), A study of residual stresses in Zircaloy-2 with rod texture, *Acta Metall. Mater.*, *42*, 4143–4153, doi:10.1016/0956-7151(94)90191-0.

Vaughan, M., J. Chen, L. Li, D. Weidner, and B. Li (2000), Use of X-ray imaging techniques at high pressure and temperature for strain measurements, in *Science and Technology of High Pressure—Proceedings of AIRAPT-17*, edited by M. H. Manghnan, W. J. Nellis, and M. F. Nicol, pp. 1097–1098, Univ. Press (India), Hyderabad, India.

Vitek, V. (1968), Intrinsic stacking faults in body-centred cubic crystals. *Philos. Mag.*, *18*, 773–786, doi:10.1080/14786436808227500.

Walker, A. M., P. Carrez, and P. Cordier (2010) Atomic scale model of dislocation cores in minerals: Progress and prospects. *Mineral. Mag.*, *74*, 381–413. doi:10.1180/minmag.2010.074.3.381.

Walte, N., F. Heidelbach, N. Miyajima, and D. Frost (2007), Texture development and TEM analysis of deformed CaIrO3: Implications for the D″ layer at the core-mantle boundary, *Geophys. Res. Lett.*, *34*, L08,306, doi:10.1029/2007GL029407.

Wang, Y., W. B. Duhram, I. C. Getting, and D. J. Weidner (2003), The deformation-DIA: A new apparatus for high temperature triaxial deformation to pressures up to 15 GPa, *Rev. Sci. Instrum.*, *74*, 3002-3011, doi:10.1063/1.1570948.

Weaver, C. W., and M. S. Paterson (1969), Deformation of cube-oriented MgO crystals under pressure, *J. Am. Ceram. Soc.*, *52*, 293–302.

Wenk, H. R., S. Matthies, R. J. Hemley, H. K. Mao, and J. Shu (2000), The plastic deformation of iron at pressures of the Earth's inner core, *Nature*, *405*, 1044–1047, doi:10.1038/35016558.

Wenk, H. R., I. Lonardelli, J. Pehl, J. Devine, V. Prakapenka, G. Shen, and H. K. Mao (2004), In situ observation of texture development in olivine, ringwoodite, magnesiowüstite and silicate perovskite at high pressure, *Earth Planet. Sci. Lett.*, *226*, 507–519, doi:10.1016/j.epsl.2004.07.033.

Wenk, H. R., I. Lonardelli, S. Merkel, L. Miyagi, J. Pehl, S. Speziale, and C. E. Tommaseo (2006), Deformation textures produced in diamond anvil experiments, analysed in radial diffraction geometry, *J. Phys. Condens. Matter*, *18*, S933–S947, doi:10.1088/0953-8984/18/25/S02.

Xu, Y., Y. Nishihara, and S.-I. Karato (2005), Development of a rotational Drickamer apparatus for large-strain deformation experiments at deep Earth conditions, in *Advances in High-Pressure Technology for Geophysical Applications*, edited by J.Chen, Y. Wang, T.S. Duffy, G. Shen, and L. F. Dobrzhinetskaya, pp. 167–182, Elsevier, Amsterdam, doi:10.1016/B978-044451979-5.50010-7.

Yamazaki, D., and S.-I. Karato (2001), High-pressure rotational deformation apparatus to 15 GPa, *Rev. Sci. Instrum.*, *72*, 4207–4211, doi:10.1063/1.1412858.

Yamazaki, D., and S. Karato (2002), Fabric development in (Mg,Fe)O during large strain, shear deformation: Implications for seismic anisotropy in Earth's lower mantle, *Phys. Earth Planet. Inter.*, *131*, 251–267, doi:10.1016/S0031-9201(02)00037-7.

Yamazaki, D., T. Yoshino, H. Ohfuji, J. Ando, and A. Yoneda (2006), Origin of seismic anisotropy in the D″ layer inferred from shear deformation experiments on post-perovskite phase, *Earth Planet. Sci. Lett.*, *252*, 372–378, doi:10.1016/j.epsl.2006.10.004.

8

Using Mineral Analogs to Understand the Deep Earth

Simon A. T. Redfern

ABSTRACT

The use of analog materials in the study of the deep Earth has a long history, as reviewed here. The most inaccessible environments of the solid Earth lie at its heart, and the use of materials as analogs for the iron alloy of Earth's inner core, for the liquid alloy of the outer core, the post-perovskite material of the base of the lower mantle, and for the most abundant terrestrial silicate, bridgmanite, are highlighted. While analogs may be imperfect in many ways, they do provide physical insight into the key processes of interest and help distinguish between potential candidate processes in minerals at extreme conditions. The development of this approach has guided mineral physics over the last few decades and appears set to continue to illuminate the challenging problems that direct approaches cannot yet address.

8.1. INTRODUCTION

The use of analogs to understand the interior of the deep Earth has a long and illustrious history. The extreme pressures and temperatures of the deep mantle and core restrict the range of experimental approaches that can be applied to the study of materials under the appropriate conditions. While structural and spectroscopic measurements can be made in the diamond anvil cell, even with laser heating, more sophisticated measurements of mineral properties are significantly complicated by the limits in size, in access, and in control of applied conditions (pressure, temperature, electric, or magnetic field) within the diamond cell.

A historical example illustrates the utility of carrying out experiments with analog materials. For at least 800 years or more it has been known that the compass needle points north. Various theories to explain the origins of

this effect have been proposed, from ideas that the pole star was attracting the lodestone needle to those of others who believed that a magnetic island located far to the north was attracting the needle. But it was only when William Gilbert carried out measurements on an "analog Earth" that the origins of Earth's magnetic field started to be understood. Gilbert, who was also physician to Queen Elizabeth I of England, carried out a series of experiments on magnetism that he described in 1600. He constructed a spherical lodestone, a single crystal of magnetite spinel, as a model Earth. Noticing the way that a compass needle pointed as it was passed across the surface of his model, he recognized that the pattern of field was similar to that seen by mariners at various latitudes and longitudes across the oceans. He concluded that Earth was the source of the magnetic field that the compass reflects and that magnetism was internal. Thus, from the application of an analog experiment a significant step forward in understanding was achieved. It should also be noted, however, that Gilbert also went on to conclude that Earth was one giant sphere of magnetite,

Department of Earth Sciences, University of Cambridge, Cambridge, UK

Deep Earth: Physics and Chemistry of the Lower Mantle and Core, Geophysical Monograph 217, First Edition.
Edited by Hidenori Terasaki and Rebecca A. Fischer.
© 2016 American Geophysical Union. Published 2016 by John Wiley & Sons, Inc.

thus simultaneously illustrating the dangers in taking the results from analog experiments too far, and too literally. Gilbert failed to recognize the analog nature of his approach, nor could he be expected to have realized this limitation given the state of understanding of the age.

Of course, the current understanding of Earth's interior means that mineral physics conclusions from the study of analogs can be developed secure in the understanding of their potential limitations and weaknesses. With this in mind, the analog approach remains a significantly useful tool in the arsenal of the deep Earth researcher since it allows the study of general phenomena, which may or may not then be translated to the real materials of the deep Earth. The modern-day equivalent of Gilbert's lodestone experiments exist as magnetohydrodynamics experiments of metallic conducting liquids, in particular the experimental investigation of rotating spheres of liquid sodium. One such is the nonlinear dynamics laboratory of Daniel Lathrop, at University of Maryland. Lathrop and co-workers have been carrying out experiments on highly turbulent magnetohydrodynamic flow in spheres up to 3 m in diameter [*Rieutord et al.*, 2012; *Triana et al.*, 2012; *Zimmerman et al.*, 2011]. Their experiments follow from similar established studies in France at the École Normale Supérieure institutes of Paris and Lyon and the French Atomic Energy Commission in Saclay [*Berhanu et al.*, 2007] as well as complementing the work of Henri-Claude Nataf in Grenoble, France [*Cabanes et al.*, 2014]. The approach provides direct analog testing of models of magnetohydrodynamic behavior that still remain intractable by computational simulation and is clearly impractical to carry out on liquid iron systems under outer core conditions. While Michael Faraday tried, and failed, to measure magnetohydrodynamic effects through the induced voltage in saltwater flowing in Earth's magnetic field during tidal movements in the river Thames, modern models of Earth's core, via laboratory analog experiments, promise to yield results directly relevant to outer core processes.

As well as models of the liquid outer core, analog experiments and approaches have been key in the development of our understanding of the nature of mineral behavior in the deep Earth and have played an important part in the historical development of our current ideas of deep Earth processes. These range from many studies of the high-temperature and high-pressure behavior of key lower mantle phases, such as perovskite-structured solids, to the nature of the liquid-solid change of state at the inner-core–outer-core boundary. The study of rheology and deformation of Earth materials in the planet's deep interior, and the mechanisms of slip and subsequent development of deformation texture, have relied heavily on studies of analog equivalents of mineral phases. The principal assumption has been that structural similarity (at the atomic scale) between analog materials and the planetary solids of interest justifies the application of results obtained to real scenarios. We shall see below, however, that not all materials isostructural with the rock-forming minerals of interest provide an accurate description of the dynamic behavior of these phases. In other contexts, especially in attempts to understand phase transition hierarchies and associated microstructral effects, analogs may provide a good insight into deep Earth behavior, however. Here, the pathway from core to mantle is charted through the application of analog materials in each context.

8.2. ANALOGS OF EARTH'S CORE MATERIALS

The conditions of Earth's core, with temperatures approaching 6000 K and pressure in excess of 360 GPa, are the most extreme within the planet. The transition from lowermost mantle to outer core, associated with a large step increase in temperature and a huge contrast in properties from solid silicate insulator (both thermally and electrically) to fluid metallic conductor, is arguably the greatest contrast in properties at any point on Earth. It certainly rivals the difference between crust and ocean or atmosphere with which we are familiar. Studying and understanding materials and their processes at these conditions challenge mineral physics, both as a computational and an experimental discipline. It is little wonder that analogs amenable to investigation at more reasonable conditions, especially for study of subtle phenomena, play an important role in understanding Earth's core.

8.2.1. Outer Core Analogs: How Aluminum Helped Computational Studies of Melting Curve of Iron

Analog methods have aided the development of computational studies of the deep Earth as well as of experimental investigations. One case in point is the evolution of estimates for the melting curve of iron in the core. For many years, the melting curve of iron at the conditions of the boundary between Earth's inner core and outer core (ICB, inner core boundary) has been the subject of considerable uncertainty, with estimates from experimental measurements using static diamond anvil cell approaches, shock experiments, and computational modeling varying widely. Even within one set of experiments using the same methods, wide disagreement was a most obvious feature.

Thus, *Yoo et al.'s* [1993] shock experiments placed the melting curve 1000 K higher than the value obtained by *Brown and McQueen* [1986]. *Boehler's* [1993] results lay 1000 K beneath the melting curve first calculated by *Alfè et al.* [2002, 2003]. Most recently, however, experiments on the melting of iron at ICB conditions, by X-ray diffraction [*Anzellini et al.*, 2013], have yielded results in good agreement with the ab initio molecular dynamics calculations carried out using co-existence methods *Alfè*, [2009]. Part of the reason for this is the improvement in experimental approaches to identify the point at which the transition from solid to liquid state takes place. In this case, the use of diffraction as a signature of solid iron was key to *Anzellini et al.'s* [2013] investigation. Alongside this are improvements in temperature characterization and stability within the complex hybrid environment within the diamond anvil cell sample chamber itself, to which *Anzellini et al.* [2013] paid careful attention. But the development of the ab initio methods of Alfè was also an important factor in the convergence of results, and this was achieved through, first, verification in analog systems.

Part of the key to improvements in calculations came from developments in the use of density functional theory combined with molecular dynamics, which *Vocadlo and Alfè* [2002] and *Alfè* [2003] first tested using the example of aluminum. While not specifically presented in the context of understanding the melting of planetary interiors, it is clear that the work on aluminum acted as a precursor for the later results on iron. The advantages of working on aluminum were twofold. First, ab initio calculations of aluminum are not complicated by the difficulties of accounting for exchange correlation, which beset the quantum mechanical computation of iron. Second, reliable and agreed experimental melting curves for aluminum existed in the literature, such that the calculations of *Vocadlo and Alfè* [2002] could be appraised and verified with confidence. In this case, the calculations and experimental results agreed well, giving support to the development of the method to predict the melting curve of iron at ICB conditions. The success of the confidence that the aluminum analog study engendered can now be recognized, given the agreement between the latest experimental results for iron and the earlier computational results of Alfè and co-workers.

8.2.2. Imitating Inner Core

In view of the existing uncertainties concerning the nature and structure of Earth's inner core, there are also limitations on how much one can attempt to apply analogs of potential relevance. A core-forming iron-nickel-based alloy with uncertain combinations of light elements may crystallize at the inner core in one of a number of structures, with face-centerd cubic (fcc), body-centerd cubic (bcc), and hexagonal close packed (hcp) being the leading contenders. Different light elements may stabilize different members of the possible polymorphs of iron, as reviewed recently by *Hirose et al.* [2013]. The principal difficulty is that the free energy surface for

iron is rather flat, with minima between polymorphs being separated by only a few hundredths of electron volt. The flat energy surface for iron is evident even at ambient pressure, with a re-entrant phase transition for bcc iron that appears at temperatures both above (δ-iron) and below (α-iron) the intermediate fcc phase (γ-iron). None the less, for pure Fe at least, all evidence points toward a hexagonal polymorph (hcp ε-iron) being stable at the pressure-temperature conditions of Earth's inner core [*Vocadlo et al.*, 1999; *Yoo et al.*, 1995]. Thus it is that hexagonal close-packed metals have been the principal focus as analogs of the solid inner core.

At atmospheric pressure, osmium has been highlighted as a dense hcp metal that shows behavior analogous to that of Fe in the inner core [*Godwal et al.*, 2012]. In addition, both Ti and Zr have an hcp structure with a *c*/*a* ratio, and relative linear compressibilities, similar to those calculated for ε-iron [*Jeanloz and Wenk*, 1988; *Mao et al.*, 1990]. *Bergman* [1998] used this as justification to employ these metals as analogs of Earth's inner core. *Bergman* [1998] found that the variation in anisotropy of these structures measured as a function of temperature is similar to that calculated for iron. From laboratory solidification studies of these analog materials, a crystal growth model was constructed, and it was this that gave the first evidence that the grain size of iron in the inner core is likely of the order of tens to hundreds of meters for the shorter grain dimension [*Bergman*, 1998].

More recently, *Bergman et al.* [2014] used directionally solidified hcp Zn-rich Sn alloys at high homologous temperatures as an analog for hcp iron at the conditions of the inner core. Here, they focus on the rheological properties and deformation mechanisms of the analog Zn-Sn alloys to arrive at an explanation of a textured inner core displaying a directionally solidified columnar structure. The limited slip systems of the hcp samples, predominantly basal slip on (0001), is linked by *Bergman et al.* [2014] to hardening that they observed in samples subjected to steady-state torsional deformation.

In fact, the use of metallurgical analogs to understand the behavior of the ICB has a rather long history. The suggestion and then widespread acceptance of the existence of a "mushy layer" at the ICB seems to have followed from observations of analog experiments on solidification of metals in the laboratory. *Loper and Roberts* [1981] and *Fearn et al.* [1981] noted that constitutional supercooling due to the rejection of impurity atoms into the molten alloy, typically results in morphological instability and the formation of a slurry or mushy zone at the interface during the casting of ingots. Testing these ideas, however, *Shimizu et al.* [2005] were careful to point out that the conditions of crystallization of Earth's inner core were far from those of casting metals in the laboratory. Growth rates at the solid-liquid interface ion in the laboratory

(a)

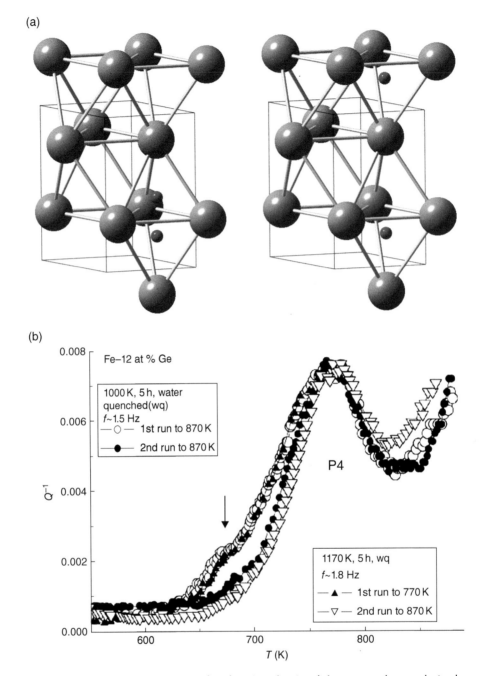

Figure 8.1 (a) Viscoelastic processes associated with pairs of point defects may play a role in the anisotropic seismic attenuation seen in Earth's inner core. Here an interstitial defect pair in hcp iron is shown: Movement of a light element from one tetrahedral interstice to another will result in relaxation of the strain dipole and generate a Zener relaxation. (b) Measuring subtle viscoelastic effects at the pressure and temperatures of the inner core is not feasible, but measurements of mechanical dissipation at seismic frequencies has already been conducted on a number of hcp alloys, including the Zener relaxation in the high-Ge iron alloy shown here (after *Golovin et al.* [2009]).

may typically be of the order of 0.1 mm/s, whereas the growth of the inner core is approximately 10 million times slower. Temperature gradients in the laboratory solidification of metals are very large, while in the core they are exceedingly small. *Shimizu et al.* [2005] quote from

Buffet et al. [2001] who warned, when reflecting on the use of analog experiments, "transferring results from one physical setting to another without proper analysis can lead to misconceptions." *Shimizu et al.* [2005] attempted such analysis, and in the end their conclusion was that the likely state at the ICB was indeed that suggested by analog experiments, but not necessarily for the same reasons.

Understanding the rheological properties of the inner core depend not only on understanding the grain microstructure (or macrostructure) but also the potential phase transitions and related interfaces, should there be a transition between the possible polymorphs that may occur. In relation to this, analogs have been employed to explore the relationship between the γ-phase of fcc iron and the hcp ε-phase, which may occur via a martensitic shear. *Poirier and Langenhorst* [2002] used iron-chromium-nickel alloys (austenitic stainless steels) quenched from high temperature to generate ε-martensite, which is metastable at atmospheric pressure. Their recovered ε-martensite, formed from austenitic stainless steel pressurized to moderate pressure and temperature in the multianvil cell, was amenable to transmission electron microscopy (TEM) analysis as well as selected area electron diffraction. They noted that the laths of ε-martensite, which lie parallel to the {111} planes of the host austenitic (fcc) phase, show basal slip as expected, as well as pyramidal slip, but that the latter ceases to operate above $0.5T_m$ (melting temperature). This led them to conclude that only basal slip would be important in the inner core, where the homologous temperature is high, but also that cobalt is a more convenient analog of inner core hcp ε-iron, for the purposes of deformation studies. More recent results on hcp iron itself have, however, called the application of these results to the core into question. *Miyagi et al.* [2008a] found that pyramidal slip dominated in their high pressure experiments on hcp iron held at high pressure in the diamond anvil cell.

Aside from the plastic deformation of the inner core, the anelastic attenuation characteristics of Earth's center have recently been interpreted in terms of results from analog systems. *Redfern* [2014] has suggested that the anisotropic attenuation seen in seismic studies of Earth's inner core can be related to a paired point defect anelastic relaxation, akin to the Zener relaxation seen in hexagonal metallic alloys. Picking up on the observations of *Mäkinen et al.* [2014] that the attenuation of P-waves in the inner core is strong along polar paths, coincident with P-wave velocity being fastest, from normal mode observations, *Redfern* [2014] has pointed out that the only anelastic process measured in alloys consistent with this observation appears to be a Zener relaxation. This would imply that the attenuation signatures of the inner core can be directly related to light element concentrations, since the Zener relaxation only occurs when solute pairs are sufficiently probable. Treating magnesium-indium hcp alloys as an analog of ε-iron containing up to 4% light element, the results of *Brozel and Leak* [1976] as well as those for hcp yttrium-hydride single crystals [*Kappesser*

et al., 1996] acquire a new significance. They demonstrate, experimentally, that the Zener relaxation is stronger along the *z* axis of an hcp crystal, the same direction that V_p is greatest, consistent with a textured inner core preferentially aligned with the hcp *z* axis parallel to Earth's rotation axis. Further experiments by *Golovin et al.* [2009] on another inner core analog, a high-Ge alloy, which is hcp at the composition of interest, showed the effect of a Zener relaxation on the inverse quality factor for an analog inner core iron alloy, measured at seismic frequencies in the laboratory (Figure 8.1).

8.3. MIMICKING LOWERMOST MANTLE

The complexity and subtleties of the lowermost mantle suggest significant structural effects that could be associated with either mineral transformations or compositional or thermal heterogeneity. Yet the difficulty in carrying out in situ studies of realistic mantle phases remains almost as challenging as those set out for the core above. For this reason, a number of researchers have adopted analogs for the study of mineral processes in the lowermost mantle, but in this case alongside (very often) complementary studies of supposed Earth-forming compositions.

8.3.1. Calcium Iridate Structured Magnesium Silicate: Post-Perovskite

The discovery of the transformation from bridgmanite (or magnesium silicate perovskite, as the mineral phase was then known) to a previously unrecognized structure (which became known as post-perovskite) at the conditions of the lowermost mantle changed views on the nature of the mantle's base [*Murakami et al.*, 2004; *Oganov and Ono*, 2004; *Iitaka et al.*, 2004]. Straight away *Oganov and Ono* [2004] identified this new dense polymorph of $MgSiO_3$ as isostructural with the layered perovskite-related compound $CaIrO_3$, first described by *Rodi and Babel* [1965] and later refined by *Hirai et al.* [2009]. The nature of the post-perovskite material, with a layered structure and anisotropic elastic properties, was quickly invoked to explain a number of the complexities associated with the D″ region above the core mantle boundary. The Clapeyron slope of the perovskite–post-perovskite transition, and predicted seismic velocities of post-perovskite, suggest that this transition occurs in the Pacific rim region of D″ [*Shim*, 2008].

Needless to say, interest in calcium iridate and its structural family blossomed with the discovery of the magnesium silicate member. Knowing that the structure itself was of geophysical interest led to a number of studies of its intrinsic behavior, best achieved by investigating the compounds that are stable (or at least metastable) at ambient conditions. The origins and nature of the anisotropy of physical properties of the post-perovskite structure was one of the early foci of investigation, with *Swainson and Yonkeu* [2007] explaining the anisotropy in terms of the connectivity of layers and freedom for rigid unit mode buckling of those layers. The thermochemical properties of $CaIrO_3$ were also subject to scrutiny, with compressibility and thermal expansion reported by *Martin et al.* [2007a,b], *Boffa Ballaran et al.* [2007], and *Aguado et al.* [2012]. The behavior of optical phonons and the mode gruneisen parameters for the structure were also measured on the $CaIrO_3$ analog material [*Hustoft et al.*, 2008].

Other analog compounds with the same structure as $CaIrO_3$ and post-perovskite were also the subject of renewed scrutiny. The large range of materials that adopt the perovskite structure (see below) made for a fertile hunting ground in which to explore the crystal chemical controls on the transition between perovskite and $CaIrO_3$-type structures. One of the earliest analog studies was that on $NaMgF_3$ fluoride perovskite, which was observed to transform to post-perovskite at more modest pressures, for example [*Liu et al.*, 2005]. Thus far, over 20 materials have been reported with the post-perovskite structure including $Ca_{1-x}Na_xIrO_3$ [*Ohgushi et al.*, 2006], $CaRhO_3$ [*Yamaura et al.*, 2009], $NaIrO_3$ [*Bremholm et al.*, 2011], $CaRuO_3$ [*Shirako et al.*, 2011], $MgGeO3$ [*Hirose et al.*, 2005], $NaZnF_3$ [*Yakovlev et al.*, 2009], and $NaCoF_3$ and $NaNiF_3$ [*Lindsay-Scott et al.*, 2014]. Phases such as $CaPtO_3$ may be synthesized at low pressure and have been investigated extensively as a function of pressure and temperature [*Hunt et al.*, 2013], but it has been suggested that the fluoride post-perovskites, although typically requiring high-pressure synthesis, may be better analogs of $MgSiO_3$ than iridate, platinate, ruthenate, or rhodinate post-perovskites [*Dobson et al.*, 2011].

The nature of the phase transition between the perovskite-structured compounds and their post-perovskite polymorphs was recognized as important not only in understanding the possible occurrence of $MgSiO_3$ post-perovskite under the conditions of the likely geotherm at the base of the mantle but also in terms of the possible origins of anisotropy, which can be linked to the seismically anisotropic properties of D″. Texture development on phase transformation has been seen in a number of the analog phases with $CaIrO_3$ structure, including $CaIrO_3$ itself. Understanding texture development and its role on lower mantle anisotropy depends upon characterizing the slip mechanisms in post-perovskite.

Deformation experiments on $MgSiO_3$ post-perovskite are experimentally challenging as it is only stable at very high pressure and is not quenchable. Initial deformation experiments were carried out on $CaIrO_3$ itself [*Miyajima et al.*, 2006b], since slip systems can be identified in the TEM. Sheared samples were also recovered and their texture analyzed by electron back-scattered diffraction (EBSD) [*Yamazaki et al.*, 2006; *Walte et al.*, 2007]. All of these investigations indicated that (010)[100] was the dominant slip system in $CaIrO_3$. Further in situ radial diffraction performed on $CaIrO_3$ held in the diamond anvil cell confirmed this conclusion [*Miyagi et al.*, 2008]. Work on other post-perovskite analog systems came up with differing conclusions, however, leading to the conclusion that the dominant slip system was a function of the chemical composition of the analog phase, rather than simply of the structure itself. Work has been conducted to characterize the slip in Mn_2O_3, $MnGeO_3$, and $MgGeO_3$ analogs in addition to $CaIrO_3$ [*Santillán et al.*, 2006; *Hirose et al.*, 2010; *Merkel et al.*, 2006]. Slip on (010) and (100) was reported for Mn_2O_3 [*Santillán et al.*, 2006] and on (001) for $MnGeO_3$ [*Hirose et al.*, 2010]. These observations demonstrate the caution with which one must approach the application of results from analog experiments. Finally, in situ diamond anvil cell (DAC) experiments by *Miyagi et al.* [2010] on strongly textured $MgSiO_3$ itself, using radial diffraction, suggested (001) [100] slip. However, even this suggestion has been thrown open to further discussion with the possibility that some of the textures observed are the result of transformation texturing rather than dislocation glide. *Miyagi et al.* [2011] explored this possibility further using $MgGeO_3$ analog post-perovskite, which showed predominant slip on (001)[100], as seen in $MgSiO_3$. They also found similar transformation texturing in $MgGeO_3$ and $MgSiO_3$, demonstrating the robustness of $MgGeO_3$ post-perovskite as an analog phase. This echoes earlier indications that germanates are good crystal chemical analogs for silicates [*Navrotsky and Ross*, 1988] as well as the most recent results for $CaGeO_3$ [*Nakatsuka et al.*, 2015]. Pleasingly, the suggested (001) slip plane for post-perovskite would explain the observed seismic anisotropy in D″ with $V_{SH} > V_{SV}$ and an anticorrelation between P and S waves [*Miyagi et al.*, 2011].

8.3.2. From Perovskite to Bridgmanite

The mineral perovskite, $CaTiO_3$, lends its name to the entire family of ABX_3 structures first determined by *Megaw* [1945]. At room temperature and ambient pressure $CaTiO_3$ perovskite is isostructural with $MgSiO_3$ bridgmanite, the principal component of the lower mantle and by far the most abundant terrestrial silicate. With *Pbnm* space group and octahedral tilts about two axes, the structure is characterized by distortion away from the cubic ABX_3 aristotype. In view of the experimental difficulties associated with the in situ study of $MgSiO_3$, an alternative approach that has long been adopted in understanding its behavior has been to investigate the high-temperature behavior of analogous phases that are more tractable experimentally. The fluoride perovskite neighborite, $NaMgF_3$, also displaying *Pbnm* symmetry, has received much attention in this regard [e.g., *Zhao et al.*, 1993]. As well as being isostructural, $CaTiO_3$ also shows the same twinned microstructures as those found in $MgSiO_3$. For $CaTiO_3$, these are thought to be associated with transitions from its *Pbnm* room temperature structure via a tetragonal intermediate to the high-temperature cubic phase, although such transitions are not expected in $MgSiO_3$, where twinning may be inherited from crystal growth or subsequent deformation. The multiple ferroelastic phase transitions seen in $CaTiO_3$ perovskite as a function of temperature [*Redfern*, 1996] are not, however, replicated in bridgmanite in the lower mantle.

The lack of identical phase transformation behavior in perovskite and bridgmanite has not, however, negated its use in other applications. There are well over 200 reported compounds in the perovskite family of structures, with another considerable selection of fluoride perovskites and most recently a growing number of organometallic halide perovskites (with applications in photovoltaics). To review all instances of the application of perovskites as analog materials to the lower mantle is beyond the scope of this chapter, but some selected examples follow.

Karato and Li [1992], for example, carried out high-temperature creep experiments on polycrystalline $CaTiO_3$ to assess the sensitivity of diffusion creep to grain size. Interestingly, they also noted that creep rate is enhanced by the transition from orthorhombic to tetragonal on heating. While this could be attributed simply to the increase in temperature needed to transform to tetragonal, it is also likely that the tetragonal, higher entropy, phase will display increased diffusion due to the Hedvall effect [*Sartbaeva et al.*, 2005]. *Karato et al.* [1995] also used rheological $CaTiO_3$ experiments to explain the absence of anisotropy in much of the lower mantle (above D″), while *Poirier et al.* [1983] had earlier employed $KZnF_3$ perovskite as an analog to estimate the material's control on the viscosity and conductivity of the lower mantle. Other early attempts to understand the rheological properties of bridgmanite include the study of ferroic titanate materials such as $BaTiO_3$ [*Beauchesne and Poirier*, 1989], $SrTiO_3$ [*Wang et al.*, 1993] as well as $CaTiO_3$ [*Li et al.*, 1996], and even aluminate $YAlO_3$ [*Wang et al.*, 1999]. More recently *Zhao et al.* [2012] investigated the rheological implication of the transformation of ringwoodite to bridgmanite plus periclase using a Co_2TiO_4 analog for magnesium silicate spinel. Their results suggested that ringwoodite breakdown results in a fine-grained intergrowth structures that flow by dislocation creep, resulting in a strengthening (potentially) in a down-going slab. The observations of deformation mechanisms on analog perovskites were reconciled to those of bridgmanite by

Cordier et al. [2004] who noted that the creep laws shear parallel to the (pseudo-)cubic <110> direction for different widely varying perovskite analogs converge into a single trend. This is assumed to reflect dislocation glide, and *Cordier et al.* [2004] proposed that titanate and niobate perovskite analogs could provide useful information about the high-temperature rheology of bridgmanite.

While analogs have been employed extensively for experimental study of plastic deformation behavior and rheological properties under high strain, perovskite analogs have also shown their utility in indicating possible origins for viscoelastic responses in the lower mantle. *Harrison and Redfern* [2002] and *Harrison et al.* [2003, 2004] identified domain wall movements as a source of anelasticity in twinned low-symmetry perovskites including $Ca_{1-x}Sr_xTiO_3$ *Pbnm* analogs. While their ambient pressure results could not be applied directly to the behavior of $MgSiO_3$ in the lower mantle, they did highlight the possibility of such effects, which are currently beyond the scope of high-P/T experimental methods. These studies also demonstrated the importance of oxygen point defect vacancies on controlling the mobility of twin domain walls, highlighting the need to better understand the activation volumes for diffusive motion of such point defects, if extrapolations to lower mantle conditions are to be made [*Goncalves-Ferreira et al.*, 2010].

Some of the earliest examples of the use of analogs in an attempt to understand the behavior of perovskite in the deep mantle include that of *Ringwood and Seabrook* [1963] who noted the value of studying germanates to replicate the high-pressure behavior of silicates. *Navrotsky and Ross* [1988] ran with this concept in their accumulation of thermochemical data on the germanate analogs. *Andrault et al.* [1996] extended that work to investigate the high-temperature behavior of $CaGeO_3$ and $SrGeO_3$ perovskites, defining the limits of their metastability under laboratory conditions. Aside from germanates, stannates have also been proposed (another group with strong isostructural resemblance to bridgmanite) as good analogs, especially $CaSnO_3$ [*Redfern et al.*, 2011; *Tateno et al.*, 2010] since it shows quasi-harmonic behavior with no indication of structural instability away from the *Pbnm* structure, akin to $MgSiO_3$. Anelastic effects in stannate perovskites were investigated by *Daraktchiev et al.* [2005] and suggest that orthorhombic twinned $MgSiO_3$ may show similar viscoelastic behavior, although the effect of pressure on the relaxation strengths and relaxation times of such processes remains unknown.

It seems likely that analogs of bridgmanite will continue to be employed to answer questions that in situ techniques cannot yet easily address. These include measurement of transport properties, understanding the role of spin transitions on conductivity (for example) within the perovskite structure, and understanding the general crystal chemical controls on properties for Earth materials with variable (and potentially heterogeneous) composition. In many cases, the results from analogs provide the first steps in understanding the general physical response of materials, of relevance to planetary interiors generally.

8.4. CONCLUSIONS

The ability to measure nonequilibrium properties such as viscoelasticity and transport properties at the conditions of the lower mantle and deeper remains challenging, if not unattainable at present. In these circumstances, although experiments on synthetic bridgmanite, for example, over a range of controlled chemistries can be made, the opportunities afforded by methods such as dielectric loss spectroscopy, mechanical spectroscopy, and more challenging techniques such as neutron inelastic scattering, demand the use of analog systems still.

A further advantage of the use of analog materials is the ability to separate out potential controlling variables and factors by isolating chemical variation (through the use of synthetic analogs), microstructural variation (for example, comparing polycrystalline samples against single crystals), and by control of external fields such as pressure, temperature, and stress in ways that are not possible under extreme conditions. In a similar manner to the use of computational methods to elucidate and solve mineral physics problems, so the use of analogs can provide fresh insights into the real natural systems of interest. Their limitations must be understood, of course, but it is likely that their use will continue for some while yet, bringing new understanding of the behavior of the deep Earth through the opportunities that they afford.

ACKNOWLEDGMENTS

The author has received funding for the study of analogs of mineral systems in the deep Earth from the UK Natural Environment Research Council, Science and Technology Facilities Council, and British Council.

REFERENCES

Aguado, F., S. Hirai, S. A. T. Redfern, and R. I. Smith (2014), Thermal expansion of CaIrO$_3$ post-perovskite by time-of-flight measurements, *J. Phys. Conf. Ser.*, *549*, 012,024.

Alfè, D. (2003), First-principles simulations of direct coexistence of solid and liquid aluminium, *Phys. Rev. B*, *68*, 064,423.

Alfè, D. (2009), Temperature of the inner-core boundary of the Earth: Melting of iron at high pressure from first-principle coexistence simulations, *Phys. Rev. B*, *79*, 060,101.

Alfè, D., G. D. Price, and M. J. Gillan (2002), Iron under Earth's core conditions: Liquid-state thermodynamics and high-pressure melting curve from ab initio calculations, *Phy. Rev. B*, *65*, 165,118.

Alfè, D., M. J. Gillan, and G. D. Price (2003), Thermodynamics from first principles; temperature and composition of the Earth's core, *Mineral. Mag.*, *67*, 113–123.

Andrault, D., J.-P. Itié, and F. Farges (1996), High-temperature structural study of germanate perovskites and pyroxenoids, *Am. Mineral.*, *81*, 822–832.

Anzellini, S., A. Dewaele, M. Mezouar, P. Loubeyre, and G. Morard (2013), Melting of iron at Earth's inner core boundary based on fast X-ray diffraction, *Science*, *340*, 464–466.

Beauchesne, S., and J. P. Poirier (1989), Creep of barium titanate perovskite: A contribution to a systematic approach to the viscosity of the mantle, *Phys. Earth Planet Inter.*, *55*, 187–199.

Bergman, M. I. (1998), Estimates of the Earth's inner core grain size, *Geophys. Res. Lett.*, *25*, 1593–1596.

Bergman, M. I., Y. Al-Khatatbeh, D. J. Lewis, and M. C. Shannon (2014), Deformation of directionally solidified alloys: Evidence for microstructural hardening of Earth's inner core? *Comptes Rendus Geosci.*, *346*, 140–147.

Berhanu, M., R. Monchaux, S. Fauve, N. Mordant, F. Pétrélis, A. Chiffaudel, F. Daviaud, B. Dubrulle, L. Marié, F. Ravelet, M. Bourgoin, Ph. Odier, J.-F. Pinton, and R. Volk (2007), Magnetic field reversals in an experimental turbulent dynamo, *Europhysi. Lett.*, *77*, 59,001.

Boehler, R. (1993), Temperatures in the Earth's core from melting-point measurements of iron at high static pressures, *Nature*, *363*, 534–536.

Boffa Ballaran, T., R. G. Trønnes, and D. J. Frost (2007), Equations of state of CaIrO$_3$ perovskite and post-perovskite phases, *Am. Mineral.*, *92*, 1760–1763.

Bremholm, M., S. E. Dutton, P. W. Stephens, and R. J. Cava (2011), NaIrO$_3$–A pentavalent post-perovskite, *J. Solid State Chem.*, *184*, 601–607.

Brown, J. M., and R. G. McQueen (1986), Phase transitions, Grüneisen-parameter, and elasticity for shocked iron between 77 GPa and 400 GPa, *J. Geophys. Res. 91*, 7485–7494.

Brozel, M. R., and G. M. Leak (1976), Zener relaxations in magnesium-indium alloys, *Il Nuovo Cimento*, *33*, 270–278.

Buffet, B., E. Garnero, and R. Jeanloz (2001), Response to Morse's "Porous sediments at the top of the Earth's core?" *Science*, *291*, 2092–2093.

Cabanes, S., N. Schaeffer, and H-C Nataf (2014), Turbulence reduces magnetic diffusivity in a liquid sodium experiment, *Phys. Rev. Lett. 113*, doi:10.1103/PhysRevLett.113.184501.

Cordier, P., T. Ungár, L. Zsoldos, and G. Tichy (2004), Dislocation creep in MgSiO$_3$ perovskite at conditions of the Earth's uppermost lower mantle, *Nature*, *428*, 837–840.

Daraktchiev, M., R. J. Harrison, E. H. Mountstevens, and S. A. T. Redfern (2005), Effect of transformation twins on the anelastic behavior of polycrystalline Ca$_{1-x}$Sr$_x$TiO$_3$ and Sr$_x$Ba$_{1-x}$SnO$_3$ perovskite in relation to the seismic properties of Earth's mantle perovskite, *Mater. Sci. Eng. A*, *442*, 199–203.

Dobson, D., S. A. Hunt, A. Lindsay-Scott, and I. G. Wood (2011), Toward better analogues for MgSiO$_3$ post-perovskite: NaCoF$_3$ and NaNiF$_3$, two new recoverable fluoride post-perovskites, *Phys. Earth Planet. Inter.*, *189*, 171–175.

Fearn, D. R., D. E. Loper, and P. H. Roberts (1981), Structure of the Earth's inner core, *Nature*, *292*, 232–233.

Godwal, B. K., J. Yan, S. M. Clark, and R. Jeanloz (2012), High-pressure behavior of osmium: An analog for iron in Earth's core, *J. Apll. Phys.*, *111*, 112,608.

Golovin, I. S., H. Neuhäuser, S. A. T. Redfern, and H.-R. Sinning (2009), Mechanisms of anelasticity in Fe-Ge-based alloys, *Mater. Sci. Eng. A*, *521*, 55–58.

Goncalves-Ferreira, L., S. A. T. Redfern, E. Artacho, E. K. H. Salje, and W. T. Lee (2010), Trapping of oxygen vacancies in the twin walls of perovskite, *Phys. Rev. B*, *81*, 024,109.

Harrison, R. J., and S. A. T. Redfern (2002), The influence of transformation twins on the seismic-frequency elastic and anelastic properties of perovskite: Dynamical mechanical analysis of single crystal LaAlO_3, *Phys. Earth Planet. Inter.*, *134*, 253–272.

Harrison, R. J., S. A. T. Redfern, and J. Street (2003), The effect of transformation twins on the seismic-frequency mechanical properties of polycrystalline $Ca_{1-x}Sr_xTiO_3$ perovskite, *Amer. Mineral.*, *88*, 574–582.

Harrison, R. J., S. A. T. Redfern, and E. K. H. Salje (2004), Dynamical excitation and anelastic relaxation of ferroelectric domain walls in LaAlO_3, *Phys. Rev. B*, *69*, 144,101.

Hirai, S., M. D. Welch, F. Aguado, and S. A. T. Redfern (2009), The crystal structure of CaIrO_3 post-perovskite revisted, *Zeits. Kristallogr.*, *224*, 345–350.

Hirose, K., K. Kawamura, Y. Ohishi, S. Tateno, and N. Sata (2005), Stability and equation of state of MgGeO_3 post-perovskite phase, *Am. Mineral.*, *90*, 262–265.

Hirose, K., Y. Nagaya, S. Merkel, and Y. Ohishi (2010), Deformation of MnGeO_3 post-perovskite at lower mantle pressure and temperature, *Geophys. Res. Lett.*, *37*, L20,302.

Hirose, K, S. Labrosse, and J. Hernlund (2013), Composition and stare of the core, *Annu. Rev. Earth Planet Sci.*, *41*, 657–691.

Hunt, S. A., A. Lindsay-Scott, I. G. Wood, M. W. Ammann, and T. Taniguchi (2013), The P-V-T equation of state of CaPtO_3 post-perovskite, *Phys. Chem. Minerals*, *40*, 73–80.

Hustoft, J., S. H. Shim, A. Kubo, and N. Nishiyama (2008), Raman spectroscopy of CaIrO_3 postperovskite up to 30 GPa, *Am. Mineral.*, *93*, 1654–1658.

Iitaka, T., K. Hirose, K. Kawamura, and M. Murakami (2004), The elasticity of the MgSiO_3 post-perovskite phase in the Earth's lowermost mantle, *Nature*, *430*, 442–445.

Jeanloz, R., and H. R. Wenk (1988), Convection and anisotropy of the inner core, *Geophys. Res. Lett.* *15*, 72–75.

Kappesser, B., H. Wipf, R. G. Barnes, and B. J. Beaudry (1996), A mechanical spectroscopy study of the anelastic relaxation of hydrogen pairs in hexagonal single crystals, *Europhys. Lett.*, *36*, 385–390.

Karato, S., and P. Li (1992), Diffusion creep in perovskite—Implications for the rheology of the lower mantle, *Science*, *255*, 1238–1240.

Karato, S., S. Q. Zhang, and H. R. Wenk (1995), Superplasticity in Earth's lower mantle—Evidence from seismic anisotropy and rock physics, *Science*, *270*, 458–461.

Lindsay-Scott, A., D. Dobson, F. Nestola, M. Alvaro, N. Casati, C. Liebske, K. S. Knight, R. I. Smith, and I. G. Wood (2014), Time-of-flight neutron powder diffraction with milligram

samples: The crystal structures of NaCoF_3 and NaNiF_3 post-perovskites, *J. Appl. Crystallogr.* *47*, 1939–1947.

Li, P., S. Karato, and Z. Wang (1996), High-temperature creep in fine-grained polycrystalline CaTiO_3, an analogue material of (Mg,Fe)SiO_3 perovskite, *Phys. Earth Planet Inter.*, *95*, 19–36.

Liu, H., J. Chen, J. Hu, C. D. Martin, D. J. Weidner, D. Häusermann, and H. K. Mao (2005), Octahedral tilting evolution and phase transition in orthorhombic NaMgF_3 perovskite under pressure, *Geophys. Res. Lett.*, *32*, L04,304.

Loper, D. E. and P. H. Roberts (1981), A study of conditions at the inner core boundary of the Earth, *Phy. Earth Planet. Inter.*, *24*, 302–307.

Mäkinen, A. M., A. Deuss, and S. A. T. Redfern (2014), Anisotropy of Earth's inner core intrinsic attenuation from seismic normal mode models, *EarthPlanet. Sci. Lett.*, *404*, 354–364.

Mao, H. K., K. Wu., L. C. Chen, J. F. Shu, and A. P. Jephcoat (1990), Static compression of iron to 300 GPa and $Fe_{0.8}Ni_{0.2}$ alloy to 260 GPa: Implications for composition of the core, *J. Geophys. Res. 95*, 21,737–21,742.

Martin, C. D., K. W. Chapman, P. J. Chupas, V. Prakapenka, P. L. Lee, S. D. Shastri, and J. B. Parise (2007a), Compression, thermal expansion, structure, and instability of CaIrO_3, the structure model of MgSiO_3 post-perovskite, *Am. Mineral. 92*, 1048–1053.

Martin, C. D., R. I. Smith, W. G. Marshall, and J. B. Parise (2007b), High-pressure structure and bonding in CaIrO3: The structural model of post-perovskite investigated with time-of-flight neutron powder diffraction, *Am. Mineral.*, *92*, 1912–1918.

Megaw, H. (1945), Crystal structure of barium titanate, *Nature*, *155*, 484–485.

Merkel, S., A. Kubo, L. Miyagi, S. Speziale, T. S. Duffy, H. K. Mao, and H. R. Wenk (2006), Plastic deformation of MgGeO_3 post-perovskite at lower mantle pressures, *Science*, *311*, 644–646.

Miyagi, L., M. Kunz, J. Knight, J. Nasiatka, M. Voltolini, and H.-R. Wenk (2008a) *In situ* phase transformation and deformation of iron at high pressure and temperature, *J. Appl. Physi.*, *104*, 103,510.

Miyagi, L., N. Nishiyama, Y. Wang, A. Kubo, D. V. West, R. J. Cava, T. S. Duffy, and H. R. Wenk (2008b), Deformation and texture development in CaIrO_3 post-perovskite phase up to 6 GPa and 1300 K, *Earth Planet. Sci. Lett.*, *268*, 515–525.

Miyagi, L., W. Kanitpanyacharoen, S. Stackhouse, B. Militzer, and H. R. Wenk (2011), The enigma of post-perovskite anisotropy: deformation versus transformation textures, *Phys. Chem. Minerals*, *38*, 665–678.

Miyajima, N., K. Ohgushi, M. Ichihara, and T. Yagi (2006), Crystal morphology and dislocation microstructures of CaIrO_3: A TEM study of an analogue of the MgSiO_3 post-perovskite phase, *Geophy. Res. Lett.*, *33*, L12,302.

Murakami, M., K. Hirose, K. Kawamura, N. Sata, and Y. Ohishi (2004), Post-perovskite phase transition in MgSiO_3, *Science 304*, 855–858.

Nakatsuka, A., S. Kuribayashi, N. Nakayama, H. Fukui, H. Arima, A. Yoneda, and A. Yoshiasa (2015), Temperature dependence of crystal structure of $CaGeO_3$ high pressure perovskite phase and experimental determination of its Debye temperatures studied by low- and high-temperature single-crystal X-ray diffraction, *Am. Mineral.*, *100*, 1190–1202.

Navrotsky, A., and N. L. Ross (1988), Study of the $MgGeO_3$ polymorphs (orthopyroxene, clinopyroxene, and ilmenite structures) by calorimetry, spectroscopy and phase equilibria, *Am. Mineral. 73*, 1355–1365.

Oganov, A. R., and S. Ono (2004), Theoretical and experimental evidence for a postperovskite phase of $MgSiO_3$ in Earth's D″ layer, *Nature*, *430*, 445–448.

Ohgushi, K., H. Gotou, T. Yagi, Y. Kiuchi, F. Sakai, and Y. Ueda (2006), Metalinsulator transition in $Ca_{1-x}Na_xIrO_3$ with post-perovskite structure, *Phys. Rev. B*, *74*, 241,104.

Poirier, J. P., and F. Langenhorst (2002), TEM study of an analogue of the Earth's inner core ε-Fe, *Phy. Earth Planet. Inter.*, *129*, 347–358.

Poirier, J. P., J. Peyronneau, J. Y. Gesland, and G. Brebec (1983), Viscosity and conductivity of the lower mantle—An experimental study on a $MgSiO_3$ perovskite analog, $KZnF_3$, *Phys. Earth Planet. Inter.*, *3*, 273–287.

Redfern, S. A. T. (1996), High-temperature structural phase transitions in perovskite $(CaTiO_3)$, *J. Phys. Condens. Matter*, *8*, 8267–8275.

Redfern, S. A. T. , C.-J. Chen, J. Kung, O. Chaix-Pluchery, J. Kreisel, and E. K. H. Salje (2011), Raman spectroscopy of $CaSnO_3$ at high temperature: A highly quasi-harmonic perovskite, *J. Phys. Conden. Matter*, *23*, 425,401.

Redfern, S. A. T. (2014), Intrinsic anisotropic anelasticity of hcp iron due to light element solute atoms, *AGU Fall Meeting Abstracts with Program*, MR12A-02, AGU, Washington, D.C.

Rieutord, M., S. A. Triana, D. S. Zimmerman, and D. P. Lathrop (2012), Excitation of inertial modes in an experimental spherical Couette flow, *Phys. Rev. E*, *86*, 026,304.

Ringwood, A. E., and M. Seabrook (1963), High-pressure phase transformation in germanate pyroxenes and related compounds, *J. Geophys. Res.*, *68*, 4601–4609.

Rodi, F., and D. Babel (1965), Erdalkaliiridium(IV)–oxide: Kristallstruktur von $CaIrO_3$, *Zeits. anorganische allgemeine Chemie*, *336*, 17–23.

Santillán, J., S. Shim, G. Shen, and V. B. Prakapenka (2006), High-pressure phase transition in Mn_2O_3: Application for the crystal structure and preferred orientation or the $CaIrO_3$ type, *Geophy. Res. Lett.*, *33*, L15,307.

Sartbaeva, A., S. A. Wells, S. A. T. Redfern, R. W. Hinton, and S. J. B. Reed (2005), Ionic diffusion in quartz studies by transport measurements, SIMS and atomistic simulations, *J. Phy. Condens. Matter*, *17*, 1099–1112.

Shim, S. (2008), The post-perovskite transition, *Annu. Rev. Earth Planet. Sci.*, *36*, 536–599.

Shimizu, H., J. P. Poirier, and J. L. Le Mouël (2005), On crystallization at the inner core boundary, *Phys. Earth Planet. Inter.*, *151*, 37–51.

Shirako, Y., H. Satsukawa, X. X. Wang, J. J. Li, Y. F. Guo, M. Arai, K. Yamaura, M. Yoshida, H. Kojitani, T. Katsumata, Y Inaguma, K. Hiraki, T. Takahashi, and M. Akaogi (2011), Integer spin-chain antiferromagnetism of the 4d-oxide

$CaRuO_3$ with post-perovskite structure, *Phys. Rev. B*, *83*, 174,411.

Swainson, I. P. and A. L. Yonkeu (2007), Tilt and buckling modes, and acoustic anisotropy in layers with post-perovskite connectivity, *Am. Mineral.*, *92*, 748–752.

Tateno, S., K. Hirose, N. Sata, and Y. Ohishi (2010), Structural distortion of $CaSnO_3$ perovskite under pressure and the quenchable post-perovskite phase as a low-pressure analogue of $MgSiO_3$, *Phys. Earth Planet. Inter. 181*, 54–59.

Triana, S. A., D. S. Zimmerman, and D. P. Lathrop. (2012), Precessional states in a laboratory model of the Earth's core, *J. Geophy. Res.*, *117*, B04,103.

Vocadlo, L., and D. Alfè (2002), *Ab initio* melting curve of the fcc phase of aluminium, *Phys. Rev. B*, *65*, 214,105.

Vocadlo, L., J. Brodholt, D. Alfè, and G. D. Price (1999), The structure of iron under the conditions of the Earth's inner core, *Geophys. Res. Lett.*, *26*, 1231–1234.

Walte, N. P., F. Heidelbach, N. Miyajima, and D. Frost (2007), Texture development and TEM analysis of deformed $CaIrO_3$: Implications for the D″ layer at the core-mantle boundary, *Geophys. Res. Lett.*, *34*, L08,306.

Wang, Z. C., S. Karato, and K. Fujino (1993), High temperature creep of single crystal strontium titanate $(SrTiO_3)$—A contribution to creep systematics in perovskites, *Phys. Earth Planet. Inter.*, *79*, 299–312.

Wang, Z. C., C. Dupas-Bruzek, and S. Karato, (1999) High temperature creep of an orthorhombic perovskite–$YAlO_3$, *Phys. Earth Planet. Inter.*, *110*, 51–69.

Yakovlev, S., M. Avdeev, and M. Mezouar (2009), High-pressure structural behavior and equation of state of $NaZnF_3$, *J. Solid State Chem.*, *182*, 1545–1549.

Yamaura, K., Y. Shirako, H. Kojitani, M. Arai, D. P Young, M. Akaogi, M. Nakashima, T. Katsumata, Y. Inaguma, and E. Takayama-Muromachi (2009), Synthesis and magnetic and charge-transport properties of the correlated 4d post-perovskite $CaRhO_3$, *J. Am. Chem. Soc.*, *131*, 2722–2726.

Yamazaki, D., T. Yoshino, H. Ohfuji, J. Ando, and A. Yoneda (2006), Origin of seismic anisotropy in the D″ layer inferred from shear deformation experiments on postperovskite phase, *Earth Planet. Sci. Lett.*, *252*, 372–378.

Yoo, C. S., N. C. Holmes, M. Ross, D. J. Webb, and C. Pike (1993), Shock temperatures and melting of iron at Earth core conditions, *Phys. Rev. Lett.*, *70*, 3931–3934.

Yoo, C. S., J. Akella, A. J. Campbell, H. K. Mao, and R. J. Hemley (1995), Phase diagram of iron by in situ X-ray diffraction: Implications for Earth's core, *Science*, *270*, 1473–1475.

Zhao, S., Z. Lin, J. Zhang, H. Xu, G. Xia, and H. W. Green (2012), Does subducting lithosphere weaken as it enters the mantle? *Geophys. Res. Lett.*, *39*, L10,311.

Zhao, Y., D. J. Weidner, J. B. Parise, and D. E. Cox (1993), Thermal expansion and structural distortion of perovskite—Data for $NaMgF_3$ perovskite: (I), *Phys. Earth Planet. Inter.*, *76*, 1–16.

Zimmerman, D. S., S. A. Triana, and D. P. Lathrop (2011), Bi-stability in turbulent, rotating spherical Couette flow, *Phys. Fluids*, *23*, 065,104.

Part III
Physical Properties of Deep Interior

9

Ground Truth: Seismological Properties of the Core

George Helffrich

ABSTRACT

Earth's core is dominantly made of iron-nickel metal alloyed with perhaps 10 wt% lighter, lower atomic number elements. The seismologically observed structures of both the solid inner core and the liquid outer core are the ground truth for the material properties of solid and liquid iron and its alloying elements. The key properties, from a materials science standpoint, are density, seismic wave speeds, and anisotropy. Perhaps surprisingly, seismology's best constraints on density and wave speeds are over radial averages as opposed to values at a particular radius. For example, the density of Earth's center is known to no better than 0.5% when averaged over 400 km and 10% if averaged over 200 km. Radial variation of wave speeds in the outer core indicate departures from a state of adiabatic self-compression, indicating compositional or thermal stratification at the top or bottom of the core. The anisotropy of the inner core does not appear to be simple: It varies laterally at the same radial level as well as varying radially. Given that anisotropy can be either intrinsic or due to material texture, it is more problematic to interpret.

9.1. INTRODUCTION

Materials scientists and mineral physicists are apt to view the main geophysical properties of Earth—density, P- and S-wave speed—as tabulated values with uncertainties given at a specified confidence level. *Dziewonski and Anderson's* [1981] preliminary reference Earth model (PREM) is one such tabulation, but another is the AK135 model of seismic wave speeds [*Kennett et al.*, 1995] augmented with a companion density model [*Montagner and Kennett*, 1996]. Yet neither of these models provide pointwise uncertainties, and they deviate significantly in parts of Earth that one would assume are well characterized: for example, the crust at 15 km ($(\rho, V_P, V_S) = (2.60, 5.80, 3.20)_{PREM}$ vs. $(2.72, 5.80, 3.46)_{AK135}$) or the top of the inner core ($(12.76, 11.02, 3.50)_{PREM}$ vs. $(12.70, 11.04, 3.50)_{AK135}$). The

discrepancies stem from not only the type of data used in the model but also the nature of the observational constraints.

The core contains both solid and a dominantly iron and nickel liquid (~90 wt%) but with a minor amount of light elements that are extremely important for the evolution of the planet and the generation of its magnetic field. The inference by *Birch* [1952] that the core was not pure iron was based on the bulk sound speed (V_P in a liquid)–density systematics on shocked materials throughout the liquid core. Subsequently the pressure-density systematics in static compression (by, e.g., diamond anvil experiments) have been invoked in support of light alloying elements in the core [*Jephcoat and Olson*, 1987]; *Hirose et al.* [2013] is a survey containing more recent developments. This is, however, a different type of constraint due to the need for a model to account for the effect of temperature and its applicability over only the more limited pressure and temperature range of the inner core itself.

Earth-Life Science Institute (ELSI), Tokyo Institute of Technology, Tokyo, Japan

Deep Earth: Physics and Chemistry of the Lower Mantle and Core, Geophysical Monograph 217, First Edition.
Edited by Hidenori Terasaki and Rebecca A. Fischer.

The inner core's anisotropy—directionally dependent wave speed—is another property that might be used to infer the composition of the inner core to solve the light element identity puzzle. The anisotropy in the core was first suggested after an analysis of travel times of inner core seismic wave arrivals by *Poupinet et al.* [1983], and later confirmed by more detailed studies [*Morelli et al.*, 1986; *Woodhouse et al.*, 1986; *Creager*, 1992; *Tromp*, 1993; *Beghein and Trampert*, 2003]. The core's anisotropy seems to be cylindrically symmetric but its intensity varies quasi-hemispherically in the inner core [*Tanaka and Hamaguchi*, 1997]. The gross pattern of anisotropy is suggestive of wave speed variation governed by symmetries due to crystal structure. Thus minerals (or solid solutions between end-member minerals) crystallizing in a particular point group might be selected for candidacy based on symmetry principles, leading to further constraints on composition.

These four themes—uncertainty, model variability, the virtues of solid vs. liquid properties, and composition—will be the focus of this contribution. Their elaboration will hopefully help present and future investigators of the physical state and composition of the core understand the problems associated with certain types of data and to focus experimental work on avenues of research providing tight observational constraints on the core's character.

9.2. MODELS AND DATA

The most often-used model for geophysical properties (density and seismic wave speeds) is the PREM [*Dziewonski and Anderson*, 1981]. This model was derived from astronomical estimates of Earth's mass and principal moment of inertia, observed eigenperiods of the free oscillations of Earth excited by large earthquakes, by long-period surface wave dispersion (frequency-dependent velocities), and to a much lesser extent by regionally averaged body wave travel times whose trends with epicentral distance, rather than their values, are fit. Thus body wave travel times are not well reproduced owing to a constant offset from the actual travel times, but the shapes of the travel time curves with epicentral distance are.

In contrast, AK135 [*Kennett et al.*, 1995] is purely a body wave model. Taking the body wave derived V_p and V_S as given, the philosophy of *Montagner and Kennett* [1996] is to find a density (and radial attenuation) profile that reproduces the free oscillation periods, which, they showed, are sensitive to these parameters. Most of the normal-mode observations may be fit this way (with a few notable exceptions). The main outcome of this approach is an unusual density trend in the shallow mantle (where the density decreases with depth in a limited range) and in the mantle above the core. Figure 9.1 shows the density and wave speed profiles for the two models, showing how

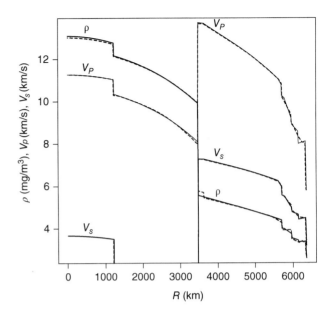

Figure 9.1 Radial dependence of density and P- and S-wave speeds in the PREM (solid) and AK135 model (dashed). Note in the shallow mantle PREM's pronounced low P-wave speeds and AK135's occasionally positive dp/dr.

emphases on different data lead to differing views of the deeper structure of Earth.

It is only the spheroidal normal modes that are sensitive to density; body wave travel times are not. Hence each model's solution for density must involve normal-mode data. A modelling philosophy that accords primary importance to the mode data will yield a density model different from one that emphasizes body wave travel times and that subsequently fits mode data.

Whatever the observable, its connection to structure—for example, the wave speed at a particular radius $v(r)$—is indirect. The relationship generally is an integral one over structure. For example, the dependence of a body wave travel time T on wave speed $v = v(r)$ is

$$T(p) = \int \left[\frac{r}{v}\right]^2 \frac{-1}{\left(r^2/v^2 - p^2\right)^{1/2}} \frac{dr}{r}, \qquad (9.1)$$

where r is radius and p is the horizontal slowness (a constant related to the take-off angle of the wave from the source to a particular receiver [*Shearer*, 2009]). While T can be observed with an uncertainty σ_T of around 10 parts per million (ppm), the uncertainty on the structural parameter σ_v is both model dependent [equation (9.1) does not depend linearly on v so is not easily solved for] and larger, around 10,000 ppm.

Masters and Gubbins [2003] showed that formal uncertainties for density, in particular, may only be applied to integral averages over a range of radii, essentially to the left-hand side of equation (9.1). The averaging lengths arise

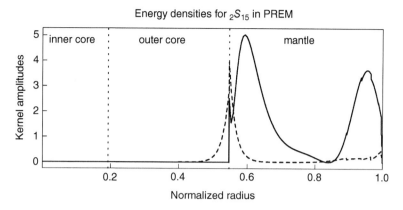

Figure 9.2 Shear energy density (solid) and compressional energy density (dashed) as a function of normalized radius (1 = surface) for the spheroidal mode $_2S_{15}$. The sensitivity peak for the compressional energy straddles the core-mantle boundary (CMB, dotted line at normalized radius ~0.55) and will therefore average structure in the outermost core as well as the lowermost mantle.

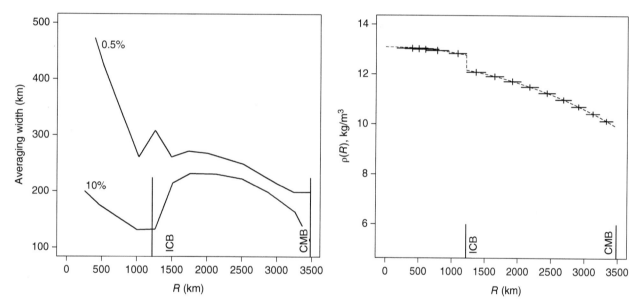

Figure 9.3 Averaging length for density in the core at the 0.5% and 10% uncertainty levels specified by *Masters and Gubbins* [2003] (left) and separated into independently resolved shells (right). On the right, the vertical tic marks show the 0.5% uncertainty at the center of the averaging width. The horizonal bars show the extent of the independent shells. Below the ICB, only two shells may be independently resolved, so average values are also plotted at 400, 500, and 600 km radius.

from the shapes of the eigenfunctions associated with each oscillation frequency (Figure 9.2). By selectively combining eigenfunctions of well-observed normal modes to forge a profile approximating a boxcar of a specific radial width, the average properties may be constrained well at a particular confidence level. *Masters and Gubbins* [2003] compiled these constraints, shown in Figure 9.3 for density; similar results are available for P-and S-wave speeds with correspondingly narrower averaging widths due to the plethora of modes (spheroidal and toroidal) that constrain them. The figure shows that

at the 0.5% uncertainty level the density is only known as an average over 400 km radius at Earth's center. On the other hand, at the 10% level, its average is known over 200 km. The figure also implies that at the 0.5% uncertainty level one only knows the average density over 300 km at the inner core boundary (ICB). Thus the precise density jump at the ICB is not known. In order to obtain this value at the 0.5% uncertainty level, one must seek an average density over a region confined to the inner core, extrapolate that average density from its midpoint to the ICB with some model, then seek an average density over a region confined to the outer core, extrapolate that from its midpoint to the ICB with some model, and then take the difference. Thus the jump will include the joint uncertainties of the individual estimates and the systematic uncertainty of the extrapolation method to either side of the ICB. The value of the density jump one effort provides is $820 \pm 180\,kg/m^3$ – a joint uncertainty of $\pm 20\%$ [*Masters and Gubbins*, 2003].

9.3. CORE WAVE SPEED MODELS

AK135 and the PREM are probably the most widely used geophysical models for Earth, but their philosophies, and thus their radial structure, are different (Figure 9.1). If we confine ourselves to the core the largest differences between them are at the top and bottom of the outer core, and within the inner core. These regions are, unfortunately, of the highest interest to contemporary investigators of deep Earth constitution and processes because their properties constrain the core's composition [*Deng et al.*, 2013; *Fiquet et al.*, 2009], the manner and rate of inner core growth [*Gubbins et al.*, 2008; *Nimmo et al.*, 2004], the energetics of driving the geodynamo [*Gomi et al.*, 2013; *Seagle et al.*, 2013; *Pozzo et al.*, 2013], and the thermal evolution of Earth [*Hirose et al.*, 2013; *Labrosse et al.*, 2007]. Thus, when choosing a model for comparison, one must have a clear justification for the preference for one over another. The PREM might be suitable for comparing mineral physics estimates of core material densities. AK135 might be suitable for comparing short-period seismic wave speeds in the inner core.

In addition to general differences between models, specific studies that target particular regions of the core show that the structure is not properly represented by either the PREM or AK135. Figure 9.4 shows a plot of wave speed differences from the PREM obtained from the results of a few detailed studies of the top and bottom of the outer core. All studies show a wave speed reduction near the top and bottom of the outer core relative to PREM, but not as much as AK135 (top) or IASP91 (bottom). What to attribute these reductions to is as yet unclear. Possibilities at the top of the core include thermal or chemical stratification [*Kaneshima and Helffrich*, 2013; *Helffrich and Kaneshima*, 2010; *Hirose et al.*, 2013; *Gubbins and Davies*, 2013], core-mantle reaction [*Komabayashi*, 2014; *Buffett and Seagle*, 2010], or relict layering from the time of Earth's accretion [*Helffrich*, 2014]. Thus deviations of observations from reference models yield key insights driving contemporary research efforts rather than simply representing differences in model-building philosophy.

9.4. ANISOTROPY AND INNER CORE

In one notable area, Earth's structural details are not represented well by any radial Earth model: the inner core. The second decade of the third millenium heralded a plethora of studies on various aspects of the inner core's structure: wave speed and attenuation (the tendency to dampen propagating waves). The picture, still blurry in some aspects but of the same basic form suggested by

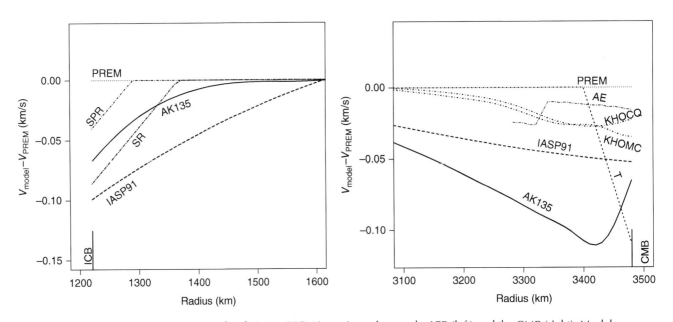

Figure 9.4 Outer core wave speeds relative to PREM in regions close to the ICB (left) and the CMB (right). Models shown are AK135, IASP91 [*Kennett and Engdahl,* 1991], SR [*Souriau and Roudil,* 1995], and SPR [*Ohtaki et al.,* 2012], AE [*Alexandrakis and Eaton,* 2010], T (model 1 of *Tanaka* [2007]), and KHOCQ and KHOMC [*Helffrich and Kaneshima,* 2010; *Kaneshima and Helffrich,* 2013].

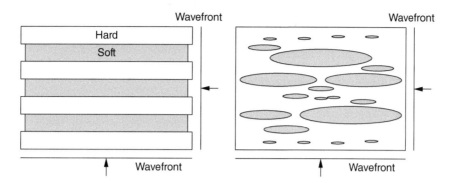

Figure 9.5 How anisotropy arises due to layering or inclusion presence. A medium made from alternating layers of hard and soft material (left) responds differently to a wavefront compressing the medium from the right and from below. The response to the leftward-moving wavefront is to transmit stress through the hard material, thus yielding faster wave speeds. The response to the upward wavefront is through the unavoidable soft material, leading to slower wave speeds. A medium with a fabric developed in soft inclusions (right) sees similar directional response to the impinging wavefronts.

early results, is of a cylindrically symmetric medium with P-wave speeds parallel to the spin axis ~3% faster than equatorial speeds [*Creager*, 1992]. However the ~1.4% hemispheric difference in wave speeds (east is 0.5% fast, west is 0.9% slow according to *Tanaka* [2012]) does not appear to be exactly hemispheric [*Irving and Deuss*, 2011a] nor does the anisotropy appear to be uniform across hemispheres [*Lythgoe et al.*, 2014]. These could represent variations in the fabric of the core as it crystallized or as it evolved with time, and if borne out by further studies the details could become robust modeling constraints for core processes.

Anisotropy, however, is fiendishly difficult to interpret because it can arise in many different ways. One mentioned earlier is intrinsic anisotropy due to the crystal structure of the solid. Other ways are compositional layering or inclusions trapped in the solid (Figure 9.5), called shape-preferred orientation (SPO). Attenuation might be related to the presence of inclusions [*Lythgoe et al.*, 2014; *Cormier and Li*, 2002], which could provide a good discriminant between anisotropy due to layering or inclusions. However, according to *Tanaka* [2012], there seems to be no clear link between the speed and the attenuation promised by some earlier studies. Thus one will be forced to invoke separate physical processes to explain wave speed variations and attenuation.

Anisotropy's clearest imprint is upon shear wave arrivals. Birefringence in the anisotropic medium imparts different wave speeds to different shear wave polarizations, leading to two distinct arrivals [*Silver and Chan*, 1991]. We already know the gross anisotropic structure of the inner core, so in principle these observations should provide architectural constraints. They already do, in their effects on normal-mode eigenfrequencies [*Woodhouse et al.*, 1986; *Tromp*, 1993; *Beghein and Trampert*, 2003]. The most recent data analyses suggest that the outermost 275 km of the inner

core is isotropic for shear waves [*Irving and Deuss*, 2011b]. Because the directional dependence of P-wave speeds should arise from the same anisotropic structure (Figure 9.5), P and S anisotropies are linked. The P anisotropy is best shown in the range of 100–500 km depth in the inner core [*Creager*, 1992]. Consequently, P anisotropy must be stronger than ~3% if confined to deeper than 275 km. This shows that there is clearly scope to incorporate body wave constraints and modern normal-mode data compilations [*Deuss et al.*, 2013] into inversions for anisotropy.

Directly observing shear waves in the inner core is difficult since they cannot pass through the liquid outer core, however. An intriguing possibility is the use of PKJKP, the inner core shear wave arrival, to constrain anisotropy [*Wookey and Helffrich*, 2008]. Observations of PKJKP are rare and technically challenging, however [*Shearer et al.*, 2011], and might never yield sufficient coverage of the inner core to provide robust constraints.

9.5. LIQUID OUTER CORE

The liquid portion of the core comprises 65% of its radius and 95% of its mass. While it seems to add needless complications to study liquid properties rather than solids—whose structures are fixed rather than fluid—the liquid portion's properties actually dominate the behavior of the core and offer a much broader pressure and temperature range for investigation. Due to the heterogeneity of the solid inner core and the problems interpreting its anisotropy, the observational liquid data provide a more straightforward target for physical property investigations.

The physical parameters governing the core's properties are bulk modulus K and density ρ; $V_P^2 = K/\rho$. *Birch* [1952] used the core's V_P and ρ to infer the presence of light alloying elements. Because the light element concentration will be on the order of 1–10%, reasonably simple liquid models may be applied to gauge deviations from a reference liquid state, presumably pure liquid Fe (or Fe+Ni), approximated by PREM (Figure 9.4). Thus one could parameterize deviations δV_p from a liquid representing PREM's outer core density and wave speed as

$$\delta V_P = s \times C(r), \qquad (9.2)$$

where C, a solution to the spherical diffusion equation, represents the radius-dependent concentration excess of a light element over the PREM and s is a scale factor converting the light element excess to wave speed. In this way *Helffrich* [2014] obtained empirical estimates of diffusion in liquid metal at core conditions and modeled light element excess at the top of the outer core [*Helffrich*, 2015]. It remains a challenge to determine the precise mixture of iron + nickel + light elements that reproduce the PREM, but it allows progress to be made on two fronts simultaneously. The light element cocktail [the $\rho(r)$ and $V_p(r)$ that match, say, the PREM] may be investigated separately [*Helffrich*, 2015] from the process that might act at the ICB or CMB to perturb wave speeds there (the deviations from the PREM; e.g., *Buffett and Seagle* [2010]).

9.6. CONCLUSIONS

Planetologists and mineral physicists seek geophysical models for the properties of planets in order to investigate their composition, mineralogy, and thermochemical evolution. The nature of the process involved in constructing a geophysical model involves subjective steps reflecting the builder's judgment. Thus a user of a model should be aware of the philosophy underlying that model before applying it to a particular purpose.

The nature of any geophysical observable is an integral over structure. The observables include seismic wave travel time, mass, and moment of inertia. While an observable may be measured to high precision, mapping the observed quantity back into structure increases its uncertainty by a few orders of magnitude. Only the integrated property is endowed with a well-defined uncertainty. This frustrates comparisons, say, of the experimentally observed density difference between solid and liquid iron to constrain the composition of the core because there is no direct comparison to a geophysical observable. Rather, comparisons to the total mass of the inner core (an integral of density structure over radius) or the average wave speed in some depth range (another integral of velocity structure over radius) will lead to better constraints on models of planetary constitution.

REFERENCES

Alexandrakis, C., and D. W. Eaton (2010), Precise seismic wave velocity atop Earth's core: No evidence for outer-core stratification, *Phys. Earth Planet. Inter.*, *180*, doi:10.1016/j.pepi.2010.02.011.

Beghein, C., and J. Trampert (2003) Robust normal mode constraints on inner-core anisotropy from model space search, *Science*, *299*, 552–555.

Birch, F. (1952), Elasticity and constitution of the Earth's interior, *J. Geophys. Res.*, *57*, 227–286.

Buffett, B. A., and C. T. Seagle (2010), Stratification of the top of the core due to chemical interactions with the mantle, *J. Geophys. Res.*, *115*, doi:10.1029/2009JB006751.

Cormier, V. F., and X. Li (2002), Frequency-dependent seismic attenuation in the inner core 2. A scattering and fabric interpretation, *J. Geophys. Res.*, *107*, doi:10.1029/2002JB001796.

Creager, K. C. (1992), Anisotropy of the inner core from differential travel-times of the phases PKP and PKIKP, *Nature*, *356*, 309–314.

Deng, L., Y.-W. Fei, X. Liu, Z. Gong, and A. Shahar (2013), Effect of carbon, sulfur and silicon on iron melting at high pressure: Implications for composition and evolution of the planetary terrestrial cores, *Geochim. Cosmochim. Acta*, *114*, 220–223.

Deuss, A., J. Ritsema, and H. van Heijst (2013), A new catalogue of normal-mode splitting function measurements up to 10 mHz, *Geophys. J. Int.*, *193*, 920–937.

Dziewonski, A., and D. L. Anderson (1981), Preliminary reference Earth model, *Phys. Earth Planet. Inter.*, *25*, 297–356.

Fiquet, G., J. Badro, E. Gregoryanz, Y.-W. Fei, and F. Ocelli (2009), Sound velocity in iron carbide (Fe3C) at high pressure: Implications for the carbon content of the Earth's inner core, *Phys. Earth Planet. Inter.*, *172*, 125–129.

Gomi, H., K. Ohta, K. Hirose, S. Labrosse, R. Caracas, M. J. Verstraete, and J. W. Hernlund (2013), The high conductivity of iron and thermal evolution of the Earth's core, *Phys. Earth Planet. Inter.*, *224*, 88–103.

Gubbins, D., and C. Davies (2013), The stratified layer at the core-mantle boundary caused by barodiffusion of oxygen, sulphur and silicon, *Phys. Earth Planet. Inter.*, *215*, 21–28.

Gubbins, D., G. Masters, and F. Nimmo (2008), A thermo-chemical boundary layer at the base of Earth's outer core and independent estimate of core heat flux, *Geophys. J. Int.*, *174*, 1007–1018.

Helffrich, G. (2014), Outer core compositional layering and constraints on core liquid transport properties, *Earth Planet. Sci. Lett.*, *391*, 256–262.

Helffrich, G. (2015), The hard sphere view of the outer core, *Earth, Planet., Space*, *67*, 73–84.

Helffrich, G., and S. Kaneshima (2010), Outer-core compositional stratification from observed core wave speed profiles, *Nature*, *468*, 807–810.

Hirose, K., S. Labrosse, and J. Hernlund (2013), Composition and state of the core, *Annu. Rev. Earth Planet. Sci.*, *41*, 657–691.

Irving, J. C. E., and A. Deuss (2011a), Hemispherical structure in inner core velocity anisotropy, *J. Geophys. Res.*, *116*, doi:10.1029/2010JB007942.

Irving, J. C. E., and A. Deuss (2011b), Stratified anisotropic structure at the top of Earth's inner core: A normal mode study, *Phys. Earth Planet. Inter., 186,* 59–69.

Jephcoat, A., and P. Olson (1987), Is the inner core of the Earth pure iron? *Nature, 325,* 332–335.

Kaneshima, S., and G. Helffrich (2013), Vp structure of the outermost core derived from analyzing large scale array data of SmKS waves, *Geophys. J. Int., 93,* 1537–1555.

Kennett, B. L. N., and E. R. Engdahl (1991), Traveltimes for global earthquake location and phase identification, *Geophys. J. Int., 105,* 429–465.

Kennett, B. L. N., E. R. Engdahl, and R. Buland (1995), Constraints on seismic velocities in the Earth from traveltimes, *Geophys. J. Int., 122,* 108–124.

Komabayashi, T. (2014), Thermodynamics of melting relations in the system Fe-FeO at high pressure: Implications for oxygen in the Earth's core, *J. Geophys. Res., 119,* 4164–4177.

Labrosse, S., J. W. Hernlund, and N. Coltice (2007), A crystallizing dense magma ocean at the base of the Earth's mantle, *Nature, 450,* 866–869.

Lythgoe, K. H., A. Deuss, J. F. Rudge, and J. A. Neufeld (2014), Earth's inner core: Innermost inner core or hemispherical variations? *Earth Planet. Sci. Lett., 385,* 181–189.

Masters, G., and D. Gubbins (2003), On the resolution of density within the Earth, *Phys. Earth Planet. Inter., 140,* 159–167.

Montagner, J.-P., and B. L. N. Kennett (1996), How to reconcile body-wave and normal-mode reference earth models, *Geophys. J. Int., 125,* 229–248.

Morelli, A., A. M. Dziewonski, and J. H. Woodhouse (1986), Anisotropy of the inner core inferred from pkikp travel-times, *Geophys. Res. Lett., 13,* 1545–1548.

Nimmo, F., G. D. Price, J. Brodholt, and D. Gubbins (2004), The influence of potassium on core and geodynamo evolution, *Geophys. J. Int., 156,* 363–376.

Ohtaki, T., S. Kaneshima, and K. Kanjo (2012), Seismic structure near the inner core boundary in the south polar region, *J. Geophys. Res., 117,* doi:10.1029/2011JB008, 717.

Poupinet, G., R. Pillet, and A. Souriau (1983), Possible heterogeneity of the Earth's core deduced from PKIKP travel times, *Nature, 305,* 204–206.

Pozzo, M., C. Davies, D. Gubbins, and D. Alfè (2013), Transport properties for liquid silicon-oxygen-iron mixtures at Earth's core conditions, *Phys. Rev. B, 87,* doi:10.1103/PhysRevB.87.014, 110.

Seagle, C., E. Cottrell, Y.-W. Fei, R. Hummer, and V. B. Prakapenka (2013), Electrical and thermal transport properties of iron and iron-silicon alloy at high pressure, *Geophys. Res. Lett., 40,* 5377–5381.

Shearer, P. M. (2009), *Introduction to Seismology,* Cambridge Univ. Press, Cambridge.

Shearer, P. M., C. A. Rychert, and Q. Liu (2011), On the visibility of the inner-core shear wave phase *PKJKP* at long periods, *Geophys. J. Int., 285,* 1379–1383.

Silver, P. G., and W. W. Chan (1991), Shear-wave splitting and subcontinental mantle deformation, *J. Geophys. Res., 96,* 16,429–16,454.

Souriau, A., and P. Roudil (1995), Attenuation in the uppermost inner-core from broad-band Geoscope PKP data, *Geophys. J. Int., 123,* 572–587.

Tanaka, S. (2007), Possibility of a low P-wave velocity layer in the outermost core from global SmKS waveforms, *Earth Planet. Sci. Lett., 259,* 486–499.

Tanaka, S. (2012), Depth extent of hemispherical inner core from PKP(DF) and PKP(Cdiff) for equatorial paths, *Phys. Earth Planet. Inter., 210–211,* 50–62.

Tanaka, S., and H. Hamaguchi (1997), Degree one heterogeneity and hemispherical variation of anisotropy in the inner core from PKP(BC)–PKP(DF) times, *J. Geophys. Res., 102,* 2925–2938.

Tromp, J. (1993), Support for anisotropy of the Earths inner-core from free oscillations, *Nature, 366,* 678–681.

Woodhouse, J. H., D. Giardini, and X. D. Li (1986), Evidence for inner core anisotropy from free oscillations, *Geophys. Res. Lett., 13,* 1549–1552.

Wookey, J., and G. Helffrich (2008), Inner-core shear-wave anisotropy and texture from an observation of PKJKP waves, *Nature, 454,* 873–877.

10

Physical Properties of the Inner Core

Daniele Antonangeli

ABSTRACT

In this chapter we will address some of the physical properties of solid iron and iron alloys at high pressure and high temperature directly relevant for the interpretations of seismic observations and for the modeling of Earth's inner core. Conceivably, density is the most important physical quantity needed for core modeling. The density difference between pure iron and seismic models is a very strong indication of the presence of light elements alloyed to iron in the inner core. Compressional and shear sound velocities are crucial as two of the few quantities directly comparable with seismic observations. Comparison of velocity vs. density systematics with inner core models provides stringent constraints on the nature and abundance of light elements. The detailed knowledge of single-crystal elasticity, together with the developed texture, is needed to understand seismic anisotropy and hence inner core structures (e.g., hemisphericity, possible existence of innermost core). Thermal conductivity is a central parameter for modeling the geodynamo and is essential to constrain the heat budget and thermal history of the core and hence its secular cooling, its age, and the crystallization rate.

Earth's inner core is the innermost portion of our planet, spanning from 5150 to 6370 km depth (corresponding to pressures between 330 and 364 GPa). Being that the direct sampling is still impossible (even though such a science fiction scenario has been envisaged [*Stevenson*, 2003]), all the available information comes from remote sensing. In this context seismology plays a primary role, providing information on the evolution with depth of compressional and shear sound velocity (respectively V_p and V_s) and density (ρ). Nowadays core seismology is so advanced not only to allow the construction of one-dimensional, radially averaged models, such as PREM (preliminary reference Earth model, [*Dziewonsky and Anderson*, 1981]), but also to permit discussion of more subtle aspects, such as lateral variation in seismic wave speed [*Niu and Wen*, 2001], and elastic anisotropy (seismic waves travel faster along the polar path than in the equatorial path [*Morelli et al.*, 1986]) or its hemispherical variation [*Deuss et al.*, 2010] (refer to Chapter 9 for more details). All these observations point out the complex nature of the inner core that, to be interpreted, calls for knowledge of the physical properties of the constituent material.

Different lines of evidence, from the analysis of meteorite compositions and Earth's differentiation models, to comparison of shock compression measurements with seismic observations, put forward the notion that iron (Fe) is main constituent of the Earth's core. Cosmochemical arguments suggest that few weight percent (wt %) nickel is likely alloyed to iron, and, since Birch's pioneering work [*Birch*, 1952] it is clear that elements lighter than Fe-Ni alloys are present in the liquid outer core. The density mismatch in the case of the solid inner core is less than for the outer core, and if by now the fact that pure Fe is too dense with respect to seismic models is well established (see Figure 10.1), the accurate

Institut de Minéralogie, de Physique des Matériaux, et de Cosmochimie (IMPMC), UMR CNRS 7590, Sorbonne Universités – UPMC, Paris, France

Deep Earth: Physics and Chemistry of the Lower Mantle and Core, Geophysical Monograph 217, First Edition.
Edited by Hidenori Terasaki and Rebecca A. Fischer.

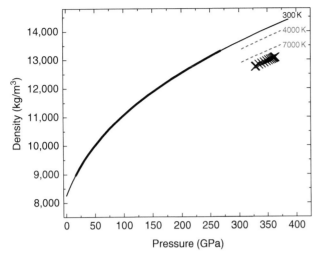

Figure 10.1 Density-pressure relationship for pure hcp-Fe in comparison with PREM. The thick solid line is the 3rd order Birch-Murnaghan fit to experimental data at high pressure and ambient temperature [*Sakai et al.*, 2014], extrapolated as thin solid line. High temperature effects (red dashed lines) are estimated applying to the 300 K data the relative correction obtained by ab initio modeling [*Dewaele et al.*, 2006]. PREM values for the Earth's inner core are plotted as crosses.

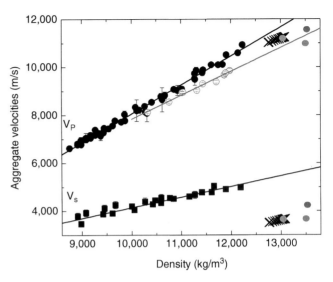

Figure 10.2 Compressional (V_p) and shear (V_s) sound velocity for pure solid hcp-Fe as a function of density. PREM is shown as crosses. Filled black points are selected experimental data obtained under static compression at ambient temperature (for an extensive discussion please refer to [*Antonangeli and Ohtani*, 2015]): solid circles - V_p [*Mao et al.*, 1998; *Crowhurst et al.*, 2004; *Antonangeli et al.*, 2004a, 2012; *Ohtani et al.*, 2013; *Decremps et al.*, 2014]; solid squares - V_s [*Mao et al.*, 1998; *Crowhurst et al.*, 2004; *Gleason et al.*, 2013; *Murphy et al.*, 2013]; open circles are results from shock-wave measurements along the Hugoniot (not reduced to 300 K) [*Brown and McQueen*, 1986]. Solid lines are linear extrapolations of the experimental results. Results of ab initio calculations at inner core density and 6700 K (but P~250 GPa) are plotted as orange dots [*Sha and Cohen*, 2010a], while calculations at 360 GPa and 7000 K (7250) are plotted as blue (dark cyan) dots [*Martorell et al.*, 2013a].

quantification of this density deficit is still the object of active research.

More generally, the most stringent constraints on the nature and constitution of Earth's core are provided by comparison of seismic observations with experimental determination, or ab initio calculations, of the physical properties of candidate materials (Fe and Fe alloys) at relevant pressure (*P*) and temperature (*T*) conditions. Inner core pressures are relatively well constrained and range from ~330 GPa at the inner core–outer core boundary (ICB) to ~364 GPa at the center of Earth. Conversely, temperature is still largely debated, with proposed values going from 4000 to 7000 K at ICB (see Chapters 1 and 2 for more extensive discussion). Recent years witnessed an impressive improvement in static high pressure, high-temperature experimentation as well as ab initio calculations (see *Hirose et al.* [2013] for a recent review). While current experimental capabilities allow to cover the *P-T* range of the entire core [*Tateno et al.*, 2010], information is basically limited to the phase diagram, and even the density determination at such extreme thermodynamic conditions is still a challenge. Only recently experimental work got to a consensus [e.g., *Tateno et al.*, 2010, 2012; *Sakai et al.*, 2011], advocating that the hexagonal closed-packed (hcp) crystalline structure is likely adopted by Fe and relevant alloys at the *P-T* conditions of the inner core (please refer to Chapter 5 for an extensive discussion).

Among the physical properties of interest, certainly sound velocity and density are of critical importance as some of the few quantities that can be directly compared with seismic observations. Selected experimental and computational results obtained for pure solid Fe are illustrated in Figure 10.2. V_p measured for hcp Fe under static compression at ambient temperature extrapolates somewhat higher than PREM, while the extrapolation of shock data obtained on solid Fe is below PREM, thus suggesting that anharmonic effects at high temperature might be significant. Indeed, within the limit of a quasi-harmonic approximation, the phonon energies (and hence the sound velocities) are expected to only depend upon volume (or density) irrespective of the specific *P-T* conditions. Eventual temperature effects at constant density are thus due to nonharmonic terms in the interatomic potential, and a quantitative assessment beyond a quasi-harmonic approach (the linear V_p-ρ relation often referred to as "Birch's law") is still the object of both experimental and theoretical study [e.g. *Lin et al.*, 2005; *Antonangeli et al.*, 2008, 2012; *Sha and Cohen*, 2010a, 2010b, *Ohtani et al.*, 2013; *Martorell et al.*, 2013a]. On qualitative grounds, high-temperature effects are expected to be more relevant for V_S than V_P and potentially become very relevant close to the melting. It is not accidental then that the difference between extrapolated measurements and PREM is much larger for V_S. However, recent calculations suggest that velocities of pure Fe might become comparable with PREM for temperatures approaching 7000 K [*Sha and Cohen*, 2010a; *Martorell et al.*, 2013a]. Light elements are still required, though, due to pressure/density mismatch. Actually, pure Fe can attain PREM values for densities and velocities, but for $P \sim 250$ GPa, so at pressures much lower than those of the core [*Sha and Cohen*, 2010a]; alternatively, core velocities can be achieved at $P \sim 360$ GPa, considering strong premelting effects [*Martorell et al.*, 2013a], but for densities much higher than PREM. Closely related to sound velocities, the elastic properties of hcp Fe, together with lattice preferred orientations, are needed to model the elastic response of the aggregate and hence interpret the observed core seismic anisotropy.

Given these premises, current research activity focuses on extending the pressure range of direct velocity determinations (currently limited to ~170 GPa [*Ohtani et al.*, 2013]), on assessing anharmonic effects at core temperatures (so far almost exclusively investigated by calculations, with experiments limited to 1700 K [*Lin et al.*, 2005]) and on quantifying effects of inclusions, be these nickel and/or light elements.

Arguments based on core differentiation models, cosmochemical abundances, chemical affinity, and volatility suggest silicon, oxygen, sulfur, carbon, and hydrogen as most likely light elements alloyed to Fe in Earth's core. Systematic X-ray diffraction studies of the pressure-volume (*P-V*) relation on Fe alloy end members [e.g., *Sata et al.*, 2010], in comparison with pure-Fe and seismic models, allow estimating effects of light element inclusion on density and compressibility, thus providing basic constraints on the potential abundance of each element. The physical properties of end members can be obtained as well by ab initio calculations [e.g., *Alfè et al.*, 1999; *Caracas and Wentzcovitch*, 2004; *Ono et al.*, 2008]. Stringent tests on these predictions come from experiments on samples of more realistic compositions [e.g., *Hirao et al.*, 2004; *Sakai et al.*, 2014] and ternary alloys [e.g., *Asanuma et al.*, 2011; *Sakai et al.*, 2012], whose treatment by ab initio methods is still a challenge. Addition of high temperature is a further, but necessary, experimental complication, and very few results exist on *P-V-T* relations on Fe alloys at megabar pressures [e.g., *Seagle et al.*, 2006, *Fischer et al.*, 2014]. The quantitative assessment of light element content on the basis of PREM density matching is currently limited by the necessary extrapolations (which in turn depend upon on the choice of the equation of state; see *Stacey and Davis* [2004] for a review on the subject) and

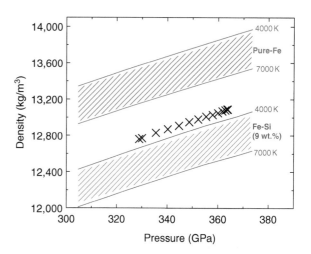

Figure 10.3 Extrapolations to core pressures of the density-pressure relationship for pure hcp-Fe (in red, after *Sakai et al.*, 2014 and *Dewaele et al.*, 2006) and Fe-Si alloy with 9 wt.% Si (in blue, after *Fischer et al.*, 2014). Dashed areas enclose density range for temperature at ICB spanning between 4000 and 7000 K. PREM values for the Earth's inner core are plotted as crosses.

from the lack of a precise knowledge of the inner core temperature. There is indeed an evident trade-off between the amount of light elements and inner core temperature, with the less impurity needed the higher is the temperature. While openly not conclusive, many studies put forward the possibility of silicon being the most abundant light element in the inner core, with a maximum amount estimated by ideal mixing models in 4–8 wt % [*Lin et al.*, 2003a; *Asanuma et al.*, 2011; *Fischer et al.*, 2014] (see Figure 10.3).

Density, however, is less precisely determined by seismic observations than sound velocities. Thus, tighter constraints on inner core compositional models can be obtained considering the velocity vs. density systematics of candidate alloys in comparison with seismic models. Sound velocity measurement of opaque, metallic samples under static compression is an experimental challenge, and very few results exist at simultaneous high pressure and high temperature. In fact, it is only with the advent of third-generation synchrotron sources and the development of inelastic scattering techniques in conjunction with the use of diamond anvil cells that systematic experiments on iron and iron alloys at megabar pressures became possible. Among the most used techniques we can mention nuclear resonant inelastic X-ray scattering (NRIXS) [*Mao et al.*, 2001] and momentum-resolved inelastic X-ray scattering (IXS) [*Fiquet et al.*, 2001]. Systematic studies on polycrystalline samples compressed in diamond anvil cells have been conducted coupling IXS for compressional velocity measurements with X-ray diffraction for density determination. Early experiments focused on end members: FeSi, FeO, FeS, FeS_2, Fe_3C, and FeH [e.g., *Fiquet et al.*, 2008; *Shibazaki et al.*, 2012]. While none of these alloys proven to be compatible with seismic observations, with pure Fe still the closest match, these data sets can be used to estimate V_p and ρ of a model alloy considering an ideal binary mixing model. The so-obtained inner core compositional model argues in favor of silicon as the most likely major light element alloyed to iron to an amount of 2–3 wt % [*Badro et al.*, 2007].

This type of analysis provided very important indications, but the resulting core models [e.g., *Badro et al.*, 2007] are based on a series of approximations: (i) the sound velocities were measured over a quite limited pressure range (up to ~100 GPa vs. 330–360 GPa of inner core) and linearly extrapolated with density; (ii) the effects of temperature were not explicitly considered, within the limit of a quasi-harmonic approximation, where sound velocities were expected to linearly depend upon density, independently of specific pressure and temperature conditions; (iii) the ideal mixing behavior to estimate the velocity as a weighted average of velocities measured on end members has never been tested; (iv) the effect of nickel inclusion was considered negligible; and (v) the comparison with seismic models was limited to the only V_p and ρ, without considering V_S.

To validate and refine these predictions we can then look at both compressional and shear sound velocity determinations, considering experiments and calculations on alloys of more realistic compositions (see Figure 10.4). Parallel velocities and density measurements up to 108 GPa on a ternary Fe-Ni-Si alloy, with 4.3 wt % Ni and 3.7 wt % Si, once extrapolated to core density and corrected for anharmonic effects at high temperature estimated by ab initio calculations, suggest that main seismic observables, V_P, V_S and ρ can be matched by an Fe alloy with 4–5 wt % Ni and about 2 wt % Si for an inner core temperature of 5000 K [*Antonangeli et al.*, 2010]. The amount of light elements is however tightly linked to temperature, with light elements generally increasing sound velocity at constant density and anharmonic high-temperature effects reducing sound velocity. Thus, a significant effort is currently devoted to extending sound velocity measurements at simultaneous high-pressure and high-temperature conditions, but so far IXS measurements are limited to 1100 K [*Antonangeli et al.*, 2012] and NRIXS measurements to 1700 K [*Lin et al.*, 2005] and results can be hardly extended to core temperatures.

Summarizing, on the basis of comparison between experiments and calculations on Fe and Fe alloys and seismic models, inner core V_P, V_S and ρ can be plausibly accounted for when considering an alloy in the Fe-Ni-Si ternary system. The exact amount of Ni is poorly constrained solely on the basis of elasticity, but inclusion around 4–5 wt % as proposed by cosmochemical and geochemical models [see e.g., *Poirier*, 1994] seems

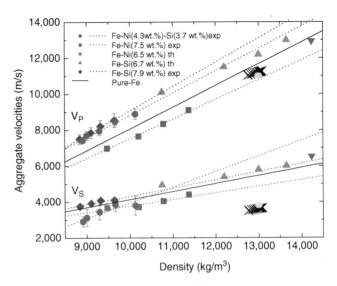

Figure 10.4 Compressional (V_p) and shear (V_S) sound velocity for Fe-Ni [measurements: *Lin et al.*, 2003b; calculations: *Martorell et al.*, 2013b], Fe-Si [measurements: *Lin et al.*, 2003b; calculations: *Tsuchiya and Fujibuchi*, 2009] and Fe-Ni-Si [measurements: *Antonangeli et al.*, 2010] alloys with hcp structure as a function of density. PREM is shown as crosses. Dotted lines are linear extrapolations of the experimental results. The solid line is a reference for pure Fe (see Figure 10.2). Grey dots are estimations for a Fe-Ni-Si alloy with 4–5 wt.% Ni and 2 wt.% Si at 5000 K [*Antonangeli et al.*, 2010].

reasonable. Tighter constraints can be placed in the maximum amount of Si, which, even when considering the trade-off between light element inclusion and core temperatures, is very unlikely to exceed 8 wt % (Fe-Si alloys with more than 8 wt % Si seem to have ρ too low and V_p too high with respect to seismic models as illustrated in Figures 10.3 and 10.4), with an expected content around 3–4 wt % at 6000–7000 K, assuming silicon the only light element in the inner core.

So far we only considered the average properties of an ideal randomly oriented aggregate. If this might be enough to constrain the bulk inner core composition, the seismic observation of lateral variation in sound velocity propagation and the seismic anisotropy call for more elaborate models, which in turn require the knowledge of (i) the single crystalline elastic tensor of core material; (ii) the texture of the aggregate, and (iii) the geodynamical mechanism producing the preferential alignment. In fact, since its original discovery [*Morelli et al.*, 1986; *Whoodhouse et al.*, 1986] seismic anisotropy was a major motivation for the study of elastic anisotropy in Fe and Fe alloys and prompted studies on the inner core dynamics proposing different generating mechanisms and textural models to explain the seismological observations. Suggestions include texturing during solidification of the inner core [*Bergman*, 1997], plastic flow induced by the magnetic field in the outer core as it solidifies at the ICB [*Karato*, 1993], large-scale convective flow in the inner core [*Yoshida et al.*, 1996; *Jeanloz and Wenk*, 1988], gravitational coupling with the mantle [*Buffet*, 1997], electromagnetic shear stresses [*Karato*, 1999; *Buffet and Wenk*, 2001], or inner core translation induced by simultaneous crystallization and melting at opposite hemispheres [*Monnereau et al.*, 2010; *Alboussiere et al.*, 2010]. In the following we will mainly focus on the single crystalline elastic tensor of core material, mostly considering the case of pure Fe in the hcp structure, and invite the reader to see Chapter 7 for more details on deformation and deformation mechanisms.

As mentioned, sound velocity measurements on metals at very high pressure and temperature are already critical in the simpler case of polycrystalline samples, and experiments on single crystals are very limited in number as well as in P-T range (~40 GPa and 1000 K) [*Antonangeli et al.*, 2004b, 2008]. From an experimental standpoint the case of iron is further complicated by the phase transition from the body-centered-cubic (bcc) to the hcp structure that iron undergoes around 13 GPa at ambient temperature, which so far has prevented experimentation on single crystals above this pressure. Therefore, all experimental investigations have been conducted on polycrystalline samples. While, in theory, this does not preclude one from extracting single-crystal tensor properties, in practice this has proven to be a very difficult inverse problem. Pioneering results obtained on Fe up to 220 GPa [*Mao et al.*, 1998] were later recognized to be incorrect [*Antonangeli et al.*, 2006] due to complications in data analysis arising from texture developed during compression, plasticity, and the approximations in the models relating radial X-ray diffraction to elasticity [*Merkel et al.*, 2006, 2009]. The developed lattice preferred orientations and the driving deformation mechanisms have been experimentally probed and modeled [e.g., *Wenk et al.*, 2000; *Merkel et al.*, 2005], but how to overcome current limitations in extracting single crystal elastic properties is still the subject of intense research activity [e.g., *Matthies et al.*, 2001; *Merkel et al.*, 2009]. The actual shape of the elastic anisotropy in hcp Fe (i.e., the dependence of the velocities on the crystal orientation) has yet to be measured *as today*, and has been only indirectly assessed by combined multitechnique approaches [*Antonangeli et al.*, 2006; *Mao et al.*, 2008]. Nonetheless, sound velocity measurements in highly textured hcp Fe show a V_p anisotropy as large as 3% already at 112 GPa [*Antonangeli et al.*, 2004a], a magnitude comparable to the cylindrical anisotropy reported by seismic observations, arguing against the need of an almost perfect alignment in the inner core invoked by early ab initio calculations [*Stixrude and Cohen*, 1995].

Ab initio methods are nevertheless the only way currently available to determine the full elastic tensor of hcp Fe.

The validity of athermal calculations of the single-crystal elastic moduli at high pressure has been proven by the overall good comparison between theoretical results and measurements on hcp crystal analogues to iron [*Steinle-Neumann et al.*, 1999; *Antonangeli et al.*, 2004b]. Indeed, calculations well reproduce the measured shape of the anisotropy, which seems to be a general feature of hcp transition metals at high pressure and is characterized by a sigmoidal form, with velocity faster along the *c* axis than in the basal plane. Calculations have however the tendency to underestimate the magnitude of the elastic anisotropy. Extension to simultaneous high-*P* and high-*T* conditions is not trivial, and the effect of temperature on the shape and magnitude of the anisotropy is still debated. Early calculations argued for an inversion in the sense of the anisotropy, with compressional acoustic wave propagation faster in the basal plane than along the *c* axis [*Steinle-Neumann et al.*, 2001] as a consequence of a significant increased c/a axial ratio at high temperature. Following calculations disproved these conclusions, showing that c/a varies only weakly with temperature [*Gannarelli et al.*, 2005]. Still, high-temperature effects, in particular approaching melting, are expected to be substantial [*Martorell et al.*, 2013a], and most recent theoretical studies point to a reduced anisotropy at core conditions, with hcp Fe possibly becoming almost isotropic [*Vočadlo et al.*, 2009; *Sha and Cohen*, 2010b] (see Figure 10.5). Provided these results are correct, it seems very difficult to reproduce the 3% cylindrical anisotropy reported in seismology

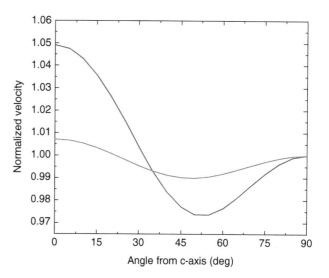

Figure 10.5 Normalized compressional sound velocities (V_p) for pure hcp-Fe as a function of propagation direction with respect to the c-axis. Orange line is after calculation at 13040 kg/m³ and 6000 K (but P~250 GPa) [*Sha and Cohen*, 2010b]. Blue line is after calculations at 360 GPa and 7000 K (but ρ ~ 13530 kg/m³), conditions at which strong premelting effects are envisaged [*Martorell et al.*, 2013a].

publications (the global inner core anisotropy is likely many times smaller than of the single crystal) [*Lincot et al.*, 2014]. Unfortunately, the necessary experimental validation of the high-temperature calculations has been so far only indirect, with measurement at relevant *P–T* conditions limited to the *c/a* axial ratio [*Tateno et al.*, 2010, 2012].

Among the other physical properties of iron and iron alloys of direct relevance for Earth's inner core we can definitively mention thermal conductivity. The estimation of inner core age, the out-coming heat flux, and the adopted structure (localized portions with homogeneous composition and crystalline structure, but radically different elastic behavior, oriented solidification, different grain size, etc.) all critically depend upon thermal conductivity of constituent alloys. Various geodynamical models consider that a thermal convection within the inner core is still active nowadays [*e.g., Deguen and Cardin*, 2011; *Cottaar and Buffett*, 2012]. Therefore, knowledge of the thermal conductivity of core material under high-pressure and high-temperature conditions is crucial for the understanding of many fundamental planetary processes, from the dynamo to the thermal history and heat budget. Unfortunately, measurements of thermal conductivity at very high pressure are technically challenging, and experimental data on iron and iron alloys, even indirect, are scant [*Konopkova et al.*, 2011]. Electronic contribution to the total thermal conductivity in metals can be obtained from electrical resistivity measurements through the Wiedemann-Franz law. While being formally a lower bound to thermal conductivity, it still provides a good approximation, as electronic heat transport is thought to be the main heat transport mechanism in metals. Electrical resistivity data for Fe and Fe-Si alloy exist up to 100 GPa [*Gomi et al.*, 2013] and serve a base to the estimation of core thermal conductivity (refer to Chapters 3 and 4 for a more extensive discussion).

In conclusion, in this chapter we briefly discussed some of the physical properties of solid iron and iron alloys at high pressure and temperature in relation to Earth's inner core. Density is possibly the most fundamental quantity needed for core modeling. The density mismatch between pure Fe and seismic models is the most striking evidence calling for light elements in the inner core. Certainly density, together with compressional and shear sound velocities is crucial for the proper interpretation of seismic data. Comparison of velocity vs. density systematics with inner core models provides the most stringent constraints on the nature and abundance of light elements alloyed to iron in Earth's core, which in turn are related to some of the most fundamental open questions in Earth science: (i) Earth's core composition and temperature profile, (ii) core-mantle differentiation, (iii) the chemical interactions at the core-mantle boundary, and (iv) the generation of Earth's magnetic field. An accurate knowledge

of single-crystal elasticity and elastic anisotropy is needed to understand seismic anisotropy and hence inner core structures (e.g., hemisphericity, possible existence of the innermost core). Thermal conductivity is a central parameter for modeling the geodynamo and is essential to constrain the heat budget and thermal history of the core and hence its secular cooling, its age, and the crystallization rate.

Arguably, knowledge of the physical properties of iron and iron alloys at pertinent pressure and temperature conditions is the master key to the ultimate comprehension of the Earth's core.

ACKNOWLEDGMENTS

I wish to thank Guillaume Morard for feedback and discussion during the long integration time of IXS measurements, which provided the quiet time needed to write this chapter.

REFERENCES

Aboussière, T., R. Deguen, and M. Melzani (2010), Melting-induced stratification above the Earth's inner core due to convective translation, *Nature, 466*, 744–747.

Alfè, D., G. D. Price, and M. J. Gillian (1999), Oxygen in the Earth's core: A first-principles study, *Phys. Earth Planet. Inter., 110*, 191–210.

Antonangeli, D., F. Occelli, H. Requardt, J. Badro, G. Fiquet, and M. Krisch (2004a), Elastic anisotropy in textured hcp-iron to 112 GPa from sound velocity wave propagation measurements, *Earth Planet. Sci. Lett, 225*, 243–251.

Antonangeli, D., M. Krisch, G. Fiquet, D.L. Farber, C.M. Aracne, J. Badro, F. Occelli, and H. Requardt (2004b), Elasticity of cobalt at high pressure studied by inelastic x-ray scattering, *Phys. Rev. Lett, 93*, 215,505.

Antonangeli, D., S. Merkel, and D.L. Farber (2006), Elastic anisotropy in hcp metals at high pressure and the sound wave anisotropy of the Earth's inner core, *Geophys. Res. Lett., 33*, L24,303.

Antonangeli, D., M. Krisch, D.L. Farber, D.G. Ruddle, and G. Fiquet (2008), Elasticity of hexagonal closed-packed cobalt at high pressure and temperature: A quasiharmonic case, *Phys. Rev. Lett., 100*, 085,501.

Antonangeli, D., J. Siebert, J. Badro, D.L. Farber, G. Fiquet, G. Morard, and F.J. Ryerson (2010), Composition of the Earth's inner core from high-pressure sound velocity measurements in Fe-Ni-Si alloys, *Earth Planet. Sci. Lett., 295*, 292–296.

Antonangeli, D., T. Komabayashi, F. Occelli, E. Borissenko, A.C. Walters, G. Fiquet, and Y. Fei (2012), Simultaneous sound velocity and density measurements of hcp iron up to 93 GPa and 1100 K: an experimental test of the Birch's law at high temperature, *Earth Planet. Sci. Lett., 331*, 210–214.

Antonangeli, D., and E. Ohtani (2015), Sound velocity of hcp-Fe at high pressure: experimental constraints, extrapolations, and comparison with seismic models, *Prog. Earth. Planet. Sci., 2*, 3.

Asanuma, H., E. Ohtani, T. Sakai, H. Terasaki, S. Kamada, N. Hirao, and Y. Ohishi (2011), Static compression of $Fe_{0.83}Ni_{0.09}Si_{0.08}$ alloy to 374 GPa and $Fe_{0.93}Si_{0.07}$ alloy to 252 GPa: Implications for the Earth's inner core, *Earth Planet. Sci. Lett.*, *310*, 113–118.

Badro, J., G. Fiquet. F. Guyot, E. Gregoryanz, F. Occelli, D. Antonangeli, and M. d'Astuto (2007), Effect of light elements on the sound velocities in solid iron: Implications for the composition of the Earth's core, *Earth Planet. Sci. Lett.*, *254*, 233–238.

Bergman, M.I. (1997), Measurements of elastic anisotropy due to solidification texturing and the implications for the Earth's inner core, *Nature*, *389*, 60–63.

Birch, F. (1052), Elasticity and constitution of the Earth's interior, *J. Geophys. Res.*, *57*, 227–286.

Brown, J.M., and R.G. McQueen (1986), Phase transition, grüneisen parameter, and elasticity for shocked iron between 77 GPa and 400 GPa, *J. Geophys. Res.*, *91*, 7485–7494.

Buffet, B.A. (1997), Geodynamic estimate of the viscosity of the Earth's inner core, *Nature*, *388*, 571–573.

Buffet, B.A., and H.R. Wenk (2001), Texturing of the inner core by Maxwellstresses, *Nature*, *413*, 60–64.

Caracas, R., and R. Wentzcovitch (2004), Equation of state and elasticity of FeSi, *Geophys. Res. Lett.*, *31*, L20,603.

Cottaar, S., and B. Buffett (2012), Convection in the Earth's inner core, *Phys. Earth Planet. Inter.*, *198–199*, 67–78.

Crowhurst, J.C., A.F. Goncharov, and J.M. Zaug (2004), Implusive stimulated light scattering from opaque materials at high pressure, *J. Phys. Condens. Matter*, *16*, S1137–S1142.

Decremps, F., D. Antonangeli, M. Gauthier, S. Ayrinhac, M. Morand, G. Le Marchand, F. Bergame, and J. Philippe (2014), Sound velocity measurements of iron up to 152 GPa by picosecond acoustics in diamond anvil cell, *Geophys. Res. Lett.*, *41*, 1459.

Deguen, R., and P. Cardin (2011), Thermochemical convection in Earth's inner core. *Geophys. J. Int.*, *187*, 1101–1118.

Deuss, A., J.C.E. Irving, and J.H. Woodhouse (2010), Regional variation of inner core anisotropy from seismic normal mode observations, *Science*, *328*, 1018–1020.

Dewaele, A., P. Loubeyre, F. Occelli, M. Mezouar, P.I. Dorogokupets, and M. Torrent (2006), Quasihy drostatic equation of state of iron above 2 Mbar, *Phys. Rev. Lett.*, *97*, 215,504.

Dziewonski, A.M., and D.L. Anderson (1981), Preliminary reference Earth model, *Phys. Earth Planet. Inter.*, *25*, 297–356.

Fiquet, G., J. Badro, F. Guyot, H. Requardt, and M. Krisch (2001), Sound velocities in iron to 110 gigapascals, *Science*, *291*, 468–471.

Fiquet, G., F. Guyot, and J. Badro (2008), The Earth's lower mantle and core, *Elements*, *4*, 177–182.

Fischer, R.A., A.J. Campbell, R. Caracas, D.M. Reaman, D.L. Heinz, P. Dera, and V. Prakapenka (2014), Equations of state in the Fe-FeSi system at high pressures and temperatures, *J. Geophys. Res. Solid Earth*, *119*, 2810–2827.

Gannarelli, C.M.S., D. Alfè, and M.J. Gillian (2005), The axial ratio of hcp iron at the conditions of the Earth's inner core, *Phys. Earth Planet. Inter.*, *152*, 67–77.

Gleason, A.E., W.L. Mao, and J.Y. Zhao (2013), Sound velocities for hexagonally close-packed iron compressed hydrostatically to 136 GPa from phonon density of states, *Geophys. Res. Lett.*, *40*, 2983–2987.

Gomi, H., K. Ohta, K. Hirose, S. Labrosse, R. Caracas, M.J. Verstraete, and J.W. Hernlund (2013), The high conductivity of iron and thermal evolution of the Earth's core, *Phys. Earth Planet. Inter.*, *224*, 88–103.

Hirao, N., E. Ohtani, T. Kondo, and T. Kinegawa (2004) Equation of state of iron-silicon alloys to megabar pressure, *Phys. Chem. Minerals*, *31*, 329–336.

Hirose, K., S. Labrosse, and J. Hernlund (2013), Composition and state of the core, *Annu. Rev. Earth Planet. Sci.*, *41*, 657–691.

Jeanloz, R., and H.R. Wenk (1988), Convection and anisotropy of the inner core, *Geophys. Res. Lett.*, *15*, 72–75.

Karato, S. (1993), Inner core anisotropy due to magnetic field-induced preferred orientation of iron, *Science*, *262*, 1708–1711.

Karato, S. (1999), Seismic anisotropy of the Earth's inner core resulting from flow induced by Maxwell stresses, *Nature*, *402*, 871–873.

Konôpkova, Z., P. Lazor, A. Goncharov, and V.V. Struzhkin (2011), Thermal conductivity of hcp iron at high pressure and temperature, *High Press. Res.*, *31*, 228–236.

Lin, J.F., A.J. Campbell, and D.L. Heinz (2003a), Static compression of iron-silicon alloys: Implications for silicon in the Earth's core, *J. Geophys. Res.*, *108*, B2045.

Lin, J.F., V.V. Struzhkin, W. Sturhahn, E. Huang, J. Zhao, M.Y. Hu, E.E. Alp, H.K. Mao, N. Boctor, and R.J. Hemley(2003b), Sound velocities of iron-nickel and iron-silicon alloys at high pressure, *Geophys. Res. Lett.*, *30*, 11.

Lin, J.F., W. Sturhahn, J. Zhao, G. Shen, H.K. Mao, and R.J. Hemley (2005), Sound velocities of hot dense iron: Birch's law revisited, *Science*, *308*, 1892–1894.

Lincot, A., R. Deguen, S. Merkel, and P. Cardin (2014), Seismic response and anisotropy of a mode hcp inner core, *C. R. Geosci.*, *346*, 148–157.

Mao, H.K., J. Shu, G. Shen, R.J. Hemley, B. Li, and A.K. Singh (1998), Elasticity and rheology of iron above 220 GPa and the nature of the Earth's inner core, *Nature*, *396*, 741–743; correction (1999), Nature, 399, 80.

Mao, H.K., J. Xu, V.V. Struzhkin, J. Shu, R.J. Hemley, W. Sturhahn, M.Y. Hu, E.E. Alp, L. Vočadlo, D. Alfè, G.D. Price, M.J. Gillian, M. Schwoerer-Böhning, D. Häusermann, P. Eng, G. Shen, H. Giefers, R. Lübbers, and G. Wortmann (2001), Phonon density of states of iron up to 153 GPa, *Science*, *292*, 914–916.

Mao, W.L., V.V. Struzhkin, A.Q.R. Baron, S. Tsutsui, C.E. Tommaseo, H.R. Wenk, M.Y. Hu, P. Chow, W. Sturhahan, J. Shu, R.J. Hemley, D.L. Heinz, and H.K. Mao (2008), Experimental determination of the elasticity of iron at high pressure, *J. Geophys. Res.*, *113*, B09,213.

Martorell, B., L. Vočadlo, J. Brodholt, and I.G. Wood (2013a), Strong premelting effect in the elastic properties of hcp-Fe under inner-core conditions, *Science*, *342*, 466–468.

Martorell, B., J. Brodholt, I.G. Wood, and L. Vočadlo (2013b), The effect of nickel on the properties of iron at the conditions of Earth's inner core: Ab initio calculations of seismic wave velocities of Fe-Ni alloys, *Earth Planet. Sci. Lett.*, *365*, 143–151.

Matthies, S., S. Merkel, H.R. Wenk, R.J. Hemley, and H.K. Mao (2001), Effects of texture on the determination of elasticity of polycrystalline ε-iron from diffraction measurements, *Earth Planet. Sci. Lett.*, *194*, 201–212.

Merkel, S., H.R. Wenk, P. Gillet, H.K. Mao, and R.J. Hemley (2004), Deformation of polycrystalline iron up to 30 GPa and 1000 K, *Phys. Earth Planet. Inter.*, *145*, 239–251.

Merkel, S., N. Miyajima, D. Antonangeli, G. Fiquet, and T. Yagi (2006), Lattice preferred orientation and stress in polycrystalline hcp-Co plastically deformed under high pressure, *J. Appl. Phys.*, *100*, 023,510.

Merkel, S., C. Tomé, and H.R. Wenk (2009), Modeling analysis of the influence of plasticity on high pressure deformation of hcp-Co, *Phys. Rev. B*, *79*, 064,110.

Monnereau, M., M. Calvet, L. Mergerin, and A. Souriau (2010), Lopsided growth of Earth's inner core, *Science*, *328*, 1014–1017.

Morelli, A., A.M. Dziewonski, and J.H. Woodhouse (1986), Anisotropy of the inner core inferred from PKIKP travel times, *Geophys. Res. Lett.*, *13*, 1545–1548.

Murphy, C.A., J.M. Jackson, and W. Sturhahn (2013), Experimental constraints on the thermodynamics and sound velocities of hcp-Fe to core pressures, *J. Geophys. Res. Solid Earth*, *118*, 1–18.

Niu, F., and L. Wen (2001), Hemispherical variations in seismic velocity at the top of Earth's inner core, *Nature*, *410*, 1081–1084.

Ohtani, E., Y. Shibazaki, T. Sakai, K. Mibe, H. Fukui, S. Kamada, T. Sakamaki, Y. Seto, S. Tsutsui, and A.Q.R. Baron (2013), Sound velocity of hexagonal close-packed iron up to core pressures, *Gephys. Res. Lett.*, *40*, 5089–5094.

Ono, S., A.R. Oganov, J.P. Brodholt, L. Vočadlo, I.G. Wood, A. Lyakhov, C.W. Glass, A.S. Côté, and G.D. Price (2008), High-pressure phase transformations of FeS: Novel phases at conditions of planetary cores, *Earth Planet. Sci. Lett.*, *272*, 481–487.

Poirier, J.P. (1994), Light elements in the Earth's outer core — A critical review, *Phys. Earth Planet. Inter.*, *85*, 319–337.

Sakai, T., E. Ohtani, N. Hirao, and Y. Ohishi (2011), Stability field of the hcp structure for Fe, Fe-Ni, and Fe-Ni-Si alloy up to 3 Mbar, *Geophys. Res. Lett.*, *38*, L09,302.

Sakai, T., E. Ohtani, S. Kamada, H. Terasaki, and N. Hirao (2012), Compression of $Fe_{88.1}Ni_{9.1}Si_{2.8}$ alloy up to the pressure of Earth's inner core, *J. Geophys. Res.*, *117*, B02,210.

Sakai, T., S. Takahashi, N. Nishitani, I. Mashino, E. Ohtani, and N. Hirao (2014), Equation of state of pure iron and $Fe_{0.9}Ni_{0.1}$ alloy up to 3 Mbar, *Phys. Earth Planet. Inter.*, *228*, 114–126.

Sata, N., K. Hirose, G. Shen, Y. Nakajima, Y. Ohishi, and N. Hirao (2010), Compression of FeSi, Fe_3C, $Fe_{0.95}O$, FeS under core pressures and implication for light element in the Earth's core,. *J. Geophys. Res.*, *115*, B09,204.

Seagle, C., A.J. Campbell, D.L. Heinz, G. Shen, and V.B. Prakapenka (2006), Thermal equation of state of Fe_3S and implications for sulfur in Earth's core, *J. Geophys. Res.*, *111*, B06,209.

Sha, X., and R.E. Cohen (2010a), First-principles thermal equation of state and thermoelasticity of hcp Fe at high pressures, *Phys. Rev. B*, *81*, 094,105.

Sha, X., and R.E. Cohen (2010b), Elastic isotropy of ε-Fe under Earth's core conditions, *Geophys. Res. Lett.*, *37*, L10,302.

Shibazaki, Y., E. Ohtani, H. Fukui, T. Sakai, S. Kamada, D. Ishikawa, S. Tsutsui, A.Q.R. Baron, N. Nishitani, N. Hirao, and K. Takemura (2012), Sound velocity measurements in dhcp-FeH up to 70 GPa with inelastic X-ray scattering: Implications for the composition of the Earth's core, *Earth Planet. Sci. Lett.*, *313–314*, 79–85.

Stacey, F.D., and P.M. Davis (2004), High pressure equations of state with applications to the lower mantle and core, *Phys. Earth Planet. Inter.*, *142*, 137–184.

Stixrude, L., and R.E. Cohen (1995), High pressure elasticity of iron and anisotropy of Earth's inner core, *Science*, *267*, 1972–1975.

Steinle-Neumann, G., L. Stixrude, and R.E. Cohen (1999), First principle elastic constants for the hcp transition metals Fe, Co and Re at high pressure, *Phys. Rev. B*, *60*, 791–799.

Steinle-Neumann, G., L. Stixrude, R.E. Cohen, and O. Gülseren (2001), Elasticity of iron at the temperature of the Earth's inner core, *Nature*, *413*, 57–60.

Stevenson, D.J. (2003), Planetary science: Mission to Earth's core — A modest proposal, *Nature*, *423*, 239–240.

Tateno, S., K. Hirose, Y. Ohishi, and Y. Tatsumi (2010), The structure of iron in Earth's inner core, *Science*, *330*, 359–361.

Tateno, S., K. Hirose, T. Komabayashi, H. Ozawa, and Y. Ohishi (2012), The structure of Fe-Ni alloy in Earth's inner core, *Geophys. Res. Lett.*, *39*, L12,305.

Tsuchiya, T., and M. Fujibuchi (2009), Effects of Si on the elastic properties of Fe at Earth's inner core pressures: First principles study, *Phys. Earth Planet. Inter.*, *174*, 212–219.

Wenk, H.R., S. Matthies, R.J. Hemley, H.K. Mao, and J. Shu (2000), The plastic deformation of iron at pressures of the Earth's inner core, *Nature*, *405*, 1044.

Whoodhouse, J.H., D. Giardini, and X.D. Li (1986), Evidence for inner core anisotropy from free oscillations, *Geophys. Res. Lett.*, *13*, 1549–1552.

Yoshida, S., I. Sumita, and M. Kumazawa (1996), Growth model of the inner core coupled with the outer core dynamics and the resulting elastic anisotropy, *J. Geophys. Res.*, *101*, 28,085–28,104.

11

Physical Properties of the Outer Core

Hidenori Terasaki

ABSTRACT

The physical properties of liquid Fe alloys are important for understanding the characteristics of the molten outer core. The possible core composition can be constrained by matching the observed seismic data with the measured sound velocity and density of liquid Fe alloys. The transport properties of liquid Fe alloys strongly influence the convection behavior of the outer core. In this chapter, we review the latest results on the elastic (density, compressibility, and sound velocity) and transport (viscosity) properties of liquid Fe alloys obtained based on experimental and numerical approaches. Combining these results, we will then discuss the influence that alloying light elements have on the properties of liquid Fe and introduce the latest models of the outer core.

11.1. INTRODUCTION

Earth's outer core is molten, as inferred from the disappearance of seismic shear waves. It is widely believed that the outer core consists of an Fe-Ni (~5 wt %) alloy [*McDonough and Sun*, 1995] and contains ~10% of lighter elements, such as S, O, Si, C, and H [e.g., *Poirier*, 1994]. However, the specifics of the light elements in the outer core are still under debate. To constrain the core composition, the density and sound velocity of various liquid Fe alloys measured at high pressures and high temperatures are necessary to compare with the observed seismic data from the outer core. The dynamics of the molten outer core are influenced largely by its elastic and transport properties, such as density, viscosity, and thermal conductivity. Convection currents in the outer core strongly influence the thermal history of Earth's interior and generate Earth's magnetic field. Hence, the nature of the light alloying elements in the core and their effect on the

physical properties are key to understanding the composition and dynamics of the outer core.

Recent technical advances can measure the elastic and transport properties of liquid materials under high pressures, and novel theoretical calculations now provide liquid properties under extreme conditions. These results enable us to discuss further the effect of pressure and composition on the physical properties of liquid Fe alloys and to obtain a more precise view of the outer core. In this chapter, the physical properties of liquid Fe alloys, especially the elastic and transport properties, are reviewed, and the proposed compositional and dynamic models of the outer core are discussed based on these properties.

11.2. DENSITY AND COMPRESSIBILITY OF LIQUID Fe ALLOYS

Three variables of pressure (P), temperature (T), and density (ρ) or specific volume (V), which are indispensable for understanding the structure and properties of Earth's interior, are linked by an equation of state (EOS). The

Department of Earth and Space Science, Osaka University, Toyonaka, Japan

Deep Earth: Physics and Chemistry of the Lower Mantle and Core, Geophysical Monograph 217, First Edition.
Edited by Hidenori Terasaki and Rebecca A. Fischer.

third-order Birch-Murnaghan (BM) EOS deduced from finite strain theory [*Birch*, 1947] is commonly used to express the compressional behavior of solid materials and can also be applied to liquid materials. We will consider the density and compressibility of liquid Fe alloys using the BM EOS in this section.

11.2.1. Density and Compressibility of Liquid Fe

The density and compressibility of liquid Fe at high pressures has been studied using shock experiments [e.g., *Brown and McQueen*, 1986; *Anderson and Ahrens*, 1994] and ab initio calculations [*Alfè et al.*, 2000; *Pozzo et al.*, 2013; *Ichikawa et al.*, 2014]. The compressibility data of liquid Fe are listed in Table 11.1. The compression curve of liquid Fe is plotted in Figure 11.1 using the BM EOS. The pressure derivative of the bulk modulus of liquid Fe ranges from 4.66 to 6.65.

The measured adiabatic bulk modulus at ambient pressure (K_{0S}) or isothermal bulk modulus at temperature T (K_{0T}) was corrected to $K_{0,1973K}$ to evaluate the reported compressibility of liquid Fe obtained over a wide pressure range. The value of K_{0S} was converted to K_{0T} using the relationship

$$K_{0T} = K_{0S} / \left(1 + \alpha\gamma_0 T\right), \qquad (11.1)$$

where γ_0 is the Grüneisen parameter at ambient pressure and α is the thermal expansivity. A value of $K_{0S} = 109.7$ GPa was obtained from shock experiments [*Anderson and Ahrens*, 1994] and was converted to $K_{0,1973K} = 83.9$ GPa using equation (11.1) with $\gamma_0 = 1.735$ and $\alpha = 0.9 \times 10^{-4}\,K^{-1}$ [*Hixson et al.*, 1990]. The isothermal bulk modulus at 7000 K, $K_{0,7000K} = 24.6$ GPa, was estimated from ab initio calculations based on the calculated density data up to 420 GPa [*Ichikawa et al.*, 2014], and gives $K_{0,1973K} = 76.9$ GPa assuming $dK_0/dT = -0.0104$ GPa/K [*Hixson et al.*, 1990]. Thus, the corrected $K_{0,1973K}$ of liquid Fe based on different methods provides comparable values (76.9–83.9 GPa). These compressibility data of liquid Fe are the base point for evaluating the effect of light elements on the elastic parameters of liquid Fe, as discussed below.

11.2.2. Effect of S, Si, and C on Density and Compressibility of Liquid Fe

The density of liquid Fe-S, Fe-Si, Fe-C, Fe-O-S at high pressures was measured using the sink/float method (Fe-S [*Balog et al.*, 2003; *Nishida et al.*, 2008], Fe-Si [*Yu and Secco*, 2008; *Tateyama et al.*, 2011]); the X-ray absorption method (Fe-S [*Sanloup et al.*, 2000; *Chen et al.*, 2005, 2014; *Nishida et al.*, 2011], Fe-Si [*Sanloup et al.*, 2004], Fe-C [*Terasaki et al.*, 2010; *Sanloup et al.*, 2011; *Shimoyama et al.*, 2013]); X-ray diffuse scattering signal

analysis (Fe-Ni-S and Fe-Ni-Si [*Morard et al.*, 2013]), and shock experiments (Fe-S [*Huang et al.*, 2013], Fe-Ni-Si [*Zhang et al.*, 2014], Fe-O-S [*Huang et al.*, 2011]). Details of the sink/float and X-ray absorption methods are described elsewhere [*Ohtani*, 2013]. The reported elastic parameters (K_{0T}, and its pressure derivative, K'_{0T}) are summarized together with measured conditions in Table 11.1. The compression curves of liquid Fe-S, Fe-Si, and Fe-C are shown in Figures 11.1a, b, and c, respectively. Although the Vinet EOS is reported to give a better fit to the measured density in the pressure range 0–3.8 GPa [*Nishida et al.*, 2011], both the BM and the Vinet EOSs applied to data obtained over wider pressure ranges provide similar elastic parameters for liquid Fe-S and FeS [e.g., *Balog et al.*, 2003; *Chen et al.*, 2014]. Hence, the BM EOS was applied here to consider the compression behavior of liquid Fe alloys.

11.2.2.1. Density

The slope of compression curve of liquid Fe-S becomes steep on the S-rich side at relatively low pressures (Figure 11.1a), suggesting that S-rich liquid Fe-S is more compressible. The density decreases nonlinearly with increasing S content at ambient pressure [*Nagamori*, 1969] and at 4 GPa [*Nishida et al.*, 2008]. This implies that the molar volume of liquid Fe-S calculated from the density deviates from ideal (linear) mixing between the molar volumes of the end-member components (Fe and FeS). The deviation of the molar volume from ideal mixing is defined as the excess molar volume.

The density of liquid Fe-Si gradually decreases up to 33 wt % Si, and then decreases markedly at concentrations above 33 wt % Si at 4 GPa and 1923 K, suggesting that liquid Fe-Si also has a nonideal mixing behavior [*Tateyama et al.*, 2011]. The excess molar volume of liquid Fe-Si is less than that of liquid Fe-S. The density of liquid Fe-C also tends to decrease nonlinearly with increasing C content [*Shimoyama et al.*, 2013]. Therefore, when we consider the effect of lighter elements on the density of liquid Fe, any nonideal mixing behavior should be taken into account, at least up to 4 GPa.

However, at higher pressures, the difference in density for different S content decreases significantly on the S-rich side, as shown in Figure 11.1a. This is because the isothermal bulk modulus, K_T, decreases with increasing S content [*Sanloup et al.*, 2000]. This may imply that the excess molar volume of liquid Fe-S decreases at higher pressure. On the other hand, the difference in density between different Si or C contents does not change much with pressure, as shown in Figures 11.1b and c. The compression curve of liquid Fe-Si (except for liquid Fe-Ni-Si) lies almost parallel to that of liquid Fe (Figure 11.1b). These behaviors are linked to the liquid compressibilities, as discussed in Section 11.2.2.2.

Table 11.1 Elastic parameters of liquid Fe alloys.

Alloy	Content[a] (wt %)	Content (at %)	P range (GPa)	T range (K)	T_0 (K)	K_{0T0} (GPa)[b]	K_{0T0}'	ρ_0 (g/cm³)[c]	EOS	Method[d]	References
Fe			275–400		1811	85.1	5.8	7.02	R-H	Shock	Anderson and Ahrens [1994]
			60–420	4000–7000	7000	24.6(6)	6.65(4)		Vinet	ab initio	Ichikawa et al. [2014]
			0.2	2125–3950							Hixson et al. [1990]
Fe-S	10	16	1.5–17.5	1773–2123	1773	63	4.8	5.50	BM	S/F	Balog et al. [2003]
						60	4.8	5.50	Vinet		
	10	16	2.1–6.0	1650–1780	1770	45(3)	4–7 (fixed)	5.20	BM	Abs	Sanloup et al. [2000]
	20	30	2.3–6.2	1600–1700	1650	29(5)	4–7 (fixed)	4.97	BM		
	27	39	1.5–4.0	1500–1530	1650	12(7)	4–7 (fixed)	4.34	BM		
	36	50	0–3.8	1500–1800	1500	2.5(3)	24(2)	4.23	Vinet	Abs	Nishida et al. [2011]
					1500	12(3)	14(3)	4.23	Vinet		
	36	50	2.1–3.8	1600–1800	1650	11(3)	5 (fixed)	4.20	B-M	Abs	Chen et al. [2014]
			3.3–5.6	1650		12(3)	5 (fixed)	4.20	Vinet		
Fe-5 wt % Ni-S	12	19	28–94	2140–2830	2600	88(7)	4 (fixed)	5.60	BM	DS	Morard et al. [2013]
					2600	65(6)	4.5 (fixed)	5.40	BM		
					2600	48(6)	5 (fixed)	5.20	BM		
Fe-Si	17	29	3–12	1773	1773	68(2)	4 (fixed)	5.88	BM	S/F	Yu and Secco [2008]
	17	29		1650	1650	76(2)	4 (fixed)	6.31	BM	Abs	Sanloup et al. [2004]
	25	40		1725	1725	73(2)	4 (fixed)	5.98	BM		
Fe-5 wt % Ni-Si	15	26	34–91	2520–3200	2850	199(16)	4 (fixed)	6.05	BM	DS	Morard et al. [2013]
					2850	174(16)	4.5 (fixed)	6.00	BM		
					2850	153(15)	5 (fixed)	5.95	BM		
Fe-C	3.5	14	1.6–6.8	1600–2200	1500	55.3(25)	5.2(15)	7.01	BM	Abs	Shimoyama et al. [2013]
	5.7	22	2.0–7.8	2273	2273			6.53	BM	Abs	Sanloup et al. [2011]
	6.7	25	3.6–9.5	1973	1973	54(3)	4 (fixed)	6.46	BM	Abs	Terasaki et al. [2010]
					1973	46(2)	7 (fixed)	6.46	BM		

[a] Content of light element.
[b] K_{0T0} represents isothermal bulk modulus at ambient pressure and T_0.
[c] ρ_0 denotes density at ambient pressure and T_0.
[d] S/F, Abs, and DS denote sink/float, X-ray absorption, and diffuse-scattering signal methods, respectively.

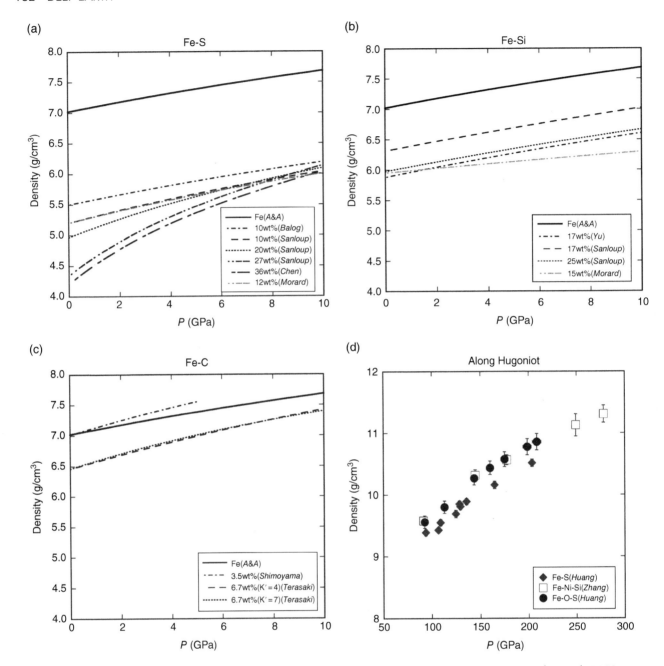

Figure 11.1 (a) Compression curves of liquid Fe [*Anderson and Ahrens*, 1994], Fe–10 wt % S [*Balog et al.*, 2003; *Sanloup et al.*, 2000], Fe–20 wt % S [*Sanloup et al.*, 2000], Fe–27 wt % S [*Sanloup et al.*, 2000], FeS [*Chen et al.*, 2014], and Fe–5 wt % Ni–12 wt % S [*Morard et al.*, 2013]. (b) Compression curves of liquid Fe, Fe–17 wt % Si [*Sanloup et al.*, 2004; *Yu and Secco*, 2008], Fe–25 wt % Si [*Sanloup et al.*, 2004], and Fe–5 wt % Ni–15 wt % Si [*Morard et al.*, 2013]. (c) Compression curves of liquid Fe, Fe–3.5 wt % C [*Shimoyama et al.*, 2013], and Fe–6.7 wt % C [*Terasaki et al.*, 2010]. The curves in (a)–(c) correspond to the isothermal compression curves at T_0 drawn using the elastic parameters listed in Table 11.1 with the third-order BM EOS. (d) Densities of liquid $Fe_{81}S_{19}$ [*Huang et al.*, 2013], $Fe_{74}Ni_8Si_{18}$ [*Zhang et al.*, 2014], and $Fe_{84.2}O_{7.2}S_{8.6}$ [*Huang et al.*, 2011] obtained by shock experiments. Measured shock wave data are aligned along the Hugoniot.

Densities of liquid Fe alloys have recently been reported at much higher pressures based on shock experiments [*Huang et al.*, 2011, 2013; *Zhang et al.*, 2014]. Figure 11.1d shows densities of Fe-S, Fe-Ni-Si, and Fe-O-S at core pressures along the Hugoniot. The density of liquid $Fe_{81}S_{19}$ is overlapped with that of liquid $Fe_{74}Ni_8Si_{18}$, suggesting that the effect of S and Si on the density of liquid Fe is similar at these conditions. On the contrary, oxygen is likely to reduce the density more effectively.

11.2.2.2. Compressibility and Liquid Structure

Figure 11.2 shows the K_{0T} values of liquid Fe-S, Fe-Si, and Fe-C as a function of light element content at 1973 K. Sulfur significantly reduces both the density and K_{0T} value of liquid Fe [*Sanloup et al.*, 2000], whereas the addition of C reduces only moderately the K_{0T} value of liquid Fe-C [*Shimoyama et al.*, 2013], and silicon does not have much effect on the K_{0T} value and only reduces the density moderately [*Sanloup et al.*, 2004]. The effect of light element content on the K_{0T} value may be explained by a difference in the local structure of the liquid. Sulfur and carbon are located at the interstitial sites in liquid Fe, while silicon substitutes at the Fe site [*Waseda*, 1980]. Interstitial sulfur modifies the local structure of liquid Fe strongly, and thus the structural ordering of liquid Fe-S is only observed over a short range [*Sanloup et al.*, 2002]. On the other hand, the local structure of liquid Fe-Si is similar to that of liquid Fe, and, thus, silicon has only a small effect on the liquid structure of Fe, leading to a minor effect on the compressibility.

There have been several reports recently on the change in elastic properties of Fe-C of around 5–6 GPa associated with a structural change in liquid Fe-C. Liquid Fe–5.7 wt % C shows an increase of compressibility above 6 GPa [*Sanloup et al.*, 2011] and liquid Fe–3.5 wt % C shows an abrupt density increase at 5 GPa [*Shimoyama et al.*, 2013]. The change in compressibility and density occurring around 5 GPa can be explained by a possible

liquid-liquid transition in the vicinity of the δ(bcc)-γ(fcc)-liquid Fe triple point occurring at 5.2 GPa [*Sanloup et al.*, 2011], or from a structural variation of liquid Fe-C associated with a change in the liquidus phase on the C-rich side from graphite to Fe_7C_3 at 5 GPa [*Nakajima et al.*, 2009; *Shimoyama et al.*, 2013]. If the abrupt change in the elastic properties of liquid Fe-C is derived from the former case (a structural transition of liquid Fe), these discontinuous changes should also be observed in other liquid Fe alloys, such as Fe-S and Fe-Si. Alternatively, if the latter case (a structural change on the Fe_3C-rich side) is more appropriate, then the abrupt change in the elastic properties will only be observed in liquid Fe-C. Since a possible structural change in the liquid phase has also been reported for other Fe alloys, such as Fe-S at 15 GPa [*Morard et al.*, 2007], discontinuity of compression curve, and a change in the elastic properties associated with such a possible structural variation should be considered when observing the compression behavior at higher pressures.

To evaluate the effect of light elements on the compressibility obtained over a much wider pressure range (near to the conditions at the core), we compared the recent results of ab initio molecular dynamic simulations on liquid Fe up to 420 GPa with data from diamond anvil cell experiments for liquid Fe-Ni-S and Fe-Ni-Si up to 94 GPa. The K_{0T} (Fe) value of 70.4 GPa at 2600 K was calculated using $K_{0,7000K} = 24.6$ GPa [*Ichikawa et al.*, 2014] with $dK_0/dT = -0.0104$ [*Hixson et al.*, 1990], and the K_{0T}(Fe–5 wt % Ni–12 wt % S) value of 48 GPa was calculated assuming $K' = 5$ at 2600 K [*Morard et al.*, 2013]. Therefore, sulfur surely reduces the K_{0T} value of liquid Fe. In contrast, the K_{0T}(Fe–5 wt % Ni–15 wt % Si) value of 153 GPa for $K' = 5$ at 2850 K (*Morard et al.*, 2013) is markedly larger than the K_{0T}(Fe) value. This needs to be clarified in future studies together with the possibility of structural transition at measured conditions.

11.3. SOUND VELOCITY OF LIQUID Fe ALLOYS

The compressional wave velocity (V_p) of a liquid is closely related to its density and elastic modulus since shear modulus of liquid can be ignored, and is expressed as,

$$V_P = \sqrt{\frac{K_S}{\rho}}, \tag{11.2}$$

where ρ is the density. K_S can be converted from K_T using equation (11.1). To consider the effect of pressure on V_p, equation (11.2) is combined with the following Murnaghan's integrated linear EOS assuming K is approximated by a linear function of P, i.e., $K \sim K_0 + K'P$,

$$\rho = \rho_0 \left(1 + \frac{K'}{K_0} P \right)^{1/K'}. \tag{11.3}$$

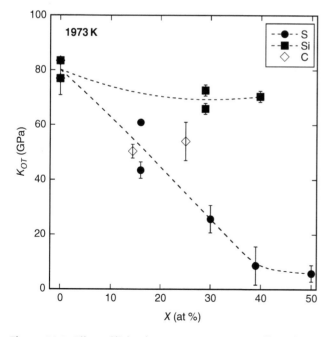

Figure 11.2 Effect of light element content on K_{0T} of liquid Fe at 1973 K. The K_{0T} values of each liquid (listed in Table 11.1) were corrected to that at 1973 K using $dK/dT = -0.0104$ GPa/K of liquid Fe [*Hixson et al.*, 1990].

Then, we can obtain

$$V_P = \left[\left(\frac{K_{0S}}{\rho_0} \right) \left(1 + \frac{K_S'}{K_{0S}} P \right)^{(1-1/K_S')} \right]^{1/2}. \quad (11.4)$$

Hence, if we obtain the density and adiabatic bulk modulus at ambient pressure and its pressure derivative, then V_P can be described as a function of pressure [*Jing et al.*, 2014]. In this section, we consider the effect of light elements (S, Si, and C) on V_P and its pressure dependence (dV_P/dP) of liquid Fe based on recent results of direct sound velocity measurements at high pressure and the calculated sound velocity from density data.

11.3.1. Sound Velocity of Liquid Fe

Direct sound velocity measurements on liquid metals were carried out for liquid Fe and Fe–Ni using ultrasonic interferometry at ambient pressure [*Nasch et al.*, 1994; *Nasch and Maghnani*, 1998] and for liquid Fe and Ni using a pulse heating technique at 0.2 GPa [*Hixson et al.*, 1990]. According to these results, the compressional wave velocity (V_P) tends to decrease almost linearly with increasing temperature ($dV_P/dT = -0.72$ m/s·K for Fe and -0.35 m/s·K for $Fe_{0.92}Ni_{0.08}$). The effect of alloying Ni

on the value of V_P of liquid Fe is not simple. It has been reported that V_P varies nonlinearly with increasing Ni content and has a minimum value around Ni = 25 wt % owing to a repulsive magnetic interaction that affects the volume and elastic modulus [*Nasch and Manghnani*, 1998].

The V_P measurements on liquid Fe at high pressure were performed under static conditions using an ultrasonic pulse-echo overlap method up to 3.6 GPa [*Jing et al.*, 2014]. The V_P measurements were also performed under dynamic conditions using shock experiments with explosive or two-stage gas guns up to 400 GPa [*Brown and McQueen*, 1986; *Nguyen and Holmes*, 2004] and with a high-power laser up to 800 GPa [*Sakaiya et al.*, 2014]. The measured V_P of liquid Fe gradually increases with pressure [*Jing et al.*, 2014] and agrees well with the calculated value of V_P from density data using equation (11.4), as shown in Figure 11.3a. In the dynamic experiments, the results were consistent with each other and the measured value of V_P increased with pressure along the Hugoniot (9.38 km/s at 206 GPa and 14.3 km/s at 810 GPa). The measured V_P were summarized in Table 11.2.

11.3.2. Effect of S, Si, and C on Sound Velocity of Liquid Fe

The V_P measurements of liquid Fe-(Ni)-S were performed using ultrasonic interferometry for Fe–5 wt %

Table 11.2 Measured P-wave velocities of liquid Fe alloys.

Alloy	Content[a] (w%)	Content (a%)	P range (GPa)	T range (K)	V_P (m/s)	Method	References
Fe			0	1823–1883	3767–3812	Ultrasonic interferometry	*Nasch et al.* [1994]
			0	1809–1950	3840–3961	Ultrasonic interferometry	*Nasch and Manghnani* [1998]
			0.2	2125–3950	3075–4035	Electrical pulse heating	*Hixson et al.* [1990]
			1.1–3.6	1873–1973	4053–4242	Ultrasonic pulse echo	*Jing et al.* [2014]
			260–400	6096–10024	6570–10910	Light gas gun (shock)	*Brown and McQueen* [1986]
			517–810		11700–14300	Laser shock	*Sakaiya et al.* [2014]
Fe-Ni-S	10	16	0	1673–1973	3150–3357	Ultrasonic interferometry	*Nasch et al.* [1997]
Fe-S	30	43	2.4–5.4	1373–1823	3105–3875	Ultrasonic pulse echo	*Nishida et al.* [2013]
	10	16	3.6–6.7	1673–1973	3466–3896	Ultrasonic pulse echo	*Jing et al.* [2014]
	20	30	2.5–8.2	1573–1773	3150–3841	Ultrasonic pulse echo	*Jing et al.* [2014]
	27	39	3.2–8.2	1573–1773	3008–3720	Ultrasonic pulse echo	*Jing et al.* [2014]
Fe-O-S	O=2.2/S=5.3	O=7.2/S=8.6	92.6–208.4		9560–10860	Light gas gun (shock)	*Huang et al.* [2011]

[a] Content of light element.

Ni–10 wt % S at ambient pressure [*Nasch et al.*, 1997] and using an ultrasonic pulse-echo overlapping method for Fe-S (S = 0–30 wt %) up to 8.2 GPa [*Nishida et al.*, 2013; *Jing et al.*, 2014] as listed in Table 11.2. The effect of temperature on the value of V_P was small [$(d \ln V_P/dT)_P = (-10\sim2) \times 10^{-5}$ K^{-1}] at high pressures [*Nishida et al.*, 2013; *Jing et al.*, 2014]. The measured and calculated values of V_P of liquid Fe-S using equation (11.4) are shown as a function of pressure in Figure 11.3a. In terms of the effect of pressure on the value of V_P, although the V_P of liquid Fe–30 wt % S has been reported to be approximated by a linear function with pressure [*Nishida et al.*, 2013], the variation of V_P with pressure for liquid Fe~Fe–27 wt % S is closer to the calculated value of V_P employing equation (11.4) [*Jing et al.*, 2014]. The calculated curves with equation (11.4) are quite similar to those using the Birch-Murnaghan EOS at the conditions up to 20 GPa, except for Fe–27 wt % S.

The V_P of liquid Fe-S decreases with increasing S content, as suggested by *Sanloup et al.* [2000], and the difference in V_P with different S contents becomes less at higher pressures (see Figure 11.3a). This behavior can be explained by the higher compressibility of the S-rich liquid. It should be noted that the V_P of liquid Fe-S is expected to become larger than that of liquid Fe at higher pressures, suggesting that the effect on V_P of S alloying is expected to reverse. This is supported by recent sound velocity measurements on liquid Fe-O-S up to 233 GPa based on shock experiments [*Huang et al.*, 2011]. In these results, the addition of S (and O) does indeed increase the bulk sound velocity of liquid Fe in the pressure range 100–200 GPa.

The calculated values of V_P of liquid Fe-Si and Fe-C are shown as a function of pressure in Figures 11.3b and c. Comparing the V_P of liquid Fe along the isentrope, the addition of Si increases the V_P of liquid Fe slightly [*Sanloup et al.*, 2004]. If we compare the V_P of liquid Fe along the isotherm, then the V_P of liquid Fe-Si is greater than that of liquid Fe up to 3 GPa at 1900 K, and this relationship is then reversed (Figure 11.3b). This tendency is opposite to the effect of S content. This is because that effect of Si on the density is larger than that on the compressibility [see equation (11.2)] up to 3 GPa. For liquid Fe-C, carbon decreases the V_P of liquid Fe moderately and the effect of pressure on the V_P value of this system is almost consistent with that of liquid Fe along the isotherm (Figure 11.3c). Consequently, the effect of light elements on the value of V_P of liquid Fe depends on the alloying light element species, and its pressure dependence also plays an important role when considering direction and magnitude of the deviation from the V_P of liquid Fe. The V_P behavior at much higher pressures, i.e., for core conditions, is considered together with the density in the next section.

11.4. CONSTRAINTS ON THE OUTER CORE COMPOSITION FROM DENSITY AND SOUND VELOCITY

11.4.1. Relationship between Density and Sound Velocity (Birch's Law)

Based on sound velocity measurements of 250 rock samples, *Birch* [1961a,b] found that the V_P of materials having a mean atomic weight (*M*) was approximated by a linear function of the density (ρ) as

$$V_P = a(M) + b\rho, \qquad (11.5)$$

where *a* and *b* are constants, and *a* is a function of *M*. This empirical linear ρ–V_P relationship is known as *Birch's law*. If Birch's law is applicable to Earth's core materials, then the measured ρ–V_P data of Fe alloys can be extrapolated to Earth's core conditions.

It has been reported that Birch's law is valid for solid Fe alloys up to Earth's core pressures based on static diamond anvil cell experiments [e.g., *Badro et al.*, 2007; *Antonangeli et al.*, 2012] and dynamic shock experiments [e.g., *Brown and McQueen*, 1986]. Although *Mao et al.* [2012] suggested that an empirical power law formulation with a temperature correction provides a better description for the ρ–V_P relationship of hcp Fe, most ρ–V_P profiles of solid Fe alloys follow Birch's law at room temperature as a first-order approximation. The addition of light elements shifts the solid Fe line to the upper left-hand side (i.e., increases V_P and decreases ρ) in a ρ–V_P plot [*Badro et al.*, 2007; *Mao et al.*, 2012]. Alloying light elements also affects the gradient ($dV_P/d\rho$) of solid Fe, and this effect depends on the light element. (For details of solid Fe alloys, see Chapter 10.)

In the case of a liquid, Birch's law is also reported to hold for liquid Fe [*Brown and McQueen*, 1986; *Hixson et al.*, 1990; *Sakaiya et al.*, 2014] and for liquid Fe-O-S [*Huang et al.*, 2011]. Figure 11.4a shows a plot of ρ–V_P of liquid Fe, Fe-S, Fe-Si, and Fe-C calculated in the range 0–20 GPa using the elastic parameters discussed in previous sections. Most of the liquids show a linear ρ–V_P relationship, except for Fe–27 wt % S, which has a slight concave curvature. If sulfur or silicon dissolves in liquid Fe, then the ρ–V_P plot will shift toward the upper left-hand side of the plot. In contrast, carbon dissolution shifts the ρ–V_P plot toward the lower side. In addition, it is noteworthy that the gradient ($dV_P/d\rho$) of liquid Fe is approximately 1.5 times steeper than that of solid Fe [*Sakaiya et al.*, 2014]. This may be closely related to the difference in elastic modulus and its pressure derivative between the solid and liquid phases.

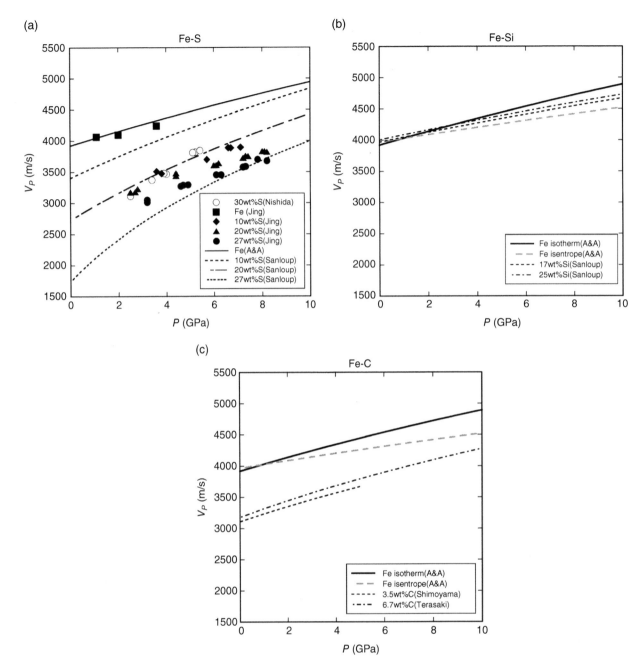

Figure 11.3 Effect of pressure on the compressional wave velocity (V_p). The curves correspond to the calculated V_p values from the density and K_{0T} values listed in Table 11.1 at 1900 K. (a) Fe-S points are from the results of Fe, Fe–10 wt % S, Fe–20 wt % S, Fe–27 wt % S [*Jing et al.*, 2014], and Fe–30 wt % S [*Nishida et al.*, 2013]. The V_p curves of Fe and Fe-S are calculated from the density based on the data of *Anderson and Ahrens* [1994] and *Sanloup et al.* [2000], respectively. (b) The V_p curves of Fe-Si calculated from the density based on the data of *Sanloup et al.* [2004]. The solid and blue dashed curves correspond to the V_p of liquid Fe [*Anderson and Ahrens*, 1994] along the isotherm at 1900 K and the isentrope, respectively. (c) The V_p curves of liquid Fe-C calculated from the density based on the data of *Terasaki et al.* [2010] and that of *Shimoyama et al.* [2013].

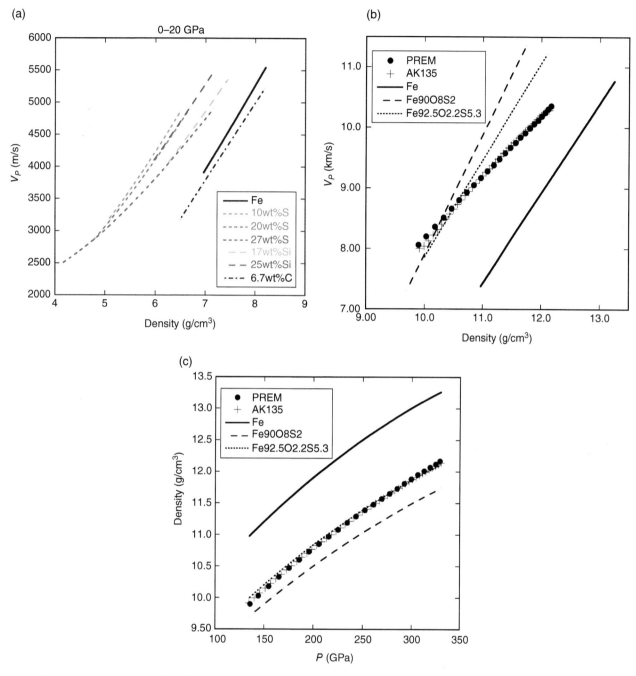

Figure 11.4 (a) A ρ–V_P plot of liquid Fe, Fe-S, Fe-Si, and Fe-C calculated in the pressure range 0–20 GPa at 1900 K using elastic parameters from the following references. Fe ρ = *Anderson and Ahrens* [1994], V_P = *Jing et al.* [2014]; Fe-S ρ = *Sanloup et al.* [2000], V_P = *Jing et al.* [2014]; Fe-Si ρ, V_P = *Sanloup et al.* [2004]; and Fe-C ρ, V_P = *Terasaki et al.* [2010]. (b) A ρ–V_P plot of the outer core inferred from seismic models (PREM, *Dziewonski and Anderson* [1981]; AK135, *Kennett et al.* [1995]) and the ρ–V_P profiles of liquid Fe [*Sakaiya et al.,* 2014] and Fe-O-S [*Huang et al.* 2011] at the outer core pressure. (c) A P–ρ plot of the outer core and the P–ρ profiles of liquid Fe and Fe-O-S. It is noted that temperature conditions of Fe–alloy data are along the Hugoniot temperatures.

11.4.2. Possible Composition of the Outer Core

To provide a constraint on the composition of Earth's outer core, the most direct way is to compare the density and sound velocity data of liquid Fe alloys under core conditions with observed seismic models, such as the preliminary reference Earth model (PREM) [*Dziewonski and Anderson*, 1981] and AK135 [*Kennett et al.*, 1995]. (For details of the seismic properties and models of the core, see Chapter 9.) The ρ–V_p behavior of liquid Fe alloys under core conditions can be estimated from measurements taken at the core conditions or by extrapolation of measured ρ–V_p data to the core conditions, assuming Birch's law holds.

Figure 11.4b shows a ρ–V_p plot of seismic model data of the outer core (PREM and AK135) and liquid Fe and Fe-O-S measured at the core conditions. For reference, the pressure–ρ relation is also shown in Figure 11.4c. The ρ–V_p profile of liquid Fe is clearly lower than the seismic models. In the range comparable with the outer core pressures, the PREM density is approximately 5%–10% [*Anderson and Ahrens*, 1994] or 6%–14% [*Sakaiya et al.*, 2014] smaller than the liquid Fe density. The PREM value of V_p is about 3% [*Anderson and Ahrens*, 1994] or 1%–4% [*Sakaiya et al.*, 2014] higher than the value of V_p of liquid Fe. Thus, a combination of light elements, which can explain both ρ and V_p gaps between the seismic models and liquid Fe, is highly likely to exist in the outer core.

The proposed composition models inferred from the physical properties are summarized in Table 11.3. If the light elements in the outer core are assumed to be S and O, then the Fe–0.5 wt % O–9.5 wt % S composition provides the best match to the ρ–V_p profile in the outer core [*Huang et al.*, 2011], even though $dV_p/d\rho$ is slightly higher than the observed seismic data. In the case of a combination of Si and S, based on the measured densities of liquid Fe-Ni-S and Fe-Ni-Si and assuming ideal mixing, Fe–5 wt % Ni–6 wt % S–2 wt % Si matches the PREM density [*Morard et al.*, 2013] although it is noted that the K_{0T} of Fe-Ni-Si reported by *Morard et al.* [2013] is significantly higher than that of Fe as mentioned in Section 11.2.2.2. Alternatively, if only Si is considered as the

light element, then the density of Fe–9 wt % Ni–10 wt % Si at the geotherm conditions agrees well with the PREM density [*Zhang et al.*, 2014].

The composition models of the outer core have been also proposed from theoretical approach on elastic properties of liquid Fe alloy. Ab initio calculation based on density functional theory provides the chemical potentials of light elements in Fe and their molar volumes [*Alfè et al.*, 2002]. By matching the calculated liquid and solid densities of Fe-S, Fe-Si, and Fe-O with observed density jump at ICB, the estimated composition of the outer core is reported to 10 ± 2.5 at % of S or Si together with 8 ± 2.5 at % of O. Recent ab initio molecular dynamic calculations on the density and bulk sound velocity of liquid Fe alloys took into account the effect of lighter elements of S, Si, O, and C [*Badro et al.*, 2014]. They suggested that the best numerical solution for the core composition is Fe–Ni–3.7 wt % O–1.9 wt % Si with no S or C. In summary, several possible core compositions (Table 11.3) can be proposed based on a geophysical approach, although these models depend on a combination of light elements. So, for the next step, a combination of light elements is required to fulfill not only the geophysical constraints but also the geochemical constraints, such as abundance of light elements in Earth's interior and their solubility in liquid Fe under the relevant core conditions. (For the core composition using a geochemical approach, see Chapter 12.)

11.5. TRANSPORT PROPERTIES OF LIQUID Fe ALLOYS

The transport properties (such as viscosity, diffusivity, electrical conductivity, and thermal conductivity) of liquid Fe alloys govern the dynamic processes of the outer core. Convectional motion of the outer core is controlled by core magnetohydrodynamics, i.e., the interaction between fluid dynamics and the magnetic field. Thus, Lorenz, Coriolis, and viscous forces contribute to the core convection and characterize the convection patterns. Viscosity and electrical conductivity are closely related to viscous and Lorenz forces,

Table 11.3 Proposed compositions of the outer core.

References	S (w%)	Si (w%)	O (w%)	C (w%)	Ni (w%)	Methods	Matching[a]
Huang et al. [2011]	9.5	–	0.5	–	–	Shock (gas gun)	ρ, V_p
Morard et al. [2013]	6	2	–	–	5	Diamond anvil cell	ρ
Zhang et al. [2014]	–	10	–	–	9	Shock (gas-gun)	ρ
Alfè et al. [2002]	6.4[b]	–	2.6	–	–	ab initio (DFT)	ρ
	–	5.6[b]	2.6	–	–	ab initio (DFT)	ρ
Badro et al. [2014]	0	1.9	3.7	0	5.6	ab initio MD	ρ, V_p

[a] Used properties to estimate the composition by matching with seismic data.
[b] S or Si is considered together with O.

respectively. In this section, we focus on the viscosity of liquid Fe alloys.

The viscosity of liquid Fe, Fe-S, and Fe-C at high pressures has been measured using the falling sphere method combined with in situ X-ray radiography [*Dobson et al.*, 2000a; *Terasaki et al.*, 2001, 2006; *Urakawa et al.*, 2001; *Rutter et al.*, 2002; *Perrillat et al.*, 2010]. The measured viscosity of Fe-S is in the range 3.2–35.6 mPa·s up to 16 GPa and 2173 K. It is known that the viscosity coefficient (η) has an Arrhenian dependence on the temperature (T)

and pressure (P), as expressed by the equation [e.g., *Poirier*, 1988]

$$\eta = \eta_0 \exp\left[\frac{Q + P\,\Delta V}{RT}\right], \qquad (11.6)$$

where η_0, R, Q, and ΔV are the pre-exponential factor, gas constant, activation energy of viscous flow, and activation volume, respectively. Thus, Q and ΔV are regarded as being indices for the effect of temperature and pressure, respectively. The value of ΔV of liquid Fe-S has been estimated to be 0.7–1.5 cm³/mol, independent of sulfur content. Liquid Fe and Fe-C show a similar ΔV value (1.2 cm³/mol) [*Rutter et al.*, 2002; *Terasaki et al.*, 2006], suggesting that all these alloy liquids have a relatively small pressure dependence on viscosity as shown in Figure 11.5. The reported Q and ΔV values of liquid Fe alloys are listed in Table 11.4. In terms of the effect of light elements on the viscosity, a weak dependence of the light element content on the viscosity of liquid Fe has been reported for liquid Fe-S and Fe-C at high pressures [*Terasaki et al.*, 2001, 2006; *Alfè and Gillan*, 1998]. This is also confirmed from Figure 11.5.

Viscosity is closely linked to diffusivity based on the Stokes–Einstein relation [e.g., *Dobson et al.*, 2001]

$$\eta = \frac{k_B T}{Da}, \qquad (11.7)$$

where k_B, D, and a correspond to Boltzmann constant, diffusion coefficient, and effective diameter of the atom or interatomic distance, respectively. Diffusivity was measured for liquid Fe, Fe-S, and Fe₃C up to 20 GPa and 2393 K on a basis of tracer (^{57}Fe) or self-diffusion [*Dobson*, 2000b, 2002; *Dobson and Widenbeck*, 2002]. Calculated viscosity using measured diffusion coefficients with equation (11.7) ranges 27–67 mPa·s for liquid Fe$_{61}$S$_{39}$ at around 5 GPa and 1293–1473 K and 15 mPa·s for liquid Fe₃C along the melting

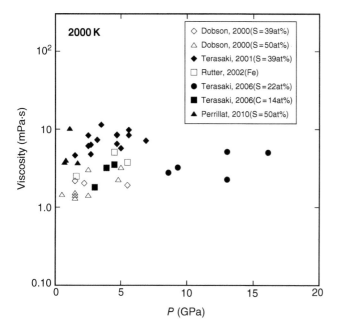

Figure 11.5 Viscosities of liquid Fe, Fe-S, and Fe-C as a function of pressure at 2000 K. The plotted viscosities are corrected to the value at 2000 K using thermodynamic parameters in Table 11.4 with equation (11.6).

Table 11.4 Viscosities and thermodynamic parameters of liquid Fe alloys.

References	Alloy	Content[a] (w%)	Content (a%)	P range (GPa)	T range (K)	η (mPa·s)	Q (kJ/mol)	ΔV (cm³/mol)
Dobson et al. [2000]	Fe-S	27	39	1.5–5.5	1423–1647	7.4–22	100(5)	
	Fe-S	36	50	0.5–5.0	1723–1980	3.8–19.5	255(18)	
Urakawa et al. [2001]	Fe-S	27	39	5.0–6.9	1333–1337	16–24		0.7
Terasaki et al. [2001]	Fe	–	–	2.8	1923	17.6		
	Fe-S	13	20	2.7	1611	13.1		
	Fe-S	20	30	2.7	1669	12.4		
	Fe-S	27	39	1.5–6.9	1253–1821	8.2–35.6	30(9)	1.5(7)
Rutter et al. [2002]	Fe	–	–	1.6–5.5	2050	2.4–4.8	40	
Terasaki et al. [2006]	Fe-S	14	22	8.6–16.1	1723–1798	3.2–8.5	30	1.46
	Fe-C	3.4	14	3.0–4.5	1605–1843	3.8–6.3	46.6	1.2
Perrillat et al. [2010]	Fe-S	36	50	0.7–1.7	1520–1600	6.3–17.9	35	

[a] Content of light element.

curve. These values are comparable with the measured viscosities listed in Table 11.4 if the values are corrected to the same P–T conditions using equation (11.6).

First-principle calculations predicted that the viscosity of Fe alloy liquids under the core–mantle boundary (CMB) conditions (135 GPa, ~4300 K) is 12–15 mPa·s [*de Wijs et al.*, 1998] and around 6.7–6.9 mPa·s for liquid Fe, $Fe_{82}Si_{10}O_8$, and $Fe_{79}Si_8O_{13}$ [*Pozzo et al.*, 2013]. These values are comparable with the lower bound of the extrapolated viscosity (3.4–468 mPa·s) under CMB conditions calculated using the thermodynamic parameters (Q and ΔV) in Table 11.4 and the Arrhenius relationship in equation (11.6). Therefore, the outer core is most likely to consist of inviscid liquid.

The importance of a viscous force in the outer core can be evaluated using the dimensionless Ekman number. The Ekman number (E) is defined as the ratio of the viscous and Coriolis forces as $E = \eta/(\rho \Omega L^2)$, where η, ρ, Ω, and L are the shear viscosity, the density, Earth's rotational velocity, and the core thickness, respectively. If the viscosity of the outer core is of the order of 10^0–10^1 mPa·s, then E is estimated to be small (~10^{-15}) [*de Wijs et al.*, 1998], suggesting that the viscous force acting on the core is negligible compared with the Coriolis force. This small value of E leads to a form of sheet plume convection rather than columnar cell convection structures, which are established with larger values of E [*Kageyama et al.*, 2008].

11.6. CONCLUSIONS

In this chapter, we reviewed the elastic and transport properties of liquid Fe alloys and discussed the effect of light elements on these properties and the proposed outer core composition models from the latest results. The main aspects are summarized below.

Sulfur significantly reduces both the density and the K_{0T} of liquid Fe, while C reduces these properties moderately, whereas Si does not have much effect on the K_{0T} and only reduces the density moderately. These effects of light element content on K_{0T} correlate with a difference in the local structure of these liquids.

The V_P of liquid Fe-S reduces with increasing S content, but the effect of S on V_P is likely to be reversed at higher pressures. The V_P of liquid Fe-Si along an isotherm is larger than that of liquid Fe at low pressures (~3 GPa). This tendency is opposite to the effect of S content.

From both measured and calculated values of ρ and V_P, most of the liquids follow a linear ρ–V_P relationship, i.e., Birch's law applies. Both S and Si dissolution shift the ρ–V_P plot toward the upper left-hand side of the plot, while the addition of C shifts the ρ–V_P plot toward the lower side of the plot. Based on matching the density and/or sound velocity with seismic data, several possible core compositions were proposed, as listed in Table 11.3.

These composition models depend on combinations of light elements.

Liquid Fe, Fe-S, and Fe-C have a relatively small pressure dependence on the viscosity. From first-principle calculations, the viscosity of liquid Fe alloys under the CMB conditions is estimated to be on the order of 10^0–10^1 mPa·s, which is consistent with the lower bound of the extrapolated values based on measured viscosities. Therefore, the outer core is likely to be composed of inviscid liquid.

ACKNOWLEDGMENT

The author thanks the two anonymous reviewers for their constructive comments and to Rebecca A. Fischer for handling the manuscript. This work was partly supported by Grants-in-Aid for scientific research from the Ministry of Education, Culture, Science, and Sport and Technology of the Japanese Government (no. 23340159 and 26247089).

REFERENCES

Alfè, D., and M. J. Gillan (1998), First-principles simulations of liquid Fe-S under Earth's core conditions, *Phys. Rev. B, 58*, 8248–8256.

Alfè, D., G. Kresse, and M. J. Gillan (2000), Structure and dynamics of liquid iron under Earth's core conditions, *Phys. Rev. B, 61*, 132–142.

Alfè, D., M. J. Gillan, and G. D. Price (2002), Composition and temperature of the Earth's core constrained by combining ab initio calculations and seismic data, *Earth Planet. Sci. Lett., 195*, 91–98.

Anderson, W. W., and T. J. Ahrens (1994), An equation of state for liquid iron and implications for the Earth's core, *J. Geophys. Res., 99*, 4273–4284.

Antonangeli, D., T. Komabayashi, F. Occelli, E. Borissenko, A. C. Walters, G. Fiquet, and Y. Fei (2012), Simultaneous sound velocity and density measurements of hcp iron up to 93 GPa and 1100 K: An experimental test of the Birch's law at high temperature, *Earth Planet. Sci. Lett., 331–332*, 210–214.

Badro, J., G. Fiquet, F. Guyot, E. Gregoryanz, F. Occelli, D. Antonangeli, and M. d'Astuto (2007), Effect of light elements on the sound velocities in solid iron: Implications for the composition of Earth's core, *Earth Planet. Sci. Lett., 254*, 233–238.

Badro, J., A. S. Cote, and J. P. Brodholt (2014), A seismologically consistent compositional model of Earth's core, *Proc. Natl. Acad. Sci., 111*, 7542–7545.

Balog, P. S., R. A. Secco, D. C. Rubie, and D. J. Frost (2003), Equation of state of liquid Fe-10 wt % S: Implications for the metallic cores of planetary bodies, *J. Geophys. Res., 108*, 2124, doi:10.1029/2001JB001646.

Birch, F. (1947), Finite elastic strain of cubic crystals, *Phys. Rev., 71*, 809–924.

Birch, F. (1961a), The velocity of compressional waves in rocks to 10 kilobars, Part 2, *J. Geophys. Res., 66*, 2199–2224.

Birch, F. (1961b), Composition of the Earth's mantle, *Geophys. J. R. Astron. Soc.*, *4*, 295–311.

Brown, J. M., and R. G. McQueen (1986), Phase transitions, Gruneisen parameter, and elasticity for shocked iron between 77 GPa and 400 GPa, *J. Geophys. Res.*, *91*, 7485–7494.

Chen, J., D. J. Weidner, L. Wang, M. T. Vaughan, and C. E. Young (2005), Density measurements of molten materials at high pressure using synchrotron X-ray radiography: Melting volume of FeS, In *Advances in High-Pressure Technology for Geophysical Applications*, edited by J. Chen, Y. Wang, T. S. Duffy, G. Shen, and L. F. Dobrzhinetskaya, 185–194, Elsevier, Amsterdam.

Chen, J., T. Yu, S. Huang, J. Girard, and X. Liu (2014), Compressibility of liquid FeS measured using X-ray radiograph imaging, *Phys. Earth Planet. Inter.*, *228*, 294–299.

De Wijs, G. A., G. Kresse, L. Vocadlo, D. Dobson, D. Alfè, M. J. Gillan, and G. D. Price (1998), The viscosity of liquid iron at the physical conditions of the Earth's core, *Nature*, *392*, 805–807.

Dobson, D. P., W. A. Crichton, L. Vocadlo, A. P. Jones, Y. Wang, T. Uchida, M. Rivers, S. Sutton, and J. P. Brodholt (2000a), In situ measurement of viscosity of liquids in the Fe-FeS system at high pressure and temperatures, *Ame. Mineral.*, *85*, 1838–1842.

Dobson, D. P. (2000b), ^{57}Fe and Co tracer diffusion in liquid Fe-FeS at 2 and 5 GPa, *Phys. Earth Planet. Inter.*, *120*, 137–144.

Dobson, D. P. (2002), Self-diffusion in liquid Fe at high pressure, *Phys. Earth Planet. Inter.*, *130*, 271–284.

Dobson, D. P., and M. Wiedenbeck (2002), Fe- and C-self diffusion in liquid Fe$_3$C to 15 GPa, *Geophys. Res. Lett.*, *29*(21), 2006, doi:10.1029/2002GL015536.

Dobson, D. P., W. A. Crichton, L. Vocadlo, A. P. Jones, Y. Wang, T. Uchida, M. Rivers, S. Sutton, and J. P. Brodholt (2000), In situ measurement of viscosity of liquids in the Fe-FeS system at high pressure and temperatures, *Am. Mineral.*, *85*, 1838–1842.

Dobson, J. P. Brodholt, L. Vocadlo, and W. A. Crichton (2001), Experimental verification of the Stokes-Einstein relation in liquid Fe-FeS at 5 GPa, *Mol. Phys.*, *99*, 773–777.

Dziewonski, A. M., and D. L. Anderson (1981), Preliminary reference Earth model, *Phys. Earth Planet. Int.*, *25*, 297–356.

Hixson, R. S., M. A. Winkler, and M. L. Hodgon (1990), Sound speed and thermophysical properties of liquid iron and nickel, *Phys. Rev. B*, *42*, 6485–6491.

Huang, H., Y. Fei, L. Cai, F. Jing, X. Hu, H. Xie, L. Zhang, and Z. Gong (2011), Evidence for an oxygen-depleted liquid outer core of the Earth, *Nature*, *479*, 513–517.

Huang, H., S. Wu, X. Hu, Q. Wang, and X. Wang (2013), Shock compression of Fe–FeS mixture up to 204 GPa, *Geophys. Res. Lett.*, *40*, 1–5, doi:10.1002/grl.50180.

Ichikawa, H., T. Tsuchiya, and Y. Tange (2014), The P-V-T equation of state and thermodynamic properties of liquid iron, *J. Geophys. Res.*, *119*, 240–252, doi:10.1002/2013JB10732.

Jing, Z., Y. Wang, Y. Kono, T. Yu, T. Sakamaki, C. Park, M. L. Rivers, S. R. Sutton, and G. Shen (2014), Sound velocity of Fe-S liquids at high pressure: Implications for the Moon's molten outer core, *Earth Planet. Sci. Lett.*, *396*, 78–87.

Kageyama, A., T. Miyagoshi, and T. Sato (2008), Formation of current coils in geodynamo simulations, *Nature*, *454*, 1106–1109.

Kennett, B. L. N., E. R. Engdahl, and R. Buland (1995), Constraints on seismic velocities in the Earth from travel times, *Geophys. J. Int.*, *122*, 108–124.

Mao, Z., J-F. Lin, J. Liu, A. Alatas, L. Gao, J. Zhao, and H-K. Mao (2012), Sound velocities of Fe and Fe-Si alloy in the Earth's core, *Proc. Natl. Acad. Sci.*, *109*, 10,239–10,244.

McDonough, W. F., and S.-S. Sun (1995), The composition of the Earth, *Chem. Geol.*, *120*, 223–253.

Morard, G., C. Sanloup, G. Fiquet, M. Mezouar, N. Rey, R. Poloni, and P. Beck (2007), Structure of eutectic Fe-FeS melts to pressures up to 17 GPa: Implications for planetary cores, *Earth Planet. Sci. Lett.*, *263*, 128–139.

Morard, G., J. Siebert, D. Andrault, N. Guignot, G. Garbarino, F. Guyot, and D. Antonangeli (2013), The Earth's core composition from high pressure density measurements of liquid iron alloys, *Earth Planet. Sci. Lett.*, *373*, 169–178.

Nagamori, M. (1969), Density of molten Ag-S, Cu-S, Fe-S, and Ni-S systems, *Trans. Metall. Soc. AIME*, *245*, 1897–1903.

Nakajima, Y., E. Takahashi, T. Suzuki, and K. Funakoshi (2009), "Carbon in the core" revisited, *Phys. Earth Planet. Inter.*, *174*, 202–211.

Nasch, P. M., and M. H. Manghnani (1998), Molar volume, thermal expansion, and bulk modulus in liquid Fe-Ni alloys at 1 bar: Evidence for magnetic anomalies?, In *Properties of Earth and Planetary Materials at High Pressure and Temperature*, M. H. Manghnani and T. Yagi, 307–317, AGU, Washington, D.C.

Nasch, P. M., M. H. Manghnani, and R. A. Secco (1994), Sound velocity measurements in liquid iron by ultrasonic interferometry, *J. Geophys. Res.*, *99*, 4285–4291.

Nasch, P. M., M. H. Manghnani, and R. A. Secco (1997), Anomalous behavior of sound velocity and attenuation in liquid Fe-Ni-S, *Science*, *277*, 219–221.

Nguyen, J. H., and N. C. Holmes (2004), Melting of iron at the physical conditions of the Earth's core, *Nature*, *427*, 339–342.

Nishida, K., H. Terasaki, E. Ohtani, and A. Suzuki (2008), The effect of sulfur content on density of the liquid Fe-S at high pressure, *Phys. Chem. Minerals.* *35*, 417–423.

Nishida, K., E. Ohtani, S. Urakawa, A. Suzuki, T. Sakamaki, H. Terasaki, and Y. Katayama (2011), Density measurement of liquid FeS at high pressures using synchrotron X-ray absorption, *Am. Mineral.*, *96*, 864–868.

Nishida, K., Y. Kono, H. Terasaki, S. Takahashi, M. Ishii, Y. Shimoyama, Y. Higo, K. Funakoshi, T. Irifune, and E. Ohtani (2013), Sound velocity measurements in liquid Fe-S at high pressure: Implications for Earth's and lunar cores, *Earth Planet. Sci. Lett.* *362*, 182–186.

Ohtani, E. (2013), Chemical and physical properties and thermal state of the core, in *Physics and Chemistry of the Deep Earth*, (edited by S. Karato, pp. 244–270, Wiley, Hoboken, N.J.

Perrillat, J-P., M. Mezouar, G. Garbarino, and S. Bauchau (2010), In situ viscometry of high-pressure melts in the Paris-Edinburgh cell: Application to liquid FeS, *High Press. Res.*, *30*, 415–423.

Poirier, J. P. (1988), Transport properties of liquid metals and viscosity of the Earth's core, *Geophys. J.*, *92*, 99–105.

Poirier, J-P. (1994), Light elements in the Earth's outer core: A critical review, *Phys. Earth Planet. Int.*, *85*, 319.

Pozzo, M., C. Davies, D. Gubbins, and D. Alfè (2013), Transport properties for liquid silicon-oxygen-iron mixtures at Earth's core conditions, *Phys. Rev. B*, *87*, 014,110.

Rutter, M. D., R. A. Secco, H. Liu, T. Uchida, M. L. Rivers, S. R. Sutton, and Y. Wang (2002), Viscosity of liquid Fe at high pressure, *Phys. Rev. B*, *66*, 060,102.

Sakaiya, T., H. Takahashi, T. Kondo, T. Kadono, Y. Hironaka, T. Irifune, and K. Shigemori (2014), Sound velocity and density measurements of liquid iron up to 800 GPa: A universal relation between Birch's law coefficients for solid and liquid metals, *Earth Planet. Sci. Lett.*, *392*, 80–85.

Sanloup, C., F. Guyot, P. Gillet, G. Fiquet, M. Mezouar, and I. Martinez (2000), Density measurements of liquid Fe-S alloys at high-pressure, *Geophys. Res. Lett.*, *27*, 811–814.

Sanloup, C., F. Guyot, P. Gillet, and Y. Fei (2002), Physical properties of liquid Fe alloys at high pressure and their bearings on the nature of metallic planetary cores, *J. Geophys. Res.*, *107*(B11), 2272, doi:10.1029/2001JB000808.

Sanloup, C., G. Fiquet, E. Gregoryanz, G. Morard, and M. Mezouar (2004), Effect of Si on liquid Fe compressibility: Implications for sound velocity in core materials, *Geophys. Res. Lett.*, *31*, doi:10.1029/2004GL019526.

Sanloup, C., W. van Westrenen, R. Dasgupta, H. Maynard-Casely, and J.-P. Perrillat (2011), Compressibility change in iron-rich melt and implications for core formation models, *Earth Planet. Sci. Lett.*, *306*, 118–122.

Shimoyama, Y., H. Terasaki, E. Ohtani, S. Urakawa, Y. Takubo, K. Nishida, A. Suzuki, and Y. Katayama (2013), Density of Fe-3.5wt%C liquid at high pressure and temperature and the effect of carbon on the density of the molten iron, *Phys. Earth Planet. Inter.*, *224*, 77–82.

Tateyama, R., E. Ohtani, H. Terasaki, K. Nishida, Y. Shibazaki, A. Suzuki, and T. Kikegawa (2011), Density measurements of liquid Fe-Si alloys at high pressure using sink-float method, *Phys. Chem. Minerals*, *38*, 801–807.

Terasaki, H., T. Kato, S. Urakawa, K. Funakoshi, A. Suzuki, T. Okada, M. Maeda, J. Sato, T. Kubo, and S. Kasai (2001), The effect of temperature, pressure, and sulfur content on viscosity of the Fe-FeS melt, *Earth Planet Sci. Lett.*, *190*, 93–101.

Terasaki, H., A. Suzuki, E. Ohtani, K. Nishida, T. Sakamaki, and K. Funakoshi (2006), Effect of pressure on the viscosity of Fe-S and Fe-C liquids up to 16 GPa, *Geophys. Res. Lett.*, *33*, L22307, doi:10.1029/2006GL027147.

Terasaki, H., K. Nishida, Y. Shibazaki, T. Sakamaki, A. Suzuki, E. Ohtani, and T. Kikegawa (2010), Density measurement of Fe$_3$C liquid using X-ray absorption image up to 10 GPa and effect of light elements on compressibility of liquid iron, *J. Geophys. Res.*, *115*, B06207, doi:10.1029/2009JB006905.

Urakawa, S., H. Terasaki, K. Funakoshi, T. Kato, and A. Suzuki (2001), Radiographic study on the viscosity of the Fe-FeS melts at the pressure of 5 to 7 GPa, *Am. Mineral.*, *86*, 578–582.

Waseda, Y. (1980), *The Structure of Non-Crystalline Materials: Liquids and Amorphous Solids*, McGraw-Hill int., New York and London.

Yu, X., and R. A. Secco (2008), Equation of state of liquid Fe-17wt%Si to 12 GPa, *High Press. Res.*, *28*, 19–28.

Zhang, Y., T. Sekine, H. He, Y. Yu, F. Liu, and M. Zhang (2014), Shock compression of Fe-Ni-Si system to 280 GPa: Implications for the composition of the Earth's outer core, *Geophys. Res. Lett.*, *41*, 4554–4559, doi:1002/2014GL06067010.

Part IV
Chemistry and Phase Relations
of Deep Interior

12

The Composition of the Lower Mantle and Core

William F. McDonough

ABSTRACT

Compositional models of the core and lower mantle are reviewed and assessed. The assumptions in the models and the constraints and uncertainties about the Earth's interior are presented. Although a compositional model for the lower mantle that matches that of the upper mantle for major elements is most compatible with observations and constraints, uncertainties are such that competing compositional models are tenable. Based on chondritic models, more than 90% of the mass for the Earth is composed of Fe, O, Mg and Si and the addition of Ni, Ca, Al and S accounts for more than 98% by mass the composition of the Earth. There is no fixed Mg/Si ratio for chondrites; variations in this ratio reflect the proportion of olivine (2:1 molar Mg/Si) to pyroxene (1:1 molar Mg/Si) accreted in the planet or chondritic parent body. Observations on active accretion disks reveal rapid grain growth in the inner disk region and spatial variations in the relative proportions of olivine to pyroxene. Compositional models for the core are constrained by limited variation in chondrites for key siderophile element ratios (e.g., Fe/Ni, Ni/Co, and Ni/Ir). The amount and relative proportions of light element(s) in the core are poorly constrained, with tradeoffs and modeling uncertainties in core temperatures and compositional space that allow for a range of model solutions. Constraints on the absolute and relative abundances of moderately volatile and volatile elements in the Earth are consistent with only ~2% by mass of sulfur and a negligible role for H, C or N in the core. There is no evidence that heat producing elements (HPE: K, Th and U) are in the core at any significant level. Geoneutrino studies are placing global scale limits on the amount of Th and U in the Earth, which in turn will constrain models for the composition of the bulk silicate Earth, the mode proportion of the Ca-bearing phase in the deep mantle, and the thermal evolution of the planet. There is no clear geochemical evidence for core-mantle exchange, but this does not preclude it from happening and not being sampled.

12.1. INTRODUCTION

The composition of the bulk Earth, the silicate Earth, and the core requires integrating the available physical and chemical data of Earth and its domains into a single coherent picture. Constraints for the composition of the deep Earth are less so than for the upper mantle and crust, and hence there is more speculation as to the nature and evolution of this remote region. There are fundamentally two significant unknowns about the lower mantle and core: (1) What element or elements comprise the light element component in the core such that it accounts for the observed lower density of the inner and outer core as compared to that of iron at these pressure and temperature conditions, and (2) is the lower mantle chemically similar

Department of Geology, University of Maryland, College Park, Maryland, USA

Deep Earth: Physics and Chemistry of the Lower Mantle and Core, Geophysical Monograph 217, First Edition.
Edited by Hidenori Terasaki and Rebecca A. Fischer.
© 2016 American Geophysical Union. Published 2016 by John Wiley & Sons, Inc.

to the upper mantle or is it distinctly different, and, if so, then the question is how different? This chapter examines the composition of the core and of the mantle and its domains, upper and lower, its physical and chemical attributes, and its evolution. I consider uncertainties in our knowledge and examine prospects of where we might make significant progress to resolve appropriately the question regarding composition of the lower and upper mantle.

Let's start, however, with fundamental definitions, particularly of what is the lower and upper mantle. The core and its inner and outer domains are physically distinct and straight forward to define as it is sharply demarcated seismically, whereas domains in the mantle are less clearly identified. A traditional, simple view is that the upper mantle is that region beneath the Moho and above the 660 km deep seismic discontinuity, which also marks a major phase change [ringwoodite disproportioning to bridgmanite (Mg-perovskite) and ferropericlase]. The lower mantle is from the 660 km deep seismic boundary to the core-mantle boundary (CMB). This definition is the one I will use throughout this discussion, thus the lower mantle includes domains that seismologists have identified, including the "D" layer, the LLSVPs (large low-shear-velocity provinces) the ULVZ (ultralow velocity zone), and the thermal boundary layer, which is the conductive interface at the base of the mantle (the equivalent to upper thermal boundary layer, the lithosphere) [*Dziewonski and Anderson*, 1981; *Garnero and McNamara*, 2008; *Lekic et al.*, 2012; *McNamara et al.*, 2010]. Seismological studies [*Fukao and Obayashi*, 2013] have also identified zones between 660 and ~1000 km depth where slabs of subducting oceanic lithosphere have turned and laid down sub-horizontally, thus potentially developing a distinctive region in the top of the lower mantle.

12.2. CHONDRITES: BUILDING BLOCKS OF TERRESTRIAL PLANETS

The inner four rocky planets are terrestrial bodies made up mostly of accreted dust as compared with the volatile gases that dominated the nebular disk. Our Sun formed in the center of this disk, and outward the protoplanetary disk gave rise to planets, moons, and icy bodies. The nebular disk, which has its origins due to the gravitational collapse of a portion of a much larger interstellar molecular cloud, contained a substantial inventory of gas and dust, with H and He dominating, hence the Sun's composition. Chemical differentiation is readily apparent in the solar system, with inner rocky planets being more refractory, and outer gas giants having a greater inventory of H, He, and the ices, compounds of H, C, N, and O. Computer simulations of accretion in the inner disk region are strongly influenced by gravitational processing, including a potential role played by Jupiter and less so Saturn, and

their inward migration during the early evolution of the protoplanetary disk. It is quite possible that different chemical components were redistributed prior, during, and subsequent to terrestrial planet formation due to inward and then outward migration of the orbits of the gas giants.

Surrounded by a thin envelope of gas and water, the Earth, a terrestrial planet, includes the present-day bulk silicate Earth (BSE, which is made up of continental and oceanic crust and the entire mantle) and the metallic core, and together this material is presumed to have a chondritic composition. Chondritic meteorites are the undifferentiated materials from the earliest days of the solar system that formed small bodies, planetismals a few to many tens of kilometers in scale that never experienced metal-silicate fractionation. The composition of chondrites has been compared to that of the sun, with the CI1 carbonaceous chondrite type having the closest match to that of the solar photosphere (Figure 12.1, top left). Other major chondrite groups also have compositions comparable to the solar photosphere, with differences in low abundance elements and the volatile elements (Figure 12.1). These compositional matches are significant given the Sun is the mass of the solar system (Jupiter, the largest of the planets is 1/1000 the mass of the Sun).

There are several different groups of chondritic meteorites that are classified by their chemical and isotopic attributes and importantly by their petrographic features, that is, the oxidation state of their minerals, the size and the abundance of chondrules (melt droplets from the nebula) relative to matrix material and other components (e.g., CAI, calcium aluminum inclusion [*MacPherson*, 2014], metal and silicate grains) [*Scott and Krot*, 2014]. The most common group is the ordinary chondrites, which are divided into three subcategories, H, L, and LL, referring to high, low, and low-low Fe contents. The ordinary chondrites represent some 80% of all meteoritic materials that fall to the Earth (considered "falls," as opposed to "finds," which are found randomly in deserts, Antarctica, and other locations worldwide). The other two major groups of chondritic meteorites are the enstatite and carbonaceous chondrites, which are minor components (2% and 4%, respectively) of the inventory of meteoritic falls.

The inventory of meteorites is restricted to material that fell recently ($<10^6$ years) to Earth, aside from rare specimens found in ancient sedimentary layers. Over the last 1100 years, there have been about 1100 meteoritic falls that have been observed and collected. Thus, our collection of meteorites might be representative of the present-day composition of the asteroid belt, or potentially biased, by the present-day orbital resonance of the Jupiter-Sun system. It is significant that the Kirkwood gaps document a gross mass reduction in the asteroid belt that reflect the effects of the Jupiter-Sun orbital

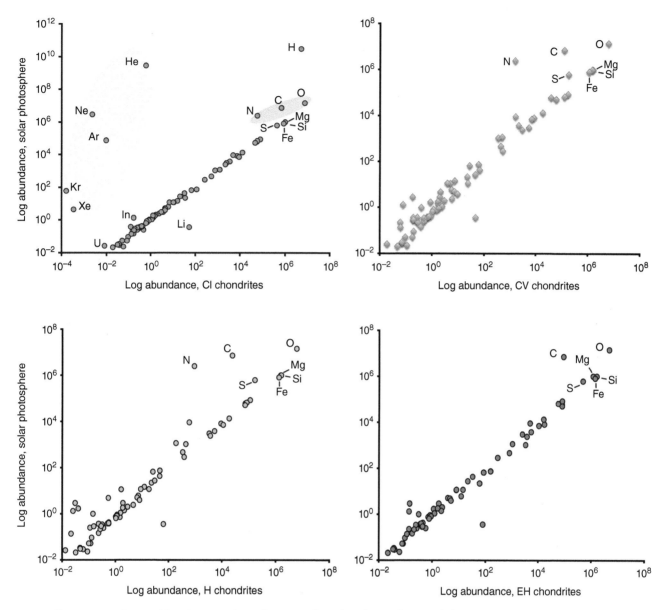

Figure 12.1 Compositional comparisons between the solar photosphere and the compositions of (top left) CI1 carbonaceous chondrite, which has the closest match to that of the Sun, (top right) CV carbonaceous chondrite, (bottom left) ordinary chondrite, and (bottom right) enstatite chondrite. Chondrite data from *Wasson and Kallemeyn* [1988] and for the Sun from *Asplund et al.* [2009] and *Lodders* [2003].

resonance and demonstrate that the asteroid belt is a processed region that does not necessarily provide an unbiased sample of the material that condensed and accreted from the nebula 4.57 billion years ago [*DeMeo and Carry*, 2014]. Thus, it is necessary to treat data from chondrites as a guide to planetary compositions and not as a requirement for specifically describing their composition [*McDonough and Sun*, 1995].

Understanding the nature and origins of the present-day positions in the solar system of asteroids and chondritic source regions, however, presents us with challenges

regarding the original distribution of components in the solar system [*DeMeo and Carry*, 2014]. Earlier ideas of a simple, organized asteroid belt with distinct compositional types separated into different regions of the belt are no longer supported, given data showing a much more mixed distribution of asteroid types throughout the belt, a consequence of the early migrational history of Jupiter and Saturn [*DeMeo and Carry*, 2014]. An enigmatic observation is that carbonaceous chondrites have the highest content of the most volatile elements (including some having abundant organic compounds), as well as the highest abundance of CAIs (i.e., in the CV carbonaceous chondrites), which is the most refractory of materials. These two observations appeared to be contradictory in that the latter material formed early in solar system history and close into the Sun, whereas the former material reflects accretion of cold, more distal, later stage components. Importantly, present-day reflectance spectroscopy and tracking of meteoroids reveal that the carbonaceous chondrites come from the inner and outer portions of the asteroid belt and beyond [*DeMeo and Carry*, 2014]. The inclination and eccentricity of the Sutter's Mill chondrite (CM type), a 2012 observed fall, placed its aphelion close to the orbit of Jupiter, consistent with the known site of some C-type asteroids [*Jenniskens et al.*, 2012]. Thus, models for planet building must recognize that during the early solar system formation (i.e., circa 10^6 years) radial inward migration of Jupiter and Saturn in the disk may have occurred (e.g., Grand Tack model, [*Walsh et al.*, 2011]), while much later after the disk is gone (e.g., several 10^8 years) gravitational scattering by these planets may have transported small bodies from the outer reaches of the solar system inward toward the rocky planets (e.g., Nice mode [*Tsiganis et al.*, 2005]). Consequently, sources of planetary building blocks and the planets themselves are products of integrated processes that required 10^6 to many 10^7 years of construction.

Elements in chondrites are classified according to two scales: their condensation behavior in the nebula and their chemical affinities during the core forming, metal-silicate separation (see Table 12.1 for element classification details). The condensation of the elements from a nebula depends on its temperature and fugacity of hydrogen gas, both of which are highly variable in time and space. It is observed that all chondrite groups inherited similar relative abundances of the refractory elements, those that have 50% condensation temperatures above 1350 K for a nebular gas of ~10^1 Pa of hydrogen [*Lodders*, 2003], but different proportions of major (Mg, Si, Fe and Ni) and volatile elements. Goldschmidt's classification of the elements (i.e., lithophile, siderophile, chalcophile, and atmophile) is based on their chemical affinities during geologic processes, and it provides predictive powers for the separation of elements between the BSE and the core. An element's behavior can, however, change depending on the oxidation/reduction potential (e.g., H or O activity). For example, under the present-day oxygen fugacity, Si in the mantle is wholly lithophile, whereas early on during core formation Si may have been partially siderophile with a portion of Earth's Si inventory being partitioned into the metallic core.

The abundances of elements in the Sun and chondrites are consistent with the theory of stellar nucleosynthesis of the elements in stars, that is, the creation of heavy nuclei from pre-existing nucleons through proton and neutron additions [*Burbidge et al.*, 1957]. The Sun is a main sequence star that is brighter than most in our Milky Way galaxy, and its metallicity is the result of a series of past stellar processes that involved element production in earlier supernovas and AGB (asymptotic giant branch) stars [*Lugaro et al.*, 2014; *Wasserburg et al.*, 1996]. Although there remains a question of whether the absolute metallicity of the Sun's core and surface are similar or different [*Haxton et al.*, 2013], the relative proportions of the major metals (Fe, O, Mg, and Si) in the Sun and the planets remain relatively constant within limits due to their universal mechanism of genesis via nucleosynthesis.

Establishing the composition of the terrestrial planets involves using a relatively standard set of model constraints and assumptions based on physical and chemical observables from the solar system. The initial boundary condition uses orbital and seismic constraints (if available, so far we have such data only for Earth and the Moon) to describe the first-order physical state of the planet (e.g., coefficient of the moment of inertia), specifically its metal to silicate fraction

Table 12.1 Classification of the elements.

>1355 K[a], **Refractory elements**	
Lithophile (not in the core, rock forming)	Be, Al, Ca, Sc, Ti, V[b], Sr, Y, Zr, Nb, Ba, REE, Hf, Ta, Th, U
Siderophile (core, affinities with Fe)	Mo, Ru, Rh, W, Re, Os, Ir, Pt
1355 – 1310 K, **Major elements**	Mg, Si, Fe, Ni
1350 – 250 K, **Moderately volatile**	
Lithophile	Li, B, Na, K, Mn[b], Rb, Cs
Siderophile	P, Co, Ga, Ge, As, Ag, Pd, Sb, Au, Tl, Bi
Chalcophile (associated with sulfide)	S[c], Cu, Se, Cd, In, Sn, Te, Hg, Pb
<250 K, **Highly volatile**	
Atmophile (in the oceans & atmosphere)	H, He, C, N, O, Ne, Ar, Kr, Xe

[a] Half-mass condensation temperature in Kelvin.

[b] V and Mn are lithophile, with considerable siderophilic affinity.

[c] The half-mass condensation temperature for S is 664 K.

(core and BSE proportions). Importantly, the mass fraction of metal to silicate in a planet defines the body's integrated oxidation state, that is, the mass ratio of metallic to oxidized iron (Figure 12.2). This ratio in combination with an understanding of the amount of iron in the silicate shell also defines the Fe/O ratio of the planet. Following this we use the chondritic meteorites as a guide to planetary composition, recognizing the compositional match between the solar photosphere and the chondrites (Figure 12.1), and that the nebular disk from which the planets accreted would have a bulk solar composition.

The chondritic constraints on planetary composition must acknowledge that there is no fixed Mg/Si ratio for chondrites. Variations in this ratio reflect the proportion of olivine (2:1 molar Mg/Si) to pyroxene (1:1 molar Mg/Si) accreted in the planet or chondritic parent body. Studies of astromineralogy increasingly reveal variations in the proportion of olivine to pyroxene in accretion disks, some with inner disk regions being richer in olivine relative to the disk wide composition [*van Boekel et al.*, 2004]. In other disks, however, the abundance of olivine is greater in the outer (versus the inner) part of the circumstellar disk, with differences in disk mineralogy relating to type of star (e.g., T Tauri vs. Herbig Ae/Be stars) [*Bouwman et al.*, 2010]. The inner disk regions (a few AU) show higher abundances of large grains and generally higher crystallinity as compared to outer disk regions, suggesting grain growth occurs more rapidly in the inner disk regions [*D'Alessio et al.*, 2005; *Sargent et al.*, 2009].

The compositions of terrestrial planets are readily described by only a few elements. Four elements, Fe, O, Mg, and Si, make up the bulk of chondrites (with the volatile rich CI1 chondrites having considerable amounts of S, CO_2, and H_2O) on a basis of either mass or atomic proportions and therefore these elements constitute ~90% the mass of the terrestrial planet (Figure 12.3 and Table 12.2). On a degassed basis (minus H_2O and CO_2), if we add five more elements to this list (S, Ca, Al, Ni, and Na),

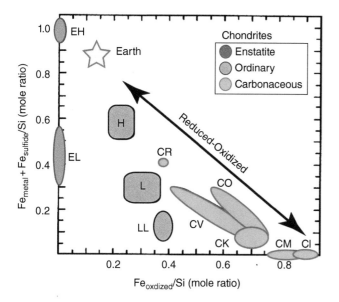

Figure 12.2 Urey-Craig diagram for chondritic meteorites, which shows the variation in oxidized Fe to silicate (x- axis) relative to the variation in reduced iron (and that in sulfide) to total silicon. The star indicates Earth's value and reflects a large metallic core.

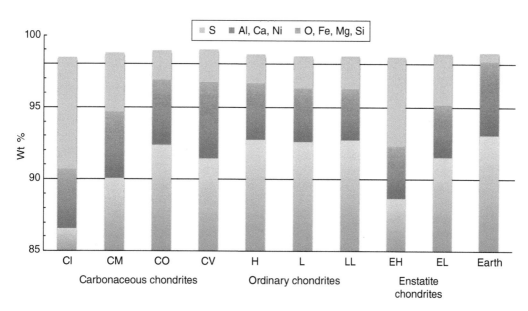

Figure 12.3 Mass proportions of the major elements (i.e., O, Fe, Mg, and Si, constituting over 85% by mass), minor (i.e., Ca, Al, and Ni, contributing 1–2 wt % each), and the volatile element S (which varies from 1 to 8 wt % in these models). These elements collectively describe >98% by mass the composition of terrestrial planets.

Table 12.2 Bulk Earth and core models based on chondritic meteorites (CI, CV, H, and EH) and on cosmochemical and geochemical grounds (Earth).

Bulk Earth models	CI	CV	H	EH	Earth
	Weight Percent				
O	31.1	35.0	35.0	28.7	29.7
Fe	26.2	24.7	27.2	30.9	31.9
Mg	13.9	15.3	13.8	11.3	15.4
Si	15.3	16.4	16.7	17.8	16.1
Al	1.2	1.8	1.1	0.9	1.6
Ca	1.3	2.0	1.2	0.9	1.7
Ni	1.5	1.4	1.6	1.9	1.8
S	7.8	2.3	2.0	6.2	0.6
	Atomic Proportions				
O	0.50	0.54	0.55	0.48	0.49
Fe	0.12	0.11	0.12	0.15	0.15
Mg	0.15	0.16	0.14	0.13	0.17
Si	0.14	0.14	0.15	0.17	0.15
Al	0.012	0.017	0.010	0.009	0.016
Ca	0.009	0.012	0.008	0.006	0.011
Ni	0.007	0.006	0.007	0.009	0.008
S	0.063	0.018	0.015	0.052	0.005

Core models	CI	CV	H	EH	Earth
	Weight Percent				
Fe	76	80	81	78	81
Ni	5.1	5.3	5.4	5.2	5.4
Si	3.5	3.5	3.5	3.5	3.5
O	7.0	7.0	7.0	7.0	7.0
S	7.8	2.3	2.0	6.2	1.9
Mean atomic no.	23.5	23.7	23.7	23.8	23.9
	Atomic Proportions				
Fe	0.604	0.664	0.670	0.623	0.670
Ni	0.038	0.042	0.042	0.040	0.043
Si	0.055	0.058	0.058	0.056	0.058
O	0.194	0.203	0.202	0.195	0.202
S	0.108	0.033	0.028	0.086	0.027

we account for more than 99% by mass of the chondrites and hence the terrestrial planets, in all cases (Figure 12.3). The amount of S in chondrites (e.g., as sulfate in CI chondrites and sulfide in EH chondrites) varies considerably. However, given Earth and the terrestrial planets have evidence for volatile depletions (e.g., low K/U and K/Th ratios), this suggests that planetary S contents are going to be on the order of 2 wt. % or less.

Once the mass fraction of the silicate shell to metallic core has been established, then four fundamental data are needed for establishing the bulk composition of this outer shell of the terrestrial planets: (1) the bulk Mg number (atomic ratio of (Mg/(Mg+Fe)) of the silicate shell, (2) the bulk ratio of olivine to pyroxene (i.e., the silicate

Mg/Si ratio), (3) the absolute abundances of the refractory lithophile elements (e.g., the amount of Al, Ca, Ti, Th), and (4) the degree of depletion of the moderately volatile elements (e.g., Na, K, Rb, S). These four parameters can provide a nearly complete compositional picture of the silicate shell of a planet. Combining the Mg number and the mass fraction of core to silicate shell provides insight into the planetary body's oxidation state and thus the amount of iron in the silicate shell. The lithophile elements are concentrated in the silicate shell and excluded from the metallic core. The refractory elements (i.e., 29 lithophile and 8 siderophile elements, see Table 12.1) are in *chondritic proportions*. Thus, if we know the absolute abundance of one of these elements, we can then determine the abundance of the rest of them assuming chondritic ratios (e.g., Al/Ca, Al/Ti, Al/Th, Al/U). Moreover, establishing the degree of depletion of moderately volatile elements in the planet (e.g., constraints from $^{40}K \rightarrow {}^{40}Ar$ in the atmosphere, $^{87}Rb \rightarrow {}^{87}Sr/{}^{86}Sr$ in rocks) can provide further insights into the relative enrichment level of the refractory elements. Collectively, these observations then place limits (by difference) on the proportion of olivine to pyroxene in the planet.

When establishing the composition of the silicate shell of the terrestrial planets, there are no unique solutions to the allowed parameter space. The four parameters listed above can be traded-off with each other, and therefore a number of competing compositional models can be permitted. Moreover, seeing through the crustal layer, the existence of layering in the mantle, hidden reservoirs, a separate sulfide layer, and continued mass transfer between the core and mantle will add further complications to the picture. Thus, remote sensing can only go so far when attempting to estimate the internal composition of a terrestrial planet.

12.3. COMPOSITIONAL MODEL FOR CORE

Emil Wiechertin 1897 appears to have been the first person to have conjectured that the Earth has a metallic core [*McDonough*, 2014]. Since then we have attempted to constrain its origin, composition, physical state, and evolution. The depth to the core-mantle boundary was established by *Beno Gutenberg* [1914], Wiechert's Ph.D. student, at 2900 km depth and is presently recognized to be 2891 ± 5 km deep [*Masters and Shearer*, 1995]. The Adams–Williamson equation [$(d\rho / dr = -GM_r\rho(r) / r^2\phi$] and the bulk sound velocity $\left[\phi = V_p^2 - \left(\dfrac{4}{3}\right)V_s^2\right]$ provide a starting density profile for the core that, in turn, is perturbed to be consistent with free oscillation frequencies [*Dziewonski and Anderson*, 1981], where ρ = density, r = radius, G = gravitational constant, M_r = mass of Earth within radius r, V_p = velocity of compressional wave, and V_s = velocity of shear wave. Beyond that we know that the outer core ($r = 3483\pm5$ km) is liquid, the inner core ($r = 1220\pm10$ km) is solid (or mostly), and the density jump across the outer-inner boundary is 0.82 ± 0.18 Mg m^{-3} [*Masters and Gubbins*, 2003]. Collectively these data demonstrate that both the outer and inner core are less dense than a compositional model of a pure Fe-Ni alloy, with the inner core having about one half the fraction of light element(s) envisaged for the outer core [*Alfè et al.*, 2000; *Birch*, 1952; *Jephcoat and Olson*, 1987; *Komabayashi and Fei*, 2010; *Masters and Gubbins*, 2003; *Stixrude et al.*, 1997].

Birch [1952] provided both a context to understand the density state of the inner and outer core, as well as the vocabulary of scientists working on this topic. Birch's admonition for the reader to be wary of claims about the nature of the deep Earth still holds today, and this includes this work too. It is fundamentally important to maintain a balance between what we know and what is possible. Foremost in discussions of Earth's core composition are the following questions:

1. What is the light element(s) in the outer (OC) and inner core (IC)?

2. How much light element component is needed to account for the density deficit in the OC and IC?

3. When and under what conditions were the light element(s) incorporated into the core component?

4. When did the inner core begin to crystallize?

5. What is the temperature at the CMB and OC-IC boundary?

6. Are there radioactive elements in the core and what are their abundances?

We do not have firm resolution to any of these questions, but a great deal of progress has been made over the last few decades, particularly with respect to both experimental and theoretical studies. The critical reviews of *Poirier* [1994] and *Anderson and Isaak* [2002] on core composition and temperature provide a fundamental framework in which to understand fully the state of unknowns and trade-offs regarding the nature of Earth's core.

This volume will provide considerable insight into the physical and chemical state of the core and lower mantle. Given the author's bias this chapter emphasizes constraints from cosmochemistry and geochemistry, and thus the conclusions reached need to be viewed in this light. The bulk Earth is modeled as a chondrite, but not a specific type known in our inventory of samples. Nonetheless, using the estimate of the primitive mantle composition (see below) and the compositional variation observed in the chondrites, the bulk Earth's composition can be predicted, and from that, the core's composition is extracted.

Compositional models for the core [*McDonough*, 2014] are presented in Table 12.2 and framed from a perspective of chondrites, the building blocks of the solar system (i.e., the chondritic ratios include Fe/Ni = 17.4 ± 0.5, Ni/Co = 20 ± 1, and Ni/Ir = $23 \pm 4 \times 10^3$). The weaknesses of the models presented in Table 12.2 are the simple assumptions (1) O, Si, and S are assumed to be present, (2) O content is arbitrarily assumed to be 7 wt. %, with the unconstrained O/Si weight ratio set at ~2 and the S content for the BSE set at 250 ppm, and (3) the core's mean atomic number is set to between 23.5 and 24.0. These models are nonunique and only meant to show one family of compositional solutions (Fe, Ni, and S); many alternative models can be generated with different proportions of O, Si, and other light elements. At present, without hard data to fix the core's light element(s) content, both the relative and absolute proportions, core compositional models remain not well constrained. Trade-offs and modeling uncertainties in core temperatures and compositional space allow for a range of model solutions [e.g., *Badro et al.*, 2014]. It is also notable that because there is no fixed Mg/Si ratio for chondrites, then one cannot make an assumption on the amount of Si in the core using chondrites and a high Mg/Si ratio of the mantle. Likewise, physical and compositional constraints on the inner core, which is ~5% the mass of the core, are weak and progress continues to find models that fit these constraints [e.g., *Chen et al.*, 2014]. We are far from understanding the details of the absolute and relative abundances of light element(s) that make up the outer core; thus, speculations on these fractions in the inner core are even less constrained. Chondritic ratios of refractory siderophile elements provide constraints for the core content of siderophile and chalcophile elements [e.g., see Figure 9 in *McDonough*, 2014] given knowledge of the BSE and bulk Earth composition. Consequently, these insights allow one to predict accurately element abundances in Earth's core down to parts per million or per billion levels.

The composition of the BSE, bulk Earth, and core, as well as the Moon's composition, all provide essential insights into how and when Earth's core formed. Constraints on the timing of Earth's core formation come from the ^{182}Hf-^{182}W isotope system [*Kleine et al.*, 2002; *Yin et al.*, 2002]. The combined modeling of inputs from large impacts during accretion and tungsten isotope data provide limits, but not a unique solution for the formational age of Earth's core, with acceptable ages being >10 Myr to ≤100 Myr after CAI formation (i.e., t_0 for the solar system) [*Kleine et al.*, 2009].

Constraints on the conditions of pressure, temperature, and fO$_2$ (i.e., fugacity of oxygen) of core formation come from knowledge of the abundances of siderophile and chalcophile elements in the BSE. For elements like W and Mo, which are refractory elements, Earth is assumed to have chondritic ratio of W/Hf and Mo/Ce (i.e., all four elements are refractory, and these ratios are relatively constant in the crust and mantle); thus these ratios in the BSE and knowledge that Hf and Ce are lithophile and excluded from the core tells us that the chalcophile element Mo and the siderophile element W were partitioned between the core and mantle in different proportion, as a function of pressure, temperature, f_{O_2}, and f_{S_2} during core formation. Likewise, the strongest constraint on Earth's high-pressure reequilibration signature comes from the experimental observations on the Ni-Co composition of the mantle [*Li and Agee*, 1996; *Siebert et al.*, 2012]. *Wood et al.* [2014] presented a recent summary of the conditions of core formation, including new insights from chalcophile elements, which involve high-pressure, high-temperature, and metal segregation under evolving f_{O_2} conditions throughout the various growth stages. Importantly, geochemical constraints only provide the integrated signal of core formation; constraints from petrology and dynamics provide the range of plausible conditions and sequences for core growth stages.

The chemical gradient in the solar system reveals that the terrestrial planets are depleted in volatiles relative to the gas giants, and more specifically Earth's sulfur budget is also depleted relative to the solar average abundance. The constraint on the estimate of sulfur content of the core comes from the observed depletion trend in the inventories of volatile elements in the silicate shell of Earth [*Dreibus and Palme*, 1996]. Likewise, Earth's overall depletion extends monotonically from the moderately volatile to highly volatile elements (e.g., C, N, and H). Many of these latter elements may likely be present in the core to some extent but are unlikely to be there in significant quantities [*McDonough*, 2014].

A question remains as to whether or not Earth's volatile budget was controlled by predifferentiated planetesimals [e.g., *Yin*, 2005] or by accretion and subsequent reequilibration of these smaller bodies in the growing Earth. The strongest constraint on Earth's depletion signature comes from the short-lived (3.7 Myr) Mn-Cr isotope system. Earth's Cr isotopic composition relative to its marked Mn depletion shows that the bulk Earth has the lowest, most depleted composition with respect to the spectrum chondritic, Mars, and the Eucrite parent bodies [*Trinquier et al.*, 2008]. The overall excesses in [53]Cr seen in chondrites coupled with the depleted composition of Earth in Mn-Cr space has been interpreted by *Palme and O'Neill* [2014] as reflecting early loss of volatiles in the pre-accreted materials that formed Earth.

Finally, in order to understand Earth's energy budget, and the power needed to drive the geodynamo [*Nimmo*, 2007], there remains a question of the presence of radioactive, heat-producing elements (i.e., HPE = K, Th, and U) in Earth's core. At present there is no compelling reason to have HPE in the core, as the energy budget does not require it. Thus, to frame the discussion, there should be no confusion between what is plausible and what is needed. *Li and Fei* [2014] and *Watanabe et al.* [2014] provide useful summaries regarding the state of the issue on the potential for K in the core, with the overall conclusion being that tens of $\mu g/g$ of K, or less, can be accommodated in the core. Studies of *Wheeler et al.* [2006] and *Bouhifd et al.* [2013] show that under core segregation conditions the geochemical behavior of U remains that of a lithophile element that is excluded from the core. *Ab initio* calculations reveal the metallization of K at pressure, due to *s*- to *d*-orbital transitions and thus allowing K to alloy with Fe [*Lee et al.*, 2004]. This plausibility argument, however, would more readily apply to Cs and Rb, larger alkali metals, which would undergo similar *s*- to *d*-orbital transitions, but at lower pressures. There is no evidence, however, that Rb is anomalously depleted in the mantle with respect to the planetary volatility curve, contrary to an expectation from the pressure-dependent metallization process. Thorium has not been found to partition into a core-forming metal phase; it is understood to be wholly lithophile. An additional constraint on the abundance and relative distribution of Th and U in the BSE comes from Pb isotope studies. The κ_{Pb} ratio (i.e., the time-integrated Th/U ratio derived from the [208*]Pb/[206*]Pb ratio) of the silicate Earth is comparable to that of chondrites at $\pm25\%$ [*Paul et al.*, 2003], consistent with negligible amounts of Th and U in Earth's core.

12.4. COMPOSITIONAL MODEL FOR BULK SILICATE EARTH (BSE) AND LOWER MANTLE

The BSE includes the present-day crust and mantle and all of their separate subdomains. The BSE has also been referred to as the primitive mantle and was the basis of the pyrolite model (pyroxene-olivine rock [*Ringwood*, 1962]). The basic question remains: Is the composition of the upper mantle, including the transition zone, and lower mantle the same (at the major element scale) or not? A range of compositional models for the mantle is reported in Table 12.3. If the composition of the upper mantle and transition zone is assumed to be approximately similar to that of *McDonough and Sun* (1995), then the composition of the lower mantle is, based on mass balance considerations, markedly different for the following models presented in Table 12.3: C1 chondrite, *Javoy et al.* [2010], and *Turcotte and Schubert* [2002]. It is often not appreciated that C1 chondritic models, so often cited for solar comparisons, set both Mg/Si and the absolute abundances of the refractory element abundances, which are different from models of the upper mantle. Likewise, a *Turcotte and Schubert* [2002] model, which assumes a high Th and U content for the BSE, must also have gross compositional layering to match the

Table 12.3 Primitive Mantle (bulk silicate earth, BSE) models.

	Cl1 chondrite	Javoy et al. [2010]	McDonough and Sun [1995]	Palme and O'Neill [2014]	*Turcotte and Shubert [2003]
			Weight Percent		
Al	1.93	1.28	2.34	2.38	3.55
Ca	2.06	1.34	2.52	2.61	3.83
Mg	21.6	22.0	22.8	22.2	21.5
Si	22.1	23.3	21.0	21.2	19.5
Fe	6.30	6.87	6.25	6.30	6.30
O	44.9	44.2	44.1	44.3	44.3
Ca/Al	1.07	1.05	1.08	1.10	1.08
Al/Si	0.087	0.055	0.111	0.112	0.182
Mg/Si	0.978	0.945	1.084	1.045	1.103
K(10^{-6} g/g)	807	144	280	260	310
Th(10^{-9} g/g)	42	42	80	85	121
U(10^{-9} g/g)	12	12	20	23	31
Power (TW)	20.3	11.2	20.5	21.8	29.8
			Atomic Proportions		
Al	0.015	0.010	0.018	0.019	0.028
Ca	0.011	0.007	0.013	0.014	0.020
Mg	0.189	0.193	0.199	0.194	0.189
Si	0.167	0.176	0.159	0.161	0.148
Fe	0.024	0.026	0.024	0.024	0.024
O	0.595	0.588	0.586	0.589	0.591
Ca/Al	0.720	0.705	0.724	0.739	0.727
Al/Si	0.091	0.057	0.116	0.117	0.189
Mg/Si	1.130	1.092	1.252	1.207	1.274

*Note: The major element abundances in this model were calculated from chondritic Ca/Al, Al/Th, and Ca/Th ratios, assumed Fe content for atomic Mg/Mg+Fe=0.89, and a mass balance.

chondritic proportions for Ca, Al, Th, and U. Collectively, these five models (Table 12.3) provide a perspective on the differences between some of the competing models for the composition of the lower mantle.

Overall, the evidence from geochemistry is that these five model compositions are permissive, regarding the chemical and mineralogical composition of the lower mantle. The upper and lower mantle can be chemically similar or different. Other data are required to resolve this issue. Commonly, there is an emphasis on combining mineral physics data with seismological constraints to find solution space that satisfies Preliminary Earth Reference Model (PREM) [*Dziewonski and Anderson*, 1981]. Increasingly, these modeling studies attempt to treat uncertainties in data and find that the lower mantle's composition is compatible with that of the upper mantle [e.g., *Zhang et al.*, 2013]. An alternative compositional model that sees distinct upper-lower mantle chemical differences has been proposed [*Murakami et al.*, 2012], however, this model has been discounted based on its incorrect Voigt-Reuss-Hill averaging of a multiphase lower mantle composition [*Cottaar et al.*, 2014]. The most compelling observation that is consistent with a relatively homogenous bulk composition for the mantle, in this author's view, is

based on seismic tomographic images of the mantle that reveal down-going slabs of oceanic lithosphere penetrating into the lower mantle (i.e., below the 660 km seismic discontinuity) [e.g., *Fukao and Obayashi*, 2013], thus requiring lower-upper mantle mass exchange and mixing.

In terms of mineral proportions, the lower mantle is, for the most part, considered to be made up of three minerals: bridgmanite (a.k.a. Mg-perovskite), Ca-perovskite, and ferropericlase, with the exception of the lowest couple hundred kilometers of the mantle seeing the phase transition of bridgmanite to post-perovskite [*Murakami et al.*, 2004]. [There is also a recent report of Mg-perovskite (bridgmanite) being unstable and disproportioning to an Fe-rich phase with a hexagonal structure (H phase) and a nearly Fe-free $MgSiO_3$ perovskite [*P Zhang et al.*, 2014].] The relative proportions of these phases in different lower mantle compositional models are shown in Figure 12.4. for the models presented in Table 12.3. The absolute mode proportion of Ca-perovskite is set by the model's predicted content of refractory lithophile elements (i.e., Ca content). The model's Mg/Si ratio sets the relative mode proportion of bridgmanite to ferropericlase, with models having higher Si contents consequently having

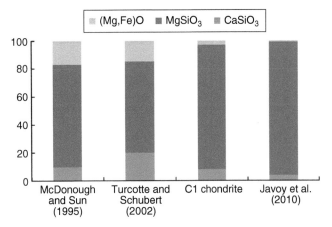

Figure 12.4 Relative proportions of the three dominant lower mantle phases in different lower mantle compositional models (see Table 12.4 for the compositional models). The minerals include bridgmanite (a.k.a. Mg-perovskite idealized as $MgSiO_3$), Ca-perovskite (idealized as $CaSiO_3$), and ferropericlase ((Mg Fe) O), The mineralogical proportions in model of *Palme and O'Neill* [2014] is the same as that in *McDonough and Sun* [1995].

lower mode proportions of ferropericlase. Based on an upper mantle composition, the lower mantle has on average by volume about 10% Ca-perovskite, 17% ferropericlase, and 73% bridgmanite (Figure 12.4).

There are many competing geochemical models that describe the composition of the BSE; these models can be viewed as being on a continuum with three dominant modes: low, intermediate, and high abundances of the refractory lithophile elements (i.e., Al, Ca, Th, U). (Th and U are mentioned here because they are also the heat producing elements, and together with K they produce >99% of Earth's radiogenic heat, with 40% contributions coming each from Th and U.) These three classes of models, labeled cosmochemical, geochemical, and geodynamic models [*Šrámek et al.*, 2013], are recognized in terms of radiogenic heat production as the low-Q (10–15 TW), medium-Q (17–22 TW), and high-Q (>25 TW) models [*Dye et al.*, 2015]. Models that conclude that the BSE has either low or high concentrations of refractory lithophile elements (low- and high-Q models) often require gross scale, chemical layering in the mantle that is often viewed as being upper and lower domains separated by the 660 km seismic discontinuity, which acts as both a phase change and a compositional boundary. Alternative layering structures can be envisaged, but their nature, origin, and mode of separation (isolation) are significant challenges for geodynamics models. Geochemical or medium-Q models typically assume a homogeneous mantle composition, at the gross scale, for the major elements (i.e., model estimates based on basalts and peridotites, with constraints from chondrites). The following will consider individually these three families of models.

12.4.1. Low Q: Cosmochemical Models

These models include the collisional erosion model [*Campbell and O'Neill*, 2012; *Caro and Bourdon*, 2010; *Jackson and Jellinek*, 2013; *O'Neill and Palme*, 2008; *Warren*, 2011], the enstatite Earth model [*Javoy*, 1995; *Javoy and Kaminski*, 2014; *Javoy et al.*, 2010; *Kaminski and Javoy*, 2013], and the non-chondritic model [*Campbell and O'Neill*, 2012]. The collisional erosion and non-chondritic models predict that the absolute abundances of refractory lithophile elements (e.g., U and Th) in Earth were initially accreted in chondritic proportions, and then in the early Hadean, Earth formed a Th-, U-, and LREE-enriched crust that was either removed during intense collisions with large impacts or the crust was subducted and permanently sequestered as a hidden reservoir at the base of the mantle. These BSE models owe their origins to the observation that Earth's ^{142}Nd isotope abundances are distinctly different from that of some chondrites [*Boyet and Carlson*, 2005; *Carlson and Boyet*, 2008], a conclusion that is at odds with the present-day range of $^{142}Nd/^{144}Nd$ isotopic compositions seen in chondrites [*Huang et al.*, 2013]. Models invoking a sequestered, hidden, Th-U-LREE-enriched reservoir at the base of the mantle are not strictly low-Q models (i.e., ≤15 TW), but in principle this hot basal layer acts like a bottom heating layer to the mantle. On a major element scale these models view the upper and lower mantle to be of similar bulk composition, or at least they do not require compositional layering.

The enstatite Earth, cosmochemical model [*Javoy and Kaminski*, 2014; *Javoy et al.*, 2010] uses the compositions of enstatite chondrites to predict the composition of the bulk Earth and recognized that the Mg/Si ratio of enstatite chondrites is markedly lower than that of the upper mantle (see geochemical models). Thus, the constructed model requires that the lower and upper mantle have distinctly different bulk compositions and predict low radiogenic heat production. In addition, these authors note that Earth and enstatite chondrites have overlapping $^{142}Nd/^{144}Nd$ isotopic compositions, and thus this shared compositional signature is a further genetic proof of the linkage between these materials.

12.4.2. Medium Q: Geochemical Models

These models [*Allegre et al.*, 1995; *Hart and Zindler*, 1986; *Jagoutz et al.*, 1979; *Lyubetskaya and Korenaga*, 2007; *McDonough and Sun*, 1995; *Palme and O'Neill*, 2014; *Ringwood*, 1975; *Wanke*, 1981] recognize the residuum-melt relationship between peridotites and basalts and seek to estimate the composition of the BSE by modeling the system to the least melt-depleted composition. These models predict a primitive composition for materials last equilibrated in the uppermost mantle and conclude that deeper materials will also have similar compositional characteristics due to whole mantle convection. Tomographic images of subducting slabs plunging, in some instances, into the deep mantle reveal that there is present-day mass transfer between the upper and lower mantle, which support the claim of whole mantle convection. Consequently, the mantle is dynamically supported to be roughly homogeneous in composition.

12.4.3. High Q: Geodynamic Models

These models [*Davies*, 2010; *Schubert et al.*, 1980; *Turcotte and Schubert*, 2002; *Turcotte et al.*, 2001] are constructed from the perspective of physics driving Earth processes and use a simple set of assumptions and equations to describe Earth. These models seek to solve the balance of mantle forces between thermal and momentum diffusivity versus viscosity and buoyancy. The models examine the thermal evolution of Earth in terms of the relative contribution of primordial to radiogenic heat, with the primordial contribution coming from the kinetic energy of accretion and the thermal energy that was converted from gravitational energy associated with core formation. Many, but not all, of these models implicitly or explicitly have chemical layering in the mantle with an upper and lower boundary at the 660 km seismic discontinuity. In addition, there are trade-offs in parameter space that can also result in alternative solutions for the rate of heat dissipation that range from high- to low-Q compositional models.

12.4.4. Thermal Earth

Understanding the heat budget and Earth's thermal evolution is key to understanding the composition of the lower mantle and core. The recent findings of the electrical and thermal conductivities of the core [*de Koker et al.*, 2012; *Gomi et al.*, 2013; *Pozzo et al.*, 2012; *Seagle et al.*, 2013], plus geodynamic considerations point to the present-day core heat flux as being high (10–16 TW). Such a high core heat flow is hard to reconcile with some models that describe the thermal evolution of the mantle, the age of inner core crystallization and low proportion of radiogenic heating. Some geodynamic models find satisfactory solutions when invoking a layered mantle scenario and/or a core containing a significant fraction of radiogenic heating [e.g., *Nakagawa and Tackley*, 2014]. Thus, if the mantle is layered, it might or might not have gross compositional differences between the upper and lower domains; however, in layered mantle scenarios the mantle geotherm necessarily has a significant step in its profile.

12.5. OUTSTANDING ISSUES AND WHAT LURKS DEEP IN MANTLE

The deep mantle is a region rich with seismologically observed structures (e.g., D″, ULVZ and LLSVP — see other chapters in this volume for additional insights), which have resulted in much speculation as to the origins and evolution of these features, with many competing hypotheses. Moreover, chemical and isotopic studies of the BSE composition have also concluded that there are inaccessible hidden layers in Earth that contain the chemical/isotopic complement of what is missing from the BSE inventory [e.g., *Boyet and Carlson*, 2006; *Jackson and Jellinek*, 2013], under the assumption of chondritic abundances of the refractory lithophile elements in the mantle. In addition, there are speculations that envisage core-mantle exchange as seen in chemical and isotopic signatures seen in basalts. From a geochemist view we can only see/sense what is down there if we can analyze it or deduce it from the consequences of an irreversible chemical fractionation event.

12.5.1. Primordial Differentiation of BSE

This scenario has common appeal given that the mantle in the early Earth, pre- and post-Moon formation, would have had a much higher temperature than today, with extensive melting of the mantle. It is also likely that the vigor of mantle convection would have been much greater and that this condition would promote global differentiation events, which may have been erased or its effects attenuated by rehomogenization due to enhanced convection. Some global differentiation scenarios might involve silicate melts sinking due to a mantle density crossover point, where melts are denser than their residuum [*Labrosse et al.*, 2007; *Lee et al.*, 2010], leading to large-scale irreversible differentiation of the mantle. Another idea is that the accessible mantle possesses a putative ^{142}Nd-isotopic anomaly and that there is a complementary reservoir hidden in the deep mantle (or that a primitive complementary crustal mass was lost to space during early Hadean impact events) [*Boyet and Carlson*, 2005; *Warren*, 2008]. The evidence for Earth being considered anomalous is no longer tenable given that Earth's ^{142}Nd-isotopic composition overlaps with that observed in chondritic meteorites [*Huang et al.*, 2013]. Importantly, all of these signatures of early global differentiation events are accounted for in the peridotites used to model the BSE composition [*McDonough and Sun*, 1995], contrary to the opposite assertions made in *Boyet and Carslon*, [2005] and *Lee et al.* [2010]. The present-day, non-primitive compositions of mantle peridotites were projected back to the intersections in chondritic-chondritic ratio space for refractory lithophile element ratios [see Figure 5 in *McDonough and Sun*, 1995] and therefore take into account all previous differentiation processes. The only scenario that cannot be modeled by the geochemist's standard toolbox is when the upper and lower mantle regions are distinct and the lower mantle is not derived in any part from the upper mantle (i.e., the two mantle domains are distinct and have no genetic relationship, physically or chemically, to each other).

12.5.2. Core-Mantle Exchange

There is the question of mass exchange between the core and the mantle, either as an undetected, but speculated process occurring at the CMB or as detected processes documented in the compositions of basalts. It is possible that there is mass exchange and/or ongoing chemical reactions occurring at the CMB, perhaps some of which are evidenced by the ULVZ [*Hernlund and Jellinek*, 2010]. However, unless these products are brought to the surface and sampled, then as geochemists we cannot validate their occurrence. On the other hand a number of suggestions have been made that the compositions of basalts record mass exchange across the CMB as evidenced by enhanced levels of key siderophile element abundances and anomalous Os isotopic compositions [e.g., *Brandon et al.*, 2003; *Herzberg et al.*, 2013; *Humayun et al.*, 2004; *Puchtel et al.*, 2005]. None of these models have withstood the scrutiny of testing, in the view of this author, as the models of *Humayun et al.* [2004] and *Herzberg et al.* [2013] are only suggestive of core mantle exchange with their observations of elevated Fe/Mn values and Ni contents, respectively. These chemical indicators, however, do not fall outside the global compositional array seen in mantle samples despite claims by these authors of the importance of their higher precision data. Moreover, the supporting evidence from W isotopic data to bolster the claim of Os isotopes revealing core-mantle exchange ended up showing the exact opposite, with Archean komatiites having higher W isotopic values than the mantle's composition as opposed to the predicted shift to lower W isotopic values [*Touboul et al.*, 2012, 2014]. In summary, there is no clear geochemical evidence for core-mantle exchange, but this does not preclude it from happening and not being sampled. However, the recent observations of W isotopic heterogeneities in Archean lavas, which appear to reflect near-primordial signatures that endure in isolation in the mantle for billions of years, present a constraint and a challenge for understanding convection in the mantle.

12.6. FUTURE PROSPECTS

The question of the composition of the lower mantle is a longstanding one and represents a major challenge to resolve. Ever improving data sets for the elastic properties of mantle minerals (including experimental studies and molecular dynamic calculations) are bringing us closer to finding acceptable solutions to defining the modal mineralogy and thermal state of the lower mantle. Comparisons of these data, when coupled to the ever improving seismological models for the mantle, will better define a coherent model for the BSE. Uncertainties remain significant, however, for the temperature and pressure derivatives of elastic moduli at lower mantle conditions.

A significant opportunity has recently presented itself with the acquisition of Earth's geoneutrino signal [*Araki et al.*, 2005; *Bellini et al.*, 2010, 2013; *Gando et al.*, 2013, 2011]. These recent experiments demonstrate that we have the capacity to directly assess the amount and distribution of Th and U inside Earth. These data will define the planetary budget of these refractory lithophile elements (defining the planet's radiogenic Q value) and hence also define the amount of Ca, Al, and Ti inside Earth (conserved refractory element ratios). These latter major element data can then be used to constrain the compositional models of the BSE and assess the mantle's Mg/Si ratio. In addition to testing models of the BSE, *Šrámek et al.* [2013] also proposed that geoneutrino studies conducted on the seafloor can be used to assess the compositional nature of the seismologically image LLSVP in the deep mantle. Their study predicted an enhanced geoneutrino flux from these two LLSVP domains under the hypothesis that they represent compositionally distinct material enriched in HPE as compared to the surrounding mantle. They presented neutrinographic images of the two LLSVP domains and set a significant research challenge for future studies of Earth's deep mantle.

REFERENCES

Alfè, D., M. J. Gillan, and G. D. Price (2000), Constraints on the composition of the Earth's core from *ab initio* calculations, *Nature*, *405*(6783), 172–175.

Allegre, C. J., J. P. Poirier, E. Humler, and A. W. Hofmann (1995), The chemical composition of the Earth, *Earth Planet. Sci. Lett.*, *134*(3–4), 515–526.

Anderson, O. L., and D. G. Isaak (2002), Another look at the core density deficit of Earth's outer core, *Phys. Earth Planet. Inter.*, *131*(1), 19–27.

Araki, T., et al. (2005), Experimental investigation of geologically produced antineutrinos with KamLAND, *Nature*, *436*(7050), 499–503.

Asplund, M., N. Grevesse, A. J. Sauval, and P. Scott (2009), The chemical composition of the sun, *Annu. Rev. Astron. Astrophy.*, *47*, 481–522.

Badro, J., A. S. Cote, and J. P. Brodholt (2014), A seismologically consistent compositional model of Earth's core, *Proc. Natl. Acad. Sci. USA.*, *111*(21), 7542–7545.

Bellini, G., et al. (2010), Observation of geo-neutrinos, *Phys. Lett. B*, *687*(4–5), 299–304.

Bellini, G., et al. (2013), Measurement of geo-neutrinos from 1353 days of Borexino, *Phys. Lett. B*, *722*(4–5), 295–300.

Birch, F. (1952), Elasticity and constitution of the Earth interior, *J. Geophys. Res.*, *57*(2), 227–286.

Bouhifd, M. A., D. Andrault, N. Bolfan–Casanova, T. Hammouda, and J. L. Devidal (2013), Metal–silicate partitioning of Pb and U: Effects of metal composition and oxygen fugacity, *Geochim. Cosmochim. Acta*, *114*, 13–28.

Bouwman, J., W. A. Lawson, A. Juhasz, C. Dominik, E. D. Feigelson, T. Henning, A. G. G. M. Tielens, and L. B. F. M. Waters (2010), The protoplanetary disk around the M4 star Recx 5: Witnessing the influence of planet formation? *Astrophys. J. Lett.*, *723*(2), L243–L247.

Boyet, M., and R. W. Carlson (2005), ^{142}Nd evidence for early (>4.53 Ga) global differentiation of the silicate Earth, *Science*, *309*(5734), 576–581.

Boyet, M., and R. W. Carlson (2006), A new geochemical model for the Earth's mantle inferred from ^{146}Sm-^{142}Nd systematics, *Earth Planet. Sci. Lett.*, *250*(1–2), 254–268.

Brandon, A. D., R. J. Walker, I. S. Puchtel, H. Becker, M. Humayun, and S. Revillon (2003), ^{186}Os-^{187}Os systematics of Gorgona Island komatiites: Implications for early growth of the inner core, *Earth Planet. Sci. Lett.*, *206*(3–4), 411–426.

Burbidge, E. M., G. R. Burbidge, W. A. Fowler, and F. Hoyle (1957), Synthesis of the elements in stars, *Rev. Modern Phys.*, *29*(4), 547–650.

Campbell, I. H., and H. S. C. O'Neill (2012), Evidence against a chondritic Earth, *Nature*, *483*(7391), 553–558.

Carlson, R. W., and M. Boyet (2008), Composition of the Earth's interior: The importance of early events, *Philos. Trans. R. Soc. A Math. Phys. Eng. Sci.*, *366*(1883), 4077–4103.

Caro, G., and B. Bourdon (2010), Non-chondritic Sm/Nd ratio in the terrestrial planets: Consequences for the geochemical evolution of the mantle crust system, *Geochim. Cosmochim. Acta*, *74*(11), 3333–3349.

Chen, B., et al. (2014), Hidden carbon in Earth's inner core revealed by shear softening in dense Fe$_7$C$_3$, *Proc. Nat. Acad. Sci.*, *111*, 17755–17758.

Cottaar, S., T. Heister, I. Rose, and C. Unterborn (2014), BurnMan: A lower mantle mineral physics toolkit, *Geochem. Geophys. Geosyst.*, *15*(4), 1164–1179.

D'Alessio, P., et al. (2005), The truncated disk of CoKu Tau/4, *Astrophys. J.*, *621*(1), 461–472.

Davies, G. F. (2010), Noble gases in the dynamic mantle, *Geochem. Geophys. Geosyst.*, *11*, Q03,005.

de Koker, N., G. Steinle-Neumann, and V. Vlcek (2012), Electrical resistivity and thermal conductivity of liquid Fe alloys at high P and T, and heat flux in Earth's core, *Proc. Natl. Acad. Sci. USA.*, *109*(11), 4070–4073.

DeMeo, F. E., and B. Carry (2014), Solar System evolution from compositional mapping of the asteroid belt, *Nature*, *505*(7485), 629–634.

Dreibus, G., and H. Palme (1996), Cosmochemical constraints on the sulfur content in the Earth's core, *Geochim. Cosmochim. Acta*, *60*(7), 1125–1130.

Dye, S., Y. Huang, V. Lekic, W. F. McDonough, and O. Šrámek (2015), Geo-neutrinos and Earth models, *Phys. Procedia*, *61*, 310–318.

Dziewonski, A. M., and D. L. Anderson (1981), Preliminary Reference Earth Model, *Phys. Earth Planet. Inter.*, *25*(4), 297–356.

Fukao, Y., and M. Obayashi (2013), Subducted slabs stagnant above, penetrating through, and trapped below the 660 km discontinuity, *J. Geophys. Res. Solid Earth*, *118*(11), 5920–5938.

Gando, A., et al. (2011), Partial radiogenic heat model for Earth revealed by geoneutrino measurements, *Nat. Geosci.*, *4*(9), 647–651.

Gando, A., et al. (2013), Reactor on-off antineutrino measurement with KamLAND, *Phys. Rev. D*, *88*(3), 033001, doi:10.1103/PhysRevD.88.033001.

Garnero, E. J., and A. K. McNamara (2008), Structure and dynamics of Earth's lower mantle, *Science*, *320*(5876), 626–628.

Gomi, H., K. Ohta, K. Hirose, S. Labrosse, R. Caracas, M. J. Verstraete, and J. W. Hernlund (2013), The high conductivity of iron and thermal evolution of the Earth's core, *Phys. Earth Planet. Inter.*, *224*, 88–103.

Gutenberg, B. (1914), Über Erdbenwellen VIIA. Beobachtungen an Registrierungen von Fernbeben in Göttingen und Folgerungen über die Konstitution des Erdkörpers, *Nachrichten der Gesellschaft der Wissenschaften zu Göttingen, Mathemathisch-Physikalische Klasse*, *25*(3), 1–52.

Hart, S. R., and A. Zindler (1986), In search of a bulk-Earth composition, *Chem. Geol.*, *57*(3–4), 247–267.

Haxton, W. C., R. G. H. Robertson, and A. M. Serenelli (2013), Solar neutrinos: Status and prospects, *Annu. Rev. Astron. Astr.*, *51*, 21–61.

Hernlund, J. W., and A. M. Jellinek (2010), Dynamics and structure of a stirred partially molten ultralow-velocity zone, *Earth Planet. Sci. Lett.*, *296*(1–2), 1–8.

Herzberg, C., P. D. Asimow, D. A. Ionov, C. Vidito, M. G. Jackson, and D. Geist (2013), Nickel and helium evidence for melt above the core-mantle boundary, *Nature*, *493*(7432), 393–397.

Huang, S. C., S. B. Jacobsen, and S. Mukhopadhyay (2013), ^{147}Sm-^{143}Nd systematics of Earth are inconsistent with a superchondritic Sm/Nd ratio, *Proc. Natl. Acad. Sci. USA.*, *110*(13), 4929–4934.

Humayun, M., L. P. Qin, and M. D. Norman (2004), Geochemical evidence for excess iron in the mantle beneath Hawaii, *Science*, *306*(5693), 91–94.

Jackson, M. G., and A. M. Jellinek (2013), Major and trace element composition of the high ^{3}He/^{4}He mantle: Implications for the composition of a nonchondritic Earth, *Geochem. Geophys. Geosyst.*, *14*(8), 2954–2976.

Jagoutz, E., H. Palme, H. Baddenhausen, K. Blum, M. Cendales, G. Dreibus, B. Spettel, H. Wanke, and V. Lorenz (1979), The abundances of major, minor and trace elements in the earth's mantle as derived from primitive ultramafic nodules, in *Proceedings of the Lunar and Planetary Science Conference*, pp. 2031–2050, Pergamon, New York.

Javoy, M. (1995), The intergral enstatite chondrite model of the Earth, *Geophys. Res. Lett.*, *22*(16), 2219–2222.

Javoy, M., and E. Kaminski (2014), Earth's uranium and thorium content and geoneutrinos fluxes based on enstatite chondrites, *Earth Planet. Sci. Lett.*, *407*, 1–8.

Javoy, M., et al. (2010), The chemical composition of the Earth: Enstatite chondrite models, *Earth Planet. Sci. Lett.*, *293*(3–4), 259–268.

Jenniskens, P., et al. (2012), Radar-enabled recovery of the Sutter's Mill meteorite, a carbonaceous chondrite regolith breccia, *Science*, *338*(6114), 1583–1587.

Jephcoat, A., and P. Olson (1987), Is the inner core of the Earth pure iron, *Nature*, *325*(6102), 332–335.

Kaminski, E., and M. Javoy (2013), A two-stage scenario for the formation of the Earth's mantle and core, *Earth Planet. Sci. Lett.*, *365*, 97–107.

Kleine, T., C. Munker, K. Mezger, and H. Palme (2002), Rapid accretion and early core formation on asteroids and the terrestrial planets from Hf-W chronometry, *Nature*, *418*(6901), 952–955.

Kleine, T., M. Touboul, B. Bourdon, F. Nimmo, K. Mezger, H. Palme, S. B. Jacobsen, Q. Z. Yin, and A. N. Halliday (2009), Hf-W chronology of the accretion and early evolution of asteroids and terrestrial planets, *Geochim. Cosmochim. Acta*, *73*(17), 5150–5188.

Komabayashi, T., and Y. W. Fei (2010), Internally consistent thermodynamic database for iron to the Earth's core conditions, *J. Geophys. Res. Solid Earth*, *115*, 1–12.

Labrosse, S., J. W. Hernlund, and N. Coltice (2007), A crystallizing dense magma ocean at the base of the Earth's mantle, *Nature*, *450*(7171), 866–869.

Lee, C.-T. A., P. Luffi, T. Hoeink, J. Li, R. Dasgupta, and J. Hernlund (2010), Upside-down differentiation and generation of a "primordial" lower mantle, *Nature*, *463*(7283), 930–933.

Lee, K. K. M., G. Steinle-Neumann, and R. Jeanloz (2004), Ab-initio high-pressure alloying of iron and potassium: Implications for the Earth's core, *Geophys. Res. Lett.*, *31*(11), 1–4.

Lekic, V., S. Cottaar, A. Dziewonski, and B. Romanowicz (2012), Cluster analysis of global lower mantle tomography: A new class of structure and implications for chemical heterogeneity, *Earth Planet. Sci. Lett.*, *357*, 68–77.

Li, J., and C. B. Agee (1996), Geochemistry of mantle-core differentiation at high pressure, *Nature*, *381*(6584), 686–689.

Li, J., and Y. Fei (2014), Experimental constraints on core composition, in *The Mantle and Core*, edited by R. W. Carlson, pp. 527–557, Elsevier, Amsterdam.

Lodders, K. (2003), Solar system abundances and condensation temperatures of the elements, *Astrophys. J.*, *591*(2), 1220–1247.

Lugaro, M., A. Heger, D. Osrin, S. Goriely, K. Zuber, A. I. Karakas, B. K. Gibson, C. L. Doherty, J. C. Lattanzio, and U. Ott (2014), Early solar system stellar origin of the [182]Hf cosmochronometer and the presolar history of solar system matter, *Science*, *345*(6197), 650–653.

Lyubetskaya, T., and J. Korenaga (2007), Chemical composition of Earth's primitive mantle and its variance: 1. Method and results, *J. Geophys. Res. Solid Earth*, *112*(B3), 21.

MacPherson, G. J. (2014), Calcium–aluminum-rich inclusions in chondritic meteorites, in *Meteorites and Cosmochemical Processes*, edited by A. Davis, pp. 139–179, Elsevier, Amsterdam.

Masters, G., and D. Gubbins (2003), On the resolution of density within the Earth, *Phys. Earth Planet. Inter.*, *140*, 159–167.

Masters, T. G., and P. M. Shearer (1995), Seismic models of the Earth: Elastic and anelastic, in *Global Earth Physics: A Handbook of Physical Constant*, edited by T. J. Ahrens, pp. 88–103, AGU, Washington, DC.

McDonough, W. F. (2014), Compositional model for the Earth's core, in *The Mantle and Core*, edited by R. W. Carlson, pp. 559–577, Elsevier, Amsterdam.

McDonough, W. F., and S. S. Sun (1995), The composition of the Earth, *Chem. Geol.*, *120*(3–4), 223–253.

McNamara, A. K., E. J. Garnero, and S. Rost (2010), Tracking deep mantle reservoirs with ultra-low velocity zones, *Earth Planet. Sci. Lett.*, *299*(1–2), 1–9.

Murakami, M., K. Hirose, K. Kawamura, N. Sata, and Y. Ohishi (2004), Post-perovskite phase transition in $MgSiO_3$, *Science*, *304*(5672), 855–858.

Murakami, M., Y. Ohishi, N. Hirao, and K. Hirose (2012), A perovskitic lower mantle inferred from high-pressure, high-temperature sound velocity data, *Nature*, *485*(7396), 90–94.

Nakagawa, T., and P. J. Tackley (2014), Influence of combined primordial layering and recycled MORB on the coupled thermal evolution of Earth's mantle and core, *Geochem. Geophys. Geosyst.*, *15*(3), 619–633.

Nimmo, F. (2007), 8.02 — Energetics of the Core, in *Treatise on Geophysics*, edited by G. Schubert, pp. 31–65, Elsevier, Amsterdam.

O'Neill, H. S. C., and H. Palme (2008), Collisional erosion and the non-chondritic composition of the terrestrial planets, *Philos. Trans. R. Soc. A Math. Phys. Eng. Sci.*, *366*(1883), 4205–4238.

Palme, H., and H. S. C. O'Neill (2014), Cosmochemical estimates of mantle composition, in *The Mantle and Core*, edited by R. W. Carlson, pp. 1–39, Elsevier, Amsterdam.

Paul, D., W. M. White, and D. L. Turcotte (2003), Constraints on the Th-232/U-238 ratio (kappa) of the continental crust, *Geochem. Geophys. Geosyst.*, *4*(12), 1525–2027, doi:10.1029/2002GC000497.

Poirier, J. P. (1994), Light-elements in the Earth's outer core — A critical review, *Phys. Earth Planet. Inter.*, *85*(3–4), 319–337.

Pozzo, M., C. Davies, D. Gubbins, and D. Alfe (2012), Thermal and electrical conductivity of iron at Earth's core conditions, *Nature*, *485*(7398), 355–358.

Puchtel, I. S., A. D. Brandon, M. Humayun, and R. J. Walker (2005), Evidence for the early differentiation of the core from Pt-Re-Os isotope systematics of 2.8-Ga komatiites, *Earth Planet. Sci. Lett.*, *237*(1–2), 118–134.

Ringwood, A. E. (1962), Model for the upper mantle, *J. Geophys. Res.*, *67*(2), 857–867.

Ringwood, A. E. (1975), *Composition and Petrology of the Earth's Mantle*, McGraw-Hill, New York.

Sargent, B. A., et al. (2009), Dust processing and grain growth in protoplanetary disks in the Taurus-Auriga star forming region, *Astrophys. J. Suppl. Ser.*, *182*(2), 477–508.

Schubert, G., D. Stevenson, and P. Cassen (1980), Whole planet cooling and the radiogenic heat-source contents of the Earth and Moon, *J. Geophys. Res. 85*(NB5), 2531–2538.

Scott, E. R. D., and A. N. Krot (2014), Chondrites and their components, in *Meteorites and Cosmochemical Processes*, edited by A. Davis, pp. 65–137, Elsevier, Amsterdam.

Seagle, C. T., E. Cottrell, Y. W. Fei, D. R. Hummer, and V. B. Prakapenka (2013), Electrical and thermal transport properties of iron and iron-silicon alloy at high pressure, *Geophys. Res. Lett.*, *40*(20), 5377–5381.

Siebert, J., J. Badro, D. Antonangeli, and F. J. Ryerson (2012), Metal-silicate partitioning of Ni and Co in a deep magma ocean, *Earth Planet. Sci. Lett.*, *321*, 189–197.

Šrámek, O., W. F. McDonough, E. S. Kite, V. Lekic, S. T. Dye, and S. J. Zhong (2013), Geophysical and geochemical constraints on geoneutrino fluxes from Earth's mantle, *Earth Planet. Sci. Lett.*, *361*, 356–366.

Stixrude, L., E. Wasserman, and R. E. Cohen (1997), Composition and temperature of Earth's inner core, *J. Geophys. Res. Solid Earth*, *102*(B11), 24729–24739.

Touboul, M., I. S. Puchtel, and R. J. Walker (2012), [182]W Evidence for long-term preservation of early mantle differentiation products, *Science*, *335*(6072), 1065–1069.

Touboul, M., J. Liu, J. O'Neil, I. S. Puchtel, and R. J. Walker (2014), New insights into the Hadean mantle revealed by [182]W and highly siderophile element abundances of supracrustal rocks from the Nuvvuagittuq Greenstone Belt, Quebec, Canada, *Chem. Geol.*, *383*, 63–75.

Trinquier, A., J. L. Birck, C. J. Allegre, C. Gopel, and D. Ulfbeck (2008), (53)Mn-(53)Cr systematics of the early Solar System revisited, *Geochim. Cosmochim. Acta*, *72*(20), 5146–5163.

Tsiganis, K., R. Gomes, A. Morbidelli, and H. F. Levison (2005), Origin of the orbital architecture of the giant planets of the Solar System, *Nature*, *435*(7041), 459–461.

Turcotte, D. L., and G. Schubert (2002), *Geodynamics*, 2nd ed., Cambridge Univ Press, Cambridge.

Turcotte, D. L., D. Paul, and W. M. White (2001), Thorium-uranium systematics require layered mantle convection, *J. Geophys. Res. Solid Earth*, *106*(B3), 4265–4276.

van Boekel, R., et al. (2004), The building blocks of planets within the "terrestrial" region of protoplanetary disks, *Nature*, *432*(7016), 479–482.

Walsh, K. J., A. Morbidelli, S. N. Raymond, D. P. O'Brien, and A. M. Mandell (2011), A low mass for Mars from Jupiter's early gas-driven migration, *Nature*, *475*(7355), 206–209.

Wanke, H. (1981), Constitution of terrestrial planets, *Philos. Trans. R. Soc. A Math. Phys. Eng. Sci.*, *303*(1477), 287–302.

Warren, P. H. (2008), A depleted, not ideally chondritic bulk Earth: The explosive-volcanic basalt loss hypothesis, *Geochim. Cosmochim. Acta*, *72*(8), 2217–2235.

Warren, P. H. (2011), Stable-isotopic anomalies and the accretionary assemblage of the Earth and Mars: A subordinate role for carbonaceous chondrites, *Earth Planet. Sci. Lett.*, *311*(1–2), 93–100.

Wasserburg, G. J., M. Busso, and R. Gallino (1996), Abundances of actinides and short-lived nonactinides in the interstellar medium: Diverse supernova sources for the r-processes, *Astrophys. J.*, *466*(2), L109–L113.

Wasson, J. T., and G. W. Kallemeyn (1988), Compositions of chondrites, *Philos. Trans. R. Soc. A Math. Phys. Eng. Sci.*, *325*(1587), 535–544.

Watanabe, K., E. Ohtani, S. Kamada, T. Sakamaki, M. Miyahara, and Y. Ito (2014), The abundance of potassium in the Earth's core, *Phys. Earth Planet. Inter.*, *237*, 65–72.

Wheeler, K. T., D. Walker, Y. W. Fei, W. G. Minarik, and W. F. McDonough (2006), Experimental partitioning of uranium between liquid iron sulfide and liquid silicate: Implications for radioactivity in the Earth's core, *Geochim. Cosmochim. Acta*, *70*(6), 1537–1547.

Wood, B. J., E. S. Kiseeva, and F. J. Mirolo (2014), Accretion and core formation: The effects of sulfur on metal–silicate partition coefficients, *Geochim. Cosmochim. Acta*, *145*, 248–267.

Yin, Q. Z. (2005), From dust to planets: The tale told by moderately volatile elements, paper presented at Chondrites and the Protoplanetary Disk, Astron. Soc. of the Pacific, San Francisco.

Yin, Q. Z., S. B. Jacobsen, K. Yamashita, J. Blichert-Toft, P. Telouk, and F. Albarede (2002), A short timescale for terrestrial planet formation from Hf-W chronometry of meteorites, *Nature*, *418*(6901), 949–952.

Zhang, L., et al. (2014), Disproportionation of $(Mg,Fe)SiO_3$ perovskite in Earth's deep lower mantle, *Science*, *344*(6186), 877–882.

Zhang, Z., L. Stixrude, and J. Brodholt (2013), Elastic properties of $MgSiO_3$-perovskite under lower mantle conditions and the composition of the deep Earth, *Earth Planet. Sci. Lett.*, *379*, 1–12.

13

Metal-Silicate Partitioning of Siderophile Elements and Core-Mantle Segregation

Kevin Righter

ABSTRACT

Compositional models of the core and lower mantle are reviewed and assessed. The assumptions in the models, and the constraints and uncertainties about the Earth's interior are presented. Although a compositional model for the lower mantle that matches that of the upper mantle for major elements is most compatible with observations and constraints, uncertainties are such that competing compositional models are tenable. Based on chondritic models, more than 90% of the mass for the Earth is composed of Fe, O, Mg and Si and the addition of Ni, Ca, Al and S accounts for more than 98% by mass the composition of the Earth. There is no fixed Mg/Si ratio for chondrites; variations in this ratio reflect the proportion of olivine (2:1 molar Mg/Si) to pyroxene (1:1 molar Mg/Si) accreted in the planet or chondritic parent body. Observations on active accretion disks reveal rapid grain growth in the inner disk region and spatial variations in the relative proportions of olivine to pyroxene. Compositional models for the core are constrained by limited variation in chondrites for key siderophile element ratios (e.g., Fe/Ni, Ni/Co, and Ni/Ir). The amount and relative proportions of light element(s) in the core are poorly constrained, with tradeoffs and modeling uncertainties in core temperatures and compositional space that allow for a range of model solutions. Constraints on the absolute and relative abundances of moderately volatile and volatile elements in the Earth are consistent with only ~2% by mass of sulfur and a negligible role for H, C or N in the core. There is no evidence that heat producing elements (HPE: K, Th and U) are in the core at any significant level. Geoneutrino studies are placing global scale limits on the amount of Th and U in the Earth, which in turn will constrain models for the composition of the bulk silicate Earth, the mode proportion of the Ca-bearing phase in the deep mantle, and the thermal evolution of the planet. There is no clear geochemical evidence for core-mantle exchange, but this does not preclude it from happening and not being sampled.

13.1. PLANETARY DIFFERENTIATION AND CORE-MANTLE SEGREGATION

Separation of a metallic core is perhaps the earliest and most significant process involved in planetary differentiation. Siderophile elements (siderophile means "iron loving") can be used to understand this physical process in early Earth. Siderophile element concentrations in the primitive

NASA Johnson Space Center, Houston, Texas, USA

upper mantle (PUM) are known to be much lower than in chondrites, potential building blocks of Earth. These low concentrations (or depletions relative to chondrites) are interpreted as due to core formation. Because the siderophile elements favor metallic liquid, formation of a core can cause depletion of siderophile elements in the mantle and the size of the depletion can potentially yield information about the conditions during core formation (Figure 13.1).

The distribution of siderophile elements between metallic liquid core and silicate mantle can be understood by

Deep Earth: Physics and Chemistry of the Lower Mantle and Core, Geophysical Monograph 217, First Edition.
Edited by Hidenori Terasaki and Rebecca A. Fischer.
© 2016 American Geophysical Union. Published 2016 by John Wiley & Sons, Inc.

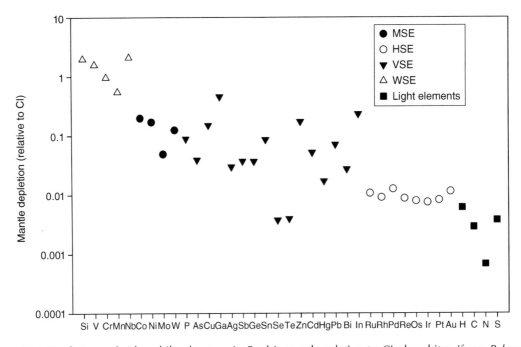

Figure 13.1 Depletions of siderophile elements in Earth's mantle relative to CI chondrites [from *Palme and O'Neill*, 2014]. Various groups of siderophile elements shown are the WSEs V, Cr, Mn, Nb, and Si, the MSEs Ni, Co, Mo, and W, the VSEs P, As, Sb, Sn, Ge, Ga, Cu, Zn, Ag, Bi, Cd, In, Te, Pb, and Se, the HSEs Re, Au, and platinum group elements Rh, Pd, Ru, Ir, Os and Pt, and the light elements S, C, H, and N. The H_2O, N, and S concentrations in the bulk silicate Earth, from Marty [2012] and *Palme and O'Neill* [2014], and bulk Earth values are from *Marty* [2012] and *Eggler and Lorand* [1993] as follows: C = 530 ppm, N = 1.68 ppm, H_2O = 2700 ppm, and S = 2 wt %.

measuring partition coefficients, D (metal/silicate) (D = concentration ratio of an element in metallic liquid and silicate melt). The siderophile elements include more than 30 elements, defined as those elements with D (metal/silicate) > 1 (at 1 bar), and are usually divided into several subclasses [e.g., *Righter*, 2003]: the weakly siderophile elements (WSEs; 1 < D < 10), moderately siderophile elements (MSEs; 10 < D < 10,000), and highly siderophile elements (HSEs; D > 10,000). In addition, we will consider some elements that are volatile and siderophile (VSE) and several light elements that are known to alloy with FeNi metallic liquids   S, Si, C, O, N, and H.

Moderately siderophile elements such as Ni, Co, Mo, and W in the upper mantle are all depleted by about a factor of approximately 10 relative to chondritic abundances [e.g., *Jagoutz et al.*, 1979; Figure 13.1]. These four elements are also refractory, thus providing tight constraints on mantle evolution. The HSEs Au and Re and the platinum group elements Pd, Pt, Rh, Ru, Ir, and Os are also in nearly chondritic ratios in primitive upper mantle but are depleted in absolute abundance by a factor of approximately 10^3 relative to chondritic values [e.g., *Becker et al.*, 2006]. The WSEs Cr, V, Mn, and Nb exhibit slight or no depletion in the PUM (Figure 13.1). In addition, these elements are known to be compatible in lower mantle minerals and thus their upper mantle deletion can be due to both core

and lower mantle partitioning [*Righter et al.*, 2011a]. For this reason, these elements hold less leverage on core formation than the moderately or highly siderophile elements. The VSEs include Ga, Ge, Cu, Zn, Sn, Sb, As, and others and have the characteristic depletion relative to chondrites but are also volatile and thus their mantle depletion must be interpreted due to pre- or postaccretional volatility combined with core formation. Finally, there are several light elements—Si, S, C, O, N, and H—that exhibit siderophile behavior. These elements can alloy with FeNi liquid and affect the metal-silicate partitioning of other siderophile metals in the MSE, HSE, WSE, and VSE groups.

All these elements encompass a wide range of partition coefficient values and can be very useful in deciphering the conditions under which a metal core may have equilibrated with a molten mantle (or a magma ocean). For example, the depletion of Ni requires a D(metal/silicate) value of approximately 30, whereas the depletion of Ru (an HSE) requires a D(metal/silicate) near 800. These required values are summarized in Figure 13.1. Because metal and silicate can equilibrate by several mechanisms, such as at the base of a deep magma ocean, or as metal droplets descending through a magma ocean, partition coefficients can potentially shed light on which mechanism may be most important, thus linking the physics and chemistry of core formation [e.g., *Rubie et al.*, 2003]. The focus of this chapter will be on the chemical equilibria, with only a brief discussion of the physical aspects of metal segregation, which will be covered in Chapter 14. In this chapter metal-silicate partitioning of siderophile elements is summarized, with brief application to understanding planetary core formation.

13.2. EXPERIMENTAL CONTROL OF P, T, f_{O_2}, AND COMPOSITION

To evaluate models for core formation and core-mantle partitioning, planetary analogue compositions are studied at a range of pressures and temperatures. At low pressures, metal-silicate systems can be studied at temperatures between 1100 and 1600°C in vertical furnaces that have large-volume hot spots allowing study of volumetrically large samples [e.g., *Newsom and Drake*, 1982]. Another advantage to low-pressure systems such as these is that the oxygen pressure (which is very low in the interior of Earth and other planets) can be controlled very precisely using gas mixing of either $CO-CO_2$ or H_2-CO_2 [*Nafziger et al.*, 1971].

Higher pressures can be achieved in the laboratory using piston cylinder techniques [0.5–4.0 GPa; *Dunn*, 1993; *Boyd and England*, 1960; *Bohlen*, 1984], multi anvil (MA) apparatuses [3.0 to ~40 GPa; *Kawai and Endo*, 1970; *Akaogi and Akimoto*, 1977; *Onodera*, 1987; *Ohtani*, 1987; *Walker et al.*, 1990; *Walker*, 1991; *Rubie*, 1999; *Ito et al.*, 2004; *Frost et al.*, 2004], or diamond anvil cell (DAC) [>10 GPa; *Bassett*, 1979; *Boehler and Chopelas*, 1991; *Campbell*, 2008] systems. Heating capabilities in these three types of apparatus are slightly different. Piston cylinder assemblies typically use an internal, electrical resistance furnace such as graphite. Heating capabilities in the multianvil are provided by graphite and also Re and $LaCrO_3$ to generate temperatures as high as 2800°C at pressures greater than the graphite-diamond transition. Samples within a DAC can be heated with a laser or with resistance heaters up to 7000 K [*Chandra Sekhar et al.*, 2003].

Study of specific compositions at high-pressure and high-temperature conditions requires careful choice of capsules or containers in the experimental assembly. Noble metal capsules (Pt and Pd) are frequently used in experimental petrology but are unsuitable for Fe metal–silicate partitioning experiments because they alloy with Fe and form eutectic melts at relatively low temperatures. As a result, studies involving Fe metal have employed refractory ceramics for capsules—MgO, Al_2O_3, and graphite. The MgO and Al_2O_3 react with the silicate melt to form MgO-rich and Al_2O_3-rich melts relative to the starting materials. Graphite is less prone to leakage of the silicate melt, but FeNi metallic liquids can alloy with carbon, resulting in as much as 6–7 wt % C in the resulting metallic liquid. Boron nitride (BN) is also a good capsule under high-*P*, *T* conditions, but most silicate melt can easily dissolve B_2O_3 (up to ~10 wt %), making BN an unattractive candidate for experiments involving silicate melts. Choice of a good capsule and an understanding of

reactivity with encapsulated melts is important for both element partitioning and isotopic fractionation. Choosing a suitable capsule material is not easy but usually there is a material that can provide good encapsulation without limiting the scope of the investigation.

The pressure of oxygen deep within planets has a fundamental control on chemical equilibria, especially whether iron is stable in the metallic or oxidized form:

$$Fe + \tfrac{1}{2}O_2 = FeO. \qquad (13.1)$$

Studies of planetary materials (meteorites, lunar samples, comet samples) have shown that metal-silicate equilibria record f_{O_2} ranging from IW-7 to IW [*Righter et al.*, 2006]. In addition, we know that Earth's mantle records nearly 10 orders of magnitude of oxygen fugacities, from just above the iron-wüstite (IW) buffer to close to the hematite-magnetite (HM) buffer [*Carmichael*, 1991]. The oxygen fugacities recorded in solar system materials (planetary and meteorite samples) thus extend across a very wide range of values and must be controlled appropriately in experimental studies. Control or monitoring of oxygen fugacity in high-pressure experimental systems is challenging but can be done by using solid buffers [e.g., Re-ReO$_2$ or Ni-NiO buffer; *Pownceby and O'Neill*, 1994; *Campbell et al.*, 2009; *Dobson and Brodholt*, 1999] or a capsule that does not react significantly with the sample of interest (e.g., graphite-lined Pt; *Holloway et al.* [1992]). The reader is referred to these references for more details of the various approaches.

The small size of the run products from MA and DAC experiments has made characterization a challenge. However, focused ion beam (FIB) sample preparation has allowed pinpoint targeting of very small features in the high-pressure samples [e.g., *Jephcoat et al.*, 2008; *Auzende et al.*, 2008; *Shofner et al.*, 2014]. Microanalytical techniques have aided in the analysis of run products [e.g., *Righter et al.*, 2010; *Bouhifd et al.*, 2013a,b]. Finally, new analytical capabilities with electron microprobe analysis (coupling with a field emission gun) have helped characterize smaller samples than ever, although correction procedures for thin samples and interferences should be improved before widespread application [*Wade and Wood*, 2012].

13.3. METAL-SILICATE PARTITIONING

13.3.1. Partitioning Behavior of Siderophile Elements: Effects of P, T, f_{O_2}, and Composition

Partitioning of siderophile elements between metal and silicate liquid can be understood in terms of simple equilibria such as.

$$
\begin{array}{ccc}
M^{n+}O_{n/2} & = & M \quad + n/4O_2 \\
\text{(silicate liquid)} & & \text{(metal)} \quad \text{(gas)}
\end{array}
\qquad (13.2)
$$

and at equilibrium:

$$-\Delta G^\circ / RT = \ln K \qquad (13.3)$$

$$-\Delta G^\circ / RT = \ln\left[(a\mathrm{M}) * \left(f_{\mathrm{O}_2}\right)^{n/4} / a\mathrm{M}^{n+}\mathrm{O}_{n/2}\right] \qquad (13.4)$$

$$-\left[\Delta H^\circ - T\,\Delta S + P\,\Delta V\right]/RT = \ln\left[\left(\gamma\mathrm{M} * x\mathrm{M}\right) * \left(f_{\mathrm{O}_2}\right)^{n/4}\right.$$
$$\left. / \left(\gamma\mathrm{M}^{n+}\mathrm{O}_{n/2} * x\mathrm{M}^{n+}\mathrm{O}_{n/2}\right)\right]$$
$$(13.5)$$

$$-\Delta H^\circ / RT + \Delta S / R - P\,\Delta V / RT - \ln\left(\gamma\mathrm{M} / \gamma\mathrm{M}^{n+}\mathrm{O}_{n/2}\right)$$
$$- n/4 \ln\left(f_{\mathrm{O}_2}\right) = \ln(x\mathrm{M}) / \left(x\mathrm{M}^{n+}\mathrm{O}_{n/2}\right)$$
$$(13.6)$$

$$\ln\left[D(\mathrm{M})\right] = -n/4 \ln\left(f_{\mathrm{O}_2}\right) - \left(\Delta H^\circ / R\right)T + \left(\Delta S / R\right)$$
$$- \left(\Delta V / R\right)P / T - \ln\left(\gamma\mathrm{M} / \gamma\mathrm{M}^{n+}\mathrm{O}_{n/2}\right)$$
$$(13.7)$$

where ΔG° is the free energy of the reaction, ΔH° is the enthalpy of the reaction, ΔS° is the entropy of the reaction, ΔV° is the volume change of the reaction, T is temperature, R is the gas constant, x is the mole fraction of the chemical species, a is the thermodynamic activity of the chemical species, f_{O_2} is the fugacity of oxygen, M is the metallic element of interest, and n is the valence in the silicate liquid [see, e.g., *Capobianco et al.*, 1993]. Such equilibria clearly would be a function of temperature, pressure, and oxygen fugacity. Additional work has demonstrated that activities of siderophile elements in metal and silicate, $a\mathrm{M}^{n+}\mathrm{O}_{n/2}$ and $a\mathrm{M}$, are dependent upon silicate liquid composition and nonmetal content of metallic liquid (e.g., S, C, or Si), respectively. All of these effects (P, T, f_{O_2}, and composition) will be discussed separately, and a framework for a comprehensive understanding guided by chemical thermodynamics will be introduced.

13.3.1.1. Oxygen Fugacity

The effect of oxygen pressure on the solubilities of siderophile elements in silicate melt has been understood for many years due to metallurgical interest and has been known to the geological community since the early work of *Palme and Ramensee* [1981] and *Newsom and Drake* [1982, 1983] on W and P. Equations (13.2)–(13.7) show that f_{O_2} dependence will depend on the valence (n) of the element in the silicate melt; specifically $\ln D(\mathrm{M})$ is dependent on $n/4 * \ln f_{\mathrm{O}_2}$. Thus a low-valence element will exhibit a weak dependence on f_{O_2}, whereas a higher valence element (W^{6+}, P^{5+}, Mo^{4+}) will exhibit a stronger f_{O_2} dependence [e.g., *Holzheid et al.*, 1994]. This f_{O_2} dependence is clear for a range of elements from In^+, Ni^{2+}, Ga^{3+}, Mo^{4+}, P^{5+}, and W^{6+} (Figure 13.2). The

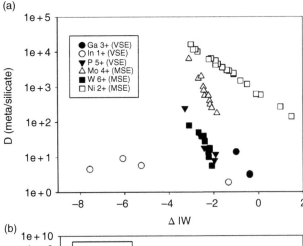

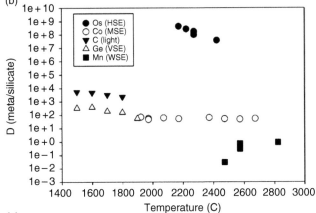

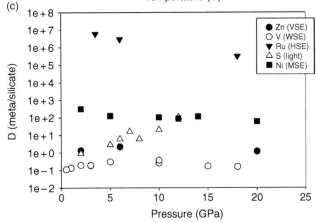

Figure 13.2 (a) Oxygen fugacity series. Data for Ga are from *Drake et al.* [1984] (at 1 bar and 1300°C), for In from *Righter et al.* [2014] (at 1 GPa and 1600°C), P from *Siebert et al.* [2011] (at 3 GPa and 1850–1900°C), for Mo and W from *Wade et al.* [2012] (at 1.5 GPa), and for Ni from *Holzheid and Palme* [1996] (at 1 bar and 1400°C). (b) Effect of temperature series. Data for Os are from *Brenan and McDonough* (2009) (2 GPa), for Co from *Chabot et al.* [2005] (at 7 GPa), for C from *Chi et al.* [2014] (at 3 GPa), for Ge from *Righter et al.* [2011b] (at 1 GPa), and for Mn from *Mann et al.* [2009] (at 18 GPa). (c) Effect of pressure series. Data for Zn are from *Mann et al.* [2009], for V from *Siebert et al.* [2011] (at 1860–1900°C), for Ru from *Mann et al.* [2012] (at 2700 K), for S from *Boujibar et al.* [2014] (at 1800–1900°C), and for Ni from *Li and Agee* [1996] (at 2000°C).

stability of high valences for several elements at high pressures (W, Mo) has been demonstrated indicating that the valences are not just a low-pressure phenomenon [*Righter et al.*, 2016; *Wade et al.*, 2013].

13.3.1.2. Temperature

A strong effect of temperature on metal-silicate partition coefficients is expected on thermodynamic grounds for those metal-silicate equilibria possessing a large enthalpy change [e.g., Capobianco et al., 1993]. In practice, some siderophile elements exhibit more temperature dependence than others [Capobianco et al., 1993; Murthy, 1991]. Several studies have shown that there is a weak temperature effect, resulting in a decrease in D (metal/silicate) at higher temperatures for Ni and Co [e.g., *Walker et al.* 1993; *Thibault and Walter*, 1995; *Chabot et al.*, 2005; Figure 13.2]. Additional studies have included multiple trace siderophile elements and have demonstrated that some D (metal/silicate) values increase at higher temperatures while some decrease [*Righter et al.*, 2010; *Siebert et al.*, 2011], thus verifying the caution of *Capobianco et al.* [1993] that one cannot generalize about siderophile elements. For example, an increase in D(Mn) with increasing temperature is in contrast to decreases in D(C), D(Ru) and D(Ge) and a weak decrease in D(Co) (Figure 13.2). Temperature effects can be difficult to disentangle from pressure effects, as discussed in the next section.

13.3.1.3. Pressure

Any effect of pressure on the magnitude of the metal-silicate partition coefficient can be related to the volumetric properties of the equilibrium represented in equations (5)–(7). The volume changes of this reaction and others like it are positive, thus indicating that siderophile elements should become more siderophile with pressure [*Righter*, 2011]. A common finding, however, is that pressure reduces the metal-silicate partition coefficients [e.g., *Li and Agee*, 1996, 2001b; *Kegler et al.*, 2008; *Bouhifd and Jephcoat*, 2003; *Cottrell et al.*, 2009, 2010]. As pressure increases (in experimental studies or in Earth's interior), so does temperature, and an increase of temperature and pressure together sometimes results in a decrease in D even though the pressure term is positive (+). Pressure causes essentially no change in D(Zn) or D(V), a slight decrease in D(Ni) and D(Ru), and an increase in D(S) (Figure 13.2).

13.3.1.4. Silicate Melt Composition

High valence elements such as Mo, W, and P can be affected significantly by melt compositional effects, whereas low-valence elements such as Ni and Co are not affected strongly by changing melt composition. Activity coefficients for siderophile metal oxides are not well known in most cases, and especially not in peridotite liquids (there are some data for basaltic and metallurgical slag systems). Therefore

proxies for activity coefficients have been used, such as basicity, *nbo/t*, or simple oxide mole fractions [*Walter and Thibault*, 1995; *O'Neill et al.*, 2008; *Righter et al.*, 2010].

The melt structural and compositional parameter *nbo/t* has been used frequently in attempts to model systematic change of solubility with silicate melt composition. The parameter *nbo/t* [*Mysen*, 1991] is the ratio of nonbridging oxygens to tetrahedrally coordinated cations in a silicate melt and is a way of estimating the degree of melt polymerization. For instance, a highly polymerized melt such as a rhyolite would have a low *nbo/t* value, close to 0.2, a basalt would have a value near 1, and depolymerized melts such as komatiite and peridotite would have values near 2 and 3, respectively. This approach gives the potential to distinguish a broad range of silicate melt compositions, but recent work suggests that such a generalization does not hold. For example, the work of *O'Neill et al.* [2008] on W shows that Ca and Mg cations in a silicate melt have different effects on the activity of W oxide dissolved in the melt, yet they would both be considered network modifiers and treated the same using an *nbo/t* approach. This drawback was noted by *Righter et al.* [2010] as well (Figure 13.3). The latter study also showed that increasing degrees of melt depolymerization can cause increases in D(metal/silicate) for some elements, which also counters the general idea that siderophile elements will be more soluble in more depolymerized melts. It is perhaps no surprise that, when considering activities in various melt compositions, each element should be considered separately because generalized assumptions about behavior may not be true in detail. For this reason, some studies have used oxide mole fractions as proxies for activity coefficients in the melt because they allow distinction between different network modifiers such as Ca, Mg, and Fe or network formers such as Si or Al.

For many elements, however, the difference between these two approaches is not significant, and the different approaches have led to similar conclusions that there is a small dependence of D(Ni) and D(Co), for example, on silicate melt composition [*O'Neill and Eggins*, 2002; *Righter and Drake*, 1999; *Righter*, 2011; *Walter and Cottrell*, 2013]. There are some studies that even choose to ignore the dependence on silicate melt composition altogether despite the abundance of evidence supporting the small effect [*Kegler et al.*, 2008; *Palme et al.*, 2011].

13.3.1.5. Metallic Liquid Composition

The variation of trace metal activities in metallic liquids can be large and not only is dependent upon the major element composition of the metal but also is influenced by the minor elements present, such as C, S, and O. The behavior of some siderophile elements can be controlled by the composition of the metal and, in particular, the Fe/Ni ratio. Because the activity coefficient of a

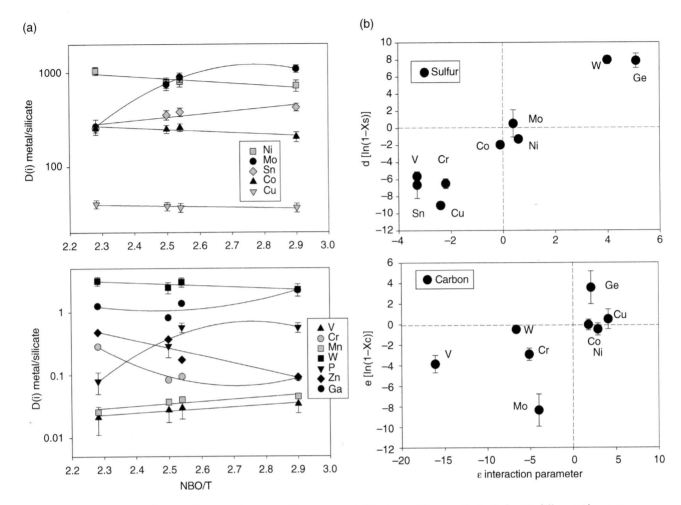

Figure 13.3 (a) Effect of silicate melt composition on metal-silicate partition coefficients for 12 different elements from the study of *Righter et al.* [2010]. Although partition coefficients for many elements decrease with silicate melt depolymerization, there are some that increase. A generalization cannot be made and the behavior of each element must be studied. *nbo/t* is calculated according to *Mysen* [1991] and corresponds to basalt values of ~1, komatiite of ~1.7 and peridotite of ~2.8. (b) Comparison of epsilon interaction parameter [from *Lupis*, 1983] and $d[\ln(1 - X_S)]$ and $e[\ln(1 - X_C)]$ terms from *Righter* [2011] and *Righter et al.* [2011b] for sulfur (top right) and carbon (bottom right). Chalcophile (S-loving) and anthracaphile (C-loving) behavior requires negative epsilon values, as demonstrated in both approaches.

siderophile element, M, may be different in a Ni-rich metal than an Fe-rich metal, an understanding of this effect is necessary before partition coefficients can be successfully applied to a natural system. A good example is Sn, which has a low activity coefficient in Fe-rich metal and a high activity coefficient in Ni-rich metal [*Capobianco et al.*, 1999]. Similarly, dissolved C and S both have large effects on the magnitude of liquid metal–liquid silicate values of *D* for many MSEs and VSEs [*Jana and Walker*, 1997a,b; *Cottrell et al.*, 2009, 2010].

Two different approaches have been taken to model these activities. One approach is to use empirical terms such as $d \ln(1 - X_S)$ that mimic the Margules parameter $\ln \gamma_2 = W_{12}(1 - X_2)^2$ (where 1 is Fe and 2 is an element such as S), terms for activity coefficients, and where coefficients for

these terms are fit by multiple linear regression. The advantage here is that the regression is carried out on data sets at the elevated pressures and temperatures and metallic liquid compositions of interest to core formation—not extrapolated from some lower pressure or temperature or unusual metal composition [*Righter and Drake*, 1999]. A potential disadvantage is that these kinds of terms are best calibrated when the coverage with mole fraction of the solute or light element is complete, but this is rarely the case. So, the assumption of linear behavior across the compositional range may not be verifiable with all data sets. However, as more data are acquired, such gaps or problems can be identified [e.g., Pd—*Wheeler et al.*, 2010].

A second approach calculates activity coefficients based on the interaction parameter approach where $\ln \gamma = \ln \gamma_0 + \sum \varepsilon_i^j X_j$

using values obtained at 1 bar and between 1400 and 1600°C, mainly from the steelmaking industry, and applies them directly (or with a temperature extrapolation) to experiments done at variable P and T [e.g., *Wade and Wood*, 2005; *Corgne et al.*, 2007]. An advantage to this approach is that there is a large body of data from the steelmaking literature [e.g., *Steelmaking Data Sourcebook*, 1988; *Elliot et al.*, 1963; *Lupis*, 1983]. The disadvantage is that there are significant secondary interactions (ρ) and temperature dependencies that are difficult to measure and therefore either do not exist or are estimated by calculation. Since these databases are tailored for steelmaking, and not core formation, they leave some large gaps in our understanding of some important elements. As such, there are no corrections for high-pressure conditions—the data are usually obtained for the low-P conditions of steelmaking. Efforts to fill gaps with new data tailored to core formation studies hold great promise [*Wood et al.*, 2014].

Each approach has strengths and weaknesses, but both can be used to quantify the effect of variable metallic liquid compositions, and agreement is typically good (Figure 13.3). With both approaches, it is important to be aware of calibration limits such as high amounts of the trace element of interest (e.g., high W in *Cottrell et al.*, 2010) or of the solute of interest such as O [*Siebert et al.*, 2013].

13.3.2. Results for Specific Groups of Siderophile Elements

13.3.2.1. Moderately Siderophile Elements

After *Murthy*'s [1991] article highlighting the issue of high-PT, metal-silicate equilibrium in the early Earth, many experimental studies focused on the MSE [*Walker et al.*, 1993; *Hillgren et al.*, 1994, 1996; *Thibault and Walter*, 1995; *Righter and Drake*, 1995; *Li and Agee*, 1996; *Righter et al.*, 1997; *Jana and Walker*, 1997a, b] and showed that the effects of temperature and pressure on D(Ni) and D(Co) are significant. Some studies showed the convergence of D(Ni) and D(Co) at higher pressures (25–30 GPa) either directly or by calculation/ extrapolation of available experimental results [*Righter et al.*, 1997; *Li and Agee*, 1996; *Ohtani et al.*, 1997; *Ito et al.*, 1998; *Bouhifd and Jephcoat*, 2003]. A systematic assessment of metal-silicate partition coefficients for the moderately siderophile elements Ni, Co, Mo, W, and P led *Righter et al.* [1997] to conclude that the excess siderophile element problem for these five elements could be resolved if their concentrations were established in a hydrous peridotite magma ocean at 25–30 GPa and 2000–2100 K. Finally, it was shown by *Righter and Drake* [1999] and *Jana and Walker* [1999] that addition of water to silicate melts has no additional or special effect on metal-silicate partition coefficient values (outside of already known effects of f_{O_2} and melt composition), thus showing that siderophile partitioning

studies done at anhydrous conditions can be applied to both hydrous and anhydrous natural systems. Subsequent work showed that many MSEs could also be explained by higher P,T conditions of 30–50 GPa and 3000–3500°C [*Li and Agee*, 2001b; *Bouhifd and Jephcoat*, 2011; *Siebert et al.*, 2012; *Kegler et al.*, 2008; *Gessmann and Rubie*, 2000; *Chabot and Agee*, 2003; *Chabot et al.*, 2005; *Wood et al.*, 2008a; *Wade et al.*, 2012]. Systematic parameterization of most MSEs has led *Righter* [2011] and *Siebert et al.* [2011] to show that MSE concentrations in the PUM can be explained by high-P,T conditions in the 30–50 GPa pressure range — these elements have been well studied and this conclusion seems incontrovertible.

13.3.2.2. Highly Siderophile Elements

HSEs have experimentally determined $D' > 10,000$, but of course the D required for equilibrium scenario are between 600 and 1000. Much of the experimental work on HSEs has been done at 1 bar conditions, where D has remained very high [10^6–10^{10}; *Fortenfant et al.*, 2003, 2006; *Borisov et al.*, 1994; *Borisov and Palme*, 1995, 1996, 1997; *Borisov and Walker*, 2000; *Lindstrom and Jones*, 1996; *Ertel et al.*, 1999, 2001; *Fleet et al.*, 1996; *Mallmann and O'Neill*, 2007; *Fonseca et al.*, 2007]. Our understanding of the effect of high-P,T conditions with peridotite and C-, S-, and Si-bearing silicate melt remains poor and incomplete. Progress in filling the gap in HSE distribution coefficient data has been difficult due to analytical problems. For example, some of these elements (and their alloys) can form submicro meter sized metallic particles that become dispersed in the melt, making analysis of just the silicate portion a major challenge [e.g., *Cottrell and Walker*, 2006]. Also, the solubilities of these elements in silicate melts are very low and therefore a highly sensitive analytical technique is required [e.g., *Ertel et al.*, 2008; *Hervig et al.*, 2004]. Despite these problems there have been a large number of studies in the last decade, and they are summarized here with an emphasis on where we still lack data.

Studies of D(Pt), D(Au), and D(Pd) at high-P,T conditions have shown that all three of these elements are less siderophile with D values close to those required for an equilibrium explanation [*Ertel et al.*, 2006; *Cottrell and Walker*, 2006; *Holzheid et al.*, 2000; *Danielson et al.*, 2005; *Righter et al.*, 2008a; *Brenan and McDonough*, 2009; *Wheeler et al.*, 2011; *Righter et al.*, 2015]. The other five HSEs—Re, Rh, Ru, Ir, and Os—are less well understood at these extreme conditions. Recently D(Ir) and D(Os) were measured at high-P,T conditions by *Brenan and McDonough* [2009] and *Yokoyama et al.* [2009], showing a significant decrease of D(Ir) and D(Os) at high temperatures. The role of higher temperature and pressure has also been examined for Ir, Re, Rh, Pt, and Ru, and the data show a decrease up to 18 GPa [*Mann et al.*, 2012]. Nonetheless, given the range (albeit incomplete) of conditions investigated, values of

D for these five elements are not lowered to those required for an equilibrium partitioning scenario (e.g., ~600 ± 200). However, extension to high-pressure conditions and to peridotites and metallic systems with C and S has not been done yet. With the exception of Au, Pd, and Pt, the distribution coefficients of the HSEs are poorly constrained at the high-P,T conditions proposed for Earth's core formation process and for the metal and silicate compositions that are relevant to that phase of Earth history.

13.3.2.3. Volatile Siderophile Elements
Until about 2008, there was very little experimental data available on the volatile siderophile elements [e.g., *Capobianco et al.*, 1999; *Newsom and Drake*, 1983; *Holzheid et al.*, 2000; *Righter and Drake*, 2000; *Schmitt et al.*, 1989]. However, since then, high-P,T studies have included Cu, Zn, P, Ge, Ga, In, Sn, Sn, and As [*Siebert et al.*, 2011; *Ballhaus et al.*, 2013; *Mann et al.*, 2009; *Corgne et al.*, 2007; *Righter et al.*, 2009, 2010, 2011b, 2014]. These new studies have allowed more thorough investigation of modeling during accretion. Studies that include only a few variables such as P and T but not f_{O_2}, metallic S, C, or Si content, or silicate melt composition [*Ballhaus et al.*, 2013] may result in a conclusion of a requirement of late accretion to explain the VSEs. But studies that have quantified all known effects result in solutions that match the observed mantle VSEs at high-P,T conditions much like the MSEs [*Righter et al.*, 2011b, 2014; *Corgne et al.*, 2007].

13.3.2.4. Light Elements
Hydrogen, O, N, C, S, and Si are all light elements that may be alloyed with FeNi metal in the core. Usually these elements are discussed in terms of their effect (separately or combined) on siderophile element partitioning because S, C, and Si are possibly in weight percent levels in the core. However, S, C, and Si are all siderophiles and their D changes with P, T, f_{O_2}, and composition, just like the MSEs, HSEs, and WSEs. Pressure and temperature have been known to affect $D(S)$ [*Li and Agee*, 2001a] and recent work has shown that when melt effects, metallic liquid composition, and f_{O_2} are also accounted for, $D(S)$ at high-P,T conditions can explain the S content of Earths primitive upper mantle [*Boujibar et al.*, 2014]. Similarly, $D(C)$ is very siderophile and recent work has shown that $D(C)$ remains below 100 at high-P,T conditions relevant to a magma ocean [*Dasgupta et al.*, 2013; *Chi et al.*, 2014]. However, *Malavergne et al.* [2008] show that $D(C)$ can be less than 100 depending on the composition (Si content) of the metallic liquid. The value $D(H)$ has been very difficult to measure, but *Okuchi* [1997] and *Okuchi and Takahashi* [1998] demonstrated that H is indeed siderophile and there may be a significant amount of H in Earth's core. Recently $D(N)$ has been studied by *Roskosz et al.* [2013] and it is also clear that N can be moderately siderophile, thus accounting for at least a portion of the N depletion in Earth's mantle.

Finally, $D(Si)$ is also variable with P, T, f_{O_2}, and metallic and silicate liquid compositions [*Ricolleau et al.*, 2011; *Kilburn and Wood*, 1997], although Si is much less siderophile than the other light elements. Clearly all of these light elements need to be understood as accretion proceeds because the core content will depend on their core-mantle distribution.

13.4. MODELING OF SIDEROPHILE ELEMENTS

13.4.1. History of Core Formation Models and Evolution of Magma Ocean Concept

A more historical treatment of core formation models is outside the scope of this review, but the reader is referred to *Brush* [1996], *Righter* [2003], or *Rubie et al.* [2007]. Many models in the 1970s to the 1990s were based on low-pressure partition coefficients for processes operating at relatively cool conditions, such as heterogeneous accretion [*Wänke*, 1981; *O'Neill*, 1991], metal-sulfide equilibrium [*Arculus and Delano*, 1981; *Brett*, 1984; *Gaetani and Grove*, 1997], and inefficient core formation [*Jones and Drake*, 1986]. When physical models accounted for the energetics of core formation [*Davies*, 1990; *Stevenson*, 1990; *Tonks and Melosh*, 1990] and dynamic modeling started to show the likelihood that early Earth experienced several giant impact events [*Cameron and Benz*, 1991; *Benz et al.*, 1989], it became clear that Earth must have experienced high temperatures where mantle melting was extensive. At that time, high-pressure and high-temperature partitioning data for siderophile elements began to be used to forward models of core formation at these more extreme conditions.

13.4.2. Physical Consideration in Magma Ocean

Models based on chemical data must also be thoroughly evaluated from a physical perspective. It is well known that metal mobility through liquid silicate is rapid [*Stevenson*, 1990; *Arculus et al.*, 1990; *Solomotov*, 2000], whereas equilibration between ponded metallic liquid and overlying convecting magma ocean is very slow [*Rubie et al.*, 2003]. *Karato and Murthy* [1997] explored the idea that metal-silicate equilibrium is attained by equilibration between metal droplets and silicate melt. Building on this idea and results of partitioning studies, *Rubie et al.* [2003] compared two modes of metal-silicate separation and equilibration: droplets and ponded metal. Rubie et al. showed that metal-silicate equilibrium can only occur in a metal droplet scenario because only that mechanism is rapid enough to allow equilibration before solidification of the magma ocean [see also *Höink et al.*, 2006]. A deep magma ocean without an insulating atmosphere will crystallize on the timescale of 1000 years [e.g., *Hofmeister*, 1983; *Davies*, 1990; *Solomotov*, 2000]. If there is a thick, hot steam atmosphere overtop the cooling rate will be slower but still rapid enough due to thermal radiation to crystallize on the order of 2×10^6 years when the temperature is well below the peridotite liquidus [e.g., *Kasting*, 1988; *Zahnle et al.*, 1988; *Zahnle*, 2006]. Equilibration of metal ponded at the base of a deep magma ocean would not occur on this timescale either. Thus, the metal droplet model seems essential for equilibrium models. If droplets descend and equilibrate rapidly, then it might suggest the P, T conditions of metal-silicate equilibrium can define the depth of the magma ocean. There are several factors, however, that complicate this simple interpretation.

First, several different settings have been identified in which droplets can form and be relevant to metal-silicate equilibrium. Because settling velocities of small iron droplets (~0.5 m/s) are much lower than typical convection velocities (~10 m/s), iron droplets may remain entrained for a significant period of time in the magma ocean. Accumulation through sedimentation at the base of the ocean will be a slow and gradual process, compared to the timescale required for chemical equilibration [e.g., *Ichikawa et al.*, 2010; *Deguen et al.*, 2011]. Furthermore, *Rubie et al.* [2007] suggest that droplets settle out in approximately the Stokes settling time but show that strong density currents develop in the droplet bearing region of the magma ocean due to density perturbations [*Golabek et al.*, 2008]. The velocities of these density currents range up to 50 m/s, which is much greater than thermal convection velocities. In addition, the gravitational energy of sinking droplets is converted into heat which raises the temperature at the base of the magma ocean by at least several hundred degrees. These observations all indicate that the temperature at the metal-silicate interface may be above the liquidus and is not necessarily constrained to be on the liquidus.

Despite the recognition that metallic droplets can equilibrate rapidly [*Rubie et al.*, 2003; *Höink et al.*, 2006; *Samuel and Tackley*, 2008; *Deguen et al.*, 2011], there are some scenarios where equilibration may not be complete. Therefore some modeling efforts have chosen to add a variable that quantifies the degree of equilibration [*Rudge et al.*, 2010; *Nimmo et al.*, 2010; *Rubie et al.*, 2011]. Modeling of the metallic core portion of a merging impactor [*Sasaki and Abe*, 2007; *Dahl and Stevenson*, 2010] shows that metallic material greater than 10 km in diameter does not equilibrate with the mantle before it merges with the core. Such scenarios indicate that the metal from an impactor may not fully equilibrate with a planet's mantle. On the other hand, *Kendall and Melosh* [2012, 2014] found evidence from modeling that equilibrium is complete due to emulsification in impact scenarios even including large impactors. If these results are broadly applicable to accretion, then equilibration seems unavoidable.

13.4.3. Quantification of P, T, f_{O_2}, and Silicate and Metallic Melt Composition Effects on D (metal/silicate)

Chemical thermodynamics can be used as a guide in assessing the relative importance of the factors identified above. Partitioning of siderophile elements (M) between metal and silicate liquid can again be understood in terms of equations (13.2) and (13.7) developed in Section 13.3, where the $\ln f_{O_2}$, $1/T$, P/T, and constant terms are related to valence, enthalpy, volume, and entropy, respectively. Addition of four terms accounting for activity coefficients in the metal and silicate results in

$$\ln D(M) = a \ln f_{O_2} + b/T + cP/T + d \ln(1 - X_S) + e \ln(1 - X_C) + f \ln(1 - X_{Si}) + g(nbo/t) + h$$

(13.8)

or

$$\ln D(M) = a \ln f_{O_2} + b/T + cP/T + \left(\frac{\gamma_{Fe}}{\gamma_M}\right) + g(nbo/t) + h,$$

(13.9)

where the activity coefficients in equation (13.9) can be calculated according to $\ln \gamma_i = \ln \gamma_{0,i} + \Sigma \varepsilon_{i,j} x_j$ (for $j = 2, \ldots, N$; e.g., *Wade and Wood* [2005]). This approach can been applied to experimental partitioning data for which temperature, pressure, oxygen fugacity, and silicate and metallic compositions are all well characterized [e.g., *Righter et al.*, 1997; *Righter*, 2011] and coefficients a to h may be determined by multiple linear regression. Enthalpies of end-member oxides and silicate melt systems (i.e., D vs. $1/T$) in general correspond well [*Wade and Wood*, 2005; *Siebert et al.*, 2011]. The slopes are usually the same, with the exception of Ge and Zn [*Siebert et al.*, 2011], which may relate to nonideal mixing in melt systems. Ultimately, the extent to which regressions can be used effectively is dependent upon the input data, and this topic is discussed more below.

Another approach that circumvents the problem of knowing f_{O_2} at high-P, T conditions is to normalize partitioning relations to the Fe + O_2 = FeO equilibrium and cast the relation in the form of an exchange equilibrium. A two-element distribution coefficient K_D based on the equilibrium

$$(n/2)\underset{\text{(metal)}}{Fe} + \underset{\text{(silicate)}}{MO_{n/2}} = (n/2)\underset{\text{(silicate)}}{FeO} + \underset{\text{(metal)}}{M}, \quad (13.10)$$

where

$$K_{D,M-Fe}^{met/sil} = \left(X_M^{met} / X_{MO_{n/2}}^{sil}\right) / \left(X_{Fe}^{met} / X_{FeO}^{sil}\right)^{n/2} \quad (13.11)$$

(n is the valence of M) is independent of oxygen fugacity, thus eliminating f_{O_2}, as a variable as well. The predictive expression thus becomes

$$\ln K_D = b/T + cP/T + d\ln(1-X_S) + e\ln(1-X_C)$$
$$+ f\ln(1-X_{Si}) + g(nbo/t) + h \tag{13.12}$$

By including activity terms for the metallic liquid (γ) and using nbo/t as a proxy for melt composition, the following expression has been used by several authors to predict exchange coefficients [*Cottrell et al.*, 2009, 2010; *Mann et al.*, 2009; *Wood et al.*, 2008a; *Kegler et al.*, 2008]:

$$\ln K_D - \mathrm{adj}(M) = b/T + cP/T + g(nbo/t) + \left(\frac{\gamma_{Fe}}{\gamma_M}\right) + h \tag{13.13}$$

In some cases the nbo/t terms have been ignored [e.g., *Kegler et al.*, 2008]. The metal activity ratios can be obtained from the literature.

In this chapter, only D will be used in the modeling that follows but K_D is introduced here because it is also encountered in the literature.

13.4.4. Linking with Accretion Models

13.4.4.1. Continuous Core Formation

With the accretion and giant impact modeling becoming mature in the late 1990s and onward continuous accretion models are more appropriate for thinking about the growth of the early Earth and the setting of core formation. Once Earth achieved the mass of Mars, it probably underwent numerous magma ocean episodes as it accreted, possibly solidifying before the next giant impact created a new magma ocean [*Stevenson*, 2008]. The calculated temperature and pressure could represent the last, largest magma ocean event [e.g., *Righter*, 2011; *Halliday*, 2008]. Alternatively, it could represent some sort of ensemble average memory of a number of magma ocean events. High-P,T conditions could also correspond to some physical boundary in the molten mantle such as the depth of equilibration of metallic droplets or a layer of entrained metallic droplets caught in convection currents. The consensus on a deep magma ocean environment being responsible for at least the moderately siderophile element fingerprint in Earth's mantle remains robust. Siderophile elements were integrated with continuous accretion models as early as *Righter and Drake* [1997], who considered the evolution of mantle Re content as Earth grew. *Righter* [2003] combined accretion models of *Agnor et al.* [1999] with W partitioning to illustrate that the Hf/W value

of the mantle would change during accretion as Earth grew in size and developed thermally. *Wade and Wood* [2005] developed this concept further and included many elements (Ni, Co, W, Nb, V, Cr) in the modeling.

13.4.4.2. Continuous Accretion and Heterogeneous Accretion

Rubie et al. [2011] showed that many elements can be fit in a model that combined aspects of continuous accretion and heterogeneous accretion. In their model, the mantle f_{O_2} is allowed to increase from initial values near IW-3.8 to final values near IW-2. Their modeling was coupled with the models of *O'Brien et al.* [2006] and also allowed for the possibility of incomplete equilibration between mantle and core. Using Fe, Ni, Co, V, W, Nb, Cr, Mn, Ga, Si, and P, *Wood et al.* [2006] and *Wood et al.* [2008a] proposed that the primitive mantle concentrations could be best matched if f_{O_2} was allowed to evolve from early reduced (ΔIW = −3.6) to late oxidized (ΔIW = −2) conditions. These are similar to those of heterogeneous accretion models of the 1970 and 1980s in which the material accreting to form Earth began reduced and ended oxidized [e.g., *Wänke*, 1981].

Additional modeling with the consideration of water sources led *Rubie et al.* [2015] to conclude that many siderophile elements (Ni, Co, V, Cr, Nb, Ta) can be explained in Earth's mantle with a multistage accretion process. Their chemically implausible core composition is modeled with a four-component system—Fe-Ni-Si-O—which excludes the important light elements S, C, and H. These three elements have a very strong influence on the partitioning of Mn, V, and Cr as well as the more robust refractory siderophile elements Mo and W, which are ignored in the modeling. Therefore, the results from this approach will be more compelling when more realistic partitioning models can be included (i.e., including the chemical effects of S and C on the partitioning of Mn, V, Cr, Mo, and W).

13.4.4.3. New Modeling: Coupling Accretion with Predictive D(Metal/Silicate) Expressions

Here we couple some recent siderophile element models with accretion models of *Agnor et al.* [1999] and *O'Brien et al.* [2014]. Simulation 1 of *Agnor et al.* [1999] produced an Earth-like planet, accreted in a timescale of 50 Ma (Figure 13.4). Simulation SA163-767-1 from *O'Brien et al.* [2014] produced an Earth-like planet with a mass of 1.1 that of Earth on a timescale of ~100 Ma and also experienced 3% of material during late accretion (Figure 13.5). For these models, we adopt the mass accreted from the growth curves and assume melting of mantle of 60% using peridotite melting relations of *Andrault et al.* [2011] and core mass fractions of ~0.3. Five elements are modeled: W, Ni, Co (MSEs), Pd (HSE)

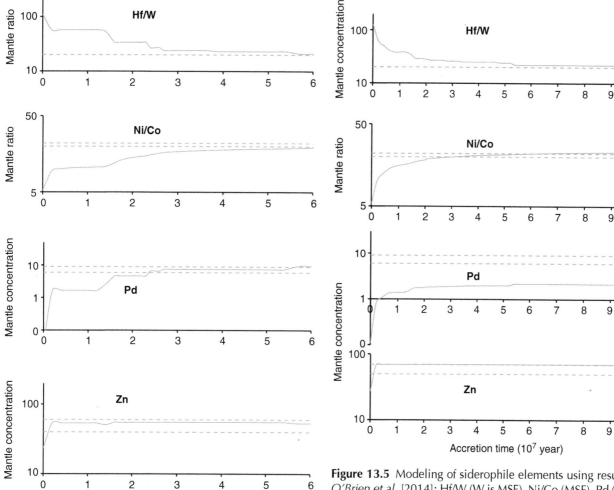

Figure 13.4 Modeling of siderophile elements using results of *Agnor et al.* [1999]: Hf/W (W is MSE), Ni/Co (MSE), Pd (HSE), and Zn (VSE). Oxygen fugacity is fixed at IW-2.4 during accretion and planet forms over 60 Ma timeframe; see text for details.

Figure 13.5 Modeling of siderophile elements using results of *O'Brien et al.* [2014]: Hf/W (W is MSE), Ni/Co (MSE), Pd (HSE), and Zn (VSE). Oxygen fugacity varies from IW-3.5 to IW-1.5 during accretion and planet forms over 100 Ma timeframe; see text for details.

and Zn (VSE); relative f_{O_2} is fixed at IW-2.4 for the *Agnor et al.* [1999] growth curve, whereas it is allowed to vary from –3.5 to –1.5 for the *O'Brien et al.* [2014] example. As these models approach the size of Earth, the chemical composition is calculated. In particular, the Ni/Co and Hf/W ratios for Earth are reproduced quite well (Figures 13.4 and 13.5), as are the Pd and Zn contents.

13.5. SUMMARY AND FUTURE

There is general agreement that Earth experienced a magma ocean condition with some disagreement about the depth and degree of equilibration between the metallic core and silicate mantle. Resolution of details of magma ocean scenarios awaits additional experimental work and calculations. There are a number of outstanding issues

that need to be addressed before we have a full understanding that will allow a distinction between various hypotheses.

13.5.1. Volatile and HSE and Unknown Effects of Si and O on D (Metal/Silicate)

First, there are many siderophile elements for which we have geochemical data on natural samples but little or no experimental data (or at least not enough to make a prediction at elevated pressure and temperature conditions). For example, we have limited experimental data for several key moderately siderophile elements such as Ag, Sb, Sn, In, and As, whose mantle depletions are well defined [e.g., *Jochum and Hofmann*, 1997; *Witt-Eickschen et al.*, 2009]. The volatile chalcophile elements such as In, Te, Pb, Se, Bi, and Cd are very poorly integrated with any accretion or core formation models. The new experiments of *Rose-Westin et al.* [2009] on Te and Se and

Mills et al. [2007] on Cs and the experiments and modeling of *Wood et al.* [2008b] for Pb have led to improvements, but again a thorough systematic understanding is lacking. The HSEs have been well defined in mantle materials [e.g., *Becker et al.*, 2006; *Fischer-Gödde et al.*, 2011], but we lack systematic high-pressure and high-temperature experimental data for most HSEs necessary to make a critical assessment of their consistency with high-pressure and high-temperature magma ocean scenarios (>20 GPa) and also the effect of alloying light elements such as C and Si.

13.5.2. Gaps in P,T Space

Second, despite publication of the results of hundreds of metal-silicate experiments, there are gaps in the experimental database on which predictive expressions are based. There are limited data at the high temperatures and pressures (most experiments are <20 GPa) at which equilibrium solutions have been proposed [see, e.g., *Righter*, 2011]. Additional experimentation in the pressure range of DAC (>30 GPa) would be particularly useful in evaluating siderophile element partitioning models as well as solid silicate and solid-liquid silicate partitioning. Moreover, an equilibrium scenario only works if the magma ocean is peridotitic, yet many of the experiments in the database have been carried out with basaltic melts. In general, the coverage of P-T-f_{O_2}s and composition is more complete for Ni, Co, W, P, Mn, V, and Cr; the partitioning of these elements has been studied across a wide range of P, T, f_{O_2}, and silicate and metal composition. Coverage for Mo, Ga, VSEs, and HSEs, however, is lacking, and much remains to be done before we have a thorough understanding of the partitioning behavior.

13.5.3. Coupling Isotopic and Experimental Studies

Several recent studies have highlighted the potential of using isotopic variations to constrain the composition of the core. In some cases, the identity of the light element is the focus, such as S [*Shahar et al.*, 2013] or Si [*Shahar et al.*, 2009; *Ziegler et al.*, 2010; *Chakrabarti and Jacobsen*, 2010]. In other cases, the isotopic fractionation associated with partitioning of a weakly siderophile element such as Cr can be assessed [*Moynier et al.*, 2011]. Although there is debate about the results obtained so far, the isotopic measurement showed significant promise for helping to constrain aspects of core formation and core-mantle equilibrium.

13.5.4. f_{O_2} in Accreting Earth: How Variable Was It?

Although some modeling argues that oxidation state increases during accretion [e.g., *Bond et al.*, 2010], experiments and calculations indicate that it may stay constant or even decrease as pressure (or depth) to the metal-silicate interface increases. *Rubie et al.* [2011] and *Wood et al.* [2008a] both carry out the modeling of siderophile elements in Earth's mantle during accretion, in which the mantle f_{O_2} evolved from IW-4 to IW-2. On the other hand, *Siebert et al.* [2013] argue that f_{O_2} of the accreting Earth may have changed from oxidized to reduced over time, based on Cr, V, O, and Si partitioning at ultrahigh pressures. Similar conclusions were reached by *Righter and Ghiorso* [2012] based on calculations of Fe-FeO equilibria to 40 GPa. However, the latter study showed that, because of uncertainties on silicate melt compressibilities at high pressures, the deep early molten Earth could exhibit either a reduced or nearly constant f_{O_2} trend with depth. Very reducing conditions are not normally considered in early Earth settings but would have a strong effect on other redox equilibria and also exert controls on the composition of the atmosphere. This is an area where more work will provide much needed constraints on the metal-silicate equilibria.

13.5.5. How Did Earth's Mantle Reach FMQ so Quickly?

If the young Earth allowed metallic liquid to pass through its mantle to the core, yet the upper mantle is not reduced enough for iron metal stability, how did Earth's mantle become oxidized to near the FMQ (fayalite-magnetite-quartz or FMQ) buffer? There has been a disconnection between the early reduced f_{O_2} required for core formation models (IW-2 to IW-3.6), compared to Earth's Archean and current upper mantle f_{O_2}, which is near FMQ [*Arculus*, 1985; *Delano*, 2001; *Trail et al.*, 2011]. Proposed oxidation mechanisms for Earth are few, perhaps not robust enough to cause the required oxidation, and have drawbacks—a satisfying explanation for this conundrum has remained elusive. For example, many have argued for late (post core formation) accretion of chondritic material to Earth's upper mantle to explain the near chondritic and elevated HSE abundances as well as the oxidation (*Chou*, 1978; *Chou et al.*, 1983). However, addition of chondritic materials (<1 mass %) to a reduced post core formation mantle will result in the reduction of those materials to a mixture of metal and silicate, and the HSEs will be partitioned into the metal and then proceed directly into the core. The oxidized late veneer has no capacity for oxidation in a large reduced planet like Earth. Another possibility may simply be that the upper mantle has become oxidized over time due to the effects of recycling and plate tectonics. However, no studies have yet revealed a secular trend of oxygen fugacity [e.g., *Eggler and Lorand*, 1995; *Canil*, 2002; *Delano*, 2001; *Yang et al.*, 2014]. A third idea is that the systematic breakdown of Fe-bearing Mg-perovskite (or bridgmanite) into Fe metal and Fe^{3+}-bearing silicates has led to natural oxidation of the upper mantle [*Frost et al.*, 2004b; *Wade and Wood*, 2005; *Williams et al.*, 2012]. Although this is an intriguing idea, and one that would occur early enough in Earth history to meet the requirements of current models, it is not without problems or questions. For example, the mantle of Mars is just as oxidized as Earth's (near FMQ) but there is not an Mg-perovskite reservoir in Mars that can produce the oxidation [*Righter et al.*, 2008b]. In addition, it is not clear if Fe_2O_3 is added to the upper mantle by Mg-perovskite dissolution; it may dissociate into FeO and Fe_2O_3 in response to the low ambient f_{O_2} set by core formation [see *Righter et al.*, 2014]. The efficiency and extent of this mechanism remain uncertain. A fourth and more likely and/or promising possibility is that the mantle was oxidized somewhat by the partitioning of H and C between the core, mantle, magma ocean, and atmosphere [*Kuramoto and Matsui*, 1996; *Okuchi and Takahashi*, 1998; *Abe et al.*, 2000]. These authors show that C prefers the core while most H prefers the silicate melt and estimate that the amount of H_2O partitioned to silicate melt is large enough to explain the amount of H_2O in the hydrosphere and mantle and provides enough oxygen to partially oxidize ferrous iron to ferric iron in the mantle. In addition, H_2 loss during accretion may provide an oxidation mechanism for the mantle that can act rapidly and extensively [*Sharp et al.*, 2013].

ACKNOWLEDGMENTS

I would like to thank R. Fischer and H. Terasaki for the invitation to contribute this volume. Reviews by J. Wade and A. Bouhifd were very helpful in improving the presentation and clarity of the chapter. This work is supported by a NASA RTOP from the Cosmochemistry Program.

REFERENCES

Abe Y., E. Ohtani, T. Okuchi, K. Righter, and M. J. Drake (2000), Water in the early Earth, in *Origin of the Earth and Moon*, edited by. R. Canup and K. Righter, 413–434 Univ. of Ariz. Press, Tucson.

Agnor, C. B., R. M. Canup, and H. F. Levison, (1999), On the character and consequences of large impacts in the late stage of terrestrial planet formation, *Icarus, 142*, 219–237.

Akaogi, M., and S. I. Akimoto, (1977), Pyroxene-garnet solid-solution equilibria in the systems $Mg_4Si_4O_{12}$-$Mg_3Al_2Si_3O_{12}$ and $Fe_4Si_4O_{12}$-$Fe_3Al_2Si_3O_{12}$ at high pressures and temperatures, *Phys. Earth Planet. Inter., 15*, 90–106.

Andrault, D., N. Bolfan-Casanova, G. Lo Nigro, M.A. Bouhifd, G. Garbarino, and M. Mezouar (2011), Solidus and liquidus profiles of chondritic mantle: Implication for melting of the Earth across its history, *Earth Planet. Sci. Lett. 304*, 251–259.

Arculus, R. J. (1985), Oxidation status of the mantle: Past and present, *Annu. Rev. Earth Planet. Sci., 13*, 75–95.

Arculus, R. J., and J. W. Delano (1981), Siderophile element abundances in the upper mantle; evidence for a sulfide signature and equilibrium with the core, *Geochim. Cosmochim. Acta, 45*, 1331–1343.

Arculus, R. J., R. D. Holmes, R. Powell, and K. Righter (1990), Metal/silicate equilibria and core formation, in *The Origin of the Earth*, edited by H. Newsom and J. H. Jones, pp. 251–271. Oxford Univ. Press, London.

Auzende, A.-L., J. Badro, F. J. Ryerson, P. K. Weber, S. J. Fallon, A. Addad, J. Siebert, and G. Fiquet (2008), Element partitioning between magnesium silicate perovskite and ferropericlase: New insights into bulk lower-mantle geochemistry, *Earth Planet. Sci. Lett. 269*, 164–174.

Ballhaus, C., V. Laurenz, C. Münker, R. O. Fonseca, F. Albarède, A. Rohrbach, M. Lagos, M. W. Schmidt, K. P. Jochum, B. Stoll, U. Weis and H. M. Helmy (2013), The U/Pb ratio of the Earth's mantle—A signature of late volatile addition, *Earth Planet. Sci. Lett., 362*, 237–245.

Bassett, W.A. (1979), The diamond cell and the nature of the Earth's mantle, *Annu. Rev. Earth Planet. Sci., 7*, 357–380.

Becker, H., M. F. Horan, R. J. Walker, S. Gao, J.-P. Lorand, and R. L. Rudnick (2006), Highly siderophile element composition of the Earth's primitive upper mantle: Constraints from new data on peridotite massifs and xenoliths, *Geochim. Cosmochim. Acta, 70*, 4528–4550.

Benz, W., A. G. W. Cameron, and H. J. Melosh (1989), The origin of the Moon and the single-impact hypothesis III, *Icarus, 81*, 113–131.

Boehler, R., and A. Chopelas (1991), A new approach to laser heating in high pressure mineral physics, *Geophys. Res. Lett., 18*, 1147–1150.

Bohlen, S. R. (1984), Equilibria for precise pressure calibration and a frictionless furnace assembly for the piston-cylinder apparatus, *Neu. Jb. Mineral. Mh., 9*, 404–412.

Bond, J. C., D. S. Lauretta, and D. P. O'Brien (2010), Making the Earth: Combining dynamics and chemistry in the Solar System, *Icarus, 205*, 321–337.

Borisov, A., and H. Palme (1995), The solubility of iridium in silicate melts: New data from experiments with $Ir_{10}Pt_{90}$ alloys, *Geochim. Cosmochim. Acta, 59*, 481–485.

Borisov, A., and H. Palme (1996), Experimental determination of the solubility of Au in silicate melts, *Mineral. Petrol., 56*, 297–312.

Borisov, A., and H. Palme (1997), Experimental determination of the solubility of platinum in silicate melts, *Geochim. Cosmochim. Acta, 61*, 4349–4357.

Borisov, A., and R. J. Walker (2000), Os solubility in silicate melts: New efforts and results. *Am. Mineral., 85*, 912–917.

Borisov, A., H. Palme, and B. Spettel (1994), Solubility of palladium in silicate melts: Implications for core formation in the Earth. *Geochim. Cosmochim. Acta, 58*, 705–716.

Bouhifd, M. A., and A. P. Jephcoat (2003), The effect of pressure on partitioning of Ni and Co between silicate and iron-rich metal liquids: A diamond-anvil cell study, *Earth Planet. Sci. Lett. 209*, 245–255.

Bouhifd, M. A., and A. P. Jephcoat (2011), Convergence of Ni and Co metal–silicate partition coefficients in the deep magma-ocean and coupled silicon–oxygen solubility in iron melts at high pressures, *Earth Planet. Sci. Lett., 307*(3), 341–348.

Bouhifd, M. A., D. Andrault, N. Bolfan-Casanova, T. Hammouda, and J. L. Devidal (2013a), Metal–silicate partitioning of Pb and U: Effects of metal composition and oxygen fugacity. *Geochim. Cosmochim. Acta, 114*, 13–28.

Bouhifd, M. A., A. P. Jephcoat, V. S. Heber, and S. P. Kelley (2013b). Helium in Earth/'s early core, *Nature Geosci.*, *6*, 982–986, doi:10.1038/NGEO1959.

Boujibar, A., D. Andrault, M. A. Bouhifd, N. Bolfan-Casanova, J. L. Devidal, and N. Trcera (2014), Metal–silicate partitioning of sulphur, new experimental and thermodynamic constraints on planetary accretion, *Earth Planet. Sci. Lett.*, *391*, 42–54.

Boyd, F. R., and J. L. England (1960), Apparatus for phase equilibrium measurements at pressures up to 50 kilobars and temperatures up to 1750°C, *J. Geophys. Res.*, *65*, 741–748.

Brenan J. M., and W. F. McDonough (2009), Core formation and metal-silicate fractionation of osmium and iridium from gold, *Nature Geosci.*, *2*, 798–801.

Brett, R. (1984), Chemical equilibration of the Earth's core and upper mantle, *Geochim. Cosmochim. Acta*, *48*, 1183–1188.

Brush, S. G. (1996), Nebulous earth. The origin of the solar system and the core of the earth from Laplace to Jeffreys, in *A History of Modern Planetary Physics*, Cambridge Univ. Press, Cambridge, 312 pp.

Cameron, A. G. W., and W. Benz (1991), The origin of the Moon and the single impact hypothesis IV, *Icarus*, *92*, 204–216.

Campbell, A. J. (2008), Measurement of temperature distributions across laser-heated samples by multispectral imaging radiometry, *Rev. Sci. Instrum.*, *79*, 015,108.

Campbell, A. J., L. R. Danielson, K. Righter, C. T. Seagle, Y. Wang, and V. B. Prakapenka (2009), High pressure effects on the iron–iron oxide and nickel–nickel oxide oxygen fugacity buffers, *Earth Planet. Sci. Lett.*, *286*, 556–564.

Canil, D. (2002), Vanadium in peridotites, mantle redox and tectonic environments: Archean to present, *Earth Planet. Sci Lett.*, *195*, 75–90.

Capobianco, C. J., J. H. Jones, and M. J. Drake (1993), Metal/silicate thermochemistry at high temperature: Magma oceans and the "excess siderophile element" problem of the Earth's Upper Mantle, *J. Geophys. Res.*, *98*, 5433–5443.

Capobianco, C. J., M. J. Drake, and J. A. DeAro (1999), Siderophile geochemistry of Ga, Ge and Sn: Cationic oxidation states in silicate melts and the effect of composition in iron-nickel alloys, *Geochim. Cosmochim. Acta*, *63*, 2667–2677.

Carmichael, I. S. E. (1991), The redox state of basic and silicic magmas: A reflection of their source regions? *Contrib. Mineral. Petrol.*, *106*, 129–141.

Chabot, N. L., and C. B. Agee (2003), Core formation in the Earth and Moon: New experimental constraints from V, Cr, and Mn, *Geochim. Cosmochim. Acta*, *67*, 2077–2091.

Chabot, N. L., D. S. Draper and C. B. Agee (2005), Conditions of core formation in the earth: Constraints from nickel and cobalt partitioning, *Geochim. Cosmochim. Acta*, *69*, 2141–2151.

Chakrabarti, R., and S. B. Jacobsen (2010), Silicon isotopes in the inner Solar System: Implications for core formation, solar nebular processes and partial melting, *Geochim. Cosmochim Acta*, *74*, 6921–6933.

Chandra Shekar, N. V., Ph. C. Sahu and K. G. Rajan (2003), Laser-heated diamond-anvil cell (LHDAC) in materials science research, *J. Mater. Sci. Technol.*, *19*, 518–525.

Chi, H., R. Dasgupta, M. Duncan, and N. Shimizu (2014), Partitioning of carbon between Fe-rich alloy melt and silicate melt in a magma ocean—Implications for the abundance and origin of volatiles in Earth, Mars, and the Moon, *Geochim. Cosmochim Acta 139*, 447–471.

Chou, C. L. (1978), Fractionation of siderophile elements in the Earth's upper mantle, in *Proceedings of the Ninth Lunar Planetary Science Conference,* 9 pp. 219–230.

Chou, C.-L., D. M. Shaw, and J. H. Crocket (1983), Siderophile trace elements in the Earth's oceanic crust and upper mantle, *J. Geophys. Res.*, *88* (Suppl. 2), A507–A518.

Corgne, A., S. Keshav, B. J. Wood, W. F. McDonough and Y. Fei (2007), Metal–silicate partitioning and constraints on core composition and oxygen fugacity during Earth accretion, *Geochim. Cosmochim. Acta*, *71*, 574–589.

Corgne A., J. Siebert and J. Badro (2009), Oxygen as a light element: A solution to single-stage core formation, *Earth Planet. Sci. Lett.*, *288*, 108–114.

Cottrell, E. and D. Walker (2006), Constraints on core formation from Pt partitioning in mafic silicate liquids at high temperatures, *Geochim. Cosmochim. Acta*, *70*, 1565–1580.

Cottrell, E., M.J. Walter, and D. Walker (2009a), Metal-silicate partitioning of tungsten at high pressure and temperature: Implications for equilibrium core formation in Earth, *Earth Planet. Sci. Lett.*, *281*, 275–287.

Cottrell, E., M. J. Walter, and D. Walker (2010), Erratum to "Metal–silicate partitioning of tungsten at high pressure and temperature: Implications for equilibrium core formation in Earth" [*Earth Planet. Sci. Lett.*, *281* (2009), 275–287], *Earth Planet. Sci. Lett.*, *289*, 631–634.

Dahl, T. W. and D. J. Stevenson (2010), Turbulent mixing of metal and silicate during planet accretion—An interpretation of the Hf–W chronometer, *Earth Planet. Sci. Lett.*, *295*, 177–186.

Danielson, L. R., T. G. Sharp, R. L. Hervig (2005), Implications for core formation of the Earth from high pressure-temperature Au partitioning experiments, *Proc. Lunar Planet. Sci. Conf.*, *34*, abstract # 1955.

Dasgupta, R., H. Chi, N. Shimizu, A. S. Buono, and D. Walker (2013), Carbon solution and partitioning between metallic and silicate melts in a shallow magma ocean: Implications for the origin and distribution of terrestrial carbon, *Geochimica Cosmochim Acta*, *102*, 191–212.

Davies, G. F. (1990), Heat and mass transport in the early Earth, in *Origin of the Earth*, edited by H. Newsom and J. H. Jones, pp. 175–194, Oxford Univ. Press, New York.

Deguen, R., P. Olsen, and P. Cardin (2011), Experiments on turbulent metal-silicate mixing in a magma ocean, *Earth Planet. Sci. Lett.*, *310*, 303–313.

Delano, J.W. (2001), Redox history of the Earth's interior since approximately 3900 Ma; implications for prebiotic molecules, in *Origins of Life and Evolution of the Biosphere, 31*, 311–341.

Dobson, D., and J. P. Brodholt (1999), The pressure medium as a solid-state oxygen buffer, *Geophys. Res. Lett.*, *26*, 259–262.

Drake, M. J., H. E. Newsom, S. J. Reed, and M. C. Enright (1984), Experimental determination of the partitioning of gallium between solid iron metal and synthetic basaltic melt: Electron and ion microprobe study, *Geochim. Cosmochim. Acta*, *48*, 1609–1615.

Dunn, T. (1993), The piston-cylinder apparatus, in *Experiments at High Pressure and Applications to the Earth's Mantle, Short Course Handbook*, edited by R. W. Luth, vol. 21, pp. 39–94, *Mineral. Assoc.*, Edmonton, Canada.

Eggler, D., and J.-P. Lorand (1993), Mantle sulfide geobarometry, *Geochim. Cosmochim. Acta, 57*, 2213–2222.

Eggler, D. H., and J. P. Lorand (1995), Sulfides, diamonds and mantle fO₂, Vol. 2, *Companhia de Pesquisa de Recursos Minerais (CPRM)*, Rio de Janeiro, in: *Proc. 5ᵗʰ International Kimberlite Conf.*, edited by H. O. A. Meyer and O. H. Leonardos, pp. 160–169.

Elliot, J. F., M. Gleiser and V. Ramakrishna (1963), *Thermochemistry for Steelmaking*, Addison Wesley, Reading, Mass.

Ertel, W., H.St.C. O'Neill, P. J. Sylvester, and D. B. Dingwell (1999), Solubilities of Pt and Rh in a haplobasaltic silicate melt at 1300°C, *Geochim. Cosmochim. Acta, 63*, 2439–2449.

Ertel, W., H.St.C. O'Neill, P. J. Sylvester, D. B. Dingwell and B. Spettel (2001), The solubility of rhenium in silicate melts: Implications for the geochemical properties of rhenium at high temperatures, *Geochim. Cosmochim. Acta, 65*, 2161–2170.

Ertel, W., M. J. Walter, M. J. Drake, and P. J. Sylvester (2006), Experimental study of platinum solubility in silicate melt to 14 GPa and 2273 K: Implications for accretion and core formation in Earth. *Geochim. Cosmochim. Acta, 70*, 2591–2602.

Ertel, W., D. B. Dingwell, and P. J. Sylvester (2008), Siderophile elements in silicate melts—A review of the mechanically assisted equilibration technique and the nanonugget issue, *Chem. Geol., 248*, 119–139.

Fischer-Gödde, M., H. Becker, and F. Wombacher (2011), Rhodium, gold, and other highly siderophile elements in orogenic peridotites and peridotite xenoliths, *Chem. Geol., 280*, 365–383.

Fleet, M. E., J. H. Crocket, and W. E. Stone (1996), Partitioning of platinum-group elements (Os, Ir, Ru, Pt, Pd) and gold between sulfide liquid and basalt melt, *Geochim. Cosmochim. Acta, 60*, 2397–2412.

Fonseca R. O. C., G. Mallmann, H.St.C. O'Neill, and I. H. Campbell (2007), How chalcophile is rhenium? An experimental study of the solubility of Re in sulphide mattes, *Earth Planet. Sci. Lett., 260*, 537–548.

Fortenfant, S. S., D. Günther, D. B. Dingwell and D. C. Rubie (2003), Temperature dependence of Pt and Rh solubilities in a haplobasaltic melt, *Geochim Cosmochim. Acta, 67*, 123–131.

Fortenfant, S. S., D. B. Dingwell, W. Ertel-Ingrisch, F. Capmas, J. L. Birck, and C. Dalpe (2006), Oxygen fugacity dependence of Os solubility in haplobasaltic melt, *Geochim. Cosmochim. Acta, 70* 742–756.

Frost, D. J., B. T. Poe, R. G. Trønnes, C. Liebske, A. Duba, D.C. Rubie (2004a), A new large-volume multi-anvil system, *Phys. Earth Planet. Inter., 143–144*, 507–514.

Frost, D. J., C. Liebske, F. Langenhorst, C. A. McCammon, R. A. Trønnes and D. C. Rubie (2004b) Experimental evidence for the existence of iron-rich metal in the Earth's lower mantle, *Nature, 428*, 409–412.

Gaetani, G. A. and T. L. Grove (1997), Partitioning of moderately siderophile elements among olivine, silicate melt, and sulfide melt: Constraints on core formation in the Earth and Mars, *Geochim. Cosmochim. Acta, 61*, 1829–1846.

Gessmann, C. K., and D. C. Rubie (2000), Experimental evidence for the origin of the Cr, V and Mn depletions in the mantle of the Earth and implications for the origin of the Moon, *Earth Planet. Sci. Lett., 184*, 95–107.

Golabek, G. J., H. Schmeling, and P. J. Tackley (2008), Earth's core formation aided by flow channelling instabilities induced by iron diapirs, *Earth Planet. Sci. Lett., 271*, 24–33.

Halliday, A. N. (2008), A young Moon-forming giant impact at 70–110 million years accompanied by late-stage mixing, core formation, and degassing of the Earth, *Philos. Trans. R. Soc. Lond. A, 366*, 4163–4182.

Hervig, R. L., F. Mazdab, L. R. Danielson, T. Sharp, A. Hamed, and P. Williams (2004), SIMS microanalyses for Au in silicates, *Am. Mineral., 89*, 498–504.

Hillgren, V. J., M. J. Drake, and D. C. Rubie (1994), High pressure and temperature experiments on core-mantle segregation in the accreting Earth, *Science, 264*, 1442–1445.

Hillgren, V. J., M. J. Drake, and D. C. Rubie (1996), High pressure and high temperature metal/silicate partitioning of siderophile elements: The importance of silicate liquid composition, *Geochim. Cosmochim. Acta, 60*, 2257–2263.

Hofmeister, A. M. (1983), Effect of a Hadean terrestrial magma ocean on crust and mantle evolution, *J. Geophys. Res., 88*, 4963–4983.

Höink, T., J. Schmalzl, U. Hansen, (2006), Dynamics of metal-silicate separation in a terrestrial magma ocean, *Geochem. Geophys. Geosyst., 7*(9), doi:10.1029/2006GC001268.

Holloway, J. R., V. Pan and G. Gudmundsson (1992), High-pressure fluid-absent melting experiments in the presence of graphite: Oxygen fugacity, ferric/ferrous ratio and dissolved CO₂, *Eur. J. Mineral., 4*, 105–114.

Holzheid, A., and H. Palme (1996), The influence of FeO in the solubilities of Co and Ni in silicate melts, *Geochim. Cosmochim. Acta, 60*, 1181–1193.

Holzheid, A., A. Borisov and H. Palme (1994), The effect of oxygen fugacity and temperature on solubilities of nickel, cobalt and molybdenum in silicate melts, *Geochim. Cosmochim. Acta, 58*, 1975–1981.

Holzheid, A., P. Sylvester, H.St C. O'Neill, D. C. Rubie, and H. Palme (2000), Evidence for a late chondritic veneer in the Earth's mantle from high-pressure partitioning of palladium and platinum. *Nature, 406*, 396–399.

Ichikawa, H., S. Labrosse, and K. Kurita (2010), Direct numerical simulation of an iron rain in the magma ocean, *J. Geophys. Res., 115*, B01,404.

Ito, E., T. Katsura and T. Suzuki (1998), Metal/silicate partitioning of Mn, Co, and Ni at high pressures and high temperatures and implications for core formation in a deep magma ocean, in *Properties of Earth and Planetary Materials at High Pressure and Temperature, Geophys. Monog. 101*, edited by M. H. Manghnani, pp. 215–225, AGU, Washington, D.C.

Ito, E., A. Kubo, T. Katsura and M. J. Walter (2004), Melting experiments of mantle materials under lower mantle conditions with implications for magma ocean differentiation, *Phys. Earth Planet. Int., 143–144*, 397–406.

Jagoutz, E., H. Palme, H. Baddenhausen, K. Blum, M. Cendales, G. Dreibus, B. Spettel, V. Lorenz, and H. Wänke (1979), The abundances of major, minor and trace elements in the earth's mantle as derived from primitive ultramafic nodules, in *Proceedings of the Tenth Lunar and Planetary Science Conference*, vol. *2*, Pergamon, New York.

Jana, D., and D. Walker (1997a), The impact of carbon on element distribution during core formation, *Geochim. Cosmochim. Acta*, *61*, 2759–2763.

Jana, D., and D. Walker (1997b), The influence of sulfur on partitioning of siderophile elements, *Geochim. Cosmochim. Acta*, *61*, 5255–5277.

Jana, D., and D. Walker (1999), Core formation in the presence of various C-H-O volatiles, *Geochim. Cosmochim. Acta*, *63*, 2299–2310.

Jephcoat, A. P., M. A. Bouhifd, and D. Porcelli (2008), Partitioning experiments in the laser-heated diamond anvil cell: Volatile content in the Earth's core, *Philos. Trans. R. Soc. Lond. A*, *306*, 4295–4314.

Jochum, K. P., and A. W. Hofmann (1997), Antimony in mantle derived rocks, *Chem. Geol.*, *139*, 39–49.

Jones, J. H., and M. J. Drake (1986), Geochemical constraints on core formation, *Nature*, *322*, 221–228.

Karato, S.-I., and V. R. Murthy (1997), Core formation and chemical equilibrium in the Earth—I. Physical considerations, *Phys. Earth Planet. Inter.*, *100*, 61–79.

Kasting, J.F. (1988), Runaway and moist greenhouse atmospheres and the evolution of Earth and Venus, *Icarus*, *74*, 472–494.

Kawai, N., and S. Endo (1970), Generation of ultrahigh hydrostatic pressures by a split sphere apparatus, *Rev. Sci. Instrum.*, *41*, 1178–1181.

Kegler, Ph., A. Holzheid, D. J. Frost, D. C. Rubie, R. Dohmen, and H. Palme (2008), New Ni and Co metal-silicate partitioning data and their relevance for an early terrestrial magma ocean, *Earth Planet. Sci. Lett.*, *268*, 28–40.

Kendall, J. D., and H. J. Melosh (2012), Fate of iron cores during planetesimal impacts, *Lunar Planet. Inst. Sci. Conf. Abstr.*, *43*, 2699.

Kendall, J., and H. J. Melosh (2014), Dispersion of planetesimal iron cores during accretional impacts, *Lunar Planet. Inst. Sci. Conf. Abstr.*, *45*, 2827.

Kilburn, M. R., and B. J. Wood (1997), Metal–silicate partitioning and the incompatibility of S and Si during core formation, *Earth Planet. Sci. Lett.* *152*, 139–148.

Kuramoto, K., and T. Matsui (1996), Partitioning of H and C between the mantle and core during the core formation in the Earth: Its implications for the atmospheric evolution and redox state of early mantle, *J. Geophys. Res.*, *101*(E6), 14,909–14,932, doi:10.1029/96JE00940.

Li, J., and C. B. Agee (1996), Geochemistry of mantle-core formation at high pressure, *Nature*, *381*, 686–689.

Li, J., and C. B. Agee (2001a), Element partitioning constraints on the light element composition of the Earth's core, *Geophys. Res. Lett.*, *28*(1), 81–84.

Li, J., and C.B. Agee (2001b), The effect of pressure, temperature, oxygen fugacity and composition on partitioning of nickel and cobalt between liquid Fe-Ni-S alloy and liquid silicate: Implications for the Earth's core formation, *Geochim. Cosmochim. Acta*, *65*, 1821–1832.

Lindstrom, D.J., and J.H. Jones (1996), Neutron activation analysis of multiple 10–100 lg glass samples from siderophile element partitioning experiments, *Geochim. Cosmochim. Acta*, *60*, 1195–1203.

Lupis, C. H. (1983), *Chemical Thermodynamics of Materials*, Elsevier Science, Amsterdam, 581 pp.

Malavergne, V., J. P. Gallien, S. B. Berthet, and H. Bureau (2008), Carbon solubility in metallic phases at high pressure and high temperature: Preliminary results and application to planetary cores, 39th LPSC, abstract #1340.

Mallmann, G., and H.St.C. O'Neill (2007), The effect of oxygen fugacity on the partitioning of rhenium between crystals and silicate melt during mantle melting, *Geochim. Cosmochim. Acta*, *71*, 2837–2857.

Mann U., D. J. Frost and D. C. Rubie (2009), Evidence for high-pressure core-mantle differentiation from the metal-silicate partitioning of lithophile and weakly-siderophile elements, *Geochim. Cosmochim. Acta*, *73*, 7360–7386.

Mann, U., D. J. Frost, D. C. Rubie, H. Becker, and A. Audétat (2012), Partitioning of Ru, Rh, Pd, Re, Ir and Pt between liquid metal and silicate at high pressures and high temperatures—Implications for the origin of highly siderophile element concentrations in the Earth's mantle, *Geochim. Cosmochim. Acta*, *84*, 593–613.

Marty, B. (2012), The origins and concentrations of water, carbon, nitrogen and noble gases on Earth, *Earth Planet. Sci. Lett.*, *313*, 56–66.

Mills, N., C. B. Agee, and D. S. Draper (2007), Metal-silicate partitioning of cesium: Implications for core formation, *Geochim. Cosmochim. Acta 71*, 4066–4081.

Moynier, F., Q. Z. Yin, and E. Schauble (2011), Isotopic evidence of Cr partitioning into Earth's core, *Science*, *331*(6023), 1417–1420.

Murthy, V. R. (1991), Early differentiation of the Earth and the problem of mantle siderophile elements: A new approach, *Science*, *253*, 303–306.

Mysen, B.O. (1991), Relations between structure, redox equilibria of iron, and properties of magmatic liquids, in *Physical Chemistry of Magmas*, edited by L. L. Perchuk and I. Kushiro, pp. 41–98, Springer-Verlag, New York.

Nafziger, R. H., G. C. Ulmer, and E. Woermann (1971), Gaseous buffering for the control of oxygen fugacity at one atmosphere, in *Research Techniques for High Pressure and High Temperature*, edited by G. C. Ulmer, pp. 9–41, Springer, Berlin and Heidelberg.

Newsom, H., and M.J. Drake (1982), The metal content of the eucrite parent body: Constraints from the partitioning behavior of tungsten, *Geochim. Cosmochim. Acta*, *46*, 2483–2489.

Newsom, H., and M.J. Drake (1983), Experimental investigations of the partitioning of phosphorus between metal and silicate phases: Implications for the Earth, Moon and eucrite parent body, *Geochim. Cosmochim. Acta*, *47*, 93–100.

Nimmo F., D. P. O'Brien, and T. Kleine (2010) ,Tungsten isotopic evolution during late-stage accretion: Constraints on Earth–Moon equilibration, *Earth Planet. Sci. Lett.*, *292*, 363–370.

O'Brien, D. P., A. Morbidelli, and H. F. Levison (2006), Terrestrial planet formation with strong dynamical friction, *Icarus*, *184*(1), 39–58.

O'Brien, D. P., K.J. Walsh, A. Morbidelli, S. N. Raymond, and A.M. Mandell, (2014), Water delivery and giant impacts in the "Grand Tack" scenario, *Icarus*, *239*, 74–84.

O'Neill, H. S. C. (1991), The origin of the Moon and the early history of the Earth—A chemical model. Part 2: The Earth, *Geochim. Cosmochim. Acta*, *55*(4), 1159–1172.

O'Neill, H. S. C., and S. M. Eggins (2002), The effect of melt composition on trace element partitioning: An experimental investigation of the activity coefficients of FeO, NiO, CoO, MoO_2 and MoO_3 in silicate melts, *Chem. Geol.*, *186*, 151–181.

O'Neill, H. S. C., A. J. Berry and S. M. Eggins (2008), The solubility and oxidation state of tungsten in silicate melts: Implications for the comparative chemistry of W and Mo in planetary differentiation process, *Chem. Geol.*, *255*, 346–359.

Ohtani, E. (1987), Ultrahigh-pressure melting of a model chondritic mantle and pyrolite compositions, in *High-Pressure Research in Mineral Physics, Geophys. Monogr. 39*, edited by M. H Manghnani and Y. Syono, pp. 87–93, AGU, Washington, D.C.

Ohtani, E., H. Yurimoto, and S. Seto (1997), Element partitioning between metallic liquid, silicate liquid, and lower-mantle minerals: Implications for core formation of the Earth, *Phys. Earth Planet. Inter.*, *100*, 97–114.

Okuchi, T. (1997), Hydrogen partitioning into molten iron at high pressure: Implications for Earth's core, *Science*, *278*, 1781–1783.

Okuchi, T., and Takahashi, E. (1998), Hydrogen in molten iron at high pressure: The first measurement, in *High Pressure-Temperature Research: Properties of Earth and Planetary Materials*, edited. by M. H. Manghnani and T. Yagi, pp. 249–260, AGU, Washington D.C.

Onodera, A. (1987), Octahedral-anvil high pressure devices, *High Temp. High Press.*, *19*, 579–609.

Palme, H., and H. S. C. O'Neill (2014), Cosmochemical estimates of mantle composition, in *Treatise on Geochemistry*, edited by Richard W. Carlson. Executive Editors: Heinrich D. Holland and Karl K. Turekian. pp. 568, Vol. 2, 2nd ed., pp. 1–39.

Palme, H., and W. Ramensee (1981), The significance of W in planetary differentiation processes: Evidence from new data on eucrites, *Lunar Planet. Sci. Conf.*, *12*, 949–964.

Palme, H., P. Kegler, A. Holzheid, D. J. Frost, and D. C. Rubie (2011), Comment on "Prediction of metal–silicate partition coefficients for siderophile elements: An update and assessment of PT conditions for metal–silicate equilibrium during accretion of the Earth" by K. Righter [*Earth Planet. Sci. Lett.*, *304* (2011), 158–167], *Earth Planet. Sci. Lett.*, *312*, 516–518.

Pownceby, M. I., and H. St. C. O'Neill (1994), Thermodynamic data from redox reactions at high temperatures. IV. Calibration of the Re-ReO_2 oxygen buffer from EMF and NiO+Ni-Pd redox sensor measurements, *Contrib. Mineral. Petrol.*, *118*, 130–137.

Ricolleau, A., Y. Fei, A. Corgne, J. Siebert, and J. Badro (2011), Oxygen and silicon contents of Earth's core from high pressure metal–silicate partitioning experiments, *Earth Planet. Science Lett.*, *310*, 409–421.

Righter, K. (2003), Metal/silicate partitioning of siderophile elements and core formation in the Early Earth, *Annu. Rev. Earth Planet. Sci.*, *31*, 135–174.

Righter, K. (2011), Prediction of metal-silicate partition coefficients for siderophile elements: An update and assessment of PT conditions for metal-silicate equilibrium during accretion of the Earth, *Earth Planet. Sci. Lett.*, *304*, 158–167.

Righter, K., and M. J. Drake (1995), The effect of pressure on siderophile element (Ni, Co, Mo, W and P) metal/silicate partition coefficients, *Meteoritics*, *30*, 565–566.

Righter, K., and M. J. Drake (1997), Metal/silicate equilibrium in a homogeneously accreting Earth: New results for Re, *Earth Planet. Sci. Lett.*, *146*, 541–553.

Righter, K., and M. J. Drake (1999), Effect of water on metal/silicate partitioning of siderophile elements: A high pressure and temperature terrestrial magma ocean and core formation, *Earth Planet. Sci. Lett.*, *171*, 383–399.

Righter, K., and M. J. Drake (2000), Metal/silicate equilibrium in the early Earth: New constraints from volatile moderately siderophile elements Ga, Sn, Cu and P. *Geochim. Cosmochim. Acta*, *64*, 3581–3597.

Righter, K., and M. S. Ghiorso (2012), Redox systematics of a magma ocean with variable pressure-temperature gradients and composition, *Proc. Nat. Acad. Sci.*, *109*, 11,955–11,960.

Righter, K., M. J. Drake and G. Yaxley (1997), Prediction of siderophile element metal–silicate partition coefficients to 20 GPa and 2800 °C: The effect of pressure, temperature, fO_2 and silicate and metallic melt composition, *Phys. Earth Planet. Inter.*, *100*, 115–134.

Righter K., M. J. Drake and E. Scott (2006), Compositional relationships between meteorites and terrestrial planets, in *Meteorites and the Early Solar System* vol. 2; edited by D. S. Lauretta and H. Y. McSween, Jr., pp. 803–828, Univ. of Ariz. Press, Tucson.

Righter, K., M. Humayun, and L.R. Danielson (2008a), Partitioning of palladium at high pressures and temperatures during core formation, *Nature Geosci. 1*, 321–323.

Righter, K., H. Yang, G. Costin, and R. T. Downs (2008b), Oxygen fugacity in the martian mantle controlled by carbon: New constraints from the nakhlite MIL03346, *Met. Planet., Sci, 43*, 1709–1723.

Righter, K., M. Humayun, A. J. Campbell, L. Danielson, D. Hill, and M. J. Drake (2009), Experimental studies of metal–silicate partitioning of Sb: Implications for the terrestrial and lunar mantles, *Geochim. Cosmochim. Acta*, *73*, 1487–1504.

Righter, K., K. Pando, L. R. Danielson, and C.-T. Lee (2010), Partitioning of Mo, P and other siderophile elements (Cu, Ga, Sn, Ni, Co, Cr, Mn, V, and W) between metal and silicate melt as a function of temperature and silicate melt composition, *Earth Planet. Sci. Lett.*, *291*, 1–9.

Righter, K., S. Sutton, L. R. Danielson, K. Pando, G. Schmidt, H. Yang, S. Berthet, M. Newville, Y. Choi, R. T. Downs, and V. Malavergne (2011a), The effect of fO_2 on the partitioning and valence of V and Cr in garnet/melt pairs and the relation to terrestrial mantle V and Cr content, *Ame. Mineral.*, *96*, 1278–1290.

Righter, K., C. King, L. Danielson, K. Pando, and C. T. Lee (2011b), Experimental determination of the metal/silicate partition coefficient of germanium: Implications for core and mantle differentiation, *Earth Planet. Sci. Lett. 304*, 379–388.

Righter, K., K. A. Pando, L. R. Danielson, and K. A. Nickodem (2014), Core-mantle partitioning of volatile elements and the origin of volatile elements in Earth and Moon. In 45th *Lunar and Planetary Science Conference*, abstract #2130.

Righter, K., L. R. Danielson, K. Pando, J. Williams, M. Humayun, R. L. Hervig, and T. G. Sharp (2015), Mantle HSE abundances in Mars due to core formation at high pressure and temperature, *Meteoritics & Planetary Science, 50*, 604–631.

Righter, K., L. R. Danielson, K. A. Pando, G. A. Shofner, S. R. Sutton, M. Newville, and C.-T. Lee (2016), Valence and metal/silicate partitioning of Mo: Implications for conditions of Earth accretion and core formation Earth and Planetary Science Letters (in press).

Rose-Weston, L., J. M. Brenan, Y. Fei, R. A. Secco, and D. J. Frost (2009), Effect of pressure, temperature, and oxygen fugacity on the metal-silicate partitioning of Te, Se, and S: Implications for earth differentiation, *Geochim. Cosmochim. Acta, 73*, 4598–4615.

Roskosz, M., M. A. Bouhifd, A. P. Jephcoat, B. Marty, and B. O. Mysen (2013), Nitrogen solubility in molten metal and silicate at high pressure and temperature, *Geochim. Cosmochim. Acta, 121*, 15–28.

Rubie, D.C. (1999), Characterising the sample environment in multi-anvil high-pressure experiments, *Phase Transitions, 68*, 431–451.

Rubie, D. C., H. J. Melosh J. E. Reid, C. Liebske, and K. Righter (2003), Mechanisms of metal/silicate equilibration in the terrestrial magma ocean, *Earth Planet. Sci Lett., 205*, 239–255.

Rubie, D. C., F. Nimmo, and H. J. Melosh (2007), Formation of the Earth's core, in *Treatise on Geophysics, vol. 9, Evolution of the Earth*, edited by D. J. Stevenson, pp. 51–90, Elsevier, Amsterdam.

Rubie, D. C., D. J. Frost, U. Mann, Y. Asahara, N. Nimmo, K. Tsuno, P. Kegler, A. Holzheid, and H. Palme (2011), Heterogeneous accretion, composition and core–mantle differentiation of the Earth, *Earth Planet. Sci. Lett., 301*, 31–42.

Rubie, D. C., S. A. Jacobson, A. Morbidelli, D. P. O'Brien, E. D. Young, J. de Vries, F. Nimmo, H. Palme, and D. J. Frost (2015), Accretion and differentiation of the terrestrial planets with implications for the compositions of early-formed Solar System bodies and accretion of water, *Icarus, 248*, 89–108.

Rudge, J. F., T. Kleine, and B. Bourdon (2010), Broad bounds on Earth's accretion and core formation constrained by geochemical models, *Nat. Geosci., 3*, 439–443.

Samuel, H. and P.J. Tackley (2008), Dynamics of core formation and equilibration by negative diapirism, *Geochem. Geophys. Geosyst., 9*, Q06,011, doi:10.1029/2007GC001896.

Sasaki, T. and Y. Abe (2007), Rayleigh-Taylor instability after giant impacts: Imperfect equilibration of the Hf-W system and its effect on the core formation age, *Earth Planets Space, 59*, 1035–1045.

Schmitt, W., H. Palme and H. Wänke (1989), Experimental determination of metal/silicate partition coefficients for P, Co, Ni, Cu, Ga, Ge, Mo, W and some implications for the early evolution of the Earth, *Geochim. Cosmochim. Acta, 53*, 173–186.

Shahar, A., K. Ziegler, E. D. Young, A. Ricolleau, E. A. Schauble, and Y. Fei (2009). Experimentally determined Si isotope fractionation between silicate and Fe metal and implications for Earth's core formation, *Earth Planet. Sci. Lett., 288*, 228–234.

Shahar, A., V. J. Hillgren, M. F. Horan, and T. D. Mock (2013). Iron Isotope Fractionation as a Probe of Core Formation and Composition, *LPI Contributions, 1768*, 8029.

Sharp, Z. D., F. M. McCubbin, and C. K. Shearer (2013), A hydrogen-based oxidation mechanism relevant to planetary formation, *Earth Planet. Sci. Lett., 380*, 88–97.

Shofner, G. A., A. J. Campbell, L. Danielson, Z. Rahman, and K. Righter (2014), Metal-silicate partitioning of tungsten from 10 to 50 GPa, *Lunar Planet. Inst. Sci. Conf. Abstr., 45*, 1267.

Siebert, J., A. Corgne, and F. J. Ryerson (2011), Systematics of metal-silicate partitioning for many siderophile elements applied to Earth's core formation, *Geochim. Cosmochim. Acta, 75*, 1451–1489.

Siebert, J., J. Badro, D. Antonangeli, and F. J. Ryerson (2012), Metal–silicate partitioning of Ni and Co in a deep magma ocean, *Earth Planet. Sci. Lett., 321*, 189–197.

Siebert, J., J. Badro, D. Antonangeli, and F. J. Ryerson (2013), Terrestrial accretion under oxidizing conditions, *Science 339*, 1194–1197.

Solomotov, V. S. (2000), Fluid dynamics of a terrestrial magma ocean. in *Origin of the Earth and Moon*, edited by R. M. Canup and K. Righter pp. 323–338, Univ. of Ariz. Press, Tucson.

Steelmaking Data Sourcebook (1988), *Part II, Recommended Values of Activity Coefficients and Interaction Parameters of Elements in Iron Alloys*, Gordon and Breach, *Science Publishers* Montreaux, pp. 273–325.

Stevenson, D. J. (1990), Fluid dynamics of core formation, in *The Origin of the Earth*, edited by H. Newsom and J. H. Jones, pp. 231–249, Oxford Univ. Press, London.

Stevenson, D. J. (2008), A planetary perspective on the deep Earth, *Nature, 451*, 261–265.

Thibault, Y., and M. J. Walter (1995), The influence of pressure and temperature on the metal/silicate partition coefficients of nickel and cobalt in a model C1 chondrite and implications for metal segregation in a deep magma ocean, *Geochim. Cosmochim. Acta, 59*, 991–1002.

Tonks, B., and H. J. Melosh (1990), The physics of crystal settling and suspension in a turbulent magma ocean, in *The Origin of the Earth*, edited by H. Newsom and J. H. Jones, pp. 151–174, Oxford Univ. Press, London.

Trail, D., E.B. Watson, and N. D. Tailby (2011), The oxidation state of Hadean magmas and implications for early Earth's atmosphere, *Nature, 480*, 79–82.

Wade, J., and B. J. Wood (2005), Core formation and the oxidation state of the Earth, *Earth Planet. Sci. Lett., 236*, 78–95.

Wade, J., and B. J. Wood (2012), Metal–silicate partitioning experiments in the diamond anvil cell: A comment on potential analytical errors, *Phys. Earth Planet. Int., 192*, 54–58.

Wade, J., B. J. Wood, and J. Tuff (2012), Metal–silicate partitioning of Mo and W at high pressures and temperatures: Evidence for late accretion of sulphur to the Earth, *Geochim. Cosmochim. Acta, 85*, 58–74.

Wade, J., B. J. Wood, and C. A. Norris (2013), The oxidation state of tungsten in silicate melt at high pressures and temperatures, *Chem. Geol., 335*, 189–193.

Walker, D. (1991), Lubrication, gasketing, and precision in multi-anvil experiments. *Am. Mineral., 76*, 1092–1100.

Walker, D., M. A. Carpenter, and C. M. Hitch (1990), Some simplifications to multi-anvil devices for high pressure experiments, *Am. Mineral., 75*, 1020–1028.

Walker, D., L. Norby and J. H. Jones (1993), Superheating effects on metal/silicate partitioning of siderophile elements, *Science, 262,* 1858–1861.

Walter, M. J., and E. Cottrell (2013), Assessing uncertainty in geochemical models for core formation in Earth, *Earth Planet. Sci. Lett., 365,* 165–176.

Walter, M. J., and Y. Thibault (1995), Partitioning of tungsten and molybdenum between metallic liquid and silicate melt, *Science, 270,* 1186–1189.

Wänke, H. (1981), Constitution of terrestrial planets, *Philos. Trans. Roy. Soc. Lond. A, 393,* 287–302.

Wheeler, K. T., D. Walker, and W. F. McDonough (2010), Pd and Ag metal-silicate partitioning applied to Earth differentiation and core-mantle exchange, *Meteor. Planet. Sci., 46,* 199–217.

Williams, H. M., B. J. Wood, J. Wade, D. J. Frost, and J. Tuff (2012), Isotopic evidence for internal oxidation of the Earth's mantle during accretion, *Earth Planet. Sci. Lett, 321,* 54–63.

Witt-Eickschen, G., H. Palme, H.St.C. O'Neill, and C. M. Allen (2009), The geochemistry of the volatile trace elements As, Cd, Ga, In and Sn in the Earth's mantle: New evidence from in situ analyses of mantle xenoliths, *Geochim. Cosmochim. Acta, 73,* 1755–1778.

Wood, B. J., E. S. Kiseeva, and F. J. Mirolo (2014), Accretion and core formation: The effects of sulfur on metal–silicate partition coefficients, *Geochim. Cosmochim. Acta, 145,* 248–267.

Wood, B. J., J. Wade, M. R. Kilburn (2008a), Core formation and the oxidation state of the Earth: Additional constraints from Nb, V and Cr partitioning, *Geochim. Cosmochim. Acta, 72,* 1415–1426.

Wood, B. J., S. G. Nielsen, M. Rehkämper, and A. N. Halliday (2008b), The effects of core formation on the Pb- and Tl-isotopic composition of the silicate Earth, *Earth Planet. Sci. Lett., 269,* 326–336.

Wood, B. J., M. J. Walter, and J. Wade (2006), Accretion of the Earth and segregation of its core, *Nature, 441,* 825–833.

Yang, X., F. Gaillard, and B. Scaillet (2014), A relatively reduced Hadean continental crust and implications for the early atmosphere and crustal rheology, *Earth Planet. Sci. Lett., 393,* 210–219.

Yokoyama, T., D. Walker, and R. J. Walker (2009), Low osmium solubility in silicate at high pressures and temperatures, *Earth Planet. Sci. Lett., 279*(3–4) 165–173.

Zahnle, K. (2006), Earth's earliest atmosphere, *Elements, 2,* 217–222.

Zahnle, K., J. F. Kasting and J. B. Pollack (1988), Evolution of a steam atmosphere during Earth's accretion, *Icarus, 74,* 62–97.

Ziegler, K., E. D. Young, E. A. Schauble, and J. T. Wasson (2010), Metal–silicate silicon isotope fractionation in enstatite meteorites and constraints on Earth's core formation, *Earth Planet. Sci. Lett., 295*(3), 487–496.

14

Mechanisms and Geochemical Models of Core Formation

David C. Rubie[1] and Seth A. Jacobson[1,2]

ABSTRACT

The formation of Earth's core is a consequence of planetary accretion and processes in Earth's interior. The mechanical process of planetary differentiation is likely to occur in large, if not global, magma oceans created by the collisions of planetary embryos. Metal-silicate segregation in magma oceans occurs rapidly and efficiently, unlike grain-scale percolation according to laboratory experiments and calculations. Geochemical models of the core formation process as planetary accretion proceeds are becoming increasingly realistic. Single-stage and continuous core formation models have evolved into multistage models that are coupled to the output of dynamical models of the giant impact phase of planet formation. The models that are most successful in matching the chemical composition of Earth's mantle, based on experimentally derived element partition coefficients, show that the temperature and pressure of metal-silicate equilibration must increase as a function of time and mass accreted and so must the oxygen fugacity of the equilibrating material. The latter can occur if silicon partitions into the core and through the late delivery of oxidized material. Coupled dynamical accretion and multistage core formation models predict the evolving mantle and core compositions of all the terrestrial planets simultaneously and also place strong constraints on the bulk compositions and oxidation states of primitive bodies in the protoplanetary disk.

14.1. INTRODUCTION

The terrestrial planets, Mercury, Venus, Earth, the Moon, and Mars, and at least some much smaller bodies in the asteroid belt (e.g., 4 Vesta), have metallic cores that are surrounded by silicate mantles. Core-mantle structures result from gravity-driven differentiation events that occurred during the early (~100 Myr) history of the Solar System. During planetary accretion, Fe-rich metal was delivered either in the form of cores of differentiated bodies (as represented by iron meteorites) or as metal

that was finely dispersed in a silicate matrix (as represented by chondritic meteorites). In both cases, given the dimensions of planetary mantles, the process of core-mantle differentiation required metal to segregate from silicate over large length scales (e.g., up to 3000 km in the case of Earth).

Here we review the mechanisms by which metal and silicate segregate to form the cores and mantles of planetary bodies. In addition, we review geochemical models of core formation and consider the implications of these for the evolution of mantle and core chemistries. Some aspects are dealt with briefly in this short review and additional sources of information are provided by *Stevenson* [1990], *Nimmo and Kleine* [2015], and *Rubie et al.* [2003, 2007, 2015a].

[1]*Bayerisches Geoinstitut, University of Bayreuth, Bayreuth, Germany*
[2]*Observatoire de la Côte d'Azur, Nice, France*

Deep Earth: Physics and Chemistry of the Lower Mantle and Core, Geophysical Monograph 217, First Edition.
Edited by Hidenori Terasaki and Rebecca A. Fischer.
© 2016 American Geophysical Union. Published 2016 by John Wiley & Sons, Inc.

14. 2. MECHANISMS OF METAL-SILICATE SEGREGATION

For metal and silicate to segregate on a planetary scale requires that at least the metal is molten [*Stevenson*, 1990]. When the silicate (which has the higher melting temperature) is in a solid state, liquid metal can potentially segregate by (a) grain-scale percolation, (b) the descent of kilometer-size diapirs, and/or (c) dyking (Figure 14.1). On the other hand, when the silicate is also largely molten and present as a global-scale magma ocean, liquid metal can segregate extremely efficiently [*Stevenson*, 1990; *Rubie et al.*, 2003]. The heat that is required to produce melting originates from the decay of short-lived isotopes (especially ^{26}Al) during the first 1–3 Myr of Solar System evolution and later from high-energy impacts between planetary bodies [*Rubie et al.*, 2007, 2015a]. The sinking of metal to the core also causes a temperature increase due to the conversion of potential energy to heat.

Many studies in recent years have concluded that core formation in Earth involved extensive chemical equilibration between metal and silicate at high pressures [e.g., *Li and Agree*, 1996]. Here we discuss two mechanisms that are consistent with such equilibration, namely grain-scale percolation and segregation in a magma ocean. Because of slow diffusion rates in crystalline silicates [e.g., *Holzapfel et al.*, 2005] and the large length scales involved, the diapir and dyking mechanisms result in insignificant chemical equilibration; these mechanisms are not discussed here but are reviewed by *Rubie et al.* [2007, 2015a]. Note also that hybrid models have been proposed, such as a combination of porous flow and diapirism [*Ricard et al.*, 2009].

14.2.1. Grain-Scale Percolation

Whether or not liquid metal can percolate through a polycrystalline silicate matrix depends on the dihedral angle θ between two solid-liquid boundaries where they intersect a solid-solid boundary at a triple junction [*von Bargen and Waff*, 1986; *Stevenson*, 1990]. This dihedral angle is controlled by the solid-solid and solid-liquid interfacial energies. When the dihedral angle is less than 60°, the liquid metal is fully connected along grain edges and can percolate efficiently through the silicate matrix. When the dihedral angle exceeds 60°, the melt forms isolated pockets when the melt fraction is low and percolation is only possible when the volume fraction of melt exceeds some critical value that ranges from 2% to 6% for dihedral angles in the range 60–85° [see also *Walte et al.*, 2007]. By measuring dihedral angles in experimentally sintered aggregates consisting of solid silicate plus liquid Fe alloy, the feasibility of percolation as a core formation mechanism can be tested.

In general, experimental studies have been performed up to pressures of 25 GPa on samples in which a few volume percent of a liquid Fe alloy are contained in a polycrystalline aggregate of olivine, ringwoodite, or bridgmanite (silicate perovskite). In general, these studies have found that dihedral angles significantly exceed 60° and are little affected by pressure, temperature, or the identity of the solid phase [e.g. *Ballhaus and Ellis*, 1996; *Minarik et al.*, 1996; *Shannon and Agee*, 1996, 1998; *Holzheid et al.*, 2000; *Terasaki et al.*, 2005, 2007, 2008). A parameter that is of considerable importance is the oxygen and/or sulfur content of the liquid metal alloy.

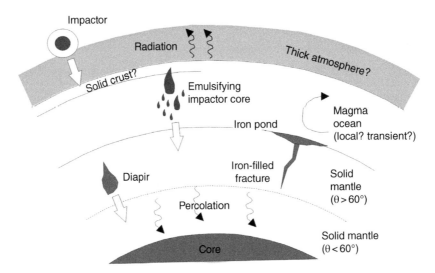

Figure 14.1 Summary of possible mechanisms by which liquid metal can segregate from silicate material during core formation; θ is the dihedral angle which controls grain-scale percolation. Note that the feasibility of percolation in the lower mantle is uncertain (see text). (Courtesy of F. Nimmo.)

The dihedral angles decrease as the concentrations of these light elements increase [*Terasaki et al.*, 2005]. In contrast, Si and C dissolved in liquid Fe have little or no effect on dihedral angles [*Mann et al.*, 2008, *Li and Fei*, 2014]. *Terasaki et al.* [2008] showed that at pressures below 2–3 GPa dihedral angles drop below 60° when the O + S content is high. Therefore, while percolation could be important during the differentiation of small bodies (planetesimals), it is unlikely to be important during core formation in larger planetary bodies.

Two studies have examined liquid Fe alloy interconnectivity in a $(Mg,Fe)SiO_3$-perovskite matrix using laser-heated diamond anvil cells (LH-DACs) up to 64 GPa. Using transmission electron microscopy to measure dihedral angles, *Takafuji et al.* [2004] found that dihedral angles apparently decrease from 94° at ~27 GPa/2400 K to 51° at ~47 GPa/3000 K, possibly as a consequence of increasing concentrations of Si and O in the metal. *Shi et al.* [2013] used in situ X-ray tomography to measure dihedral angles and to generate three-dimensional (3-D) images of metal distribution. They concluded that dihedral angles decrease from 72° at 25 GPa to 23° at 64 GPa and that the liquid Fe metal forms interconnected networks at pressures above 40–50 GPa. Although both studies concluded that percolation could have been an efficient segregation mechanism in Earth's lower mantle, there are caveats to consider concerning the experimental technique. In particular, a sample that is pressurized in the LH-DAC is subjected to an extremely high differential stress that may have a major influence on microstructure, dihedral angles, and connectivity. In fact, the tomographic images of *Shi et al.* [2013] suggest strongly that the interconnected metal in their samples is present along conjugate shear zones that result from high differential stress. Furthermore, they reported dihedral angles as low as 12° at 52 GPa, which raises the question as to whether textural equilibrium under hydrostatic stress conditions was attained.

Two additional factors to consider when discussing grain-scale percolation are the effects of (a) the presence of a small fraction of silicate melt and (b) deformation. Contrary perhaps to expectations, the presence of small fractions (e.g., 2–8 vol %) of silicate melt does not facilitate the percolation of Fe or FeS melt through largely crystalline silicate. Thus high volume fractions of silicate melt are required for metallic liquids to segregate efficiently [*Yoshino and Watson*, 2005; *Bagdassarov et al.*, 2009; *Holzheid*, 2013]. Several studies have shown that deformation can enhance percolation in high-dihedral-angle systems, especially when strain rates are high [e.g., *Hustoft and Kohlstedt*, 2006; *Walte et al.*, 2011]. However, the process is inefficient because small fractions of metallic liquid are always left stranded in the silicate matrix. In addition, it is unlikely that deformation at low strain rates

(characteristic of mantle convection) enhances percolation [*Walte et al.*, 2011].

Cerantola et al. [2015] studied the effects of deformation on FeS segregation in a polycrystalline olivine matrix that also contains a small percentage of basaltic liquid. The results show that the presence of silicate liquid actually inhibits sulfide melt segregation by reducing its connectivity.

If grain-scale segregation does occur (e.g., during differentiation of planetesimals), it seems likely that chemical equilibration would be fast because of large surface-to-volume ratios, even though diffusion in solid silicates might be slow. To our knowledge, the extent and efficiency of such equilibration has never been modeled quantitatively.

14.2.2. Metal-Silicate Segregation in Magma Oceans

The late stages of Earth accretion involved collisions with smaller planetesimals and embryos that culminated in the Moon-forming giant impact [*Hartmann and Davis*, 1975; *Cameron and Ward*, 1976; *Chambers and Wetherill*, 1998; *Agnor et al.*, 1999]. As well as delivering Fe metal to Earth, such impacts provided sufficient energy to cause extensive melting and deep magma ocean formation [*Tonks and Melosh*, 1993; *Rubie et al.*, 2007, 2015a]. The delivery of energy is localized around the impact site and likely results in a roughly spherical melt pool (Figure 14.2) that could extend to the core-mantle boundary in the case of a Mars-size impactor, as suggested for the Moon-forming event [*Canup and Asphaug*, 2001; *Ćuk and Stewart*, 2012]. Isostatic readjustment then results in the spreading out of the magma to form a global magma ocean hundreds of kilometers deep (e.g., Figure 14.1) but on a timescale that is uncertain [*Reese and Solomatov*, 2006]. The lifetime of the global magma ocean is also very uncertain and could be short (e.g., several 1000 years) or long (~100 Myr) (especially in the case of a "shallow" magma ocean, which can have a basal pressure of up to 40 GPa), depending on the absence or presence of a dense insulating atmosphere [*Abe*, 1997; *Solomatov*, 2000]. In the case of short-lived magma oceans, convection is turbulent with a Rayleigh number on the order of 10^{27}–10^{32} and convection velocities of several meters per second [*Solomatov*, 2000; *Rubie et al.*, 2003].

A magma ocean provides an environment in which metal-silicate segregation can occur rapidly and efficiently due to (a) the large density contrast between these materials and (b) the very low viscosity of ultramafic silicate liquids at high pressure [*Liebske et al.*, 2005]. Many impacting bodies are likely to have been already differentiated [*Urey*, 1955; *Hevey and Sanders*, 2006] and their metallic cores would plunge through the magma ocean (Figure 14.2). Within an end-member scenario, known as "core merging," impactor cores remain intact and merge directly with Earth's

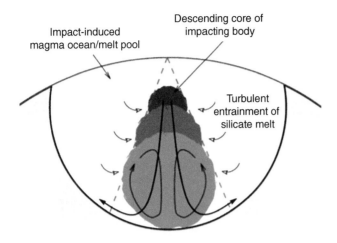

Figure 14.2 Fluid dynamical model of the descent of an impactor's iron core through an impact-induced semispherical magma ocean/melt pool. As the core sinks, silicate liquid is turbulently entrained in a descending metal-silicate plume which broadens with depth. Only the silicate liquid that is entrained in the plume equilibrates chemically with the metal. After *Deguen et al.* [2011].

protocore [e.g., *Halliday*, 2004]. Within the other end-member scenario, impactor cores emulsify completely into small droplets as they sink through Earth's molten mantle [*Stevenson*, 1990; *Rubie et al.*, 2003]. In the first case there would be very limited chemical equilibration between metal and silicate at high pressure whereas in the second case equilibration would be complete [*Rubie et al.*, 2003]. The mechanical behavior of impactor cores as they sink through a magma ocean thus determines the partitioning of siderophile elements between the core and mantle (as discussed below) and is also critical for estimating the timescale of core formation from tungsten isotope anomalies [e.g., *Nimmo et al.*, 2010].

Liquid metal sinking through silicate liquid tends to form droplets of a stable size that is controlled especially by the metal-silicate interfacial energy. Large metal blobs tend to break up as they sink due to mechanical instabilities whereas very small droplets coalesce to reduce interfacial energy [*Rubie et al.*, 2003]. The stable droplet size for typical magma ocean properties is ~1 cm diameter with a settling velocity of ~0.5 m/s [*Rubie et al.*, 2003]. The exact values depend on the Si-O-S content of the Fe alloy because this affects the metal-silicate interfacial energy [*Terasaki et al.*, 2012].

Understanding the extent to which impactor cores emulsify into centimeter-size droplets in a magma ocean is a challenging problem. Emulsification is difficult to study numerically because relevant length scales vary over many orders of magnitude, from hundreds of kilometers to centimeters. Alternatively, the problem can be approached through fluid dynamics experiments on analog liquids, although high impact velocities cannot then be taken into account. Currently it seems likely that planetesimal cores emulsify completely, whereas for much larger embryo cores the answer is still uncertain [*Rubie et al.*, 2003; *Olson and Weeraratne*, 2008; *Dahl and Stevenson*, 2010; *Deguen et al.*, 2011; *Kendall and Melosh*, 2014, 2016; *Samuel*, 2012; *Landeau et al.*, 2014]. Based on siderophile element partitioning and Earth's mantle tungsten isotope anomaly, estimates of the fraction of core-forming metal that equilibrated with Earth's mantle include 36% [*Rudge et al.*, 2010] (note that this estimate increases to 53% when the new Hf/W measurements of *König et al.* [2011] are included), 30–80% [*Nimmo et al.*, 2010] and 70–100% [*Rubie et al.*, 2015b].

14.3. GEOCHEMICAL MODELS OF CORE FORMATION

14.3.1. Element Partitioning and Oxygen Fugacity

Earth's mantle is depleted in siderophile (metal-loving) elements, relative to chondritic and Solar System abundances, because of extraction into the core. The degree of depletion depends on the metal-silicate partition coefficient, which for element M is described as

$$D_M^{met\text{-}sil} = \frac{C_M^{met}}{C_M^{sil}}, \quad (14.1)$$

where C_M^{met} and C_M^{sil} are the molar concentrations of M in metal and silicate respectively. Moderately siderophile elements (e.g., Ni, Co, W, Mo) are defined as having $D_M^{met\text{-}sil}$ values $<10^4$ at 1 bar, highly siderophile elements (e.g., Ir, Pt, Pd, Ru) have $D_M^{met\text{-}sil}$ values $>10^4$, and values for lithophile elements (e.g., Al, Ca, Mg) are <1. In general, partition coefficients depend on pressure, temperature, and in some cases the compositions of the metal and silicate phases [*Rubie et al.*, 2015a]. In addition, oxygen fugacity (f_{O_2}) is a critical controlling parameter, as shown by the equilibrium

$$\underset{\text{metal}}{M} + (n/4)O_2 = \underset{\text{silicate liquid}}{MO_{n/2}}, \quad (14.2)$$

where n is the valence of element M when dissolved in silicate liquid. At high oxygen fugacities M is concentrated more in silicate liquid, whereas at low f_{O_2} it is concentrated more in the metal. In addition, the dependence of partitioning on oxygen fugacity is strong in the case of high–valence elements and weak for low-valence elements. Oxygen fugacity is determined relative to the iron-wüstite buffer from the Fe content of metal and the FeO content of coexisting ferropericlase or silicate, both of which give comparable results [e.g., *Asahara et al.*, 2004; *Mann et al.*, 2009].

There have been numerous experimental studies of the dependence of siderophile element partitioning on pressure (P), temperature (T), and f_{O_2} with the aim of determining the conditions of metal-silicate equilibration during core formation [e.g., *Rubie et al.*, 2015a, Table 3]. In most studies the aim has been to reproduce core-mantle partition coefficients $D_M^{core-mantle} = C_M^{core} / C_M^{mantle}$ [e.g., *Wood et al.*, 2006]. Values of C_M^{mantle} are based on estimated primitive mantle concentrations [e.g., *Palme and O'Neill*, 2014] and C_M^{core} values are estimated by mass balance assuming a given Earth bulk composition [*McDonough*, 2003]. Several types of core formation models have been fit to partitioning results, as reviewed below.

14.3.2. Single-Stage Core Formation

In the simplest and most commonly applied model of core formation, chemical equilibration between the mantle and core at a single set of P-T-f_{O_2} conditions is assumed [e.g., *Li and Agee*, 1996; *Corgne et al.*, 2009]. This has led to a great variety of P-T estimates, ranging from 25 to 60 GPa and 2200 to 4200 K, mostly assuming an f_{O_2} about two log units below the iron-wüstite buffer [*Rubie et al.*, 2015a, Table 3]. A typical conclusion of such studies is that metal-silicate equilibration occurred at the base of a magma ocean at an equilibration pressure corresponding to the ocean's depth (i.e., in the range 700–1500 km). In this case, the equilibration temperature should lie close to the peridotite liquidus at the equilibration pressure. However, *Wade and Wood* [2005] found that $D_M^{core-mantle}$ values for Ni, Co, V, Mn, and Si could be matched with an equilibration pressure of 40 GPa but only when the equilibration temperature exceeds the peridotite liquidus by ~700 K, which is physically unlikely.

Righter [2011] determined a P-T estimate of 27–46 GPa and 3100–3600 K based on element partitioning data and argued that this represents the conditions of a final equilibration event that occurred at the culmination of Earth's accretion and growth. This seems to imply that a large fraction of the metal of the core equilibrated with a large fraction of the silicate of the mantle *in a single event* at pressure-temperatures conditions that correspond to a shallow midmantle depth.

14.3.3. Continuous Core Formation and Evolution of Oxidation State

In the "continuous" core formation model of *Wade and Wood* [2005], Earth accretion and the concurrent delivery of core-forming metal occur in small steps of 1% mass

[see also *Wood et al.*, 2006, 2008]. Each batch of metal equilibrates with the silicate magma ocean at its base, the depth of which increases as Earth grows. Thus metal-silicate equilibration pressures and temperatures (the latter defined by the peridotite liquidus) increase during accretion. In order to reproduce core-mantle partition coefficients of Ni, Co, Cr, V, Nb, and Si, a magma ocean thickness corresponding to 35% of mantle depth is required and P-T equilibration conditions reach a maximum of 40–50 GPa and ~3250 K at the end of accretion. However, reproduction of mantle concentrations was not possible when f_{O_2} remains constant and, instead, conditions need to become increasingly oxidizing, by ~2 log units, during accretion [see also *Rubie et al.*, 2011].

For f_{O_2} to increase significantly during accretion requires the FeO content of the mantle to increase [e.g. from <1 to 8 wt %], for which there are two viable mechanisms [*Rubie et al.*, 2011]. First, when Si partitions into the core, the FeO content of the mantle increases by the reaction

$$2Fe + SiO_2 \rightarrow 2FeO + Si$$
$$\text{core} \quad \text{mantle} \quad \text{mantle} \quad \text{core} \quad (14.3)$$

Thus for every mole of Si that partitions into the core, 2 moles of FeO are added to the mantle, which means that this mechanism is very effective at oxidizing the mantle. Second, the accretion of relatively oxidized material during the later stages of accretion can also significantly increase the mantle FeO content. Note that the "oxygen pump" or "self-oxidation" mechanism proposed by *Wade and Wood* [2005] increases Fe^{3+} but not the FeO content of the mantle: It is therefore not a viable oxidation mechanism in the present context [*Rubie et al.*, 2015a].

It has also been proposed that core formation may occur under initially oxidizing conditions [*Rubie et al.*, 2004; *Siebert et al.*, 2013]. This model is based on Earth's mantle/magma ocean initially containing ~20 wt % FeO. The resulting high f_{O_2} causes FeO to partition into the core so that the mantle FeO is progressively reduced during accretion to its current value of 8 wt %. However, if a small amount (e.g., 2–3 wt %) of Si also partitions into the core, which is inevitable at high temperatures [*Siebert et al.*, 2013], this model fails based on mass balance. This is because the initial ~20 wt % FeO is reduced only slightly during accretion (to 17–18 wt %) because of the production of FeO by reaction (14.3) [*Rubie et al.*, 2015b].

14.3.4. Multistage Core Formation

Earth accreted through a series of high-energy impacts with smaller bodies consisting of kilometer- to multi-kilometer-size planetesimals and Moon- to Mars-size embryos [e.g. *Chambers and Wetherill*, 1998]. Such impacts, as well as delivering energy that caused extensive melting, added Fe-rich metal which segregated to Earth's protocore. Thus core formation was multistage and was an integral part of the accretion process.

A preliminary model of multistage core formation is based on an idealized accretion scenario in which Earth accretes through impacts with differentiated bodies that have a mass ~10% of Earth's mass at the time of each collision [*Rubie et al.*, 2011]. The metal of the impacting bodies equilibrates, partially or completely, in a magma ocean at a pressure that is a constant fraction of Earth's core-mantle boundary pressure and at a temperature close to the corresponding peridotite liquidus. Thus, as in the model of continuous core formation, metal-silicate equilibration pressures increase as Earth grows in size. In contrast to previous studies, this model is not based on assumptions about oxygen fugacity and its evolution. Instead, the bulk compositions of the accreting bodies are defined in terms of nonvolatile elements, which are assumed to be present mostly in Solar System (CI chondritic) relative abundances. The oxygen content is the main compositional variable that enables a wide range of compositions between two extreme end members to be defined: At low oxygen content all Fe is present as metal, whereas at high oxygen content all Fe is present as FeO in the silicate.

The compositions of equilibrated metal and silicate liquids at a given *P-T* are expressed as

$$\left[\underbrace{(FeO)_x (NiO)_y (SiO_2)_z (Mg_u \ Al_m \ Ca_n)O}_{\text{silicate liquid}} \right] + \underbrace{\left[Fe_a \ Ni_b \ Si_c \ O_d \right]}_{\text{metal liquid}}.$$

(14.4)

The indices *u*, *m*, and *n* are determined from the bulk composition alone because Mg, Al, and Ca do not partition into the metal. The other seven indices, *x*, *y*, *z*, *a*, *b*, *c*, and d, are determined by simultaneously solving four mass balance equations (for Fe, Si, O, and Ni), two partitioning expressions for Si and Ni, and a model of oxygen partitioning [*Frost et al.*, 2010] by an iterative process that is described in detail by *Rubie et al.* [2011, (supplementary data)]. Trace elements have little effect on the mass balance and concentrations in metal and silicate are based on partitioning alone. Using this approach, the core of each impactor is equilibrated (fully or partially) with the magma ocean at high *P-T* and the resulting metal is added to the protocore. Thus the evolution of mantle and core compositions is modeled throughout Earth's accretion [*Rubie et al.*, 2011, Figure 4].

For simplicity, two bulk compositions are used in this model of Earth accretion: (1) a highly reduced composition in which 99.9% of Fe and ~20% of available Si are present as metal and (2) a relatively oxidized composition in which ~60% of Fe is present as metal and ~40% as FeO (the initial metal contents of these two compositions are 36 and 20 wt%, respectively). These compositional parameters, together with equilibration pressures and extent of metal reequilibration, are refined by a least squares minimization in order to fit the final mantle concentrations of FeO, SiO$_2$, Ni, Co, W, Nb, V, Ta, and Cr to those of Earth's primitive mantle. Best results are obtained when the initial 60%–70% of Earth accretes from the reduced composition and the final 30%–40% from the more oxidized material, with equilibration pressures that are 60%–70% of core-mantle boundary pressures at the time of each impact. In addition, during the final few impacts, only a limited fraction of the metal equilibrates with silicate. Note that, apart from the need for high-pressure metal-silicate equilibration, this model has similarities to early models of heterogeneous accretion [*Wänke*, 1981; *O'Neill*, 1991].

14.3.5. Combined Accretion and Core-Mantle Differentiation Model

The multistage core formation model described above has recently been extended by combining it with *N*-body accretion simulations [*Rubie et al.*, 2015b]. The latter study concentrates on Grand Tack accretion models because of their success in reproducing the masses and orbital characteristics of the terrestrial planets especially Earth and Mars [*Walsh et al.*, 2011; *O'Brien et al.*, 2014; *Jacobson and Morbidelli*, 2014]. Using the mass balance/partitioning approach described above, the compositions of primitive bodies in the solar nebula are defined in terms of oxidation state and water content as a function of heliocentric distance by least squares regressions. This is done by adjusting the fitting parameters in order to produce an Earth-like planet with a mantle composition identical to (or close to) that of Earth's primitive mantle. The only composition-distance model that provides acceptable results involves bodies close to the Sun (<0.9–1.2 AU) having a highly reduced composition and those from further out being increasingly oxidized (Figure 14.3a). Beyond the giant planets, all bodies are completely oxidized (i.e., with no metallic cores) and contain 20 wt % H$_2$O, which results in ~1000 ppm H$_2$O in Earth's mantle. Note that in the Grand Tack model the C-complex asteroids, which are thought to be carbonaceous chondrite parent bodies containing water that matches that of Earth [*Morbidelli et al.*, 2000; *Alexander et al.*, 2012], are delivered to Earth and the outer Main Belt from their initial locations at the inner edge of an extended outer disk (i.e., beyond Saturn) by the outward migration of the giant planets [*Walsh et al.*, 2011; *O'Brien et al.*, 2014]. The bodies originally located between the

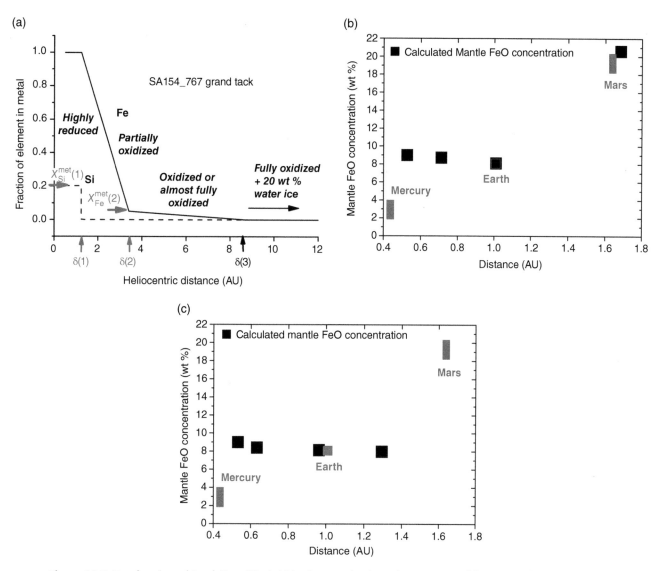

Figure 14.3 Results of combined Grand Tack *N*-body accretion/core formation models. (a) Best-fit composition-distance model for primitive bodies at heliocentric distances from 0.7 to 12 AU for *N*-body simulation SA154_767. Oxidation state is defined by the fractions of total Fe and Si that are initially present as metal. For example, in the highly reduced composition, 99.9% of bulk Fe and 20% of bulk Si are present as metal The four parameters labeled in red, together with equilibrium pressure, were refined by least squares minimization to fit Earth's mantle composition. (b,c) Calculated FeO contents of the mantles of Mercury, Venus, Earth, and Mars in simulations SA154_767 and 2:1-0.25-10A, respectively. Actual mantle compositions of Mercury, Earth, and Mars are shown as red symbols.

snow line and 6–8 AU were swept up by Jupiter and Saturn and were not accreted to the terrestrial planets. (The snow line is the minimum distance from the Sun in the protoplanetary disk at which temperatures were low enough for water ice to condense. It was likely located at ~3 AU.)

Hundreds of impacts and associated core formation events are simulated for all embryos, and thus the final terrestrial planets, simultaneously. For the first time, metallic cores of impacting bodies are modeled to only equilibrate with a small fraction of the target's mantle/magma ocean (Figure 14.2); this is in contrast to all previous studies in which it has been assumed that the entire magma ocean equilibrates with the metal. One consequence is that metal-silicate equilibration pressures need to be relatively high [*Rubie et al.*, 2015b].

Earth's mantle composition is perfectly reproduced in both simulations shown in Figure 14.3b,c. The Martian mantle composition is correctly predicted in the simulation shown in Figure 14.3b but is much too FeO poor in the simulation of Figure 14.3c — this is because the embryo that formed Mars originated too close to the Sun and therefore under conditions that are too reducing in simulation 2:1-0.25-10A (for further discussion see *Rubie et al.* [2015b]). Mercury's FeO-poor mantle is not reproduced in either simulation because, although it starts with an appropriately-reduced composition, Mercury subsequently accretes too much oxidized material in both simulations.

The ability to predict mantle and core compositions of Mercury, Venus, and Mars, in addition to that of Earth, adds a new constraint to both accretion and core formation models. For example, in terms of reproducing the composition of the Martian mantle, the accretion model of Figure 14.3b is more successful than that of Figure 14.3c.

14.4. CONCLUSIONS AND OUTLOOK

Accretion and core-mantle differentiation of the terrestrial planets can now be modeled by combining astrophysical simulations of planetary accretion with geochemical models of core formation [*Rubie et al.*, 2015b]. Such models are becoming increasingly sophisticated and realistic in terms of incorporating new constraints on both physical and chemical processes. The approach enables the compositions and mass ratios of the cores and mantles of all the terrestrial planets to be modeled simultaneously. By fitting the mantle compositions of model Earth-like planets (e.g., located at ~1 AU and of ~1 Earth mass), fitting parameters can be refined by least squares minimization. In addition, the bulk compositions of primitive bodies at heliocentric distances ranging from ~0.7 to 10 AU can be defined in terms of oxidation state and water content. These combined models can place new constraints on our understanding of both planetary accretion and core-mantle differentiation processes.

Currently only non-volatile elements and H_2O have been considered in the combined model. However, there is scope for also including a range of other elements, such as S and other volatile elements, highly siderophile elements, and both stable and radiogenic isotopes.

A number of caveats should be mentioned (see *Rubie et al.* [2015b] for a detailed discussion). For example, element partitioning data are often extrapolated from <25 GPa to pressures as high as 80 GPa. In order to reduce the uncertainties, careful but challenging experimental studies using the DAC are required. For example, the partitioning of Si, O, Ni, Co, V, and Cr between metal and silicate has been studied recently to 100 GPa and 5500 K [*Fischer et al.*, 2015]. In addition, partition coefficients are often a function of the concentrations of elements such as Si, O, and S in the liquid metal, although for most of the elements considered so far the effects are small [e.g., *Tuff et al.*, 2011]. Finally, accretion models are currently based on the assumption of 100% efficiency, whereas in reality material may often be ejected and lost during accretional impacts.

ACKNOWLEDGMENTS

D.C.R. and S.A.J. were supported by the European Research Council (ERC) Advanced Grant "ACCRETE" (contract number 290568). We thank Francis Nimmo for kindly providing Figure 14.1; David O'Brien, Alessandro Morbidelli, Herbert Palme, and Ed Young for discussions; and Kevin Righter and an anonymous referee for helpful reviews.

REFERENCES

Abe, Y. (1997), Thermal and chemical evolution of the terrestrial magma ocean, *Phys. Earth Planet. Inter.*, *100*, 27–39.

Agnor, C. B., R. M. Canup, and H. F. Levison (1999) On the character and consequences of large impacts in the late stage of terrestrial planet formation, *Icarus*, *142*, 219–237.

Alexander, C. M. O'D., R. Bowden, M. L. Fogel, K. T. Howard, C. D. K. Herd, and L. R. Nittler (2012), The provenances of asteroids, and their contributions to the volatile inventories of the terrestrial planets, *Science*, *337*, 721–723.

Asahara, Y., T. Kubo, and T. Kondo (2004), Phase relations of a carbonaceous chondrite at lower mantle conditions, *Phys. Earth Planet. Inter.*, *143–144*, 421–432.

Bagdassarov, N. G. Solferino, G. J. Golabek, M. W. Schmidt (2009), Centrifuge assisted percolation of Fe–S melts in partially molten peridotite: Time constraints for planetary core formation, *Earth Planet. Sci. Lett.*, *288*, 84–95.

Ballhaus, C., and D. J. Ellis (1996), Mobility of core melts during Earth's accretion, *Earth Planet. Sci. Lett.*, *143*, 137–145.

Cameron, A. G. W. and W. R. Ward (1976), The origin of the Moon, *Lunar Sci.*, *7*, Abstract 1041.

Canup R. M., and E. Asphaug E, (2001), Origin of the Moon in a giant impact near the end of the Earth's formation, *Nature*, *412*, 708–712.

Chambers, J. E., and G. W. Wetherill (1998), Making the terrestrial planets: *N*-body integrations of planetary embryos in three dimensions, *Icarus*, *136*, 304–327.

Cerantola, F., N. Walte, and D. C. Rubie (2015), Deformation of a crystalline olivine aggregate containing two immiscible liquids: Implications for early core-mantle differentiation, *Earth Planet. Sci. Lett.*, *417*, 67–77.

Corgne, A., J. Siebert, and J. Badro (2009), Oxygen as a light element: A solution to single-stage core formation, *Earth Planet. Sci. Lett.*, *288*, 108–114.

Ćuk, M., and S. T. Stewart (2012), Making the Moon from a fast-spinning Earth: A giant impact followed by resonant despinning, *Science*, *338*, 1047–1052.

Dahl, T. W., and D. J. Stevenson (2010), Turbulent mixing of metal and silicate during planet accretion and interpretation of the Hf-W chronometer, *Earth Planet. Sci. Lett.*, *295*, 177–186.

Deguen, R., P. Olson, and P. Cardin (2011), Experiments on turbulent metal-silicate mixing in a magma ocean, *Earth Planet. Sci. Lett.*, *310*, 303–313.

Fischer, R. A., Y. Nakajima, A. J. Campbell, D. J. Frost, D. Harries, F. Langenhorst, N. Miyajima, K. Pollock, and D. C. Rubie (2015), High pressure metal-silicate partitioning of Ni, Co, V, Cr, Si and O, *Geochim. Cosmochim. Acta*, *167*, 177–194.

Frost, D. J., Y. Asahara, D. C. Rubie, N. Miyajima, L. S. Dubrovinsky, C. Holzapfel, E. Ohtani, M. Miyahara, and T. Sakai (2010), The partitioning of oxygen between the Earth's mantle and core, *J. Geophys. Res.*, *115*, B02,202, doi:10.1029/2009JB006302.

Halliday, A. N. (2004), Mixing, volatile loss and compositional change during impact-driven accretion of the Earth, *Nature*, *427*, 505–509.

Hartmann, W. K., and D. R. Davis (1975), Satellite-sized planetesimals and lunar origin, *Icarus*, *24*, 504–515.

Hevey, P., and I. Sanders (2006), A model for planetesimal meltdown by [26]Al and its implications for meteorite parent bodies, *Meteorit. Planet. Sci.*, *41*, 95–106.

Holzapfel, C., D. C. Rubie, D. J. Frost, and F. Langenhorst (2005), Fe-Mg interdiffusion in (Mg,Fe)SiO$_3$ perovskite and lower mantle reequilibration, *Science*, *309*, 1707–1710, doi:10.1126/science.1111895.

Holzheid, A. (2013), Sulphide melt distribution in partially molten silicate aggregates: Implications to core formation scenarios in terrestrial planets, *Eur. J. Mineral*, *25*, 267–277.

Holzheid A., M. D. Schmitz, and T. L. Grove (2000), Textural equilibria of iron sulphide liquids in partly molten silicate aggregates and their relevance to core formation scenarios, *J. Geophys. Res.*, *105*, 13,555–13,567.

Hustoft J. W., and D. L. Kohlstedt (2006), Metal-silicate segregation in deforming dunitic rocks, *Geochem. Geophys. Geosyst.*, *7*, Q02,001, doi:10.1029/2005GC001048.

Jacobson, S. A., and A. Morbidelli (2014), Lunar and terrestrial planet formation in the Grand Tack scenario, *Philos. Trans. R. Soc. A*, *372*, doi: 10.1098/rsta.2013.0174.

Kendall, J. D., and H. J. Melosh (2014), Dispersion of planetesimal iron cores during accretional impacts, *LPSC Conf.*, *45*, Abstract 2827.

Kendall, J. D., and H. J. Melosh (2016), Differentiated planetesimal impacts into a terrestrial magma oceans: Fate of the iron core., *Earth Planet. Sci. Lett.*, submitted.

König, S., C. Münker, S. Hohl, H. Paulick, A. R. Barth, M. Lagos, J. Pfänder, and A. Büchl. (2011), The Earth's tungsten budget during mantle melting and crust formation, *Geochim. Cosmochim. Acta*, *75*, 2119–2136.

Landeau, M., R. Deguen, and P. Olson (2014), Experiments on the fragmentation of a buoyant liquid volume in another liquid, *J. Fluid Mech*, *749*, 478–518, doi:10.1017/jfm.2014.202.

Li, J., and C. B. Agee (1996), Geochemistry of mantle-core differentiation at high pressure, *Nature*, *381*, 686–689.

Li, J., and Y. Fei (2014), Experimental constraints on core composition, in *The Mantle and Core*, edited by R. W. Carlson, vol. *2, Treatise on Geochemistry*, pp. 527–557, edited by. H. D. Holland and K. K. Turekian, Elsevier-Pergamon, Oxford.

Liebske, C., B. Schmickler, H. Terasai, B. T. Poe, A. Suzuki, K. Funakoshi, R. Ando, and D. C. Rubie (2005), Viscosity of peridotite liquid up to 13 GPa: Implications for magma ocean viscosities, *Earth Planet. Sci. Lett.*, *240*, 589–604.

Mann, U., D. J. Frost, and D. C. Rubie (2008), The wetting ability of Si-bearing Fe-alloys in a solid silicate matrix — Percolation during core formation under reducing conditions? *Phys. Earth Planet. Inter.*, *167*, 1–7.

Mann, U., D. J. Frost, and D. C. Rubie (2009), Evidence for high-pressure core-mantle differentiation from the metal-silicate partitioning of lithophile and weakly siderophile elements, *Geochim. Cosmochim. Acta*, *73*, 7360–7386, doi:10.1016/j.gca.2009.08.006.

McDonough, W. F. (2003), Compositional model for the Earth's core, in The Mantle and Core, edited by R. W. Carlson, Vol. 1, *Treatise on Geochemistry*, edited by H. D. Holland and K. K. Turekian, pp. 547–568, Elsevier-Pergamon, Oxford.

Minarik, W. G., F. J. Ryerson, and E. B. Watson (1996), Textural entrapment of core-forming melts, *Science*, *272*, 530–533.

Morbidelli, A., J. Chambers, J. I. Lunine, J. M. Petit, F. Robert, G. B. Valsecchi, and K. E. Cyr (2000), Source regions and timescales for the delivery of water to Earth, *Meteor. Planet. Sci.*, *35*, 1309–1320.

Nimmo, F., and T. Kleine (2015), Early differentiation and core formation: Processes and timescales, in *The Early Earth, AGU Geophys. Monogr. 212*, edited by J. Badro and M. Walter, AGU, Washington, D.C., 83–102.

Nimmo, F., D. P. O'Brien, and T. Kleine (2010), Tungsten isotopic evolution during late-stage accretion: Constraints on Earth-Moon equilibration, *Earth Planet. Sci. Lett.*, *292*, 363–370.

O'Brien, D. P., K. J. Walsh, A. Morbidelli, S.N. Raymond, and A. M. Mandell (2014), Water delivery and giant impacts in the "Grand Tack" scenario, *Icarus*, *239*, 74–84, doi:10.1016/j.icarus.2014.05.009.

Olson, P., and D. Weeraratne (2008), Experiments on metal-silicate plumes and core formation, *Philos., Trans. R. Soc. A*, *366*, 4253–4271.

O'Neill, H. S. C. (1991), The origin of the moon and the early history of the Earth—A chemical model. Part 2: the Earth, *Geochim. Cosmochim. Acta*, *55*, 1159–1172.

Palme, H., and H. St. C. O'Neill (2014), Cosmochemical estimates of mantle composition, in *The Mantle and Core*, edited by R. W. Carlson, Vol. *2, Treatise on Geochemistry*, edited by H. D. Holland and K. K. Turekian, pp. 1–39, Elsevier-Pergamon, Oxford.

Reese, C. C., and V. S. Solomatov (2006), Fluid dynamics of local martian magma oceans, *Icarus*, *184*, 102–120, doi:10.1016/j.icarus.2006.04.008.

Ricard, Y., O. Šrámek, and F. Dubuffet (2009), A multi-phase model of runaway core-mantle segregation in planetary embryos, *Earth Planet. Sci. Lett.*, *284*, 144–150.

Righter, K. (2011), Prediction of metal–silicate partition coefficients for siderophile elements: An update and assessment of PT conditions for metal–silicate equilibrium during accretion of the Earth, *Earth Planet. Sci. Lett.*, *304*, 158–167.

Rubie, D. C., H. J. Melosh, J. E. Reid, C. Liebske, and K. Righter (2003), Mechanisms of metal-silicate equilibration in the terrestrial magma ocean, *Earth Planet. Sci. Lett.*, *205*, 239–255.

Rubie, D. C., C. K. Gessmann, and D. J. Frost (2004), Partitioning of oxygen during core formation on the Earth and Mars, *Nature*, *429*, 58–61.

Rubie, D. C., F. Nimmo, and H. J. Melosh (2007), Formation of the Earth's core, in Evolution of the Earth, edited by D. Stevenson, Vol. 9, Treatise on Geophysics, edited by G. Schubert, pp. 51–90, Elsevier, Amsterdam.

Rubie, D. C., D. J. Frost, U. Mann, Y. Asahara, K. Tsuno, F. Nimmo, P. Kegler, A. Holzheid, and H. Palme (2011), Heterogeneous accretion, composition and core-mantle differentiation of the Earth, *Earth Planet. Sci. Lett.*, *301*, 31–42, doi:10.1016/j.epsl.2010.11.030.

Rubie, D. C., F. Nimmo, and H. J. Melosh (2015a), Formation of the Earth's core, in Evolution of the Earth, edited by D. Stevenson, Vol. 9, Treatise on Geophysics, 2nd Edition, edited by G. Schubert, pp. 43–79, Elsevier, Amsterdam.

Rubie, D. C., S. A. Jacobson, A. Morbidelli, D. P. O'Brien, E. D. Young, J. de Vries, F. Nimmo, H. Palme, and D. J. Frost (2015b), Accretion and differentiation of the terrestrial planets with implications for the compositions of early-formed Solar System bodies and accretion of water, *Icarus*, *248*, 89–108, doi: 10.1016/j.icarus.2014.10.015.

Rudge, J. F., T. Kleine, B Bourdon (2010), Broad bounds on Earth's accretion and core formation constrained by geochemical models, *Nature Geosci.*, *3*, 439–443.

Samuel, H. (2012), A re-evaluation of metal diapir breakup and equilibration in terrestrial magma oceans, *Earth Planet. Sci. Lett.*, *313–314*, 105–114.

Shannon, M. C., and C. B. Agee (1996), High pressure constraints on percolative core formation, *Geophys. Res. Lett*, *23*, 2717–2720.

Shannon M. C., and C. B. Agee (1998), Percolation of core melts at lower mantle conditions, *Science*, *280*, 1059–1061.

Shi, C. Y., L. Zhang, W. Yang, Y. Liu, J. Wang, J. Meng, J. C. Andrews, and W. L. Mao (2013), Formation of an interconnected network of iron melt at Earth's lower mantle conditions, *Nature Geosci.*, *6*, 971–975, doi:10.1038/ngeo1956.

Siebert, J., J. Badro, D. Antonangeli, and F. J. Ryerson (2013), Terrestrial accretion under oxidizing conditions, *Science*, *339*, 1194–1197.

Solomatov, V. S. (2000), Fluid dynamics of a terrestrial magma ocean, in *Origin of the Earth and Moon*, edited by R. M. Canup and K. Righter, pp. 323–338, Univ. of Ariz. Press, Tucson.

Stevenson, D. J. (1990), Fluid dynamics of core formation, in *Origin of the Earth*, edited by Newsom H. E. and J. H. Jones, pp. 231–250, Oxford Univ. Press.

Takafuji N., K. Hirose, S. Ono, F. Xu, M. Mitome, and Y. Bando (2004), Segregation of core melts by permeable flow in the lower mantle, *Earth Planet. Sci. Lett.*, *224*, 249–257.

Terasaki, H., D. J. Frost, D. C. Rubie, and F. Langenhorst (2005), The effect of oxygen and sulphur on the dihedral angle between Fe-O-S melt and silicate minerals at high pressure: Implications for Martian core formation, *Earth Planet. Sci. Lett.*, *232*, 379–392, doi:10.1016/j.epsl.2005.01.030.

Terasaki, H., D. J. Frost, D. C. Rubie, and F. Langenhorst (2007), The interconnectivity of Fe-O-S liquid in polycrystalline silicate perovskite at lower mantle conditions, *Phys. Earth Planet. Int.*, *161*, 170–176.

Terasaki, H., D. J. Frost, D. C. Rubie, and F. Langenhorst (2008), Percolative core formation in planetesimals, *Earth Planet. Sci. Lett.*, *273*, 132–137.

Terasaki, H., S. Urakawa, D. C. Rubie, K. Funakoshi, T. Sakamaki, Y. Shibazak, S. Ozawa, and E. Ohtani (2012), Interfacial tension of Fe–Si liquid at high pressure: Implications for liquid Fe-alloy droplet size in magma oceans, *Phys. Earth Planet. Int.* *174*, 220–226.

Tonks, W. B., and H. J. Melosh (1993), Magma ocean formation due to giant impacts, *J. Geophys. Res*, *98*, 5319–5333.

Tuff, J., B .J. Wood, and J. Wade (2011), The effect of Si on metal–silicate partitioning of siderophile elements and implications for the conditions of core formation, *Geochim. Cosmochim. Acta*, *75*, 673–690.

Urey, H. C. (1955), The cosmic abundances of potassium, uranium, and thorium and the heat balance of the Earth, the Moon, and Mars, *Proc. Natl. Acad. Sci.*, *41*, 127–144.

von Bargen N., and H. S. Waff (1986), Permeabilities, interfacial areas and curvatures of partially molten systems: Results of numerical computations of equilibrium microstructures, *J. Geophys. Res*, *91*, 9261–9276.

Wade, J., B. J. Wood, (2005). Core formation and the oxidation state of the Earth, *Earth Planet. Sci. Lett.*, *236*, 78–95.

Walsh, K. J., A. Morbidelli, S. N. Raymond, D. P. O'Brien, and A. M. Mandell (2011), A low mass for Mars from Jupiter's early gas-driven migration, *Nature*, *475*, 206–209.

Walte, N. P., J. K. Becker, P. D. Bons, D. C. Rubie, and D. J. Frost (2007), Liquid distribution and attainment of textural equilibrium in a partially-molten crystalline system with a high-dihedral-angle liquid phase, *Earth Planet. Sci. Lett.*, *262*, 517–532.

Walte, N. P., D. C. Rubie, P. D. Bons, and D. J. Frost (2011), Deformation of a crystalline aggregate with a small percentage of high-dihedral-angle liquid: Implications for core–mantle differentiation during planetary formation, *Earth Planet. Sci. Lett.*, *305*, 124–134.

Wänke, H. (1981), Constitution of terrestrial planets, *Philos. Trans. R. Soc*, *303*, 287–302.

Wood B. J., M. J. Walter, and J. Wade (2006), Accretion of the Earth and segregation of its core, *Nature*, *441*, 825–833, doi:10.1038/nature04763.

Wood, B. J., J. Wade, and M. R. Kilburn (2008), Core formation and the oxidation state of the Earth: Additional constraints from Nb, V and Cr partitioning, *Geochim. Cosmochim. Acta* *72*, 1415–1426.

Yoshino, T., and E. B. Watson (2005), Growth kinetics of FeS melt in partially molten peridotite: An analog for core-forming processes, *Earth Planet. Sci. Lett.*, *235*, 453–468.

15

Phase Diagrams and Thermodynamics of Core Materials

Andrew J. Campbell

ABSTRACT

The density contrast at Earth's inner core boundary (ICB) can act as an important constraint on the chemical and thermal structure of the core if the melting relationships of relevant Fe-rich alloys were accurately known. Currently, there are no experimental constraints on these solid-melt phase loops at appropriately high-pressure, high-temperature conditions. However, a simplified thermodynamic analysis, using available melting curves for Fe-rich binaries and their endmember alloys, suggests that high-pressure melting relations can be reasonably approximated using an entropy change of melting of $\Delta S_m = 0.70$–$0.75R$ for these systems. With this parameter, and extrapolated melting curves, multicomponent phase diagrams are calculated to ICB conditions. The phase relations so calculated are compatible with geophysical constraints on the alloy in Earth's core.

15.1. INTRODUCTION

The differentiation of Earth into its metallic core and rocky mantle and crust is one of the most significant events in the planet's formation and evolution. The composition and dynamics of the core today carry the imprint of those early differentiation processes, and a closer understanding of the core's current composition and thermal state would provide insight into the chemical and physical conditions relevant to the birth of our planet. Cosmochemical abundances and the seismological structure of Earth's deep interior together indicate that the core's main constituent is iron, alloyed with ~5 wt % nickel and perhaps ~10 wt % of lighter elements [*Birch*, 1952]. This light element alloying component is regarded to be mainly S, O, Si, and/or C in uncertain relative importance, and other elements including H and Mg are also sometimes considered potentially important. An important research goal in high-pressure mineral physics is to better define the light element component of Earth's core, whose identity will improve our understanding of Earth's formation and current thermal structure.

Mineral physics investigations into the core's light element component rest on the comparison between measured or computed properties of a candidate alloy with the seismologically determined density and velocity structure of the core, which provide several key constraints. First, the density and velocity of the outer core must be satisfied by the properties of the alloy. This is an important and widely used constraint but ultimately limited because the trade-offs between compositions and temperature leave the problem underdetermined. An important additional constraint is the density (and velocity) jump at the inner core boundary (ICB), which is a consequence of crystallization of a denser solid inner core from the liquid, light-element-rich outer core. This 7% density contrast [*Masters and Gubbins*, 2003] represents a compositional tie line in the phase relations that describe the properties of the core's alloy composition at the ICB pressure (330 GPa) and temperature

Department of the Geophysical Sciences, University of Chicago, Chicago, Illinois, USA

Deep Earth: Physics and Chemistry of the Lower Mantle and Core, Geophysical Monograph 217, First Edition.
Edited by Hidenori Terasaki and Rebecca A. Fischer.
© 2016 American Geophysical Union. Published 2016 by John Wiley & Sons, Inc.

(uncertain, but in the range of 5000 K). A successful candidate core composition will match this solid-liquid density contrast on its liquidus at the ICB temperature, and moreover, an adiabat for the model composition anchored near that ICB temperature defines, through the alloy's equation of state, the density and velocity profiles that must match those of the preliminary reference Earth mode (PREM) [*Dziewonski and Anderson*, 1981] or any similar seismological model for the core.

There has been considerable progress defining the equations of state and velocities for various iron alloy compositions at high pressures and temperatures. Both static and dynamic experimental studies, as well as ab initio investigations, have provided a useful set of equations of state constrained to core pressures for Fe [e.g., *Dewaele et al.*, 2006], FeO [e.g., *Jeanloz and Ahrens*, 1980; *Campbell et al.*, 2009; *Fischer et al.*, 2011], FeS and Fe$_3$S [e.g., *Brown et al.*, 1984; *Seagle et al.*, 2006; *Fei et al.*, 2000; *Kamada et al.*, 2014b], Fe$_3$C [*Sata et al.*, 2010; *Litasov et al.*, 2013], and Fe-Si alloys [*Fischer et al.*, 2012, 2014]. In addition, velocity data are available for many iron-rich alloys to high pressures based on inelastic X-ray scattering measurements [e.g., *Badro et al.*, 2007; *Antonangeli et al.*, 2010; *Mao et al.*, 2012; *Kamada et al.*, 2014a; *Chen et al.*, 2014]. Improvement in this direction is still needed, of course—particularly in obtaining data from appropriate liquid alloys [*Morard et al.*, 2013] and higher experimental *PT* conditions covering the entire range of Earth's core—but the available data are adequate to resolve density and velocity comparisons to the core.

In contrast, the phase diagrams of candidate core-forming alloys are much more poorly resolved, and this is nonetheless the subject of the present chapter. The tools that have permitted impressive growth in our understanding of the physical properties of iron-rich alloys—namely synchrotron X-ray scattering methods in diamond anvil cells and dynamic compression studies—are not as well suited for detailed chemical investigations. Prospects for the near future are good though, because improvements in sample recovery methods (principally focused ion beam micromachining) promise to allow petrological studies of these and other geologically relevant systems to core conditions. Recent studies have reported high-resolution electron microscopy of samples recovered from outer core conditions [e.g., *Nomura et al.*, 2014], and one can anticipate that similar studies on candidate core compositions will soon follow. In the meantime, however, there is lamentably poor understanding of Fe-rich phase diagrams at the *PT-X* conditions relevant to Earth's ICB. There is no experimental information on solid-melt partitioning in Fe alloys at the conditions of Earth's core, much less at the more extreme conditions of the ICB. Consequently there are no data with which to benchmark

the accuracy of the handful of ab initio studies that exist on this subject [e.g., *Alfè et al.*, 2007; *Zhang and Yin*, 2012]. This chapter reviews the essential thermodynamics of the melting relations in core-forming alloys and applies these principles to the limited available data in the hope of outlining where future experimental studies can best be applied to constrain the composition of Earth's core.

15.2. THERMODYNAMIC BASIS

Here we consider the thermodynamics of phase loops during eutectic melting of Fe-rich alloys relevant to studies of Earth's core. The binary alloy endmembers (FeO, FeSi, Fe$_3$S, and Fe$_3$C) are described here using an associated solution model, applied previously for this use by *Stevenson* [1981] and *Svendsen et al.* [1989]. In associated Fe-Z solutions, the solute Z is present as monomers in addition to associated Fe$_i$Z$_j$ complexes [*Prigogine and Defay*, 1954]. Although binary compounds are sometimes taken as the solute (e.g., FeO in *Komabayashi* [2014]), associated solution models successfully describe both alloy solutions [e.g., *Sharma and Chang*, 1979; *Chuang and Chang*, 1982] and silicate melts [*Wen and Nekvasil*, 1994; *Ghiorso*, 2004]. The purpose of revisiting this approach is that there is now more detailed information available for not only the melting curves of some endmembers but also the binary eutectics between them. This information can be used, in the context of the associated solution model, to estimate the shapes of the liquidus curves, including the eutectic compositions. Furthermore, we extend this model here to multicomponent Fe-rich alloy systems, whereas the earlier analyses were restricted to binary systems because of the limited experimental data available.

For endmember compositions the chemical potential of the liquid (μ^L_0) can be related to that of the solid (μ^S_0) to first order by

$$\mu^L_0 = \mu^S_0 - \Delta S_m (T - T_m), \tag{15.1}$$

where ΔS_m is the entropy change upon melting and T_m is the melting temperature of the pure substance. Here, differences in heat capacity between solid and liquid are assumed negligible near the melting point. Accordingly, in a solid–liquid phase loop the activities (a) of this end-member component are related to one another by

$$\ln \frac{a^L}{a^S} = \frac{\Delta S_m}{RT} (T - T_m) \tag{15.2}$$

because $\mu^L_i = \mu^S_i$ for each component i in the solutions and $\mu = \mu_0 + RT \ln a$. This approach was used, for example, by *Williams and Jeanloz* [1990] to estimate the eutectic from melting curves of Fe and FeS, assuming a value for ΔS_m. Below we will build upon recent measurements of eutectic temperatures to evaluate ΔS_m at high pressures for different binary systems, allowing us to extend the existing data to ICB conditions with greater confidence in our assumptions.

For binary compounds Fe_iZ_j, where Z represents any candidate light element in the core, we model both the solid and liquid phases as fully associated mixtures such that FeO, Fe_3S, etc., are distinct, energetically favored species in both phases: $iFe + jZ \leftrightarrow Fe_iZ_j$. This leads to [Prigogine and Defay, 1954]

$$\mu_{Fe_iZ_j} = i\mu_{Fe} + j\mu_Z = i\mu_{0,Fe} + iRT \, ln a_{Fe} + j\mu_{0,Z}$$
$$+ jRT \, ln \, a_Z, \qquad (15.3)$$
$$\mu_{Fe_iZ_j} = \mu_{0,Fe_iZ_j} + RT \, ln a_{Fe_iZ_j},$$

where μ_0 for Fe_iZ_j refers to the stoichiometric endmember, with mole fraction $X_Z = j/(i+j)$. From (3),

$$a_{Fe_iZ_j} = a_{Fe}^i a_Z^j e^{-\Delta\mu/RT} \qquad (15.4)$$

with the definition $\Delta\mu = \mu_{0,Fe_iZ_j} - i\mu_{0,Fe} - j\mu_{0,Z}$. This term can be obtained by considering that $a_{Fe_iZ_j} = 1$ at $X_{Fe} = i/(i+j)$, $X_Z = j/(i+j)$. Assuming ideal solution behavior for each element, $e^{-\Delta\mu/RT} = (i+j)^{i+j}/(i^ij^j)$ and (4) becomes

$$a_{Fe_iZ_j} = \frac{(i+j)^{i+j}}{i^i j^j} X_{Fe}^i X_Z^j \qquad (15.5)$$

Equations (15.2) and (15.5) describe the phase relations between liquid and solid alloy at a given pressure. In the following section these will be compared to experimental melting curves, including eutectic melting temperatures in binary systems, to show that the ΔS_m term can be reasonably estimated as $R \, ln \, 2$ per mole atoms at high-pressure conditions. This will serve as a basis for estimating multicomponent (ternary) phase diagrams relevant to Earth's core at 330 GPa.

15.3. APPLICATION TO EXPERIMENTAL MELTING TEMPERATURES

Fischer (Chapter 1 in this volume) has summarized experimental constraints on the melting curve of Fe at high pressures and also binary eutectic temperatures in various Fe-rich systems. In addition, melting curves are available for endmember compounds FeO [Fischer and Campbell, 2010], Fe_3C [Lord et al., 2009], and FeSi [Lord et al., 2010; Fischer et al., 2013]. These melting temperatures can be applied to equation (2) to obtain estimates of ΔS_m or of solid-melt partitioning.

As an example, consider the Fe-FeO system. Available evidence from X-ray diffraction experiments indicates that there is very limited departure in the endmembers of this binary system from pure Fe and FeO, so the temperature along the two branches of the liquidus can be calculated from equations (2) and (5) as

$$T = \frac{T_m^{Fe}}{1 - (R/\Delta S_m^{Fe}) ln(X_{Fe}^L)} \quad \text{and}$$
$$T = \frac{T_m^{FeO}}{1 - (R/\Delta S_m^{FeO}) ln(4X_{Fe}^L X_O^L)}, \qquad (15.6)$$

where the left equation describes the liquidus on the Fe-rich side of the eutectic composition and the right equation describes the liquidus on the FeO-rich side. At the eutectic both equations describe the same temperature (T_{eut}), and the eutectic composition can be computed from this, using also $X_{Fe}^L + X_O^L = 1$. For comparison, at 80 GPa the melting points of Fe, FeO, and their eutectic are approximately 3500, 3200, and 2600 K, respectively [Anzellini et al., 2013; Fischer and Campbell, 2010; Seagle et al., 2008]. (See also Chapter 1). A pressure of 80 GPa is practical for this comparison because it is near the upper limit of experimentally determined melting points for FeO and the eutectic, and it is a sufficient pressure to be considered the "high-pressure" regime insofar as the entropy change upon melting is concerned. These experimental melting points can be reconciled with equation (15.6) if the entropy change $\Delta S_m = 0.75R$ per mole of atoms for both Fe and FeO (Figure 15.1). This produces a eutectic composition at 80 GPa of 7.8 wt % oxygen, slightly lower than the ~10.5 wt % O reported by Seagle et al. [2008] but within 2σ uncertainties of those experiments.

In the high-pressure limit, when $\Delta V_m \rightarrow 0$, the entropy change of melting for simple substances is expected to approach $\Delta S_m \rightarrow R ln2$ [Stishov et al., 1973; Stishov, 1988]. More exactly [Tallon, 1980],

$$\Delta S_m = R \, ln \, 2 + \alpha K_T \Delta V_m$$
$$= R \, ln \, 2 + \gamma C_V \frac{\Delta V_m}{V}, \qquad (15.7)$$

where $\gamma = \alpha K_T V/C_V$ is the Grüneisen parameter of the substance. Reasonable estimates for iron alloys in Earth's core might be $\gamma \approx 1.3$, $C_V \approx 5R$ (including an approximate electronic heat capacity at high T) [Brown and McQueen, 1986], and $\Delta V_m/V \approx 1.5\%$ [Laio et al., 2000], producing $\Delta S_m \approx 0.79R$, close to the value of $0.75R$ obtained from analysis of the Fe-FeO system. Ab initio and molecular dynamic studies have produced higher values of $\Delta S_m = 0.86R$ to $1.05R$ for Fe at 330 GPa [Laio et al., 2000; Alfè et al., 2002; Zhang et al., 2015]. As Zhang et al. [2015]

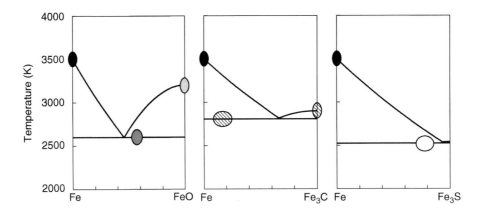

Figure 15.1 Binary phase diagrams for Fe-FeO, Fe-Fe$_3$C, and Fe-Fe$_3$S at 80 GPa calculated from experimental melting temperatures and equations (15.6), (15.8), and (15.9). Black ovals [*Anzellini et al.*, 2013], dark gray oval [*Seagle et al.*, 2008], light gray oval [*Fischer and Campbell*, 2010], hatched ovals [*Lord et al.*, 2009], white oval [*Kamada et al.*, 2012].

noted however, these methods also overestimate ΔS_m for Fe at 1 bar by a similar amount.

The Fe-Fe$_3$C provides another useful case to consider, because the melting temperatures in this system have also been determined experimentally [*Lord et al.*, 2009; *Anzellini et al.*, 2013]. In this case the analog to equation (15.6) becomes

$$T = \frac{T_m^{Fe}}{1 - \dfrac{R}{\Delta S_m^{Fe}} \ln\left(X_{Fe}^L\right)} \quad \text{and}$$

$$T = \frac{T_m^{Fe_3C}}{1 - \dfrac{R}{\Delta S_m^{Fe_3C}} \ln\left(\dfrac{256}{27} X_{Fe}^{L3} X_C^L\right)}. \quad (15.8)$$

Here again it has been assumed that the solid end-members Fe and Fe$_3$C are effectively stoichiometric, although this is a less well justified assumption than in the Fe-FeO system, as solid iron can contain up to 2 wt % C at 1 bar and probably to some degree at high pressures too. Nevertheless, this is a minor correction. Using melting points at 80 GPa of 3500, 2900, and 2800 K for Fe, Fe$_3$C, and their eutectic, respectively [*Lord et al.*, 2009; *Anzellini et al.*, 2013], one finds that equation (15.8) is satisfied with $\Delta S_m = 0.71R$ per mole of atoms for both Fe and Fe$_3$C (Figure 15.1). Again this value is slightly greater than the high-pressure limit of $R\ln2$, indicating that a value of $\Delta S_m = 0.70$–$0.75R$ is a reasonable approximation with which to estimate the thermodynamics of melting at high pressures comparable to those in Earth's core.

An important difficulty in the Fe-C system is that at higher pressures the relevant carbide endmember is probably Fe$_7$C$_3$, not Fe$_3$C. Less is known about the physical properties of Fe$_7$C$_3$, but it is evident from the limited

melting data on this phase by *Lord et al.* [2009] that its melting point is much higher than that of Fe$_3$C, perhaps 3500 K at 80 GPa. This strongly suggests that the eutectic melting point depression in Fe-Fe$_7$C$_3$ is somewhat less than that of Fe-Fe$_3$C, but there are no experimental data yet to verify this. In this chapter Fe$_3$C will be used as the relevant carbide phase, acknowledging that an important improvement will be to measure the iron-carbide eutectic at sufficiently high pressure (~120 GPa [*Lord et al.*, 2009]) that Fe$_7$C$_3$ is the solidus carbide phase.

The converse problem presently exists with the high-pressure melting in the iron–sulfide system. Experimental measurements of Fe-Fe$_3$S eutectic melting have been accomplished up to 175 GPa [*Kamada et al.*, 2012], with remarkable consistency among results from different authors [*Campbell et al.*, 2007; *Chudinovskikh and Boehler*, 2007; *Stewart et al.*, 2007; *Morard et al.*, 2008; *Kamada et al.*, 2010; 2012]. However, the melting curve of the endmember Fe$_3$S has not been measured. Using $\Delta S_m = 0.70$–$0.75R$ as above and the reported melting point depression for the Fe-Fe$_3$S system at high pressures (1050 K from Chapter 1), the melting point of Fe$_3$S can be estimated using an analysis similar to that in equation (15.8):

$$T = \frac{T_m^{Fe}}{1 - \dfrac{R}{\Delta S_m^{Fe}} \ln\left(X_{Fe}^L\right)} \quad \text{and}$$

$$T = \frac{T_m^{Fe_3S}}{1 - \dfrac{R}{\Delta S_m^{Fe_3S}} \ln\left(\dfrac{256}{27} X_{Fe}^{L3} X_C^L\right)} \quad (15.9)$$

The resulting melting point at 80 GPa is ~2450 K (Figure 15.1). This is indistinguishable from T_{eut} and correspondingly places the eutectic composition near

16 wt % S, which is the sulfur content in Fe_3S. Although not impossible, this is higher than the eutectic composition of 12.5% S at 85 GPa reported by *Kamada et al.* [2012] and likely reflects some nonideality in the Fe-S system persisting to high pressure. Here we have assumed again that any deviation from ideal Fe_3S stoichiometry is negligible, although there is evidence that this is not strictly correct [*Li et al.*, 2001].

The Fe-Si binary system is even more challenging to treat in the absence of Si partitioning measurements between solid and melt. Using again 80 GPa as an example condition to compare to experimental results, the sub-solidus phase diagram shows eutectic melting between face-centered cubic (*fcc*)-Fe and a B2-structured Si-rich alloy, although at pressures >90 GPa the Fe-rich phase becomes hexagonal close packed (*hcp*) [*Fischer et al.*, 2013; *Anzellini et al.*, 2013]. At 80 GPa the melting temperature of Fe is approximately 3500 K. There is extensive solid solution of Si into both the fcc and hcp structures of Fe, plus various intermetallics at low pressures, and also evidence for extensive solid solution in the B2 phase of FeSi, perhaps to as low as ~25 mol % Si [*Fischer et al.*, 2013]. Consequently, analysis of this phase diagram is not as easily simplified as the Fe-FeO, Fe-Fe_3C, and Fe-Fe_3S systems above. Nonetheless, equation (15.2) provides some insight into the Si partitioning behavior when the high-pressure melting behavior of Fe-Si alloys is considered. As summarized in Chapter 1, the melting point depression in Fe-Si alloys at high pressure is 200 ± 200 K, which is a relatively small but uncertain degree. Using these values with $\Delta S_m = 0.70$–$0.75R$ as before, equation (15.2) yields solid-liquid partitioning of Fe between metal and melt of $X^S_{Fe}/X^L_{Fe} = 1.00$–1.10 at the eutectic temperature, consistent with the narrow fcc+melt phase loops at 80 GPa described by *Fischer et al.* [2013] on the basis of X-ray diffraction data. Additional constraints from future work will be required to obtain the compositions of the coexisting phases. Nevertheless, it is evident from this analysis that the partitioning of Si between solid metal and melt is weak and not by itself likely to produce a large density contrast like the 7% change observed at the ICB [*Masters and Gubbins*, 2003].

15.4. PHASE RELATIONS AT ICB CONDITIONS

A principal result from the previous section is that the available experimental data on high-pressure melting of Fe-rich alloys supports an estimate of $\Delta S_m = 0.70$–$0.75R$ for their entropy of melting. This value of ΔS_m is slightly higher than the $\Delta V_m \rightarrow 0$ limit of $R\ln2$ [*Stishov*, 1988], consistent with the observation that the Fe melting curve remains positive to core pressures and has a ΔV_m of 1%–2% [*Laio et al.*, 2000; *Alfè et al.*, 2002]. The difference between 0.70 and 075 has an effect of only ~100 K in

calculated melting temperatures, and going forward we will assume that $\Delta S_m = 0.75R$ is a suitable estimate for any Fe-rich alloy composition. In this section, the analysis of melting phase relations will be extended to 330 GPa, the pressure at Earth's ICB, to predict the chemistry of melting in Fe-rich systems under conditions at which no such experimental data exist. The results will be speculative, as they are based on assumptions of ideal mixing and extrapolations of melting curves, but they can serve as a roadmap for future experimental exploration into these phase diagrams.

The melting curve of Fe has been extensively studied with static and dynamic experimental methods, as well as ab initio calculations, because of its key importance to the thermal structure of Earth's interior. Following Chapter 1, in this chapter the melting curve of *Anzellini et al.* [2013] will be used as a reference for pure Fe (i.e., 6200 K at 330 GPa), mainly for consistency as the experimental techniques used in that study (X-ray diffraction in a laser heated diamond anvil cell) were similar to those used in several studies of eutectic melting in Fe-binary systems [e.g., *Campbell et al.*, 2007; *Seagle et al.*, 2008; *Morard et al.*, 2008; *Asanuma et al.*, 2010; *Kamada et al.*, 2010, 2012; *Fischer et al.*, 2012, 2013]. Likewise, for convenience the extrapolation of binary eutectic temperatures to the ICB pressure will follow the summary of *Fischer* in Chapter 1. As observed in Chapter 1, the eutectic depressions in the Fe-FeO, Fe-Fe_3S, Fe-Fe_3C, and Fe-FeSi systems remain approximately constant, within experimental uncertainty, over the high-pressure ranges in which they have been investigated, and therefore their extrapolations to 330 GPa can simply track the *Anzellini et al.* [2013] melting curve of Fe. Constant eutectic depressions of 900 K (Fe-FeO), 700 K (Fe-Fe_3C), 1050 K (Fe-Fe_3S), and 200 K (Fe-FeSi) will be applied for the respective binary systems, and the thermodynamic framework of the previous section will be used to calculate phase relations at the ICB.

As in the previous section, equations (15.6), (15.8), and/or (15.9) can be used to calculate liquidus phase relations coexisting with solid phases Fe, FeO, Fe_3C, and Fe_3S. For now we can approximate these phases as pure, i.e., $a^S_i = 1$ in equation (15.2) for each of these components. The phase diagrams produced here are restricted to the Fe-O-C-S system, because the known solid solution in Fe-Si alloys is so extensive that without further information it is pointless to project phase relations with an Fe-Si component to inner core conditions. It is also assumed that the relevant endmembers are FeO, Fe_3C, and Fe_3S, although it is known that Fe_7C_3 plays an important role at high pressures and Fe_3S dissociates to an unidentified sulfide at $P > 250$ GPa [*Ozawa et al.*, 2013].

For each ternary system, cotectic curves and eutectic points are obtained by simultaneous solution of equations

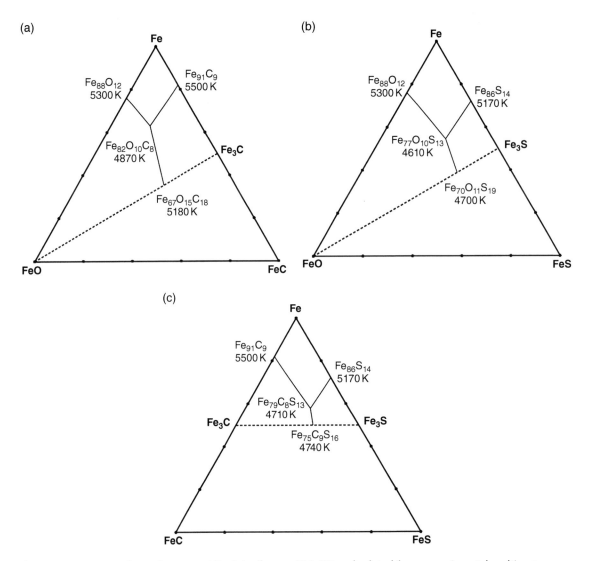

Figure 15.2 Ternary phase diagrams of Fe-rich alloys at 330 GPa calculated from experimental melting temperatures extrapolated to ICB conditions: (a) Fe-FeO-Fe$_3$C ternary; (b) Fe-FeO-Fe$_3$S ternary; (c) Fe-Fe$_3$C-Fe$_3$S ternary. Binary and ternary eutectic compositions (mole fractions) and temperatures are labeled. The melting temperature of Fe is 6200 K [*Anzellini et al.*, 2013], and those of FeO (8370 K), Fe$_3$C (6310 K), and Fe$_3$S (5460 K) were calculated from equations (15.6), (15.8), and (15.9), using eutectic melting point depressions summarized by *Fischer* Chapter 1 and a value of $\Delta S_m = 0.75R$ per mole atoms.

(15.6), (15.8), and/or (15.9), with the additional constraint $X_{Fe} + X_O + X_C + X_S = 1$. The results are shown in Figure 15.2. In the 330 GPa ternary systems, the eutectic temperatures lie in the range 4600–4900 K, which are not unreasonable lower bounds on the ICB temperature. The eutectic compositions in the Fe-O-S and Fe-C-S ternaries contain 10–12 wt % light element, which is consistent with that needed to explain the core density deficit, but the Fe-O-C eutectic is significantly more Fe-rich (5.5 wt % O + C), suggesting that sulfur may be important to the solid-melt phase relations to obtain a core composition that is consistent with the geophysical constraint that the liquid is ~7% less dense than the coexisting solid [*Masters and*

Gubbins, 2003]. The same analysis can be applied to the Fe-O-C-S quaternary system, from which one obtains a eutectic temperature $T_{eut} = 4290$ K at $X_O = 0.08$, $X_C = 0.07$, $X_S = 0.13$. These multicomponent eutectic compositions are compatible with the 7% density contrast observed at the ICB.

The binary eutectic melting point depressions, adapted here from Chapter 1, are critical input to the calculated phase diagrams in Figure 15.2, and uncertainty in these melting points propagate significantly to the liquidus phase boundary compositions. To give an example, increasing the Fe-O eutectic depression from $\Delta T_{eut} = 900$ K to $\Delta T_{eut} = 1200$ K would shift the calculated Fe-O eutectic composition at 330 GPa from $X_O = 0.08$ to $X_O = 0.12$, with corresponding shift in the neighboring cotectic curves illustrated in Figure 15.2. *Komabayashi* [2014] obtains a much greater oxygen content in the 330 GPa Fe-O eutectic (9 wt %) by invoking $\Delta T_{eut} = 2000$ K, which is not supported by experiments at lower pressures [*Seagle et al.*, 2008] (see also Chapter 1). This example underscores the importance of obtaining greater experimental constraints on the melting curves of Fe-alloys at pressures approaching those of Earth's ICB.

There are very few multicomponent data at sufficiently high pressure with which to compare such calculations, but *Terasaki et al.* [2011] report Fe-O-S eutectic melting temperatures that are indistinguishable from the Fe-S binary eutectic [*Kamada et al.*, 2012] up to 140 GPa. This result does not match well the analysis presented here (~500 K difference between the Fe-O-S and Fe-S eutectics) and further highlights the need for more experimental investigations to provide a larger foundation upon which thermodynamic models of candidate core-forming alloys can be improved.

15.5. OUTLOOK

The calculated phase diagrams in Figure 15.2 serve as examples of how reasonable approximations (an ideal associated solution model; an entropy change of melting for all alloys $\Delta S_m = 0.75$ R, slightly above the $\Delta V_m \rightarrow 0$ limit; binary eutectic melting point depression unchanging with pressure) allow the construction of useful Fe-X liquidus diagrams using only melting curves as experimental input. Obviously this approach has limitations, but these can be overcome with continued effort from both experimental and ab initio methods. Of primary importance will be solid-liquid partitioning experiments at the ICB pressure, 330 GPa. These are essential to understand Fe-Si or other systems in which significant solid solution exists. Experiments like these have not yet been reported, but continuing advances in both high-pressure generation [e.g., *Tateno et al.*, 2010; *Dubrovinsky et al.*, 2012] and petrological examination of recovered diamond anvil samples [e.g., *Siebert et al.*, 2012; *Nomura et al.*, 2014; *Ozawa et al.*, 2013] indicate they are on the horizon. Even a relatively small number of phase equilibria experiments would allow important refinement of the calculated phase diagrams in Figure 15.2, and subsolidus experiments can also provide useful constraints on the extent of solid solution expected in metal coexisting with melt. Additionally, extending the melting curves to higher pressures will allow more accurate phase diagram calculations from the data. Not only eutectic melting but also endmember melting curves are useful input to a thermodynamic framework in which the chemistry of the ICB can be understood.

REFERENCES

Alfè, D., G. D. Price, and M. J. Gillan (2002), Iron under Earth's core conditions: Liquid-state thermodynamics and high-pressure melting curve from ab initio calculations, *Phys. Rev. B, 65*, 165,118.

Alfè, D., M. J. Gillan, and G. D. Price (2007), Temperature and composition of the Earth's core, *Contemp. Phys., 48*, 63–80.

Antonangeli, D., J. Siebert, J. Badro, D. L. Farber, G. Fiquet, G. Morard, and F. J. Ryerson (2010), Composition of the Earth's inner core from high-pressure sound velocity measurements in Fe-Ni-Si alloys. *Earth Planet. Sci. Lett., 295*, 292–296.

Anzellini, S., A. Dewaele, M. Mezouar, P. Loubeyre, and G. Morard (2013), Melting of iron at Earth's inner core boundary based on fast X-ray diffraction, *Science, 340*, 464–466.

Asanuma, H., E. Ohtani, T. Sakai, H. Terasaki, S. Kamada, T. Kondo, and T. Kikegawa (2010), Melting of iron–silicon alloy up to the core–mantle boundary pressure: Implications to the thermal structure of the Earth's core, *Phys. Chem. Minerals, 37*, 353–359.

Badro, J., G. Fiquet, F. Guyot, E. Gregoryanz, F. Occelli, D. Antonangeli, and M. d'Astuto (2007), Effect of light elements on the sound velocities in solid iron: Implications for the composition of Earth's core, *Earth Planet. Sci. Lett., 254*, 233–238.

Birch F. (1952), Elasticity and composition of earth's interior, *J. Geophys. Res., 57*, 227-286.

Brown, J. M., and R. G. McQueen (1986), Phase transitions, Grüneisen parameter, and elasticity for shocked iron between 77 GPa, and 400 GPa. *J. Geophys. Res., 91*, 7485–7494.

Brown, J. M., T. J. Ahrens, and D. L. Shampine (1984), Hugoniot data for pyrrhotite and the Earth's core, *J. Geophys. Res., 89*, 6041–6048.

Campbell, A. J., C. S. Seagle, D. L. Heinz, G. Shen, and V. B. Prakapenka (2007), Partial melting in the iron-sulfur system at high pressure: A synchrotron X-ray diffraction study, *Phys. Earth Planet. Inter., 162*, 119–128.

Campbell, A. J., L. Danielson, K. Righter, C. T. Seagle, Y. Wang, and V. B. Prakapenka (2009), High pressure effects on the iron-iron oxide and nickel-nickel oxide oxygen fugacity buffers, *Earth Planet. Sci. Lett., 286*, 556–564.

Chen, B., Z. Li, D. Zhang, J. Liu, M. Y. Hu, J. Zhao, W. Bi, E. E. Alp, Y. Xiao, P. Chow, and J. Li (2014), Hidden carbon in Earth's inner core revealed by shear softening in dense Fe_7C_3, *Proc. Natl. Acad. Sci., 111*, 17,755–17,758.

Chuang, Y.-Y., and Y. A. Chang (1982), Extension of the associated solution model to ternary metal-sulfur melts: Cu-Ni-S, *Metall. Trans. B, 13*, 379–385.

Chudinovskikh, L., and R. Boehler (2007), Eutectic melting in the system Fe-S to 44 GPa, *Earth Planet. Sci. Lett., 257*, 97–103.

Dewaele, A., P. Loubeyre, F. Occelli, M. Mezouar, P. I. Dorogokupets, and M. Torrent (2006), Quasihydrostatic equation of state of iron above 2 Mbar, *Phys. Rev. Lett., 97*, 215,504.

Dubrovinsky, L., N. Dubrovinskaia, V. B. Prakapenka, and A. Abakumov (2012), Implementation of micro-ball nano-diamond anvils for high-pressure studies above 6 Mbar, *Nature Comm., 3*, 1163.

Dziewonski, A. M., and D. L. Anderson (1981), Preliminary reference Earth model, *Phys. Earth Planet. Inter.*, *25*, 297–356.

Fei, Y., J. Li, C. M. Bartka, and C. T. Prewitt (2000), Structure type and bulk modulus of Fe$_3$S, a new iron-sulfur compound, *Am. Mineral.*, *85*, 1830–1833.

Fischer, R. A., and A. J. Campbell (2010), High pressure melting of wüstite. *Am. Mineral.*, *95*, 1473–1477.

Fischer, R. A., A. J. Campbell, G. A. Shofner, O. T. Lord, P. Dera, and V. B. Prakapenka (2011), Equation of state and phase diagram of FeO, *Earth Planet. Sci. Lett.*, *304*, 496–502.

Fischer, R. A., A. J. Campbell, R. Caracas, D. M. Reaman, P. Dera, and V. B. Prakapenka (2012), Equation of state and phase diagram of Fe-16Si alloy as a candidate component of Earth's core. *Earth Planet. Sci. Lett.*, *357–358*, 268–276.

Fischer, R. A., A. J. Campbell, D. M. Reaman, N. A. Miller, D. L. Heinz, P. Dera, and V. B. Prakapenka (2013), Phase diagrams in the Fe-FeSi system at high pressures and temperatures, *Earth Planet. Sci. Lett.*, *373*, 54–64.

Fischer, R. A., A. J. Campbell, R. Caracas, D. M. Reaman, D. L. Heinz, P. Dera, and V. B. Prakapenka (2014), Equations of state in the Fe-FeSi system at high pressures and temperatures, *J. Geophys. Res.*, *119*, 2810–2827.

Ghiorso, M. S. (2004), An equation of state for silicate melts. I. Formulation of a general model, *Am. J. Sci.*, *304*, 637–678.

R. Jeanloz, and T. J. Ahrens (1980), Equations of state of FeO and CaO, *Geophys. J. R. Astron. Soc.*, *62*, 505–528.

Kamada, S., H. Terasaki, E. Ohtani, T. Sakai, T. Kikegawa, Y. Ohishi, N. Hirao, N. Sata, and T. Kondo (2010), Phase relationships of the Fe-FeS system in conditions up to the Earth's outer core, *Earth Planet. Sci. Lett.*, *294*, 94–100.

Kamada, S., E. Ohtani, H. Terasaki, T. Sakai, M. Miyahara, Y. Ohishi, and N. Hirao (2012), Melting relationships in the Fe-Fe$_3$S system up to the outer core conditions, *Earth Planet. Sci. Lett*, *359–360*, 26–33.

Kamada, S., E. Ohtani, H. Fukui, T. Sakai, H. Terasaki, S. Takahashi, Y. Shibazaki, S. Tsutsui, A. Q. R. Baron, N. Hirao, and Y. Ohishi (2014a), The sound velocity measurements of Fe$_3$S, *Am. Mineral.*, *99*, 98–101.

Kamada, S., E. Ohtani, H. Terasaki, T. Sakai, S. Takahashi, N. Hirao, and Y. Ohishi (2014b), Equation of state of Fe$_3$S at room temperature up to 2 megabars, *Phys. Earth Planet. Inter.*, *228*, 106–113.

Komabayashi, T. (2014), Thermodynamics of melting relations in the system Fe-FeO at high pressure: Implications for oxygen in the Earth's core, *J. Geophys. Res.*, *119*, doi:10.1002/2014JB010980.

Laio, A., S. Bernard, G. L. Chiarotti, S. Scandolo, and E. Tosatti (2000), Physics of iron at Earth's core conditions, *Science*, *287*, 1027–1030.

Li, J., Y. Fei, H. K. Mao, K. Hirose, and S. R. Shieh (2001), Sulfur in the Earth's inner core, *Earth Planet. Sci. Lett*, *193*, 509–514.

Litasov, K. D., I. S. Sharygin, P. I. Dorogokupets, A. Shatskiy, P. N. Gavryushkin, T. S. Sokolova, E. Ohtani, J. Li, and K. Funakoshi (2013), Thermal equation of state and thermodynamic properties of iron carbide Fe$_3$C to 31 GPa and 1473 K, *J. Geophys. Res.*, *118*, 1–11, doi: 10.1002/2013JB010270.

Lord, O. T., M. J. Walter, R. Dasgupta, D. Walker, and S. M. Clark (2009), Melting in the Fe-C system to 70 GPa, *Earth Planet. Sci. Lett.*, *284*, 157–167.

Lord, O. T., M. J. Walter, D. P. Dobson, L. Armstrong, S. M. Clark, A. Kleppe (2010), The FeSi phase diagram to 150 GPa, *J. Geophys. Res.*, *115*, B06,208.

Mao, Z., J.-F. Lin, J. Liu, A. Alatas, L. Gao, J. Zhao, and H.-K. Mao (2012), Sound velocities of Fe and Fe-Si alloy in the Earth's core, *Proc. Natl. Acad. Sci.*, *109*, 10,239–10,244.

Masters, G. and D. Gubbins (2003), On the resolution of density within the Earth, *Phys. Earth Planet. Inter.*, *140*, 159–167.

Morard, G., D. Andrault, N. Guignot, C. Sanloup, M. Mezouar, S. Petitgirard, and G. Fiquet (2008), In situ determination of Fe-Fe$_3$S phase diagram and liquid structural properties up to 65 GPa, *Earth Planet. Sci. Lett*, *272*, 620–626.

Morard, G., J. Siebert, D. Andrault, N. Guignot, G. Garbarino, F. J. Guyot, and D. Antonangeli (2013), The Earth's core composition from high pressure density measurements of liquid iron alloys, *Earth Planet. Sci. Lett.*, *373*, 169–178.

Nomura, R., K. Hirose, K. Uesugi, Y. Ohishi, A. Tsuchiyama, A. Miyake, and Y. Ueno (2014), Low core–mantle boundary temperature inferred from the solidus of pyrolite, *Science*, *343*, 52–525.

Ozawa, H., K. Hirose, T. Suzuki, Y. Ohishi, and N. Hirao (2013), Decomposition of Fe$_3$S above 250 GPa, *Geophys. Res. Lett.*, *40*, 4845–4849.

Prigogine, I., and R. Defay (1954), *Chemical Thermodynamics*, Longmans Green, London.

Sata, N., K. Hirose, G. Shen, Y. Nakajima, Y. Ohishi, and N. Hirao (2010), Compression of FeSi, Fe$_3$C, Fe$_{0.95}$O, and FeS under the core pressures and implication for light element in the Earth's core, *J. Geophys. Res.*, *115*, B09,204, doi:10.1029/2009jb006975.

Seagle, C. T., A. J. Campbell, D. L. Heinz, G. Shen and V. B. Prakapenka (2006), Thermal equation of state of Fe$_3$S and implications for sulfur in Earth's core, *J. Geophys. Res.*, *111*, B06,209.

Seagle, C. S., D. L. Heinz, A. J. Campbell, V. B. Prakapenka, and S. T. Wanless (2008), Melting and thermal expansion in the Fe-FeO system at high pressure, *Earth Planet. Sci. Lett.*, *265*, 655–665.

Sharma, R. C., and Y. A. Chang (1979), Thermodynamics and phase relationships of transition metal-sulfur systems: Part III. Thermodynamics properties of the Fe-S liquid phase and calculation of the Fe-S phase diagram, *Metall. Trans. B*, *10*, 103–108.

Siebert J., J. Badro, D. Antonangeli, and R. J. Ryerson (2013), Terrestrial accretion under oxidizing conditions. *Science*, *339*, 1194–1197.

Stevenson, D. J. (1981), Models of the Earth's core, *Science*, *214*, 611–619.

Stewart, A. J., M. W. Schmidt, W. van Westrenen, and C. Liebske (2007), Mars: A new core-crystallization regime, *Science*, *316*, 1323–1325.

Stishov, S. M. (1988), Entropy, disorder, melting, *Sov. Phys. Usp.*, *31*, 52–67.

Stishov, S. M., I. N. Makarenko, V. A. Ivanov, and A. M. Nikolaenko (1973), On the entropy of melting, *Phys. Lett. A*, *45*, 18.

Svendsen, B., W. W. Anderson, T. J. Ahrens, and J. D. Bass (1989), Ideal Fe-FeS, Fe-FeO phase relations and Earth's core, *Phys. Earth Planet. Inter*, *55*, 154–186.

Tallon, J. L. (1980), The entropy change on melting of simple substances, *Phys. Lett.*, *76A*, 139–142.

Tateno, S., K. Hirose, Y. Ohishi, and Y. Tatsumi (2010), The structure of iron in Earth's inner core, *Science*, *330*, 359–361.

Terasaki, H., S. Kamada, T. Sakai, E. Ohtani, N. Hirao, and Y. Ohishi (2011), Liquidus and solidus temperatures of a Fe-O-S alloy up to the pressures of the outer core: Implications for the thermal structure of Earth's core, *Earth Planet. Sci. Lett.*, *304*, 559–564.

Wen, S., and H. Nekvasil (1994), Ideal associated solutions: Application to the system albite quartz H_2O, *Am. Mineral.*, *79*, 316–331.

Williams, Q., and R. Jeanloz (1990), Melting relations in the iron–sulfur system at ultra-high pressures: Implications for the thermal state of the Earth, *J. Geophys. Res.*, *95*, 19,299.

Zhang, W.-J., Z.-Y. Liu, Z.-L. Liu and L.-C. Cai (2015), Melting curves and entropy of melting of iron under Earth's core conditions, *Phys. Earth Planet. Inter.*, *244*, 69–77.

Zhang, Y., and Q.-Z. Yin (2012), Carbon and other light element contents in the Earth's core based on first-principles molecular dynamics, *Proc. Natl. Acad. Sci.*, *109*, 19579–19583.

16

Chemistry of Core-Mantle Boundary

John W. Hernlund

ABSTRACT

Earth's core-mantle boundary (CMB) is a chemically reactive interface between a mantle and core that are, on average, far from equilibrium. The chemical processes taking place at the CMB depend critically on the transport of reactants into this region, the degree of mixing and mass exchange that occurs while at the CMB, and subsequent removal of reaction products in order to vacate space for subsequent reactants. Thus the subject of CMB chemistry is a problem of core and mantle dynamics in addition to thermodynamical and transport properties of materials at the extreme pressure-temperature conditions of Earth's deep interior. Unfortunately, little is known about the particular phase equilibria between metals and silicates at the CMB, although the gross behavior of some simple systems is understood well enough that hypotheses can be proposed and tested using geophysical observations. Integrative studies that link CMB chemistry to a diverse range of observable phenomena offer the best hope for progress in the future.

16.1. INTRODUCTION

The subject of core-mantle boundary (CMB) chemistry is a rich problem, touching upon numerous subjects intimately connected to the composition, structure, and dynamics of the core and deep mantle. As with any geological process, the relics left behind as a consequence of CMB chemistry in principle leave behind a record of the evolution of the deep Earth since formation. Our challenge is to understand the nature of these processes and how to interpret the results of nature's experiment in order to learn more about the workings of the deep interior of our planet through deep time.

Experimental and theoretical constraints upon CMB chemistry have been slow to progress owing to the extreme pressure (~136 GPa) and temperatures (~3500–5000 K)

Earth-Life Science Institute, Tokyo Institute of Technology, Meguro, Japan

that have prevailed in this region over the past 4.5 Gyr of Earth's history. Methods such as dynamic shock compression [e.g., *Tan and Ahrens*, 1990] are useful for constraining the equation of state of materials at extreme conditions but are practically useless for determining chemical equilibria. Large-volume multianvil experiments are the best tool for studying chemical equilibria in Earth's interior; however, they are still limited to ~100 GPa [e.g., *Yamazaki et al.*, 2014] and temperatures are limited to ~2000 K. Small-volume diamond-anvil cells can readily achieve CMB conditions when heated by a laser and the tiny samples can be analyzed using modern microtechnology; however, due to the small size, these exhibit very high intrinsic temperature gradients (e.g., 10^3 K over 10^{-4} m is ~10^7 K/m). Such a large temperature gradient tends to drive the system away from equilibrium owing to the Soret effect [*Platten*, 2005; *Sinmyo and Hirose*, 2010]. It is hoped that continued methodological developments will circumvent these difficulties.

Deep Earth: Physics and Chemistry of the Lower Mantle and Core, Geophysical Monograph 217, First Edition.
Edited by Hidenori Terasaki and Rebecca A. Fischer.
© 2016 American Geophysical Union. Published 2016 by John Wiley & Sons, Inc.

Regardless of the challenges faced in understanding high *P-T* behavior, much effort has been made in recent years to begin establishing a zero-order understanding of the nature of CMB chemistry, which is the subject of the remainder of this chapter. The presentation begins with a discussion of the initial conditions for CMB chemistry established during core formation. This is followed by an overview of chemical equilibria between metals and silicates at high *P-T*, with a focus on the role of major elements. The connection between chemical reactions and dynamical processes in the CMB region is explored next. The chapter concludes that much work is yet to be done on the problem of CMB chemistry, along with an optimistic assessment that much progress may be made in coming years by leveraging multiple perspectives and constraints.

16.2. INITIAL CONDITIONS FOR CMB CHEMISTRY

The planets are thought to have been formed by gradual accretion of mass over a period of ~10 Myr from a disk of dust and gas remaining in orbit around the newborn Sun. These were supernova remnants representing several generations of cosmic nucleosynthetic cycles and thus containing a significant fraction of heavy elements. Although oxygen is abundant in the solar system, it was not so prevalent that no unoxidized metals were available for terrestrial planets to form as a mixture of oxides and metals (thus permitting Earth to have a core). According to modern models of planetary formation, relatively larger "embryos" formed on the order of ~1 Myr timescales and some rapidly accumulated gas to become proto-gas giants at an early stage [e.g., *Kokubo and Ida*, 1998]. The embryos continued to accrete smaller planetesimals and other debris, dynamically interacting in their evolving orbits and sometimes merging with other embryos in giant collisions like the one that is thought to have formed Earth's Moon.

The initial conditions for CMB chemistry were established during accretion and core formation [*Stevenson*, 1981; *Murthy*, 1991]. Equilibration is thought to have been reached by mixing between metallic and silicate liquids inside the molten cavities produced by energetic impacts as planetesimals and embryos collided with the growing proto-Earth. Significant mixing should have occurred as liquid metal sank through silicate melt under the action of gravity [e.g., *Deguen et al.*, 2014]. Elemental partitioning of moderately siderophile elements between these immiscible liquids left a fingerprint in the silicate Earth whose calibration in pressure-temperature and composition $(P - T - X)$ using laboratory experiments suggests equilibration in ~1000 km deep magma oceans [*Li and Agee*, 1996; *Righter*, 2003]. High-density metal liquids presumably sank to the

bottom of these magma oceans and temporarily ponded above a deeper layer of mostly solid rock. Depending on the transport mechanism and rheological properties of the proto-mantle, subsequent descent of the metal through the rocky layer to the core may have taken place over time scales of hours to millions of years, with the possibility of further chemical interactions on the way down [*Stevenson*, 1990]. Although some elements of this narrative are well-established, existing experimental and observational constraints still permit a wide range of scenarios, and thus there are many unresolved questions about how this process occurred and no clear consensus regarding what is uniquely required by partitioning constraints and observed elemental abundances. For example, the data alone cannot distinguish between single-stage and multiple-stage core formation [*Walter and Cottrell*, 2013], although there are very good physical reasons for adopting a multistage view. Another source of uncertainty is temperature and oxygen fugacity (f_{O_2}): Ni and Co partitioning constraints can be satisfied at high temperatures at present-day oxygen fugacity or at lower temperatures with an evolving oxygen fugacity [*Wade and Wood*, 2005].

Chemical exchange between metals and silicates during core formation established the bulk composition of both the core and mantle, thus setting the large-scale context for subsequent interactions at the CMB. Because silicate-metal equilibration is sensitive to $P-T$ and was initially established at pressures (and probably temperatures) less than those that have prevailed at the CMB since Earth's formation, the CMB itself is undoubtedly a chemically reactive frontier. This is true even if metallic liquid were to completely re-equilibrate with the solid silicate proto-mantle upon descent into the core, since the end result would still represent a convolution of incremental accretion events taking place at ever greater $P - T$ conditions as the protoplanet grew in size.

The disequilibrium between the mantle and core at the largest scales means that there is a gradient in chemical potential(s) between the core and mantle. Some species become more soluble in metallic liquid iron at CMB relative to the conditions of core formation while some become less soluble, and thus the nature of CMB chemical disequilibrium since formation should be representative of these relative states, in addition to evolving conditions of the CMB through geological time. If average mantle material were circulated to the CMB and placed adjacent to average core material, the disequilibrium between them would drive chemical reactions. If any of the reacted material were later transported away from the CMB, it would exhibit a chemical composition distinct from the average mantle or core and could in principle carry a distinct fingerprint of core-mantle interaction.

Thus far there is still no clear evidence that any material has been borne up to Earth's surface following CMB interactions. It has been proposed that some material has been transported from the CMB bearing Os isotopic signatures reflecting inner core crystallization [*Brandon and Walker*, 2005]. Hot-spot lavas exhibiting elevated Fe/Mn have also been attributed to CMB interactions and subsequent carriage of products upward in mantle plumes [*Humayun et al.*, 2004]. Diamonds bearing extremely iron-enriched inclusions have also been suggested to represent a chemical signature of the CMB region [*Wirth et al.*, 2014]. While these are all exciting possibilities, such claims remain controversial and are not unique interpretations of the data. For example, high Fe/Mn could be explained by a Si-rich source rock [*Sobolev et al.*, 2005], and other processes could lead to Fe-enrichment besides core-mantle interaction (e.g., subduction of Fe-rich rocks). In principle, it may be possible to discover a distinct chemical signature of processes that can only take place at CMB conditions, and further research into these possibilities may yet enable us to uniquely identify a physical sample that has interacted with the core.

16.3. METAL-SILICATE EQUILIBRIA

The best way to predict the nature of core-mantle reactions through time is to directly measure the outcome of metal/silicate partitioning of all relevant species as a function of $P-T-X$. However, as mentioned previously, the elevated $P-T$ conditions at the CMB present significant technical challenges for producing quality measurements of this kind. Nevertheless, many interesting insights have been gained from such experiments, and it is useful to review some of the basic results and recent progress.

Of the major elements in Earth, Fe, Si, and O are likely to play the most important roles in the dynamics of chemical reactions at the CMB. For Earth's average composition, Fe exhibits a preference for the metallic core while Si and O favor incorporation into mineral phases or melts in the mantle. However, Fe, Si, and O could all be stable in concentrations of ~10% or more in either the metallic core or silicate/oxide mantle. Iron, silicon, and oxygen exert a strong influence upon phase equilibria and physical properties of both oxides and metals, and variations in their concentrations may produce dramatic density variations in both the core and mantle that exert a strong influence on their respective dynamics. Therefore, constraining the Fe-Si-O system is a key ingredient to constraining the chemistry and associated dynamics of the CMB.

An important study for assessing CMB chemistry in the system Fe-Si-O using the laser-heated diamond anvil cell was carried out by *Knittle and Jeanloz* [1991]. They mixed iron-bearing (Fe,Mg)SiO$_3$ perovskite (now known as the mineral "bridgmanite," Br) with metal (M) iron, compressed the sample to mid-mantle pressures, and following laser-heating they found iron-free Br, FeO, and

FeSi in the quench products. The simplest interpretation of this result suggests the following reaction:

$$(Mg,Fe)SiO_3^{Br} + Fe^M \rightarrow MgSiO_3^{Br} + FeO^? + FeSi^?.$$
(16.1)

The phase associated with each chemical species is denoted using a superscript. A question mark is deliberately written for the phase of FeO and FeSi to represent the uncertainty in the stable phase that hosted them at elevated P-T prior to quenching. While this experiment suggested a preference for dissolution of the FeSiO$_3$ component of Br into metal, the interpretation of the coexisting FeO and FeSi is less clear. Shock experiments on FeO wüstite (Wu) indicated that it becomes electrically conducting at high pressures [*Knittle et al.*, 1986], suggesting that metallic phases of FeO and FeSi might be produced by the reaction (16.1) at pressures much lower than the CMB. Indeed, the tendency for pure FeO to become metallic at lowermost mantle conditions has been supported by more recent diamond-anvil cell experiments and theoretical calculations [*Fischer et al.*, 2011; *Ohta et al.*, 2012]. The separation into O- and Si-bearing phases observed in the study of Knittle and Jeanloz was thought to represent two stable solid phases; however, it is not a unique possibility. Such an outcome could also result from immiscibility of O - and Si-bearing metals induced by quenching from high to low temperature [e.g., *Tsuno et al.*, 2013]. For example, Fe-O-Si may be stable as a continuous miscible solution in liquid metal at high P-T for a broad range of O and Si concentrations, but they unmix at lower temperatures to form a mixture of FeO and FeSi.

The results of *Knittle and Jeanloz* [1991] stirred much debate and prompted new models of the CMB region. For example, *Kellogg and King* [1993] considered the possibility of accumulating a FeO-/FeSi-rich dense layer formed by such a reaction on the dynamics of mantle plumes, and *Manga and Jeanloz* [1996] discussed the influence of a corresponding high thermal conductivity layer at the CMB. One important question is whether the kind of reaction observed by Knittle and Jeanloz should proceed if the iron metal initially contained enough dissolved Si and O to account for Earth's outer core density deficit. *Buffett et al.* [2000] proposed that the Knittle-Jeanloz-type reaction could run backward if the amount of Si and/or O in the outer core is on the other side of the equilibrium. If this occurred, resultant underplating of the CMB by a mushy electrically conducting layer might help to provide a Maxwell stress couple to the solid mantle for core-mantle angular momentum exchange on ~10 yr timescales, potentially helping to explain observed variations in Earth's rotation rate.

Another issue concerns the pressure-dependence of reaction (16.1), which is a function of the volume difference between products and reactants. Clearly, higher pressures will favor the side of the reaction occupying the smaller volume. The volume change of oxygen dissolution in metal was not initially well-constrained, and some have argued for negligible changes based on analogue systems [*Walker et al.*, 2002] while behaviors extrapolated from lower pressures suggested that the reaction would be impeded in the deep mantle [e.g., *Rubie et al.*, 2004]. The recent expansion of experimental constraints and models to higher pressures appears to indicate that reaction (16.1) is more sensitive to temperature than pressure [*Asahara et al.*, 2007; *Ozawa et al.*, 2008, 2009; *Frost et al.*, 2010; *Tsuno et al.*, 2013], although the precise temperature-dependence is still poorly constrained because experimental data have not been acquired over a sufficiently broad range of temperatures.

Also important is the coupled solubility when both Si and O are present. Strongly variable behavior can develop even in the simple ternary Fe-Si-O system, depending on the relative behavior of Si and O in the metal. This becomes more clear if we write the relevant chemical equilibria in terms of component oxides,

$$FeO^{Ox} \rightarrow Fe^{M} + O^{M} \qquad (16.2)$$

and

$$SiO_2{}^{Ox} \rightarrow Si^{M} + 2O^{M}. \qquad (16.3)$$

If Si and O exhibit similar tendencies toward dissolution and/or become mutually compatible, then one might expect to find both Si and O in roughly equal proportions in the metal and coexisting rocks, as proposed by *Tsuno et al.* [2013] for temperatures exceeding 3000 K at lower mantle pressures. But the general behavior could be unbalanced in either direction if the reaction does not occur at conditions in which they are mutually compatible. For example, FeO could more readily decompose to the metallic state [i.e., forward reaction (16.2)] while Si exhibits a relatively greater preference for the oxide state [i.e., reverse reaction (16.3)], thus leaving a metal enriched in oxygen alongside a rock depleted in FeO and enriched in SiO_2. Conversely, Si may be more energetically favorable in metal than is oxygen, in which case reaction (16.3) proceeds while excess oxygen is returned to the oxide as FeO [i.e., reverse reaction (16.3)], leaving a silicon-rich metal alongside a relatively FeO-rich and SiO_2-depleted rock. This range of possible behaviors illustrates why it is important to consider the entire system of relevant components, rather than focusing only on the behavior of a singe reaction or component.

One important constraint on mutual Si-O solubility in metallic iron can be understood by writing the reaction for dissolution of silica into metal as

$$SiO_2{}^{Ox} + 2Fe^{M} \rightarrow Si^{M} + 2\xi FeO^{M} + 2(1-\xi)FeO^{Ox},$$
$$(16.4)$$

where ξ is the mole fraction of FeO that goes into the metal phase and can be expressed in terms of the equilibrium constant $K_{FeO}^{M/Ox}$ for the reaction $FeO^{Ox} \rightarrow FeO^{M}$ as

$$\xi = \frac{K_{FeO}^{M/Ox}}{1 + K_{FeO}^{M/Ox}}. \qquad (16.5)$$

Thus for example, if FeO were present in roughly similar concentrations in metal and oxide phases at equilibrium at the CMB, then $\xi \sim 0.5$. This means that about half of the oxygen involved in any further silica dissolution into the core is partitioned into the metal, while the other half contributes to enrichment of the oxide phase in FeO. In fact, so long as $\xi < 1$, any Si dissolution into the core inevitably enriches the residual mantle reaction product in FeO.

The above conclusion is not restricted to the Fe-Si-O system. Incorporation of any species from an oxide state into core metal will enrich the residual rock in FeO in precisely the same manner as SiO_2 incorporation. For example, if the magma ocean(s) in which the core was originally forged contained significant H_2O, it could have reacted with iron to leave an H-enriched metal and FeO-enriched oxide as products [e.g., *Okuchi*, 1997]. Such tendencies are usually exacerbated by higher temperatures, which drive reactions such as equation (16.4) further to the right. Unfortunately, the reaction (16.4) is still not well-constrained at CMB conditions, particularly the strength of the temperature-dependence, and this could have a major influence on many hypotheses. For example, it was recently proposed that Earth could have accreted from relatively oxidized materials with excess oxygen partitioning into the core [*Siebert et al.*, 2013]; however, such a mechanism might be untenable if a significant amount of SiO_2 also dissolves into the core (thus driving O back out of the core as FeO).

The Fe-S-O system is another interesting ternary system for CMB chemistry. The coexistence of two immiscible liquids (one S-rich and one O-rich) is well-documented at ambient pressure and is important in the smelting of iron. *Helffrich and Kaneshima* [2004] developed a thermodynamical model using parameters obtained at ambient pressure and suggested that Earth's outer core could contain two immiscible liquids for some ranges of bulk core S and O concentrations. They sought evidence for this phenomena using seismic waves that should produce detectable reflections from any immiscible fluid-fluid interface and did not find any such signatures. The lack of observation of seismic immiscibility thus suggested unique constraints on the relative abundances of O and S in the core. However, it is important to note that some experiments show that miscibility gaps such as the one in the Fe-FeO system can disappear at high pressures [e.g., *Tsuno et al.*, 2007], and thus it is not straightforward to predict behavior at the CMB without experimental constraints at the appropriate *P-T* conditions.

One might hope that the present experimental constraints would allow us to determine the direction of CMB reactions for proposed "typical" mantle or core compositions, and some steps have been made in this direction. For simple systems such as Fe-FeO it is predicted that typical mantle such as pyrolite yields a metal in equilibrium which hosts more oxygen than is allowed for by the core density deficit [*Asahara et al.*, 2007; *Ozawa et al.*, 2008, 2009; *Frost et al.*, 2010]. Thus, if we limit our view to only include the behavior of oxygen, then the inevitable conclusion is that either the base of the mantle is depleted in FeO [*Asahara et al.*, 2007] and/or the top of the core is enriched in oxygen [*Buffett and Seagle*, 2010]. Seismic observations of objects such as ultralow-velocity zones [*Garnero and Helmberger*, 1996] or core-rigidity zones [e.g., *Rost and Revenaugh*, 2001] at the CMB suggest that the bottom of the mantle is enriched (not depleted) in FeO, thus favoring O-enrichment on top of the core. However, one must be careful to note that the chemistry may be strongly affected by the presence of other chemical species on both sides of the CMB, and the story according to the Fe-FeO system may not capture the full behavior of the CMB chemical system.

16.4. DYNAMICAL CONSIDERATIONS OF CMB REACTIONS

Although typical mantle and typical core are out of equilibrium with one another at the CMB, it does not necessarily follow that the CMB itself represents a discontinuity in chemical potential(s). Instead, chemical reactions will occur locally at the frontier, changes in the composition of materials on either side will result, and the discontinuity in chemical potentials is transformed into a gradient with a length scale that is dictated by transport styles and properties on either side of the CMB. If there is any significant residence time of reacted material on either side of the CMB, then the CMB itself is not *a priori* expected be a pristine interface. There may also be significant lateral variations along the CMB driven by chemical reactions, since mantle materials with significantly different bulk composition [e.g., midocean ridge basalt (MORB) vs. harzburgite] may be transported to the CMB and brought into contact with the top of the core in different locations.

The outcome of CMB chemical reactions is strongly influenced by material transport properties, such as diffusivity and viscosity, which exhibit dramatic changes at the CMB. The equilibration time between a piece of mantle with length scale L_m and a piece of core with length scale L_c is dictated by the slowest diffusivity of species

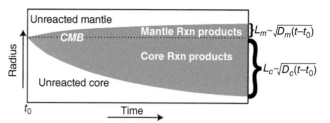

Figure 16.1 Spatial evolution of core-mantle reaction products with time at a stagnant (nonflowing) CMB. The reaction proceeds by diffusion of species alone over length scales governed by the limiting diffusion timescales L_m^2/D_m in the mantle and L_c^2/D_c in the core. For Earth's CMB, L_c is $\sim 10^3$–10^4 times larger than L_m; the difference is much larger than can be represented on this figure.

involved in the mantle contribution to the reaction D_m and the core contribution to the reaction D_c. The time-scale for equilibration is L_m^2/D_m for the mantle bit and L_c^2/D_c for the core bit, and when placed together the relative length scales over which chemical reactions take place over a given time span are $L_c/L_m = \sqrt{D_c/D_m}$. If one adopts typical values at temperatures in the range of 3000–4000 K, such as $D_c \sim 10^{-8}$ m²/s [*Vočadlo et al.*, 2003] and $D_m \sim 10^{-16}$–10^{-14} m²/s [*Van Orman and Crispin*, 2010], then $L_c/L_m \sim 10^3$–10^4. Therefore the higher diffusivity of reacting species in core material permits simultaneous equilibration of regions significantly larger in the core than in the mantle (Figure 16.1).

Convection of material through the CMB region is also an essential ingredient in determining the length scales of processes driven by CMB reactions. Mantle material circulating through a partial arc of the CMB at a tectonic rate of ~ 10 cm/yr would have a residence time of ~ 50 Myr at the CMB such that the mantle side of the CMB could be replaced by sustained mantle convection roughly 100 times in Earth's history. For outer core flows of order 10^{-4} m/s consistent with the present secular variation of the geomagnetic field, the residence time is just ~ 1000 yr for the same traverse across the CMB, sufficient to refresh the core side of the CMB millions of times throughout Earth's history. In ~ 50 Myr a diffusion-limited chemical reaction on the mantle side would influence a region only 10–1000 cm thick, and after refreshing the CMB 100 times through Earth's history the amount of reacted material would amount to only 10–100 ppm of the total mantle. For the core, the diffusion-limited reaction boundary layer for residence times of 1000 yr is 10 m, and if this were re-circulated and mixed into the deeper core, it would change the composition of the entire core on 10 Myr timescales. Therefore, so long as material is free to convect in and out of the CMB region from the mantle and core sides, then CMB reactions will have little effect on the mantle but can bring the entire outer core into equilibrium with the CMB over relatively short geological timescales.

The above discussion regarding convection of material through the CMB is incomplete because it neglects the important dynamical feedback that the chemical reactions themselves should exert upon convective transport. For example, if the density of mantle material increases by ~1% or more as a consequence of iron-enrichment following chemical reactions, it will become heavy enough to compete against the thermal buoyancy that drives mantle convection, and thus reacted material may tend to accumulate at the CMB. The ability for reacted mantle to resist immediate viscous entrainment is enhanced if the reacted material is rheologically weak, perhaps as a consequence of the reaction itself. Likewise, if the density of outer core material is reduced by $~10^{-8}$ or more by addition of light components to the heavy metal owing to CMB reactions, then it may be buoyant enough to overcome thermal buoyancy forces in the core and would thus resist downward entrainment into the deeper core, accumulating instead just beneath the CMB. On the other hand, if mantle material becomes less dense and/or core material more dense owing to CMB reactions, then their respective buoyant transport rates away from the CMB will be accelerated. In these latter cases, reacted material would be transported away and refreshed more quickly and thus the CMB itself would be kept sharp and pristine. However, if heterogeneous dense matter accumulates at the base of the mantle and/or buoyant metal accumulates on top of the core, chemical reactions can produce chemically distinct domains with significantly longer residence times at the CMB. Thus, if chemical reactions were the only mechanism capable of generating strong heterogeneity at the CMB, then it would be possible to predict much about the nature of CMB chemistry using geophysical observations alone. Unfortunately, however, there are many other factors that may influence CMB structures.

Further complexity is added to the context of CMB chemistry owing to dramatic differences in the rheology of the core and mantle. If dense matter accumulates at the base of the mantle as a result of CMB reactions, then so long as it continues to behave as a creeping viscous solid, it will still be subject to viscous entrainment into overlying mantle flows and will be gradually eroded away [Sleep, 1988]. This is true even for density anomalies as large as ~10%. In the liquid outer core, on the other hand, viscous forces are virtually absent and the only straightforward mechanism available for downward mixing of a buoyant metallic rind produced by ingestion of light alloys is turbulent entrainment. However, energy considerations suggest that the amount of material susceptible to turbulent entrainment is very small, even for modest stabilizing density variations of ~1% [Hernlund and McNamara, 2014].

Dynamics can also play an important role in increasing the degree of mixing between mantle and core material at local scales. If the CMB is a sharp interface between pure rock above and pure metal below, then the volume that can be reacted is strictly limited by diffusion across the interface. Experiments suggest that liquid Fe can migrate through iron-free MgO periclase via a proposed "morphological instability" [Otsuka and Karato, 2012]; however, it is not clear whether the requisite conditions would have existed in the context of Earth's long-term CMB evolution. In particular, the mechanism may rely upon strong local chemical disequilibrium (e.g., between Fe metal and Fe-free oxide) but may not be as vigorous with lesser degrees of disequilibrium (or smaller chemical potential gradients). Alternatively, it may be possible for iron liquid to simply migrate between grain boundaries of solid rock or "wet" the grain boundaries. While metal has a high surface tension at lower pressures and temperatures that can inhibit grain boundary wetting, experimental evidence has shown that this behavior may change at high P-T [Sakai et al., 2006]. The metal liquid, being relatively heavy, could not percolate significantly into the solid unless there is an external driving force. For example, it is expected that the CMB is depressed by up to ~1 km beneath mantle downwellings owing to the dynamic topography effect. In this case a fluid outer core material that wets grain boundaries could be driven up into the depressed portions of the mantle by simple hydrostatic pressure gradients [Kanda and Stevenson, 2006]. For the mantle side this can result in a significantly larger reaction zone that would otherwise be limited by the small elemental diffusivity of species in oxide phases. As the reacted material is pushed away from centers of mantle downwelling, the dynamic viscous force causing its depression will diminish and much of the reacted zone could be buoyed upward, with the metallic liquid inside the interstices draining back into the core. It is also possible that a small fraction of iron liquid would remain trapped in the rock at volume fractions of order ~1% [Sakai et al., 2006].

If buoyant O-rich and/or Si-rich material accumulates atop the core as a consequence of CMB reactions that endow it with sufficient buoyancy to resist entrainment and mixing, then other processes could be triggered as a consequence. For example, if the composition of fluid at the very top of the core is close to stoichiometric FeO owing to CMB reactions, then core cooling to the FeO melting temperature will result in crystallization of FeO solid. Current experimental constraints [Fischer and Campbell, 2010] and uncertainties in CMB temperature are probably enough to admit this scenario. If the liquid metal layer at the very top of the core has a density similar to FeO liquid (or less, owing to additional O or other light alloys), then the crystal would sink into the core, perhaps remelting at depth (depending upon the pressure dependence of the FeO melting curve and the geotherm). This kind of process could enhance downward transport of CMB reaction products deeper into the core. On the other hand, if the liquid is heavier than the FeO solid, then the snow would float upward and become inherited as a solid metallic layer underplating the mantle, a mechanism that is not dissimilar to the kind of process proposed by *Buffett et al.* [2000]. This would have the effect of helping to maintain the composition of the CMB near the stoichiometry of FeO. FeO snow is only one possibility; other phases with different stoichiometry could also crystallize from a light alloy enriched layer on top of the core.

The dynamical scenarios discussed above are relatively speculative in nature and are only useful insofar as they can explain observations. There is much uncertainty about the degree of chemical heterogeneity in the deep mantle, the nature of deep-mantle convection, the existence of stratified layers on top of the core, and many other features. We also do not know the stable phase assemblages at the base of the mantle for the full range of compositions that satisfy seismic observations. Models of coupled core-mantle thermal evolution require the ancient CMB to be above the melting temperature of any plausible composition, suggesting that the base of the mantle was once extensively molten [*Labrosse et al.*, 2007]. The existence of such a basal magma ocean and the secular changes driven by fractional crystallization would exert a profound influence on the early history of CMB chemical evolution, such as enriching the melt in iron [*Nomura et al.*, 2011] and hence driving more oxygen into the top of the core. The uncertainties invoked by this scenario alone are sufficient to argue that we do not have a firm grasp of CMB chemistry at the present, and further progress must await future developments and discoveries.

16.5. SUMMARY

CMB chemistry is a rich subject and is important for interpreting geophysical observations of the CMB region. However, future progress requires better and more diverse experimental constraints on the behaviors of materials at the relevant *P-T* conditions. Seismological surveys of the CMB region have been very useful to illuminate some of its structures, but there is still no firm link to the ways in which observed features are linked to chemical processes. The best way forward is to continue considering a very wide range of possibilities, performing experimental work that will have the greatest impact on present uncertainties, and connecting a more diverse range of observations to processes that may be related to CMB chemistry (including core formation). In principle, both existing and emerging technologies should permit us to tackle these problems with more confidence in the coming decade.

REFERENCES

Asahara, Y., D. J. Frost, and D. C. Rubie (2007), Partitioning of FeO between magnesiowüstite and liquid iron at high pressures and temperatures: Implications for the composition of the Earth's outer core, *Earth Planet. Sci. Lett.*, *257*, 435–449, doi:10.1016/j.epsl.2007.03.006.

Brandon, A. D., and R. J. Walker (2005), The debate over core-mantle interaction, *Earth Planet. Sci. Lett.*, *232*, 211–225.

Buffett, B. A., and C. T. Seagle (2010), Stratification of the top of the core due to chemical interactions with the mantle, *J. Geophys. Res.*, *115* B04,407, doi:10.1029/2009JB006751.

Buffett, B. A., E. J. Garnero, and R. Jeanloz (2000), Sediments at the top of the Earth's core, *Science*, *290*, 1338–1342.

Deguen, R., M. Landeau, and P. Olson (2014), Turbulent metalsilicate mixing, fragmentation, and equilibration in magma oceans, *Earth Planet. Sci. Let.*, *391*, 274–287.

Fischer, R. A., and A. J. Campbell (2010), High pressure melting of wüstite, *Am. Mineral.*, *95*, 1473–1477.

Fischer, R. A., A. J. Campbell, O. T. Lord, G. A. Shofner, P. Dera, and V. B. Prakapenka (2011), Phase transition and metallization of FeO at high pressures and temperatures, *Geophys. Res. Lett.*, *38*, L24,301, doi:10.1029/2011GL049800.

Frost, D. J., Y. Asahara, D. C. Rubie, N. Miyajima, L. S. Dubrovinsky, C. Holzapfel, E. Ohtani, M. Miyahara, and T. Sakai (2010), Partitioning of oxygen between the Earth's mantle and core, *J. Geophys. Res.*, *115*, B02,202, doi:10.1029/2009JB006302.

Garnero, E. J., and D. V. Helmberger (1996), Seismic detection of a thin laterally varying boundary layer at the base of the mantle beneath the central-Pacific, *Geophys. Res. Lett.*, *23*, 977–980.

Helffrich, G., and S. Kaneshima (2004), Seismological constraints on core composition from Fe-O-S liquid immiscibility, *Science*, *306*, 2239–2242.

Hernlund, J. W., and A. K. McNamara (2014), Dynamics of the core-mantle boundary region, *Treatise on Geophysics*, 2nd Edition, edited by G. Schubert and D. Bercovici, Elsevier, Amsterdam, *7*, 461–519, 2015.

Humayun, M., L. Qin, and M. D. Norman (2004), Geochemical evidence for excess iron in the mantle beneath hawaii, *Science*, *306*, 91, doi:10.1126/science.1101050.

Kanda, R.V. S., and D. J. Stevenson (2006), Suction mechanism for iron entrainment into the lower mantle, *Geophys. Res. Lett.*, *33*, L02,310, doi:10.1029/2005GL025009.

Kellogg, L. H., and S. D. King (1993), Effect of mantle plumes on the growth of D″ by reaction between the core and mantle, *Geophys. Res. Lett.*, *20*, 379–382.

Knittle, E., and R. Jeanloz (1991), The Earth's core-mantle boundary: Results of experiments at high pressures and temperatures, *Science*, *251*, 1438–1443.

Knittle, E., R. Jeanloz, A. C. Mitchell, and W. J. Nellis (1986), Metallization of $Fe_{0.94}O$ at elevated pressures and temperatures observed by shock-wave electrical resistivity, *Solid State Commun.*, *59*, 513.

Kokubo, E., and S. Ida (1998), Oligarchic growth of protoplanets, *Icarus*, *131*, 171–178.

Labrosse, S., J. W. Hernlund, and N. Coltice (2007), A crystallizing dense magma ocean at the base of the Earth's mantle, *Nature*, *450*, 866–869.

Li, J., and C. B. Agee (1996), Geochemistry of mantle-core differentiation at high pressure, *Nature*, *381*: 686–689, doi:10.1038/381686a0.

Manga, M., and R. Jeanloz (1996), Implications of a metal-bearing chemical boundary layer in D″ for mantle dynamics, *Geophys. Res. Lett.*, *23*, 3091–3094.

Murthy, V. R., (1991), Early differentiation of the earth and the problem of mantle siderophile elements: A new approach, *Science*, *253*, 303–306, doi:10.1126/science.253.5017.303.

Nomura, R., H. Ozawa, S. Tateno, K. Hirose, J. Hernlund, S. Muto, H. Ishii, and N. Hiraoka (2011), Spin crossover and iron-rich silicate melt in the Earth's deep mantle, *Nature*, *473* (7346), 199–202, doi:10.1038/nature09940.

Ohta, K., R. E. Cohen, K. Hirose, K. Haule, K. Shimizu, and Y. Ohishi (2012), Experimental and theoretical evidence for pressure-induced metallization in FeO with the rock-salt type structure, *Phys. Rev. Lett.*, *108*, 026,403, doi:10.1103/PhysRevLett.108.026403.

Okuchi, T., (1997), Hydrogen partitioning into molten iron at high pressure: Implications for Earth's core, *Science*, *278*, 1781–1784, doi:10.1126/science.278.5344.1781.

Otsuka, K., and S. Karato (2012), Deep penetration of molten iron into the mantle caused by a morphological instability, *Nature*, *492*, 243–246, doi:10.1038/nature11663.

Ozawa, H., K. Hirose, M. Mitome, Y. Bando, N. Sata, and Y. Ohishi (2008), Chemical equilibrium between ferropericlase and molten iron to 134 GPa and implications for iron content at the bottom of the mantle, *Geophys. Res. Lett.*, *35*, L05308, doi:10.1029/2007GL032648.

Ozawa, H., K. Hirose, M. Mitome, Y. Bando, N. Sata, and Y. Ohishi (2009), Experimental study of reaction between perovskite and molten iron to 146 GPa and implications for chemical equilibrium at the core-mantle boundary, *Phys. Chem. Minerals*.

Platten, J. K., (2005), The soret effect: A review of recent experimental results, *J. Appl. Mech*, *73*, 5–15, doi:10.1115/1.1992517.

Righter, K., (2003), Metal-silicate partitioning of siderophile elements and core formation in the early Earth, *Annu. Rev. Earth Planet. Sci.*, *31*(1), 135–174, doi:10.1146/annurev.Earth.31.100901.145451.

Rost, S., and J. Revenaugh (2001), Seismic detection of rigid zones at the top of the core, *Science*, *294*: 1911–1914.

Rubie, D. C., C. K. Gessmann, and D. J. Frost (2004), Partitioning of oxygen during core formation on the Earth and Mars, *Nature*, *429*: 58–61.

Sakai, T., T. Kondo, E. Ohtani, H. Terasaki, N. Endo, T. Kuba, T. Suzuki, and T. Kikegawa (2006), Interaction between iron and post-perovskite at core-mantle boundary and core signature in plume source region, *Geophys. Res. Lett.*, *33*, L15,317, doi:10.1029/2006GL026868.

Siebert, J., J. Badro, D. Antonangeli, and F. J. Ryerson (2013), Terrestrial accretion under oxidizing conditions, *Science*, *339*: 1194–1197, doi:10.1126/science.1227923.

Sinmyo, R., and K. Hirose (2010), The Soret diffusion in laser-heated diamond-anvil cell, *Phys. Earth Planet. Inter.*, *180*: 172–178, doi:10.1016/j.pepi.2009.10.011.

Sleep, N. H. (1988), Gradual entrainment of a chemical layer at the base of the mantle by overlying convection, *Geophys. J.*, *95*, 437–447.

Sobolev, A. V., A. W. Hofmann, S. V. Sobolev, and I. K. Nikogosian (2005), An olivine-free mantle source of Hawaiian shield basalts, *Nature*, *434*, 590–597, doi:10.1038/nature03411.

Stevenson, D. J. (1981), Models of the Earth's core, *Science*, *214*, 611–619.

Stevenson, D. J. (1990), Fluid dynamics of core formation, in *Origin of the Earth*, edited by H. E. Newsom and J. H. Jones, pp. 231–249, Oxford Univ. Press.

Tan, H., and T. J. Ahrens (1990), Shock temperature measurements for metals, *High Pressure Res.*, *2*, 145–157.

Tsuno, K., E. Ohtani, and H. Terasaki (2007), Immiscible two-liquid regions in the Fe-O-S system at high pressure: Implications for planetary cores. *Phys. Earth. Planet. Inter.*, in press, *160*(1), 75–85, doi:10.1016/j.pepi.2006.09.004.

Tsuno, K., D. J. Frost, and D.C. Rubie (2013), Simultaneous partitioning of silicon and oxygen into the earth's core during early earth differentiation, *Geophys. Res. Lett.*, *40*, 66–71, doi:10.1029/2012GL054116.

Van Orman, J. A., and K. L. Crispin (2010), Diffusion in oxides, *Rev. Mineral. Geochem.*, *72*, 757–825.

Vočadlo, L., D. Alfè, M. J. Gillan, and G. D. Price (2003), The properties of iron under core conditions from first principles calculations, *Phys. Earth Planet. Inter.*, *140*, 101–125.

Wade, J., and B. J. Wood (2005), Core formation and the oxidation state of the Earth, *Earth Planetary Sci. Lett.*, *236*(1–2), 78–95, doi:10.1016/j.epsl.2005.05.017.

Walker, D., S.M. Clark, and L. M. D. Cranswick (2002), O₂ volumes at high pressure from KClO₄ decomposition: D″ as a siderophile element pump instead of a lid on the core, *Geochem. Geophys. Geosys.*, *3*, 1070.

Walter, M. J., and E. Cottrell (2013), Assessing uncertainty in geochemical models for core formation in earth, *Earth Planet. Sci. Lett.*, *365*, 165–176.

Wirth, R., L. Dobrzhinetskaya, B. Harte, A. Schreiber, and H.W. Green (2014), High-Fe (Mg,Fe)O inclusion in diamond apparently from the lowermost mantle, *Earth Planet. Sci. Lett.*, *404*, 365–375, doi:10.1016/j.epsl.2014.08.010.

Yamazaki, D., E. Ito, T. Yoshino, N. Tsujino, A. Yoneda, X. Guo, F. Xu, Y. Higo, and K. Funakoshi (2014), Over 1 Mbar generation in the Kawai-type multianvil apparatus and its application to compression of $(Mg_{0.92}Fe_{0.08})SiO_3$ perovskite and stishovite, *Phys. Earth Planet. Inter.*, *228*: 262–267.

17

Phase Transition and Melting in the Deep Lower Mantle

Kei Hirose[1,2]

ABSTRACT

The discovery of post-perovskite (ppv) first provided a likely explanation for the shear velocity increase at the top of the D″ layer. Other seismic anomalies such as strong seismic anisotropy and anticorrelation between the anomalies of shear and bulk sound velocities are also, at least in part, reconciled with the presence of ppv in the lowermost mantle. However, the compositional effect on the pressure range of the ppv transition and its sharpness remain controversial, which makes the interpretation of complex seismic structures difficult. In addition, recent experiments show melting-phase relations and partial-melt compositions formed under deep lower mantle pressures. Melting as well as deep subduction of dense materials over the history of the Earth contributes to complex seismic structures in D″.

17.1. INTRODUCTION

Seismology shows strong anomalies in the deepest part of the mantle called the D″ layer, including abrupt shear wave velocity increase a few hundred kilometers above the core-mantle boundary (CMB) (the D″ discontinuity). Since the origins of such anomalies are difficult to explain with the known properties of $MgSiO_3$ perovskite (now formally called bridgmanite), the D″ layer has long been the most enigmatic region inside Earth. *Sidorin et al.* [1999] speculated a solid-solid phase transition to reconcile the D″ seismic discontinuity, but a specific phase transition in major lower mantle minerals had not been identified until 2004. Therefore, the D″ discontinuity has been often attributed to be thermal or chemical boundary [e.g., *Wysession et al.*, 1998; *Lay and Garnero*, 2004].

The year 2014 was the 10th anniversary of the discovery of perovskite (pv) to post-perovskite (ppv) phase transition [*Murakami et al.*, 2004; *Oganov and Ono*, 2004; *Tsuchiya et al.*, 2004]. After the first report on pure $MgSiO_3$ pv, it has been observed to occur in a variety of simple and multicomponent systems such as natural pyrolite [*Murakami et al.*, 2005; *Ono and Oganov*, 2005; *Ohta et al.*, 2008; *Grochorlski et al.*, 2012] and midoceanic ridge basalt (MORB) materials [*Hirose et al.*, 2005; *Ohta et al.*, 2008; *Grocholski et al.*, 2012]. Physical properies of ppv, such as electrical and thermal conductivity, viscosity, and deformation mechanism, have also been extensively studied in the last ten years (see the most recent review by *Hirose et al.* [2015]), and it has been argued that a majority of seismic anomalies observed in the D″ region, such as D″ discontinuity, shear wave polarization anisotropy, and anticorrelation between the anomalies of shear and bulk sound velocities, may be explained by the occurrence of ppv.

[1] Earth-Life Science Institute (ELSI), Tokyo Institute of Technology, Meguro, Tokyo, Japan

[2] Laboratory of Ocean-Earth Life Evolution Research, Japan Agency for Marine-Earth Science and Technology, Yokosuka, Kanagawa, Japan

Deep Earth: Physics and Chemistry of the Lower Mantle and Core, Geophysical Monograph 217, First Edition.
Edited by Hidenori Terasaki and Rebecca A. Fischer.
© 2016 American Geophysical Union. Published 2016 by John Wiley & Sons, Inc.

Recent seismological observations, however, revealed more complex structures in the D″ layer, such as reflectors with strong topography over short distances [*Hutko et al.*, 2006], which are not reconciled with the pv-ppv transition. While the ppv transition causes velocity increase and decrease for S and P waves, respectively, both S and P waveforms display a positive polarity in Siberia and the Central Pacific [*Cobden and Thomas*, 2013]. Since the D″ layer is a convection boundary at the bottom of the mantle, one can naturally expect rich geological structures there. The complex structure may be caused not only by deep subduction of oceanic lithosphere but also by melting/crystallization in a deep lower mantle over the history of Earth [e.g., *Labrosse et al.*, 2007]. In order to understand such geological structures, one needs to distinguish anomalies of chemical origin from those of phase transition origin and therefore needs to better understand the compositional effects on the pressure range of the ppv phase transition. Such compositional effects, however, still remain controversial [*Ohta et al.*, 2008; *Grocholski et al.*, 2012]. In addition, the dissociation of pv into almost iron-free pv and iron-rich hexagonal "H phase" has been reported to occur in a deep lower mantle [*Zhang et al.*, 2014], which could be an additional ingredient to cause complex seismological structure.

This chapter reviews previous experimental results on the ppv phase transition. In addition, recent studies are introduced on melting-phase relations and partial-melt compositions under deep lower mantle pressures. The presence of ppv and the possible scenarios on the formation of dense materials in the lowermost mantle are discussed. Recently, silicate pv has been given an official name of bridgmanite, but this chapter will use "post-perovskite," not "post-bridgmanite," because (1) bridgmanite is a name for Mg-rich perovskite (Mg must be a dominant cation) and therefore post-bridgmanite cannot be applied to Fe-rich post-perovskite that is important in the lowermost mantle and (2) we have a custom of using the name of the crystal structure in relevant cases; for example, the dissociation of Mg_2SiO_4 ringwoodite is called, not post ringwoodite, but post-spinel transformation.

17.2. POST-PEROVSKITE PHASE TRANSITION

17.2.1. Experimental Difficulty

Extensive efforts have been paid to determine the ppv phase transition boundary, since it is key to understanding the nature of the D″ region. However, experiments under relevant high-pressure and high-temperature (*P-T*) conditions are not easy, and experimental results using laser-heated diamond anvil cell (DAC) techniques include uncertainty. It is mainly because (1) the absolute pressure scale is not available at lowermost mantle *P-T* conditions, (2) the temperature gradient is relatively large in laser-heated sample, and (3) chemical segregation occurs due to a large temperature gradient, often called Soret diffusion, leading to strong chemical heterogeneity in a sample, in particular for iron. These could be the main sources of discrepancy between DAC experimental results obtained by different groups, although additional factors may also be involved.

17.2.1.1. Pressure Determinations

In recent high-*P*, high-*T* experiments with in situ X-ray diffraction measurements, pressure is determined at high temperature from the volume of a pressure marker using its known equation of state (EoS). The accuracy of such a high-temperature EoS has been a matter of extensive debate in the last 15 years [e.g., *Fei et al.*, 2004]. Gold is often used as a pressure standard for experiments on Fe-bearing samples because it is believed to be much less chemically reactive with iron than platinum and MgO. The EoSs of gold reported in the literature give different pressures, as much as 15 GPa at 110 GPa and 2400 K (Figure 17.1) [*Hirose et al.*, 2008a].

The high-temperature EoS of MgO is less controversial, in part because electronic thermal pressure is not necessary to consider for nonmetals. Indeed, Mg_2SiO_4 post-spinel transformation and $MgSiO_3$ ppv phase transition boundaries match the 660 km and the D″ seismic discontinuities, respectively, when using the EoS of MgO proposed by *Speziale et al.* [2001] or *Wu et al.* [2008]. The EoSs of gold proposed by *Fei et al.* [2007] and *Hirose et al.* [2008a] are based on the MgO pressure scale by *Speziale et al.* [2001]. Note that these two give the highest pressure at the high temperatures of the lower mantle among existing EoSs of gold (Figure 17.1).

17.2.1.2. Soret Diffusion

Soret diffusion is diffusion of elements driven by a temperature gradient. In gases, heavier elements or elements with larger ionic radii migrate from hot to cold regions, and lighter elements move opposite. It occurs also in liquids and solids, although it is more complex. *Lesher and Walker* [1991] reported that the Soret diffusion in silicate melt is controlled by the *Z/r* ratio of cations rather than its mass and volume (*Z* is the cation charge, *r* is the ionic radii, and *Z/r* is the proxy of cation field strength).

The Soret diffusion is known to occur often in a laser-heated DAC sample, depending on a thermal gradient in a laser-heated sample that is controlled by laser-heating optics and sample configuration. Both iron and magnesium migrate from the high-temperature part toward the cold edge of a laser-heated area; In contrast, silicon is enriched at a hot region (Figure 17.2) [*Ozawa et al.*, 2009; *Sinmyo and Hirose*, 2010]. Note that the samples shown in Figure 17.2 were not molten because iron was enriched at the outer margin of a laser-heated area (partial melt,

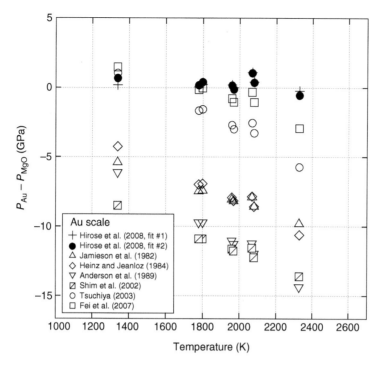

Figure 17.1 Comparison of pressures determined from existing Au pressure scales and those obtained by the MgO scale of *Speziale et al.* [2001] based on simultaneous volume measurements around 110 GPa at high temperatures [from *Hirose et al.*, 2008a].

if any, forms a melt pool at the hottest part of a sample and is always enriched in iron compared to residual solid). It leads to a chemically heterogeneous sample during laser heating, which could be a main source of inconsistencies among previous experimental studies on iron-bearing samples. *Nomura et al.* [2011] found that the Soret diffusion is very fast at relatively high temperatures; strong iron depletion was observed in 5 at 3500 K and 77 GPa under a subsolidus condition. This was a serious problem in past laser-heating DAC experiments, in which the chemical characterization of a sample was not performed after heating in most of the cases. It could also be a source of discrepancy found in previous experimental results on iron-bearing systems [e.g., *Mao et al.*, 2004; *Murakami et al.*, 2005; *Dorfman et al.*, 2013].

Sinmyo and Hirose [2010] reported that the magnitude of Soret diffusion depends strongly on a sample configuration. It can be avoided when the sample is covered with a thin film of metal, which provides relatively homogeneous temperature distributions and helps minimize the Soret diffusion (Figure 17.2d). On the other hand, laser heating caused strong chemical heterogeneity when the sample was mixed with fine metal powder; metal powder moved to form aggregates, and the heterogeneous distributions of the aggregates caused a complex and large temperature gradient (Figure 17.2c). It is therefore very important to check the chemical homogeneity of a sample after laser-heating experiments.

17.2.2. Pure MgSiO$_3$ ppv

Murakami et al. [2004] first reported MgSiO$_3$ ppv but did not tightly constrain the *P-T* location of the phase transition boundary. Later, on the basis of both forward (transition from pv to ppv) and backward (reversal) experiments in wide *P-T* space up to 171 GPa and 4400 K, *Tateno et al.* [2009] found that the ppv phase transition boundary in pure MgSiO$_3$ is located at 121 GPa and a plausible deep lower mantle temperature of 2400 K, with a Clapeyron slope of +13.3 MPa/K (Figure 17.3b). Such pressure corresponds to 2650 km depth, matching the depth of the D″ seismic discontinuity. These results are based on the MgO pressure scale proposed by *Speziale et al.* [2001].

On the other hand, when using a different pressure marker or EoS, the same experimental data give very different pressure ranges of transition and Clapeyron slope. With gold as a pressure marker [*Tsuchiya*, 2003], *Hirose et al.* [2006] (*Tateno et al.* [2009] is an extension of this work) demonstrated that the ppv phase transition occurs at 113 GPa and 2400 K with the Clapeyron slope of +4.7 MPa/K. The experiments performed by *Ono and Oganov* [2005], based on the platimun pressure scale [*Holmes et al.*, 1989], showed the boundary at 129 GPa and 2400 K and a slope of +7.0 MPa/K.

The Soret diffusion can occur in pure MgSiO$_3$ sample (Figure 17.2), possibly leading to dissociation into MgO

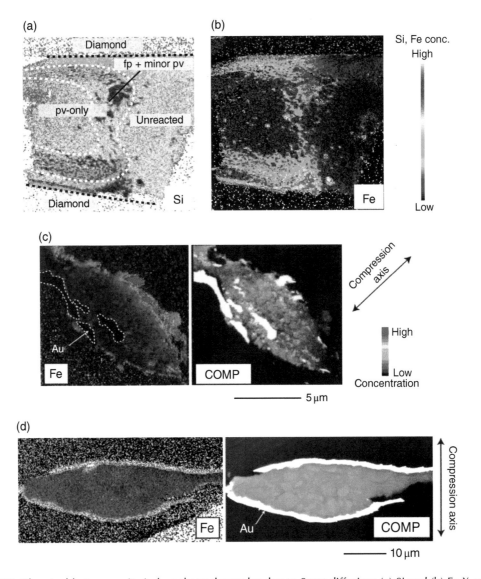

Figure 17.2 Chemical heterogeneity in laser-heated samples due to Soret diffusion: (a) Si and (b) Fe X-ray maps of $(Mg_{0.9}Fe_{0.1})_2SiO_4$ sample after laser heating [from *Ozawa et al.*, 2009]. Silicon migrated toward the center of the hot spot, leading to the formation of pv only in the absence of fp. On the contrary, iron was segregated to the low-temperature part, resulting in strong enrichment in fp. Note that iron enrichment occurs in partial melt formed at the hottest area when heated to melting temperature (see Figure 17.4). X-ray maps of Fe and backscattered electron images of pyrolitic sample (c) mixed with Au powder and (d) coated with Au after laser heating [from *Sinmyo and Hirose*, 2013]. The Au powder moved to form large aggregates, causing heterogeneous temperature distributions and resulting in the strong heterogeneity in Fe. On the other hand, sample remained chemically homogeneous when it was coated with metal.

and SiO_2. Such dissociation, however, does not affect the results of phase equilibria studies introduced above. In addition, both low- and high-pressure phases have identical chemical composition in a simple system such as $MgSiO_3$, which is also a great advantage to perform reversal experiments, unlike the case for pv and ppv formed in pyrolite. Moreover, the first-order structural transition is a discontinuous reaction in single-component systems (two phases coexist only on a univariant curve), and thus determination of the boundary is

much more straightforward than that in multicomponent systems, in which the transition occurs in some pressure range.

The overall uncertainty in experimental determinations of the ppv phase transition boundary may be ±5–10 GPa, mainly due to the lack of accurate pressure scale at high temperature. This is a similar magnitude of uncertainty to that in high-temperature abinitio calculations. The experimental results of the ppv transition pressure by *Tateno et al.* [2009] are between those calculated using

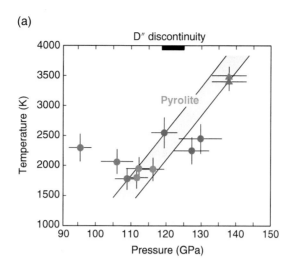

(a)

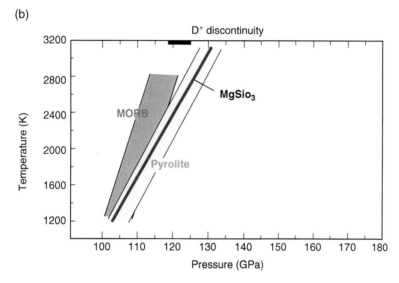

(b)

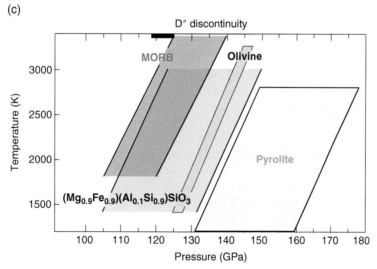

(c)

Figure 17.3 Experimentally determined ppv phase transition boundaries. All pressures were recalculated to be consistent with the MgO pressure scale [*Speziale et al.*, 2001]. (a) Stabilities of pv (red), ppv (blue), and pv+ppv (green) in a pyrolitic lower mantle reported by *Murakami et al.* [2005] and *Ohta et al.* [2008] (circles) and *Fiquet et al.* [2010] (triangles). (b) Pressure ranges of the ppv transition (pv+ppv two-phase coexisting region) reported by Tokyo Tech and French groups; pyrolite from (a) and MORB from *Hirose et al.* [2005] and *Ohta et al.* [2008]. The ppv transition boundary in pure $MgSiO_3$ is also shown [*Tateno et al.*, 2009]. (c) Ppv transition boundaries obtained by the MIT group; pyrolite, olivine, and MORB from *Grocholski et al.* [2012] and $(Mg_{0.9}Fe_{0.1})(Si_{0.9}Al_{0.1})O_3$ from *Catalli et al.* [2009].

local density approximation (LDA) and generalized gradient approximation (GGA) [*Tsuchiya et al.*, 2004; *Oganov and Ono*, 2004]. As shown above, the Clapeyron slope of the ppv boundary in pure $MgSiO_3$ obtained by experiments is strongly dependent on a pressure scale, ranging from +4.7 to +13.3 MPa/K. Theory predicted +7.5 MPa/k [*Tsuchiya et al.*, 2004] and +10 MPa/K [*Oganov and Ono*, 2004].

17.2.3. Ppv in Pyrolitic Mantle

Experimental determination of the ppv phase transition boundary in multicomponent systems is much more complicated that that for pure $MgSiO_3$ for a number of reasons. First, Soret diffusion easily causes strong chemical heterogeneity (Figure 17.2). Therefore, long heating duration in laser-heated DAC experiments does not necessarily help the chemical equilibrium in multicomponent systems. As was reported by *Sinmyo and Hirose* [2010], Soret diffusion causes the migration of both Fe and Al toward the low-temperature part of a sample, leading to the formation of (Fe,Al)-depleted $MgSiO_3$ phase at hot areas and Fe-rich $MgSiO_3$ and Al-rich $MgSiO_3$ phases at cold portions (Fe- and Al-rich areas are likely separated). In such a situation, both pv and ppv coexist in an apparently very wide pressure range, because aluminum is known to significantly expand the stability of pv with respect to ppv [*Tateno et al.*, 2005; *Akber-Knutson et al.*, 2005] although uncertainty remains [*Tsuchiya and Tsuchiya*, 2008]. Note that the effect of Soret diffusion is largest when metal powder is mixed with sample powder (Figure 17.2c).

Second, the sample must be amorphous before heating. The first heating of such homogeneous sample will give an equilibrium-phase assemblage and composition. Nevertheless, once three solid phases are formed in pyrolite at the first heating, subsequent heating at different *P-T* conditions hardly provides chemical reequilibrium, because the lattice diffusion of elements is known to be very slow, at least in pv (~100 nm/h) [*Holzapfel et al.*, 2005]. Chemical disequilibrium within the sample could lead to a disequilibrium-phase assemblage such as the coexistence of pv and ppv. Considering that typical grain size in a pyrolitic mantle material is 100 nm in laser-heated DAC when heated at ~2000 K [*Sinmyo et al.*, 2008], it is necessary to heat samples at least for 1 h when reheating is made with changing *P-T* conditions. It is, however, difficult to heat sample for 1 h without suffering Soret diffusion. This means that chemical equilibrium is obtained only when first heating the pyrolitic material X-ray diffraction (XRD) and practically we could obtain a single datum in each separate run [note that the absence of peaks does not necessarily mean that the sample is still amorphous after the first heating, because the XRD

measurements do not detect small crystals probably less than ~10 nm]. It also means that reversal experiments are virtually impossible.

Third, a two-phase coexisting region is overestimated by laser-heated DAC experiments. Apparently a wide two-phase region can be caused by (1) chemical heterogeneity induced by Soret diffusion, (2) large temperature variations in the X-ray-probed area in both radial and axial directions, and (3) compositional zoning in each phase caused by multiple heating cycles. Note that the Soret effect causes iron enrichment in the low-temperature portion, including the area near the diamonds in the X-ray spot. Soret diffusion does not help sharp transition. Note also that Soret diffusion occurs even in the presence of pressure medium, although it reduces the thermal gradient within a sample [*Sinmyo and Hirose*, 2010].

The first work on the ppv phase transition in a pyrolitic lower mantle was by *Murakami et al.* [2005] (Figure 17.3a). Murakami and others employed a starting material coated with a thin film of gold in order to make a homogeneous temperature distribution in a laser-heated sample. *Sinmyo and Hirose* [2010] later demonstrated that such a metal coating helps reduce the effect of Soret diffusion (Figure 17.2d). *Murakami et al.* [2005] reported that the ppv phase transition occurs around 113 GPa and 2500 K based on the Au scale by *Tsuchiya* [2003], which is recalculated to 120 GPa when using the Au scale of *Fei et al.* [2007], which is based on the MgO scale of *Speziale et al.* [2001]. They did not observe the coexistence of pv and ppv.

Ohta et al. [2008] augmented experimental data from *Murakami et al.* [2005] to constrain the sharpness of the boundary (Figure 17.3a). They used gold powder mixed with sample powder, which may have caused extensive Soret diffusion, but the results of *Ohta et al.* [2008] were fairly consistent with those by *Murakami et al.* [2005]. The Au scale by *Hirose et al.* [2008a] was employed, which is based on the Mgo scale of *Speziale et al.* [2001] and thus is close to the Fei et al., scale (Figure 17.1). Ohta and others concluded that the transition occurs over the 5 GPa pressure range around 120 GPa at 2500 K. Note that both *Murakami et al.* [2005] and *Ohta et al.* [2008] performed laser heating at a single *P-T* condition of interest in each run in order to avoid kinetic hindering of phase transition and chemical disequilibrium, both of which cause the apparent coexistence of two phases (pv and ppv).

Subsequently, *Fiquet et al.* [2010] observed a drastic change in the XRD pattern of pyrolite with increasing temperature from 3400 to 3500 K at 138 GPa, indicating an abrupt phase transition from ppv to pv (Figure 17.3a). Fiquet and others used the EoS of pv [*Ricolleau et al.*, 2009] for pressure determinations, which is based on the Fei et al., Au scale and thus the Speziale et al., MgO scale.

All of these three results are consistent with each other (Figure 17.3a). Combining them, the ppv phase transition takes place in pyrolite rather sharply over 5 GPa between 120 (ppv in) and 125 (pv out) GPa with Clapeyron slope of +13.6 MPa/K, primarily based on the MgO pressure scale by *Speziale et al.* [2001]. Note that the boundary is close to that in pure $MgSiO_3$. The location of the boundary matches the depth of the D″ seismic discontinuity (~2600 km depth).

Ono and Oganov [2005] performed similar XRD measurements to determine the ppv transition boundary in pyrolite. They showed that the transition took place at 124 GPa and 2500 K based on the Au scale proposed by *Jamieson et al.* [1982], which is equivalent to 131 GPa by the *Fei et al.*, Au scale. Note that they repeated a number of heating cycles with changing *P-T* conditions in a single experiment and therefore their results may have suffered both kinetic hindering of phase transition and compositional segregation due to the Soret effect. They did not mention the coexistence of pv and ppv.

More recent experiments performed by *Grocholski et al.* [2012] reported the ppv transition in a pyrolitic lower mantle between 140 and 169 GPa at 2500 K with Clapeyron slope of +5.6 MPa/K based on the Au scale of *Tsuchiya* [2003]'s, which are converted into 146 and 175 GPa and +12 MPa/K with the Fei et al. Au scale (Figure 17.3c). They repeated heating with changing *P-T* conditions, but the results of the first heating cycles were consistent with those obtained by subsequent heating cycles. Compared to the earlier experiments by *Murakami et al.* [2005], *Ohta et al.* [2008], and *Fiquet et al.* [2010] using the same pressure scale (Figure 17.3b), Grocholski and others reported (1) much higher pressure range (120–125 GPa vs. 146–175 GPa at 2500 K) and (2)a much wider pv + ppv two-phase coexisting region (5 GPa vs. ~30 GPa).

Such higher pressure range of the ppv phase transition in pyrolite (140–169 GPa) is, however, not consistent with results on the similar transition in $(Mg_{0.9}Fe_{0.1})(Al_{0.1}Si_{0.9})O_3$ bulk composition (112–139 GPa at 2500 K) determined by the same group [*Catalli et al.*, 2009] (Figure 17.3b). The chemical composition of pv formed in a pyrolitic lower mantle is similar to $(Mg_{0.9}Fe_{0.1})(Al_{0.1}Si_{0.9})O_3$ [*Sinmyo and Hirose*, 2013], and therefore the ppv transition should occur in a very similar pressure range. *Grocholski et al.* [2012] argued that aluminum is responsible for such high ppv transition pressure in pyrolite, but the experiments by *Catalli et al.* [2009] demonstrated that the ppv transition occurs at similar pressure ranges in $(Mg_{0.9}Fe_{0.1})(Al_{0.1}Si_{0.9})O_3$ and $(Mg_{0.91}Fe_{0.09})SiO_3$. The main difference between natural pyrolite and $(Mg_{0.9}Fe_{0.1})(Al_{0.1}Si_{0.9})O_3$ is the presence of ferropericlase (fp) in the former, which narrows the width (pressure range) of the transition but does not increase the average transition pressure [*Sinmyo et al.*, 2011].

It is certainly true that the incorporation of iron and aluminum expands the pv + ppv two-phase coexisting region. Such effects of iron and aluminum could be large [*Mao et al.*, 2004; *Tateno et al.*, 2005; *Catalli et al.*, 2009; *Andrault et al.*, 2010; *Grocholski et al.*, 2012] but may be small [*Hirose et al.*, 2006; *Ohta et al.*, 2008; *Fiquet et al.*, 2010], in particular in the presence of fp [*Sinmyo et al.*, 2011]. As mentioned earlier, several technical issues, such as large temperature variations and the Soret effects are involved in laser-heated DAC experiments, and therefore this issue has been controversial quantitatively (Figure 17.3).

These discussions are based on the Speziale et al. MgO pressure scale (and the Fei et al. Au scale based on it), but the accuracy of high-temperature EoS of MgO by *Speziale et al.* [2001] is not clear. If *Grocholski et al.* [2012] employs the Au scale by *Shim et al.* [2002], the transition pressure is recalculated to be 131–160 GPa (Figure 17.1), indicating that the ppv transition starts in a pyrolitic material within the pressure range of Earth's mantle. The inconsistency between *Murakami et al.* [2005], *Ohta et al.* [2008], *Fiquet et al.* [2010], and *Grocholski et al.* [2012] is not clear (Figures 17.3b, c), but it is certainly important to assess the chemical homogeneity of the laser-heated DAC sample (in other words, the effect of Soret diffusion) by ex situ sample characterization (Figure 17.2). Indeed, inconsistency is often found in DAC experiments on iron-bearing systems. A large part of such inconsistency may be reconciled with the strong segregation of iron in laser-heated sample due to the Soret effect.

In order to clarify the pressure range of the ppv transition in pyrolite, better high-temperature generation on samples using multianvil press or resistance-heated DAC may be required. The recent multianvil experiments by *Yamazaki et al.* [2014] were performed up to 106 GPa and 1200 K, close to the ppv boundary at such a low temperature. Furthermore, a resistive internally heated DAC experiment has been carried out up to 1650 K around 50 GPa [*Komabayashi et al.*, 2009]. Temperature generation by a resistive externally heated DAC has been limited to ~1000 K, but recent experiments were conducted up to 1300 K [*Du et al.*, 2013] or 1460 K [*Umemoto et al.*, 2014]. Spatial and temporal temperature fluctuations in these resistance-heated experiments are much less than those in laser-heated DAC experiments, which can avoid Soret diffusion and help in the accurate determination of sample temperature.

17.2.4. Ppv in Depleted Mantle

Subducting oceanic plates are mainly composed of about 60 km thick depleted peridotite layer, from which 10–20% partial melts were extracted under midoceanic ridges. Considering the average oceanic crust production rate during the Mesozoic to the present time (about 25 km³ per year) [*Reymer and Schubert*, 1984] and assuming this production (subduction) rate for the last 4 billion years, the total amount of MORB crust produced corresponds to ~10 vol % of the Earth's mantle, suggesting that all mantle materials once experienced partial melting on average. Therefore, the depleted peridotitic materials are not only a main constituent of subducting slabs but also possibly a primary constituent of Earth's mantle.

The ppv phase transition has been examined not in such depleted peridotite but in $(Mg,Fe)_2SiO_4$ olivine bulk composition, which is the "end-member" depleted peridotite. The experiments performed by *Hirose et al.* [2006] on $(Mg_{0.89}Fe_{0.11})_2SiO_4$ olivine bulk composition observed pv + fp assemblage at 106 GPa and ppv + fp at 118 GPa and ~2400 K based on the *Tsuchiya* [2003]'s Au scale, which corresponds to 112 and 124 GPa by the MgO scale. It is consistent with the transition pressure in pure $MgSiO_3$.

More recent experiments by *Grocholski et al.* [2012] demonstrated that the ppv transition occurs in $(Mg_{0.89}Fe_{0.11})_2SiO_4$ olivine bulk composition between 131 and 134 GPa at 2400 K by the *Tsuchiya et al.*, Au scale, which are recalculated to be 137 and 140 GPa by the MgO scale. They again reported higher pressure range for the phase transition than that by *Hirose et al.* [2006], but the narrow two-phase region is reasonable considering the low iron concentrations in pv and ppv when coexisting with fp.

17.2.5. Ppv in MORB Material

The ppv phase transition has been observed in MORB materials as well (Figures 17.3b and c). Combining with *Hirose et al.* [2005], *Ohta et al.* [2008] reported that the transition occurred in a pressure range between 112 and 117 GPa at 2500 K using the Au scale of *Hirose et al.* [2008a], which is equivalent to the MgO scale. Similar experiments by *Grocholski et al.* [2012] also demonstrated the ppv transition in MORB between 108 and 122 GPa at 2500 K based on the Au scale of *Tsuchiya et al.*, which corresponds to 114 and 128 GPa by the MgO scale. These two studies are broadly consistent with each other, unlike the cases for pyrolite and olivine compositions (see above).

Such transition pressure in MORB is lower by 4 GPa [*Ohta et al.*, 2008] or approximately by 30 GPa [*Grocholski et al.*, 2012] than that in a pyrolitic mantle. The releatively large stability of ppv in the MORB materials may be due to the effect of large concentrations of sodium [*Hirose et al.*, 2005] and iron, although the effect of iron has been very controversial. The stabilization of ppv by iron was suggested by a series of theoretical predictions [e.g., *Caracas and Cohen*, 2005; *Stackhouse et al.*, 2006] as well as experiments by *Mao et al.* [2004] and *Dorfman et al.* [2013], but the opposite effects were observed by *Murakami et al.* [2005], *Tateno et al.* [2007], *Hirose et al.* [2008b], and *Andrault et al.* [2010]. However, considering that $MgSiO_3$-rich pv formed in MORB materials is

strongly enriched in aluminum [*Hirose et al.*, 1999, 2005; *Ricolleau et al.*, 2010] that is known to increase the ppv transition pressure [*Tateno et al.*, 2005; *Akber-Knutson et al.*, 2005], the reduction in ppv transition pressure by 30 GPa seems too large [*Grocholski et al.*, 2012].

17.3. H-PHASE?

Very recently, *Zhang et al.* [2014] reported on the basis of synchrotron XRD measurements that $(Mg,Fe)SiO_3$ pv disproportionates into nearly pure $MgSiO_3$ pv and Fe-rich phase with a hexagonal structure above 95 GPa and 2200 K, corresponding to ~2200 km depth. Similar disproportionation was observed in Al-bearing $(Mg,Fe)SiO_3$ pv as well. The latter new phase was called H phase (it is different from phase H, which is the recently discovered dense hydrous magnesium silicate [*Nishi et al.*, 2014]).

Such disproportionation and H phase have not been observed by a great number of earlier experimental works conducted at similar *P-T* ranges. For example, *Mao et al.* [2004] observed the coexistence of pv and ppv in $(Mg,Fe)SiO_3$ with 20–40 mol % $FeSiO_3$ (Fs# 20–40) at 100–108 GPa and 2000 K. The experiments performed by *Tateno et al.* [2007] showed that $(Mg_{0.5}Fe_{0.5})SiO_3$ pv is stable as a single phase to 108 GPa and 2330 K and the pv-to-ppv transition occurs at higher pressures. Previous works on $(Mg_{0.91}Fe_{0.09})SiO_3$ and $(Mg_{0.9}Fe_{0.1})(Al_{0.1}Si_{0.9})O_3$ by *Catalli et al.* [2009] and on (Fe,Al)-bearing $MgSiO_3$ by *Andrault et al.* [2010] also did not show evidence of the dissociation of pv. The H phase has never been observed in natural mantle materials under deep lower mantle conditions to >3000 K [*Ohta et al.*, 2008; *Fiquet et al.*, 2010; *Grocholski et al.*, 2012]. The recent work on $(Mg,Fe)SiO_3$ by *Dorfman et al.* [2013] also explored the relevant *P-T* range but did not report any signs related to the H phase.

Zhang et al. [2014] emphasized the importance of very stable and uniform heating for long duration above 2200 K in order to observe the H phase. However, as described above, it is very likely that long laser heating at such high temperatures causes Soret diffusion, which produces Fe-poor and Fe-rich portions within a sample. It is possible that Soret diffusion might be involved in the formation of the H phase. It is also possible that the H phase is a quench crystal formed upon rapid temperature quenching of a partial melt, which is not thermodynamically stable.

17.4. MELTING UNDER DEEP LOWER MANTLE PRESSURES

17.4.1. Melting of Pyrolitic Mantle

Melting temperature, supersolidus-phase relations, and partial-melt compositions of pyrolitic [*Fiquet et al.*, 2010; *Nomura et al.*, 2014; *Tateno et al.*, 2014] and chondritic [*Andrault et al.*, 2012] lower mantle materials have been

88GPa

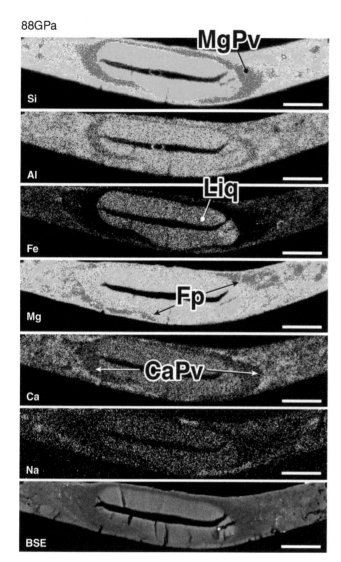

Figure 17.4 Backscattered electron image and X-ray maps for quenched partially molten pyrolite sample at 88GPa [*Tateno et al.*, 2014]. Phase segregation occurred due to thermal gradient with partial melt at the center. Such texture suggests that MgSiO₃-rich pv (MgPv) is a liquidus phase, followed by (Mg,Fe)O fp and CaSiO₃ pv (CaPv). Scale bar, 10 μm. Unlike Soret diffusion (Figure 17.2), iron is enriched in a quenched partial-melt pool formed at the center.

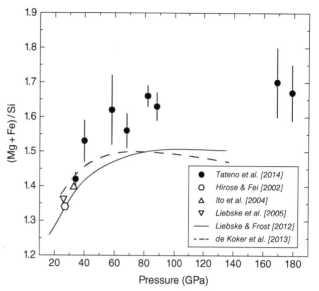

Figure 17.5 Variations in the (Mg+Fe)/Si molar ratio in melt formed by partial melting of pyrolite as a function of pressure [from *Tateno et al.*, 2014]. Solid and dashed curves represent change in the Mg/Si ratio of eutectic melt composition in the MgO-MgSiO₃ binary system obtained by thermodynamic [*Liebske and Frost*, 2012] and ab initio [*de Koker et al.*, 2013] calculations, respectively.

the lower mantle remains highly controversial [e.g., *Dixon et al.*, 2002; *Bolfan-Casanova et al.*, 2003; *Marty*, 2012].

X-ray maps of a sample section obtained by melting experiments are given in Figure 17.4. Unlike Soret diffusion, iron enrichment occurs at the hottest part of the laser-heated sample, which represents partial melt. These melting experiments consistently showed that Mg-rich pv is a liquidus phase in pyrolite (the first crystallizing phase from liquid pyrolite), followed by CaSiO₃ pv and then fp (Figure 17.4) [*Fiquet et al.*, 2010; *Tateno et al.*, 2014], suggesting that incipient partial melt is enriched in the (Mg,Fe)O component. Tateno and others demonstrated that melt formed by ~50% partial melting (in other words 50% solidification) of pyrolite exhibits (Mg+Fe)/Si = 1.65 (molar ratio) at the CMB (Figure 17.5), remarkably higher than (Mg+Fe)/Si = 1.47 of pyrolite (KLB-1 peridotite). Since pv is the liquidus phase, melts formed by lower degrees of partial melting (higher degrees of solidification) should have (Mg+Fe)/Si ratio even higher than 1.65. The experiments by *Nomura et al.* [2011] reported the (Mg+Fe)/Si ratio of melt formed from $(Mg_{0.89}Fe_{0.11})_2SiO_4$ to be 2.2 at the CMB.

The eutectic melt composition in the SiO₂-MgO binary system has been determined experimentally up to 25 GPa and extrapolated thermodynamically to CMB pressure [*Liebske and Frost*, 2012], showing that the Mg/Si molar ratio of the eutectic melt increases to 1.5 with increasing pressure to 80 GPa and remains constant at higher pressures. The first-principles calculations by

examined by recent DAC studies. Both *Fiquet et al.* [2010] and *Andrault et al.* [2011] based their work on in situ XRD measurements and reported a solidus temperature to be about 4200 ± 150 K at the CMB pressure. On the other hand, the more recent study by *Nomura et al.* [2014], using ex situ textural and chemical characterization of recovered samples, demonstrated that the solidus temperature of pyrolite is as low as 3570 ± 200 K at the CMB, substantially lower than earlier results. Their sample included ~400 ppm H₂O, which partly explains such lower melting temperature. The typical water concentration in

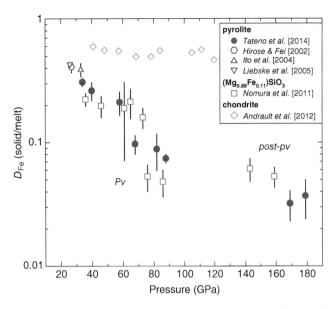

Figure 17.6 Iron partition coefficient between pv and partial melt, $D_{Fe} = [Fe^{pv}]/[Fe^{melt}]$, obtained in a pyrolitic mantle (red) [*Tateno et al.*, 2014] and olivine bulk compositions (blue) [*Nomura et al.*, 2011] based on ex situ microprobe analyses. The data reported by *Andrault et al.* [2012] in a chondritic mantle composition using in situ XRF analyses (green) showed much less iron enrichment in partial melt.

de Koker et al. [2013] also predicted a very similar change in the Mg/Si ratio of the eutectic melt. The higher (Mg+Fe)/Si ratio observed in partial melts from pyrolite and olivine composition [*Nomura et al.*, 2011; *Tateno et al.*, 2014] is likely due to the effect of iron (Figure 17.5).

Fiquet et al. [2010] reported the liquidus temperature of pyrolite to be 5300 ± 150 K at the CMB pressure, which is higher by >1000 K than the solidus temperature. The experiments on a chondritic mantle by *Andrault et al.* [2011] provided the liquidus temperature of 4725 ± 150 K, about 600 K higher than the solidus temperature. The latter study employed thermal insulation layers and thus the axial temperature gradient was smaller. In any case, however, it is rather difficult to determine the liquidus temperature by laser-heated DAC experiments, because the solid phase could remain at a low-temperature area whose temperature is hard to estimate. On the other hand, both the thermodynamic modeling by *Liebske and Frost* [2012] and the first-principles calculations by *de Koker et al.* [2013] have demonstrated that the eutectic composition in the MgO-SiO$_2$ binary system is similar to that of pyrolite at the CMB pressure. Liebske and Frost therefore argued that the difference between the liquidus and solidus temperatures of a deep lower mantle is narrow, only about 300 K.

17.4.2. Melting of Subducted Crustal Materials

Solidus temperatures of subducted former crustal materials are likely to be less than that of pyrolite in a deep lower mantle. *Hirose et al.* [1999] originally argued that the melting temperature of MORB is lower by a few hundred kelvins, suggesting that the ultralow-velocity zone (ULVZ) is caused by partial melting of MORB materials. More recent experiments performed by *Andrault et al.* [2014] confirmed the low solidus temperature of MORB (3800 K at the CMB) compared to that of the chondritic mantle (4150 K). They also reported that the partial-melt composition becomes progressively more SiO$_2$ rich with increasing pressure. One of the grounds for the low melting temperature of MORB is higher concentrations of water. The subducted MORB crust includes a substantial amount (15–25 wt %) of Al-bearing SiO$_2$ phase [*Hirose et al.*, 2005; *Ricolleau et al.*, 2010], which may possess 1000–3000 ppm H$_2$O [*Lakshtanov et al.*, 2007]. Indeed, the MORB sample used by *Andrault et al.* [2014] included 2700 ppm H$_2$O.

The chemical composition of melt formed by partial melting of MORB is closely related to its melting-phase relations. The multianvil experiments by *Hirose and Fei* [2002] reported that partial melts at 27.5 GPa have much lower SiO$_2$ and CaO and higher FeO and MgO contents than their source MORB, reflecting that Ca-rich pv and SiO$_2$ stishovite are the first and second liquidus phases (disappear first and second with increasing temperature), respectively. On the other hand, the DAC study by *Andrault et al.* [2014] performed at deep lower mantle pressures demonstrated that the SiO$_2$ phase disappears at temperatures lower than those for Mg-rich pv and Ca-rich pv, suggesting that the partial-melt composition becomes SiO$_2$ rich with increasing pressure.

In addition to normal oceanic plates, banded-iron formations (BIFs), which consist of thin bands of SiO$_2$ and Fe$_3$O$_4$/Fe$_2$O$_3$, may have subducted into a deep mantle in the past, at 2.8–1.8 Ga [*Klein*, 2005; *Dobson and Brodholt*, 2005]. Such Fe$_3$O$_4$/Fe$_2$O$_3$ could be reduced into FeO in the mantle. The melting curve of FeO has been determined to 60 GPa, and its extrapolation shows that FeO melts at 3690 K at the CMB [*Fischer and Campbell*, 2010]. The eutectic temperature in the SiO$_2$-FeO binary system must be lower than that of the FeO end member, suggesting that the subducted BIFs may have undergone melting near the base of the mantle. Indeed, since the SiO$_2$ + FeO mixture is much denser than the surrounding mantle, it is very likely that the BIFs subducted to the very bottom of the mantle. The recent experiments by *Kato et al.* [2013] demonstrated that eutectic melts in the SiO$_2$-FeO binary system are strongly enriched in FeO.

17.5. GEOPHYSICAL IMPLICATIONS

17.5.1. Origin of D″ Discontinuity

The topography of the D″ layer is consistent with a solid-solid phase transition with a large positive Clapeyron slope, as originally argued by *Sidorin et al.* [1999] before the discovery of ppv. The pv-ppv transition is a unique major mantle phase change that occurs in the relevant pressure range (SiO$_2$ also undergoes a phase transition in the lowermost mantle but is known to cause negative shear wave velocity change, which is not consistent with observations). While compositional effects on the pressure range and the sharpness of the ppv transition remain controversial [e.g., *Hirose et al.*, 2006; *Andrault et al.*, 2010; *Grocholski et al.*, 2012] (see below), the ppv transition seems the best way to explain the origin of D″ seismic discontinuity [*Hirose et al.*, 2015], no matter what materials constitute the lowermost part of the mantle. Indeed, the ppv transition (or the occurrence of ppv) may account not only for the velocity increase at the top of the D″ region but also for the positive correlation between the anomalies of shear and compressional waves ($\delta \ln V_s/d \ln V_p$) and the negative correlation between those of shear and bulk sound velocities ($d \ln V_s/d \ln V\phi$) [*Wentzcovitch et al.*, 2006].

As described in Section 17.2, the experiments by the Tokyo Tech group demonstrated that the ppv transition in a variety of mantle-related compositions (MgSiO$_3$, iron-bearing olivine, and pyrolite) occurs at similar pressure ranges [e.g., *Hirose et al.*, 2006; *Ohta et al.*, 2008; *Tateno et al.*, 2009], suggesting that the compositional effect on the ppv transition is not large (Figure 17.3b). Their results are consistent with those of *Fiquet et al.* [2010] performed at higher temperatures and CMB pressures. On the other hand, the results by the MIT group show (1) strong compositional dependence and (2) higher transition pressure (Figure 17.3c) [*Catalli et al.*, 2009; *Grocholski et al.*, 2012]. The origin of such discrepancy is not clear, but a number of small differences, such as the condition of starting material, stress state, and Soret effect, may have been involved.

The sharpness of the phase transition boundary is key to be observed by seismology. A velocity increase over a transition width of 90 km or larger may be detected, but existing data suggest much narrower features [*Lay*, 2008]. Both *Catalli et al.* [2009] and *Andrault et al.* [2010] observed that the ppv transition occurs in a wide pressure range about 30 GPa in (Fe,Al)-bearing SiO$_3$ bulk compositions (Figure 17.3c), although the sharpness of the boundary is usually overestimated by laser-heated DAC experiments due to temperature variations the in X-ray-probed area, in particular when the sample was scanned

by laser [*Andrault et al.*, 2010] because each specific part of the sample experienced both high and low temperatures during scanning. Nevertheless, *Sinmyo et al.* [2011] argued that coexistence with fp in pyrolite sharpens the ppv transition boundary (see Figure 7 in their paper). The more recent work by *Grocholski et al.* [2012], however, again demonstrated that pv and ppv coexist over 30 GPa in pyrolite.

The experiments by *Ohta et al.* [2008], with the particular aim of constraining the sharpness of the ppv boundary in pyrolite, showed that it occurs over 5 GPa (90 km) in a pyrolitic mantle. While it could be overestimated, it is broader than the upper bound for the sharpness of the D″ discontinuity (30 km, 2 GPa by *Lay* [2008]). It is possible that the proportion between pv and ppv may change not linearly with increasing pressure (depth) but rather abruptly in a small pressure range, as discussed previously on the sharpness of the 410 km discontinuity [*Stixrude*, 1997]. Alternatively, the D″ discontinuity might be caused by the sharper ppv transition in depleted peridotitic (harzburgitic) material [*Grocholski et al.*, 2012] (Figure 17.3b) (see Section 17.5.2).

Ammann et al. [2010] argued that S-wave velocity increase and the small reduction in P-wave velocity observed at the top of D″ could be produced by the onset of strong deformation. If this is the case, the ppv transition must start at shallower depths. Some experiments as well as theoretical calculations supported the stability of Fe-bearing ppv at low pressures [e.g., *Mao et al.*, 2004; *Zhang and Oganov*, 2006; *Dorfman et al.*, 2013], but the effect of iron on the stability of ppv still remains controversial [*Hirose et al.*, 2008b; *Andrault et al.*, 2010]. As introduced above, ppv occurs in MORB materials at relatively shallow depths [*Ohta et al.*, 2008; *Grocholski et al.*, 2012]. The phase proportion of ppv is, however, limited to less than 40%, and thus it may be difficult to cause strong seismic anisotropy in MORB materials. The slip system and weakness of other constituent phases in MORB remain to be examined.

17.5.2. Global Occurrence of ppv?

The D″ discontinuity is observed primarily in regions with fast-velocity anomalies. The lack of seismic discontinuity, however, does not necessarily mean the absence of the ppv phase transition. Recent seismological study by *Cobden et al.* [2012] argued on the basis of >10,000 $d \ln V_S$ and $d \ln V_P$ data near the CMB that ppv is present globally, whether or not it coexists with pv, even underneath the Central Pacific.

The D″ seismic discontinuity is possibly caused by the sharp ppv transition in depleted peridotitic (harzburgitic) materials, which may be piled in the fast-velocity regions of the lowermost mantle. Outside such an area, namely the transitional region between the fast- and slow-velocity anomaly regions, might be dominated by pyrolite, in which the ppv transition may be too broad to be observed as a reflector. Or, as was simulated by *Nakagawa and Tackley* [2005], such a transitional region could be a mixture of MORB and harzburgitic materials but enriched in the former more than the fast-velocity region, because subducted MORB is denser by a few percent than the average lower mantle and thus accumulates at the bottom of the mantle [*Hirose et al.*, 2005; *Ricolleau et al.*, 2010]. If this is the case, since the pressure ranges of the ppv transition in these two materials are different (Figure 17.3), the boundary may be intermittent and therefore hard to be observed as reflectors. The slow-velocity region, called large low-shear-velocity province (LLSVP), is possibly composed of a variety of dense materials in addition to MORB [*Deschamps et al.*, 2012] (see below).

The occurrence of ppv is strongly dependent on temperature at the mantle side of the CMB. Recent experimental work by *Nomura et al.* [2014] on the basis of textural and chemical characterization of DAC samples demonstrated that the solidus temperature of a pyrolitic mantle is as low as 3570 ± 200 K at the CMB. It may provide the upper bound for the CMB temperature because the mantle side of the CMB is not globally molten (ultralow-velocity zone is only local). *Fiquet et al.* [2010] reported that the transition from ppv to pv occurs in pyrolite over a small temperature range above 3400 K at 138 GPa (Figure 17.3a). These suggest that the stability of pv right above the CMB is limited, supporting the wide occurrence of ppv.

17.5.3. Origin of ULVZ

The ULVZ most likely indicates the presence of partial melt in a thermal boundary layer above the CMB [*Williams and Garnero*, 1996]. Such ULVZ is not globally observed, suggesting that melting occurs only in materials with low melting temperatures, because temperature at the CMB must be isothermal. Indeed, partial melting of an ambient lower mantle does not produce a density of "mush" comprising melt and residual solid comparable to the ULVZ density induced by seismology [*Thomas and Asimow*, 2013]. Alternatively, it has been repeatedly proposed that the melting temperature of subducted MORB crust is lower by a few hundred kelvins than that of surrounding mantle and its partial melting near the CMB is responsible for the ULVZ [*Hirose et al.*, 1999; *Andrault et al.*, 2014]. Nevertheless, it is not clear whether the MORB crust reaches the bottom of the mantle, and many alternative scenarios have been proposed.

As described in Section 17.4.2, subducted BIFs, consisting of thin SiO_2 and FeO bands, have low melting (eutectic) temperature, likely less than 3500 K at the CMB. Because of their much larger density than that of surrounding mantle, they most likely reached the CMB and underwent partial melting there. In addition, the partial melt formed from such BIFs would remain molten at present, unless it dissolved into molten core.

It has also been proposed that melts present near the CMB are the remnants of a basal magma ocean (BMO) [*Labrosse et al.*, 2007; *Nomura et al.*, 2011]. The chemical composition of such residual magma after extensive crystallization may be inferred from the eutectic melt composition in the SiO_2-MgO binary system (Mg/Si = 1.5 at 136 GPa) [*Liebske and Frost*, 2012; *de Koker et al.*, 2013]. On the other hand, recent DAC experiments on pyrolite reported that the chemical compositions of melts formed by ~50% solidification have (Mg+Fe)/Si = 1.65 in the deep lower mantle [*Tateno et al.*, 2014]. Since pv is the first phase to crystallize from such partial melts, the residual magma would evolve into even more (Mg,Fe)O-rich and SiO_2-poor composition with further solidification. It is known that melt is denser with increasing Mg/Si ratio at a given Fe/Mg ratio [*Funamori et al.*, 2010; *Thomas et al.*, 2012; *de Koker et al.*, 2013]. In addition to the Mg/Si ratio, iron partitioning is key for melt to become denser than solid, but it still remains controversial (Figure 17.6). The high-pressure X-ray fluorescence study performed by *Andrault et al.* [2012] on a chondritic mantle demonstrated the iron partition coefficient between pv and melt, D_{Fe} (pv/melt) = 0.5 under the lowermost mantle pressure, suggesting that partial melt is buoyant and thus could not cause the ULVZ. In contrast, more recent work by *Tateno et al.* [2014] on the basis of ex situ microprobe analysis of a section of DAC sample showed much stronger iron enrichment in partial melts formed in a pyrolitic mantle; D_{Fe} (pv/melt) = 0.08 at ~100 GPa. The residual magma should therefore eventually become denser than solids during solidification of the BMO [*Funamori and Sato*, 2010; *Thomas et al.*, 2012].

Alternatively, sound velocity measurements at high pressure suggested that the ULVZ is attributed not to melt but to iron-rich ppv [*Mao et al.*, 2006] or (Mg,Fe)O magnesiowüstite [*Wicks et al.*, 2010]. The formations of these iron-rich phases, in particular iron-rich magnesiowüstite, may be possible as a consequence of fractional crystallization in a hypothetical BMO [*Labrosse et al.*, 2007; *Nomura et al.*, 2011]. However, iron-rich magnesiowüstite has a low melting temperature depending on the Fe/Mg ratio; the Fe end-member FeO would melt at 3690 K at the CMB [*Fischer and Campbell*, 2010]. Such magnesiowüstite would not occur as a single phase but most likely coexists with pv or ppv, which further reduces solidus temperature. Similarly, *Dobson and Brodholt* [2005] proposed that subducted BIFs by itself are responsible for the ULVZ, but the eutectic temperature in the SiO_2-FeO binary system is also likely to be low [*Kato et al.*, 2013]. These suggest that the presence of partial melt is more likely a mechanism to cause the ULVZ.

On the other hand, the iron-rich silicate and oxide are unlikely to form in the mantle side of the CMB as a consequence of the core-mantle chemical reaction [*Takafuji et al.*, 2005]. The core-mantle chemical reaction experiments by *Ozawa et al.* [2009] showed that the (Si,O)-bearing outer core coexists with almost iron-free $(Mg_{0.99}Fe_{0.01})SiO_3$ pv at the CMB. *Frost et al.* [2010] argued on the basis of the partitioning of oxygen between fp and molten iron that $(Mg_{0.86}Fe_{0.14})O$, typical lower mantle fp, is in equilibrium with the core containing 11 wt % O at 136 GPa and 3500 K, while the core density deficit is reconciled only with ~8 wt % O [*Li and Fei*, 2007; *Sata et al.*, 2010]. These studies suggest that (1) the iron content in pv and fp in equilibrium with the outer core depends on the oxygen concentration in the core and (2) the mantle side of the CMB is likely depleted in iron unless the top of the core is strongly enriched in oxygen. Seismology suggests that the top few hundred kilometers of the core exhibits slow velocity and is therefore chemically distinct from the bulk outer core [*Helffrich and Kaneshima*, 2010]. Nevertheless, simple addition of oxygen increases sound velocity [*Badro et al.*, 2014], inconsistent with the observations.

17.6. FUTURE PERSPECTIVE

Post-perouskite was discovered in 2004, and since then, with knowledge of the detailed properties of ppv in the last 10 years, the lowermost mantle has been studied extensively not only by mineral physics but also by seismology and geodynamics. Both experimental and theoretical studies of ppv have examined not only its stability but also its elasticity, electrical and thermal conductivity, viscosity, and element partitioning [see a recent review by *Hirose et al.*, 2015]. These works have shed light on the D'' layer that has long been the most enigmatic region inside Earth.

On the other hand, one can naturally expect rich "geology" in the lowermost mantle, simply because it is a convection boundary layer. We often speculate on materials that possibly exist near the base of the mantle but are still far away from identifying them with confidence. In order to interpret seismological data and reveal geological structures there, more knowledge of the compositional effects on the pv-ppv transition and sound velocity of ppv, including its anisotropy, is highly demanded.

Experimental studies of ppv have been carried out mainly with laser-heated DAC techniques at the lowermost mantle *P-T* conditions. Indeed, the laser-heated DAC has developed very rapidly in the last decade or so

in combination with synchrotron X-ray measurements. While only few laser-heated DAC experiments had been made at the lowermost mantle conditions (>120 GPa and >2000 K) before 2004, a number of groups are now working on synchrotron XRD measurements under such *P-T* conditions. In addition, combined with in situ [e.g., *Andrault et al.*, 2012, 2014] and ex situ chemical characterizations [e.g., *Fiquet et al.*, 2010; *Nomura et al.*, 2011; *Tateno et al.*, 2014], recent melting experiments have reported not only melting temperature but also melting-phase relations and melt/solid element partitioning under deep lower mantle conditions.

As introduced in this chapter, laser-heated DAC experiments include uncertainty, mainly because laser heating causes much larger temporal and spatial temperature fluctuations than resistance heating. The effect of Soret diffusion is possibly a source of discrepancy among existing DAC experimental data on iron-bearing systems (see Figure 17.3, for example), but other issues such as hydrostaticity and reaction with a pressure medium might also be involved. It is therefore key to develop new techniques other than laser-heated DAC. Both multianvil press and resistance-heated DAC provide much better high-temperature environments for samples. Recent multianvil experiments were conducted above 100 GPa and high temperature [*Yamazaki et al.*, 2014]. Temperature generation by resistive externally heated DAC has been traditionally limited to ~1000 K but recently extended to ~1500 K [*Du et al.*, 2013; *Umemoto et al.*, 2014]. Such efforts are precious and help significantly in elucidating the nature of complexities in the D'' layer with robust experimental results.

ACKNOWLEDGMENTS

Discussions with C. Thomas, S. Tateno, J. Hernlund, and C. Houser were helpful. The review comments by Denis Andrault and an anonymous referee helped improve the manuscript.

REFERENCES

Akber-Knutson, S., G. Steinle-Neumann, and P. D. Asimow (2005), Effect of Al on the sharpness of the $MgSiO_3$ Pv to Ppv phase transition, *Geophys. Res. Lett.*, 32, L14,303, doi:10.1029/2005GL023192.

Ammann, M. W., J. P. Brodholt, J. Wookey, and D. P. Dobson (2010), First-principles constraints on diffusion in lower-mantle minerals and a weak D'' layer, *Nature*, 465, 462–465, doi:10.1038/nature09052.

Andrault, D., M. Muñoz, N. Bolfan-Casanova, N. Guignot, J-P. Perrillat, G. Aquilanti, and S. Pascarelli (2010), Experimental evidence for Pv and Ppv coexistence throughout the whole D'' region, *Earth Planet. Sci. Lett.*, 293, 90–96.

Andrault, D., N. Bolfan-Casanova, G. L. Nigro, M. A. Bouhifd, G. Garbarino, and M. Mezouar (2011), Solidus and liquidus profiles of chondritic mantle: Implication for melting of the Earth across its history, *Earth Planet. Sci. Lett.*, *304*, 251–59.

Andrault, D., S. Petitgirard, G. L. Nigro, J-L Devidal, G. Veronesi, G. Garbarino, and M. Mezouar (2012), Solid–liquid iron partitioning in Earth's deep mantle, *Nature*, *487*, 354–357, doi:10.1038/nature11294.

Andrault, D., G. Pesce, M. A. Bouhifd, N. Bolfan-Casanova, J.M. Hénot, and M. Mezouar (2014), Melting of subducted basalt at the core-mantle boundary, *Science*, *23*, 892–895, doi:10.1126/science.1250466.

Badro, J., A. S. Coté, and J. P. Brodholt (2014), A seismologically consistent compositional model of Earth's core, *Proc. Natl. Acad. Sci.*, *111*, 7542–7545.

Bolfan-Casanova, N., H. Keppler, and D. C. Rubie (2003), Water partitioning at 660 km depth and evidence for very low water solubility in magnesium silicate perovskite, *Geophys. Res. Lett.*, *30*, 1905, doi:10.1029/2003GL017182.

Caracas, R., and R. E. Cohen (2005), Effect of chemistry on the stability and elasticity of the Pv and Ppv phases in the $MgSiO_3$–$FeSiO_3$–Al_2O_3 system and implications for the lowermost mantle, *Geophys. Res. Lett.*, *32*, L16,310, doi:10.1029/2005GL023164.

Catalli, K., S. H. Shim, and V. Prakapenka (2009), Thickness and Clapeyron slope of the Ppv boundary, *Nature*, *462*, 782–785.

Cobden, L., I. Mosca, J. Trampert, and J. Ritsema (2012), On the likelihood of post-perovskite near the core–mantle boundary: A statistical interpretation of seismic observations, *Phys. Earth Planet. Int.*, *210–211*, 21–35.

Cobden, L., and C. Thomas (2013), The origin of D″ reflections: a systematic study of seismic array data sets, *Geophys. J. Int.*, *194*, 1091–1118.

de Koker, N., B. B. Karki, and L. Stixrude (2013), Thermodynamics of the MgO–SiO_2 liquid system in Earth's lowermost mantle from first principles, *Earth Planet. Sci. Lett.*, *361*, 58–63.

Deschamps, F., L. Cobden, and P. J. Tackley (2012), The primitive nature of large low shear-wave velocity provinces, *Earth Planet. Sci. Lett.*, *349–350*, 198–208.

Dixon, J. E., L. Leist, C. Langmuir, and J-G. Schilling (2002), Recycled dehydrated lithosphere observed in plume-influenced mid-ocean-ridge basalt, *Nature*, *420*, 385–389.

Dobson, D. P. and J. P. Brodholt (2005), Subducted banded iron formations as a source of ultralow-velocity zones at the core–mantle boundary, *Nature*, *434*, 371–374.

Dorfman, S. M., Y. Meng, V. B. Prakapenka, and T. S. Duffy (2013), Effects of Fe-enrichment on the equation of state and stability of (Mg,Fe)SiO_3 perovskite, *Earth Planet. Sci. Lett.*, *361*, 249–257.

Fei, Y., J. Li, K. Hirose, W. Minarik, J. V. Ormana, C. Sanloup, W. van Westrenen, T. Komabayashi, and K. Funakoshi (2004), A critical evaluation of pressure scales at high temperatures by in situ X-ray diffraction measurements, *Phys. Earth Planet. Inter.*, *143–144*, 515–526.

Fei, Y., A. Ricolleau, M. Frank, K. Mibe, G. Shen, and V. Prakapenka (2007), Toward an internally consistent pressure scale, *Proc. Natl. Acad. Sci. USA*, *104*, 9182–9186.

Fiquet, G., A. L. Auzende, J. Siebert, A. Corgne, H. Bureau, H. Ozawa, and G. Garbarino (2010), Melting of peridotite to 140 gigapascals, *Science*, *329*, 1516–18.

Fischer, R. A. and A. J. Campbell (2010), High pressure melting of wüstite, *Am. Mineral.*, *95*, 1473–1477.

Frost, D. J., Y. Asahara, D.C. Rubie, N. Miyajima, L.S. Dubrovinsky, C. Holzapfel, E. Ohtani, M. Miyahara, and T. Sakai (2010), Partitioning of oxygen between the Earth's mantle and core, *J. Geophys. Res.*, *115*, B02,202, doi:10.1029/2009JB006302.

Funamori, N., and T. Sato (2010), Density contrast between silicate melts and crystals in the deep mantle: An integrated view based on static-compression data, *Earth Planet. Sci. Lett.*, *295*, 435–440, doi:10.1016/j.epsl.2010.04.021.

Grocholski, B., K. Catalli, S-H. Shim, and V. Prakapenka (2012), Mineralogical effects on the detectability of the Ppv boundary, *Proc. Natl. Acad. Sci. USA*, *109*, 2275–2279.

Helffrich, G., and S. Kaneshima (2010), Outer-core compositional stratification from observed core wave speed profiles, *Nature*, *468*, 807–810.

Hirose, K., and Y. Fei (2002), Subsolidus and melting phase relations of basaltic composition in the uppermost lower mantle, *Geochim. Cosmochim. Acta*, *66*, 2099–2108.

Hirose, K., Y. Fei, Y. Ma, and H. K. Mao (1999), The fate of subducted basaltic crust in the Earth's lower mantle, *Nature*, *397*, 53–56.

Hirose, K., N. Takafuji, N. Sata, and Y. Ohishi (2005), Phase transition and density of subducted MORB crust in the lower mantle, *Earth Planet. Sci. Lett.*, *237*, 239–251.

Hirose, K., R. Sinmyo, N. Sata, and Y. Ohishi (2004), Determination of Ppv phase transition boundary in $MgSiO_3$ using Au and MgO internal pressure standards, *Geophys. Res. Lett.*, *33*, L01,310, doi:10.1029/2005GL024468.

Hirose K., N. Sata, T. Komabayashi, and Y. Ohishi (2008a), Simultaneous volume measurements of Au and MgO to 140 GPa and thermal equation of state of Au based on the MgO pressure scale, *Phys. Earth Planet. Inter.*, *167*, 149–154.

Hirose K., N. Takafuji, K. Fujino, S. R. Shieh, and T. S. Duffy (2008b), Iron partitioning between Pv and Ppv: A transmission electron microscope study, *Am. Mineral. 93*, 1678–1681, doi:10.2138/am.2008.3001.

Hirose, K., R. Wentzcovitch, D. A. Yuen, and T. Lay (2015), Mineralogy of the Deep Mantle – The Post-Perovskite Phase and its Geophysical Significance. in *Gerald Schubert (editor-in-chief) Treatise on Geophysics, 2nd edition, 2*. pp. 85–115, Oxford, Elsevier.

Holmes, N. C., J. A. Moriarty, G. R. Gathers, and W. J. Nellisn (1989), The equation of state of platinum to 660 GPa (6.6 Mbar). *J. Appl. Phys.*, *66*, 2962–2967.

Holzapfel, C., D. C. Rubie, D. J. Frost, and F. Langenhorst (2005), Fe-Mg Interdiffusion in (Mg,Fe)SiO3 perovskite and lower mantle reequilibration, *Science*, *309*, 1707–1710.

Hutko, A., T. Lay, E. J. Garnero, and J. S. Revenaugh (2006), Seismic detection of folded, subducted lithosphere at the core-mantle boundary, *Nature*, *441*, 333–336.

Jamieson, J. C., J. N. Fritz, and M. H. Manghnani (1982), Pressure measurement at high temperature in x-ray diffraction studies: Gold as a primary standard, in *High-pressure Research in Geophysics*, edited by S. Akimoto and M. H. Manghnani, pp. 27–48, CAPJ, Tokyo.

Kato, C., R. Nomura, and K. Hirose (2013), Melting of FeO-SiO$_2$ system at high pressure and the fate of subducted banded iron formations, *Mineral. Mag.*, *77*(5), 1436.

Klein, C. (2005), Some Precambrian banded iron-formations (BIFs) from around the world: Their age, geologic setting, mineralogy, metamorphism, geochemistry, and origin, *Am. Mineral.*, *90*, 1473–1499.

Komabayashi, T., Y. Fei, Y. Meng, and V. Prakapenka (2009), In-situ X-ray diffraction measurements of the γ-ε transition boundary of iron in an internally-heated diamond anvil cell, *Earth Planet. Sci. Lett.*, *282*, 252–257.

Labrosse, S., J. W., Hernlund, and N. Coltice (2007), A crystallizing dense magma ocean at the base of the Earth's mantle, *Nature*, *450*, 866–869.

Lakshtanov, D. L., K. D. Litasov, S. V. Sinogeikin, H. Hellwig, J. Li, E. Ohtani, and J. D. Bass (2007), Effect of Al^{3+} and H$^+$ on the elastic properties of stishovite, *Am. Mineral.*, *92*, 1026–1030.

Lay, T. (2008), Sharpness of the D″ discontinuity beneath the Cocos Plate: Implications for the Pv to Ppv phase transition, *Geophys. Res. Lett.*, *35*, L03304, doi:10.1029/2007GL032465.

Lay, T., and E. J., Garnero (2004), Core-mantle boundary structures and processes, in *The State of the Planet: Frontiers and Challenges in Geophysics*, edited by R. S. J. Sparks and C. J. Hawkesworth, pp. 25–42, AGU, Washington, D.C.

Lesher, C. E., and D. Walker (1991), Thermal diffusion in petrology, in *Diffusion, Atomic Ordering and Mass Transport*, edited by J. Ganguly, S. Saxena, and S. K. Saxena, pp. 396–451, Springer Verlag, New York.

Li, J., and Y. Fei (2007), Experimental constraints on core composition, in *Treatise on Geochemistry*, vol. 2, *The Mantle and Core*, edited by R. W. Carlson, pp. 1–31, Elsevier, Amsterdam.

Liebske, C., and D. J. Frost (2012), Melting phase relations in the MgO–MgSiO$_3$ system between 16 and 26 GPa: Implications for melting in Earth's deep interior, *Earth Planet. Sci. Lett.*, *345–348*, 159–170.

Mao, W. L., G. Shen, V. B. Prakapenka, Y. Meng, A. J. Cambell, D. Heinz, J. Shu, R. J. Hemley, and H. K. Mao (2004), Ferromagnesian Ppv silicates in the D″ layer of the earth, *Proc. Natl. Acad. Sci. USA*, *101*, 15,867–15,869.

Mao, W. L., H. K. Mao, W. Sturhahn, J. Zhao, V. B. Prakapenka, Y. Meng, J. Shu, Y. Fei, and R. J. Hemley (2006), Iron-rich post-perovskite and the origin of ultralow-velocity zones, *Science*, *312*, 564–565, doi:10.1126/science.1123442.

Marty B. (2012), The origins and concentrations of water, carbon, nitrogen and noble gases on Earth, *Earth Planet. Sci. Lett.*, *313–314*, 56–66.

Murakami, M., K. Hirose, K. Kawamura, N. Sata, and Y. Ohishi (2004), Post-perovskite phase transition in MgSiO$_3$, *Science*, *304*, 855–858.

Murakami, M., K. Hirose, N. Sata, and Y. Ohishi (2005), Post-perovskite phase transition and crystal chemistry in the pyrolitic lowermost mantle, *Geophys. Res. Lett.*, *32*, L03,304, doi:10.1029/2004GL021956.

Nakagawa, T., and P. J. Tackley (2005), The interaction between the Ppv phase change and a thermochemical boundary layer near the core-mantle boundary, *Earth Planet. Sci. Lett.*, *238*, 204–216.

Nishi, M., T. Irifune, J. Tsuchiya, Y. Tange, Y. Nishihara, K. Fujino, and Y. Higo (2014), Stability of hydrous silicate at high pressures and water transport to the deep lower mantle, *Nature Geosci.*, *7*, 224–227.

Nomura, R., H. Ozawa, S. Tateno, K. Hirose, J. Hernlund, S. Muto, H. Ishii, and N. Hiraoka (2011), Spin crossover and iron-rich silicate melt in the Earth's deep mantle, *Nature*, *473*, 199–202.

Nomura, R., K. Hirose, K. Uesugi, Y. Ohishi, A. Tsuchiyama, A. Miyake, and Y. Ueno (2014), Low core-mantle boundary temperature inferred from the solidus of pyrolite, *Science*, *343*, 6170, 522–525, doi:10.1126/science.1248186.

Oganov, A. R., and S. Ono (2004), Theoretical and experimental evidence for a Ppv phase of MgSiO$_3$ in Earth's D″ layer, *Nature*, *430*, 445–448.

Ohta, K., K. Hirose, T. Lay, N. Sata, and Y. Ohishi (2008), Phase transitions in pyrolite and MORB at lowermost mantle conditions: Implications for a MORB-rich pile above the core-mantle boundary, Earth Planet. *Sci. Lett.*, *267*, 107–117.

Ono, S., and A. R. Oganov (2005), In situ observations of phase transition between Pv and CaIrO$_3$-type phase in MgSiO$_3$ and pyrolitic mantle composition, *Earth Planet. Sci. Lett.*, *236*, 914–932.

Ozawa, H., K. Hirose, M. Mitome, Y. Bando, N. Sata, and Y. Ohishi (2009), Experimental study of reaction between perovskite and molten iron to 146 GPa and implications for chemically distinct buoyant layer at the top of the core, *Phys. Chem. Minerals*, *36*, 355–363, doi:10.1007/s00269-008-0283-x.

Reymer, A., and G. Schubert (1984), Phanerozoic addition rates to the continental crust and crustal growth, *Tectonics*, *3*, 63–77.

Ricolleau, A., Y. Fei, E. Cottrell, H. Watson, L. Deng, L. Zhang, G. Fiquet, A-L. Auzende, M. Roskosz, G. Morard, and V. Prakapenka (2009), Density profile of pyrolite under the lower mantle conditions, *Geophys. Res. Lett.* *36*, L06,302, doi:10.1029/2008GL036759.

Ricolleau, A., J. P. Perrillat, G. Fiquet, I. Daniel, J. Matas, A. Addad, N. Menguy, H. Cardon, M. Mezouar, and N. Guignot (2010), Phase relations and equation of state of a natural MORB: Implications for the density profile of subducted oceanic crust in the Earth's lower mantle, *J. Geophys. Res.*, *115*, B08,202, doi:10.1029/2009JB006709.

Sata, N., K. Hirose, G. Shen, Y. Nakajima, Y. Ohishi, and N. Hirao (2010), Compression of FeSi, Fe$_3$C, Fe$_{0.95}$O, and FeS under the core pressures and implication for light element in the Earth's core, *J. Geophys. Res.*, *115*, B09204.

Shim, S., T. S. Duffy, and K. Takemura (2002), Equation of state of gold and its application to the phase boundaries near 660 km depth in the Earth's mantle, *Earth Planet. Sci. Lett.*, *203*, 729–739.

Sidorin, I., M. Gurnis, and D. V. Helmberger (1999), Evidence for a ubiquitous seismic discontinuity at the base of the mantle, *Science*, *286*, 1326–1329.

Sinmyo, R., and K. Hirose (2010), The Soret diffusion in laser-heated diamond-anvil cell, *Phys. Earth Planet. Inter.*, *180*, 172.

Sinmyo, R., and K. Hirose (2013), Iron partitioning in pyrolitic lower mantle, *Phys. Chem. Miner.*, *40*, 107–113, doi:10.1007/s00269-012-0551-7.

Sinmyo, R., K. Hirose, D. Nishio-Hamane, Y. Seto, K. Fujino, N. Sata, and Y. Ohishi (2008), Partitioning of iron between perovskite/post-perovskite and ferropericlase in the lower mantle, *J. Geophys. Res.*, *113*, B11,204.

Sinmyo R., K. Hirose, S. Muto, Y. Ohishi, and A. Yasuhara (2011), The valence state and partitioning of iron in the Earth's lowermost mantle, *J. Geophys. Res.*, *116*, B07,205, doi:10.1029/2010JB008179.

Speziale, S., C. Zha, T. S. Duffy, R. J. Hemley, and H. K. Mao (2001), Quasi-hydrostatic compression of magnesium oxide to 52 GPa: Implications for the pressure-volume-temperature equation of state, *J. Geophys. Res.*, *106*, 515–528.

Stackhouse, S., J. P. Brodholt, and G. D. Price (2006), Elastic anisotropy of $FeSiO_3$ end-members of the Pv and Ppv phases, *Geophys. Res. Lett.*, *33*, L01,304, doi:10.1029/2005GL023887.

Stixrude, L. (1997), Structure and sharpness of phase transitions and mantle discontinuities, *J. Geophys. Res.*, *102*, 14,835–14,852, doi:10.1029/97JB00550.

Takafuji, N., K. Hirose, M. Mitome, and Y. Bando (2005), Solubilities of O and Si in liquid iron in equilibrium with $(Mg,Fe)SiO_3$ perovskite and the light elements in the core, *Geophys. Res. Lett.*, *32*, L06,313.

Tateno, S., K. Hirose, N. Sata, and Y. Ohishi (2005), Phase relations in $Mg_3Al_2Si_3O_{12}$ to 180 GPa: Effect of Al on Ppv phase transition, *Geophys. Res. Lett.*, *32*, L15,306, doi:10.1029/2005GL023309.

Tateno, S., K. Hirose, N. Sata, and Y. Ohishi (2007), High solubility of FeO in Pv and the Ppv phase transition, *Phys. Earth Planet. Inter.*, *160*, 319–325.

Tateno, S., K. Hirose, N. Sata, and Y. Ohishi (2009), Determination of Ppv phase transition boundary up to 4400 K and implications for thermal structure in D″ layer, *Earth Planet. Sci. Lett.*, *277*, 130–136.

Tateno, S., K. Hirose, and Y. Ohishi (2014), Melting experiments on peridotite to lowermost mantle conditions, *J. Geophys. Res. Solid Earth*, *119*, 4684–4694, doi:10.1002/2013JB010616.

Thomas, C. W. and P. D. Asimow (2013), Direct shock compression experiments on premolten forsterite and progress toward a consistent high-pressure equation of state for $CaO-MgO-Al_2O_3-SiO_2-FeO$ liquids, *J. Geophys. Res.*, *118*, 5738–5752.

Thomas, C. W., Q. Liu, C. B. Agee, P. D. Asimow, and R. A. Lange (2012), Multi-technique equation of state for Fe_2SiO_4 melt and the density of Fe-bearing silicate melts from 0 to 161 GPa, *J. Geophys. Res.*, *117*, B10,206, doi:10.1029/2012JB009403.

Tsuchiya, J., and T. Tsuchiya (2008), Postperovskite phase equilibria in the $MgSiO_3$-Al_2O_3 system, *Proc. Natl. Acad. Sci. USA*, *105*, 19,160–19,164.

Tsuchiya, T. (2003), First-principles prediction of the P-V-T equation of state of gold and the 660-km discontinuity in Earth's mantle, *J. Geophys. Res.*, *108*(B10), 2462, doi:10.1029/2003JB002446.

Tsuchiya, T., J. Tsuchiya, K. Umemoto, and R. M. Wentzcovitch (2004), Phase transition in $MgSiO_3$ Pv in the Earth's lower mantle, *Earth Planet. Sci. Lett.*, *224*, 241–248.

Umemoto, K., K. Hirose, S. Imada, Y. Nakajima, T. Komabayashi, S. Tsutsui, and A. Q. R. Baron (2014), Liquid iron-sulfur alloys at outer core conditions by first-principles calculations, *Geophys. Res. Lett.*, *41*, doi:10.1002/2014GL061233.

Wentzcovitch, R. M., T. Tsuchiya, and J. Tsuchiya (2006), $MgSiO_3$ ppv at D″ conditions, *Proc. Natl. Acad. Sci. USA*, *103*, 543–546.

Wicks, J. K., J. M. Jackson, and W. Sturhahn (2010), Very low sound velocities in iron-rich (Mg,Fe)O: Implications for the core-mantle boundary region, *Geophys. Res. Lett.*, *37*, L15,304, doi:10.1029/2010GL043689.

Williams, Q., and E. J. Garnero (1996), Seismic evidence for partial melt at the base of Earth's mantle, *Science*, *273*, 1528–1530.

Wu, Z., R. M. Wentzcovitch, K. Umemoto, B. Li, and K. Hirose (2008), PV-T relations in MgO: An ultra-high P-T scale for planetary sciences applications, *J. Geophys. Res.*, *113*, B06,204, Correction, *J. Geophys. Res.*, *115*, B05,201, 2010.

Wysession, M. E., T. Lay, J. Revenaugh, Q. Williams, E. Garnero, R. Jeanloz, and L. Kellogg (1998), The D″ discontinuity and its implications, in *The Core-Mantle Boundary Region*, edited by M. Gurnis, M. E. Wysession, E. Knittle, and B. A. Buffett, pp. 273–297, AGU, Washington, D. C.

Yamazaki, D., E. Ito, T. Yoshino, N. Tsujino, A. Yoneda, X. Guo, F. Xu, Y. Higo, and K. Funakoshi (2014), Over 1 Mbar generation in the Kawai-type multianvil apparatus and its application to compression of $(Mg_{0.92}Fe_{0.08})SiO_3$ perovskite and stishovite, *Phys. Earth Planet. inter.*, *228*, 262–267.

Zhang F. and A. R. Oganov (2006), Valence state and spin transitions of iron in Earth's mantle silicates. *Earth Planet. Sci. Lett.* *249*(3–4), 436–443, doi:10.1016/j.epsl.2006.07.023.

Zhang, L., Y. Meng, W. Yang, L. Wang, W. L. Mao, Q-S. Zeng, J. S. Jeong, A. J. Wagner, K. A. Mkhoyan, W. Liu, R. Xu, and H. K. Mao (2014), Disproportionation of $(Mg,Fe)SiO_3$ perovskite in Earth's deep lower mantle, *Science*, *344*, 877–882.

18

Chemistry of the Lower Mantle

Daniel J. Frost and Robert Myhill

ABSTRACT

Various lines of evidence have been used to argue that the lower mantle is chemically distinct in major elements from the upper mantle. One of the few approaches that may clarify this would be to compare observed seismic velocities with estimates computed from mineral physical models. This requires a method for determining the chemistry and modes of mantle minerals for a given bulk composition as a function of pressure and temperature. Aside from phase changes near the core-mantle boundary, the main change taking place throughout the bulk of an isochemical mantle will be the exchange of Fe between the principal minerals ferropericlase and bridgmanite. Here we review experimental data for this exchange and present simple models to describe this behavior. All experimental data to date seem to indicate that bridgmanite will contain subequal proportions of both Fe^{2+} and Fe^{3+} throughout much of the lower mantle. If the entire mantle is isochemical, the low Fe^{3+} contents of the upper mantle imply that metallic Fe should exist in the lower mantle. An alternative hypothesis is that the lower mantle is oxygen enriched but becomes reduced by the oxidation of species such as carbon during upwelling.

18.1. INTRODUCTION

Earth's lower mantle, extending from 660 to 2891 km depth, contains ~70% of the mass of the mantle and is ~50% of the mass of the entire planet [*Dziewonski and Anderson*, 1981]. Due to its size, the lower mantle has a large potential to bias the composition of the bulk silicate Earth (BSE), which in major element terms is generally assumed to be the same as the upper mantle [*Allegre et al.*, 1995; *McDonough and Sun*, 1995; *O'Neill and Palme*, 1998]. Any chemical differences between the upper and lower mantle would have major implications for our understanding of the scale of mantle convection, the origin of the terrestrial heat flux, and aspects such as the volatile content of the interior. Furthermore, many constraints on the processes involved in the accretion and differentiation of Earth would be lost if element concentrations in the upper mantle could not be reliably assumed to reflect the mantle as a whole.

Ultimately the best prospect for determining whether the lower mantle has the same major element composition as the upper mantle is to compare seismic reference models for shear and longitudinal wave velocities in the lower mantle with mineral physical estimates for what these velocities should be if the mantle is isochemical [*Birch*, 1952; *Anderson*, 1968; *Jackson*, 1983; *Stixrude and Jeanloz*, 2007; *Murakami et al.*, 2012; *Cottaar et al.*, 2014; *Wang et al.*, 2015]. Such comparisons require not only high-pressure and high-temperature equation-of-state data for mantle minerals but also knowledge of the proportion and composition of minerals in the lower mantle for a given bulk composition. Significant uncertainties exist for both types of data. It must also be recognized that any fit to observed

Bayerisches Geoinsitut, University Bayreuth, Bayreuth, Germany

Deep Earth: Physics and Chemistry of the Lower Mantle and Core, Geophysical Monograph 217, First Edition.
Edited by Hidenori Terasaki and Rebecca A. Fischer.

lower mantle velocities in terms of temperature and composition is unlikely to be unique [*Bina and Helffrich*, 2014]. This approach may nevertheless be effective at excluding some hypotheses concerning the composition and thermal structure of the lower mantle.

Here arguments for plausible bulk chemical compositions of the lower mantle are briefly reviewed, followed by a discussion of how the mineralogy of the lower mantle and in particular the chemical compositions of minerals may change with depth. Determining a mineralogical model of this type is a key step toward the goal of calculating precise densities and seismic velocities for plausible lower mantle compositions.

18.2. EVIDENCE FOR CHEMICALLY DISTINCT LOWER MANTLE

The lack of earthquakes at depths greater than 700 km was one of the earliest lines of evidence used to argue that the subducted lithosphere did not penetrate into the lower mantle [*Oliver and Isacks*, 1967]. This combined with observations of a global seismic discontinuity at similar depths and early interpretations of the density through the lower mantle formed the basis for the hypothesis that the lower mantle could be chemically isolated and not involved in the convective processes responsible for plate tectonics [*Anderson*, 1968; *Richter and Johnson*, 1974; *Richter and McKenzie*, 1978; *Anderson*, 1979]. The principal evidence against this has come from seismic tomography that shows that some slabs stagnate at 660 km [see *Fukao et al.*, 2001] while others penetrate directly into the lower mantle [*Creager and Jordan*, 1984; *van der Hilst et al.*, 1997; *Grand*, 2002; *Fukao and Masayuki*, 2013]. This has provided the most compelling evidence for whole-mantle convection but it does not necessarily mean that the lower mantle has the same composition as the upper mantle, as it remains difficult to quantify the total mass exchange between the upper and lower mantle over the last 4.5 Gyr.

Geodynamic simulations have also been used to argue for an isolated lower mantle. At the top of the lower mantle (660 km) the breakdown reaction of the mineral ringwoodite into an assemblage of ferropericlase and perovskite-structured $(Mg,Fe)SiO_3$, now called bridgmanite [*Tschauner et al.*, 2014], appears to have a negative Clausius-Clapeyron slope [*Liu*, 1974, 1975, 1976; *Ito and Takahashi*, 1989; *Ito et al.*, 1990]. Models indicate that a large negative slope of between −4 and −8 MPa/K would cause a local inversion in buoyancy that could induce two-layered mantle convection [*Christensen and Yuen*, 1985]. While experimental estimates for the slope of the transformation are closer to −3 MPa/K [*Ito and Takahashi*, 1989; *Ito et al.*, 1990] and may be as low as

−2 to −0.4 MPa/K [*Katsura et al.*, 2003], it should be noted that recent experimental studies also predict transition pressures for the transformation that are too low to explain the discontinuity at all. The absolute pressure and Clapeyron slope determined for the transition may well be unreliable due to a lack of a rigorously calibrated high-temperature pressure marker. Models suggest that an increase in viscosity around the upper-lower mantle boundary would also inhibit whole-mantle convection. An ~30 times increase in viscosity would cause slabs to buckle within the mantle transition zone [*Gaherty and Hager*, 1994], in agreement with tomographic images [e.g., *Li and van der Hilst*, 2010] and analysis of focal mechanisms and earthquake locations [*Myhill*, 2013]. *Ringwood* [1982, 1991], recognizing that complete two-layered mantle convection may be unlikely, proposed that subducted material may tend to accumulate at 660 km but after sufficient thermal equilibration material would avalanche through into the lower mantle. Three-dimensional (3D) geodynamic simulations appear to support this scenario [*Tackley et al.*, 1993; *Solheim and Peltier*, 1994], and it has been proposed that such avalanches could trigger superplume events, which may have been responsible for the episodic growth of the continental crust [*Condie*, 1998].

Chemical differences between the upper and lower mantle have long been invoked by geochemists to explain the isotopic and trace element heterogeneity displayed by ocean island basalts (OIBs), which are often considered to be the product of plumes rising from the lower mantle. The OIB source appears nominally "primitive" and unaffected by the partial-melting events linked to the formation of the continental crust, which has depleted the source of midocean ridge basalt (MORB) in incompatible elements [see *Hofmann*, 2014]. This depletion is consistent with a small-degree melting event that would not have significantly influenced the concentrations of major elements but appears, for example, to have left the MORB source too depleted in heat-producing elements (U, Th, K) to account for the expected radiogenic component of Earth's surface heat flow. This and similar arguments involving rare gases have led many studies to conclude that the lower mantle holds an undepleted and undegassed reservoir [*Jochum et al.*, 1983; *Albarède and van der Hilst*, 2002; *Arevalo et al.*, 2009]. Early mass balance estimates indicated that the depleted MORB source may be ~30% of the mass of the mantle [*Jacobsen and Wasserburg*, 1979], which leaves the primitive reservoir with the same mass as the lower mantle. More recent estimates for the mass of the MORB source reservoir are much larger, however, [*Jackson and Jellinek*, 2013; *Hofmann*, 2014] and imply that the primitive reservoir may be only a fraction of the lower mantle. Isotopic evidence for the presence of ancient recycled crust in the

OIB source [*Lassiter and Hauri*, 1998; *Sobolev et al.*, 2000] is similarly at odds with the perception of a mainly isolated primitive lower mantle. Convection in the lower mantle may, however, favor reentrainment of recycled material into plumes at the base of the mantle, leaving a sizable proportion of the lower mantle chemically isolated [*Solomatov and Reese*, 2008; *Campbell and O'Neill*, 2012].

18.3. POTENTIAL ORIGINS OF CHEMICALLY DISTINCT LOWER MANTLE

Although it is unlikely that Earth formed from any single class of meteorite, major element ratios estimated for the BSE support Earth's similarity with carbonaceous chondrites [*O'Neill and Palme*, 1998]. Table 18.1 shows a range of silicate Earth compositions determined from bulk analyses of the major chondritic meteorite groups [*Wasson and Kallemeyn*, 1988]. The compositions were calculated by assuming a mantle FeO concentration of 8 wt % and separating the remaining Fe and Ni into the core, which is assumed to contain no light element. No other elements were considered and the initial Na concentrations were normalised to 20% of the chondritic

value to account for volatility. The primitive Earth's mantle composition determined by *Palme and O'Neill* [2004] is shown for comparison. Chondrite-based estimates of silicate Earth compositions have Mg/Si ratios that are lower than any petrological estimates for this ratio within the BSE or upper mantle [*Allegre et al.*, 1995; *McDonough and Sun*, 1995; *Workman and Hart*, 2005; *O'Neill and Palme*, 1998]. The resulting upper mantle normative mineral contents are olivine poor and in the case of high enstatite chondrites (EHS) become quartz normative. None of the compositions approach fertile mantle xenoliths, which have normative olivine contents of ~60 wt %. CV and CO chondrite compositions match many of the major element concentrations but are still ~3–4 wt % SiO_2 enriched/MgO depleted relative to the BSE.

The apparently high Mg/Si ratio of the upper mantle raises three possibilities that are not mutually exclusive; that is, the bulk Earth has a superchondritic Mg/Si ratio, significant Si entered the Earth's core, or some region of the mantle has a lower Mg/Si ratio than the upper mantle. The first possibility, that the Earth has a higher Mg/Si ratio than any chondrite or the solar photosphere, is quite plausible given that this ratio varies among chondrites.

Table 18.1 Chondrite-based BSE compositions, upper and lower mantle mineralogy, and core mass.

	Earth	CI	CM	CO	CV	H	L	LL	EH	EL
SiO_2	45.0	49.5	49.1	48.0	47.8	51.9	52.5	52.6	57.9	54.4
TiO_2	0.2	0.2	0.2	0.2	0.2	0.1	0.1	0.1	0.1	0.1
Cr_2O_3	0.38	0.8	0.8	0.7	0.8	0.8	0.8	0.7	0.7	0.6
Al_2O_3	4.45	3.5	4.0	3.8	4.7	3.1	3.1	2.9	2.5	2.7
FeO	8.05	8.1	8.1	8.1	8.1	8.1	8.1	8.1	8.1	8.1
MgO	37.8	34.8	34.5	36.0	34.4	33.3	32.8	33.0	28.5	32.0
CaO	3.55	2.8	3.2	3.1	3.8	2.5	2.4	2.4	1.9	1.9
Na_2O[a]	0.36	0.3	0.2	0.2	0.1	0.3	0.3	0.2	0.3	0.2
Total	99.8	100.0	100.0	100.0	100.0	100.0	100.0	100.0	100.0	100.0
Upper mantle normative mineralogy										
Olivine	54	31	30	38	38	18	15	15	Quartz	6
Opx	23	53	51	44	44	68	73	73	Norm.	86
Cpx	20	14	15	15	15	11	10	9		6
Spinel	3	3	3	3	3	2	2	2		2
Lower mantle normative mineralogy										
Bdg	80	90	89	86	85	96	95	95	85	92
Fper	13	5	6	8	8	0	0	0	0	0
Ca-Pv	6	5	5	5	7	4	4	4	3	3
SiO_2	0	0	0	0	0	0	1	1	12	5
Core	32	26	25	23	23	26	19	16	30	20

All values in weight percent. Primitive Earth's mantle composition from *O'Neill and Palme* [1998], chondrite compositions from *Wasson and Kallemeyn* [1988]. Abbreviations: Bdg, bridgmanite; Fper, ferropericlase; Ca-Pv, $CaSiO_3$ perovskite. Except for Earth the calculated size of the core is solely based on the available Fe and Ni in the meteorite composition and ignores any light element.

[a] Initially normalized to 20% of the chondrite concentrations.

Redistribution of forsterite (Mg_2SiO_4) that had partially condensed from solar nebula gas has been one mechanism proposed to explain this [*Larimer and Anders*, 1970]. Earth could have formed from material that was, on average, enriched in Mg_2SiO_4 compared to any chondrite. As this fractionation may have only involved forsterite, an otherwise quite pure mineral, it would be difficult to then constrain the bulk Earth Mg/Si ratio via element concentrations calculated from cosmochemical observations. However, it has recently been proposed that concurrent isotopic fractionation between condensing forsterite and nebula gas occurred, which caused variations in the $^{30}Si/^{28}Si$ ratio observed among terrestrial planets and meteorites [*Dauphas et al.*, 2015]. *Dauphas et al.* [2015] provide good evidence for this by demonstrating a clear and predictable correlation between the $^{30}Si/^{28}Si$ and Mg/Si ratios of chondritic meteorites. They constrain the bulk Earth Mg/Si ratio using this trend from the seemingly constant $^{30}Si/^{28}Si$ ratio of the mantle and accounting for the fact that some Si likely also entered the core. The resulting bulk Earth Mg/Si ratio is higher than any chondrite, in agreement with petrological estimates.

Silicon partitioning into the core has often been proposed to explain the apparent superchondritic Mg/Si ratio of the mantle [*Allegre et al.*, 1995; *McDonough and Sun*, 1995; *O'Neill and Palme*, 1998] and it is in line with geophysical observations that the core is 5–10% less dense than expected for pure iron-nickel liquid [*Birch*, 1952; *Poirier*, 1994]. In order to obtain an Earth-like Mg/Si ratio from the chondrite-based mantle compositions shown in Table 18.1 would require between 5 wt % (CV) and 20 wt % (EH) Si to have separated into the core. The higher bulk Earth Mg/Si ratio determined by *Dauphas et al.* [2015] from the BSE $^{30}Si/^{28}Si$ ratio implies that ~3.6 wt % Si exists in the core. Although the uncertainties on this estimate are high (+6.0/–3.6 wt %), this relatively low value is in much better agreement with more recent mineral physics estimates for the Si content based on the core's density and seismic velocity [*Antonangeli et al.*, 2010; *Badro et al.*, 2014].

It would seem, from the above arguments, that models requiring a lower Mg/Si ratio, in part or all of the lower mantle, to explain the upper mantle superchondritic Mg/Si ratio are becoming increasingly redundant. Given the large uncertainties, however, in both mineral physics estimates for the silicon content of the core and cosmochemical estimates for the bulk earth Mg/Si ratio, such models cannot yet be completely refuted. If the Mg/Si ratio of the lower mantle were lower, this would be manifest as a raised bridgmanite to ferropericlase ratio, as shown by the normative lower mantle mineralogies calculated in Table 18.1. One quite appealing process to achieve this would be as a result of fractional crystallization and settling of bridgmanite from a global magma

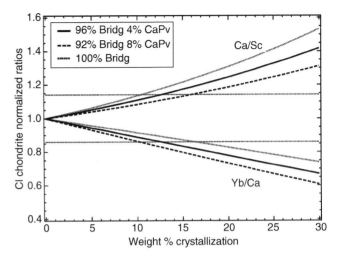

Figure 18.1 Chondrite-normalized Ca/Sc and Yb/Ca ratios of the residual liquid as a function of crystallization of bridgmanite and $CaSiO_3$ perovskite from a magma ocean. Horizontal lines indicate the uncertainty on both element ratios in the upper mantle. Solid, dashed, and dotted lines show the effect of different proportions of bridgmanite and $CaSiO_3$ perovskite in the fractionating assemblage.

ocean during accretion [*Agee and Walker*, 1988]. The formation of the moon testifies to the fact that toward the end of accretion the Earth was involved in at least one giant impact, which would have melted a large proportion, if not all, of Earth's mantle [*Canup*, 2004; *Carlson et al.*, 2014]. As the magma ocean cooled, bridgmanite would have been the liquidus phase throughout much of the lower mantle [*Ito et al.*, 2004; *Walter et al.*, 2004; *Andrault et al.*, 2011; *Liebske and Frost*, 2012]. Crystallization and settling of bridgmanite from an initially chondritic magma ocean could, in principle, have driven up the MgO content of the residual liquid, which would ultimately crystallize to form the upper mantle. In detail, however, mass balance calculations based on experimentally determined liquidus phase compositions fail to find a crystallization sequence that can derive the composition of the upper mantle in this way [*Walter et al.*, 2004; *Jackson et al.*, 2014]. Furthermore, a number of well-constrained chondritic element ratios apparent in the upper mantle would not have been preserved if more than ~10% bridgmanite fractionation occurred from an initially chondritic magma ocean, as shown in Figure 18.1 [*Kato et al.*, 1988; *Walter et al.*, 2004; *Corgne et al.*, 2005; *Liebske et al.*, 2005]. It should be noted, however, that both the liquidus-phase composition and trace element partitioning arguments are based around experiments performed at conditions that correspond only to the very top of the lower mantle.

A number of seismological studies have argued for the existence of large-scale (>1000 km) heterogeneities in the mantle at >2400 km depths [*Su et al.*, 1994;

Li and Romanowicz, 1996; *Ni et al.*, 2002]. Due to their seismic properties, these heterogeneities have been termed large low-shear-velocity provinces (LLSVPs). Two antipodal provinces have been identified beneath Africa and the Pacific, and based on paleomagnetic reconstructions, it has been proposed [*Torsvik et al.*, 2008] that the majority of hot spots with a suggested deep-mantle origin derive from the edges of these anomalies [but see also *Austermann et al.*, 2014]. The relatively small size of these anomalies means that they could have originally formed by fractional crystallization of bridgmanite, with or without Ca-perovskite, from a global magma ocean without upsetting known chondritic ratios in the residual liquid that formed the upper mantle [*Jackson et al.*, 2014]. However, if LLSVPs have remained at the base of the mantle for the timescales proposed by plume reconstructions, then they are probably composed of material denser than the average mantle at these depths. The most plausible explanation for this would be as a result of iron enrichment, but as the available evidence indicates that the magma ocean itself should be more enriched in iron than the crystallizing phases [*Andrault et al.*, 2011, 2012; *Nomura et al.*, 2011], it probably makes more sense if LLSVPs formed from the crystallization of a dense residual basal magma ocean, rather than crystal settling from a global magma ocean.

In summary, although the hypothesis that the lower mantle may be different in major element composition from the upper mantle has been eroded in recent years by geophysical observations of slabs entering the lower mantle, the prospect cannot be excluded and there are mechanisms by which a chemically distinct lower mantle could have been created and at least partially preserved over geological history. As proposed, the only way to categorically exclude this possibility is to compare seismic velocities for the lower mantle with mineral physical models based on a given chemical composition. Recently such an approach was used to argue that the lower mantle is indeed chemically distinct, with a Mg/Si ratio closer to that of carbonaceous chondrites [*Murakami et al.*, 2012]. If such comparisons are to be meaningful, however, some account must be taken of the fact that the elasticity of mantle minerals changes as a function of their composition. Building a realistic model for the effects of composition on the seismic velocities of lower mantle minerals remains a key goal in mineral physics.

18.4. MINERALOGY AND CRYSTAL CHEMISTRY OF THE LOWER MANTLE

If the entire mantle is isochemical, then the lower mantle should be composed of bridgmanite, ferropericlase, and $CaSiO_3$ perovskite in the approximate proportions given for the Earth model in Table 18.1 Although there are a large number of estimates for the composition of

the BSE based on peridotite rocks and mantle melting scenarios such as the pyrolite model [*Ringwood*, 1975; *McDonough and Sun*, 1995; *O'Neill and Palme*, 1998; *Walter*, 2003], across the range of these ultramafic compositions the predicted proportion of bridgmanite in the lower mantle changes only between 80 and 82 wt %.

$CaSiO_3$ perovskite exsolves from garnet at depths between 500 and 660 km in the transition zone [*Irifune and Tsuchiya*, 2007; *Saikia et al.*, 2008]. Bridgmanite is formed through two reactions. The first is dissociation of $(Mg,Fe)_2SiO_4$ ringwoodite into ferropericlase and bridgmanite and is responsible for the 660 km seismic discontinuity. Due to the similarity in Fe-Mg partitioning across the transformation, it should take place over a narrow pressure interval equivalent to <2 km in depth [*Ito and Takahashi*, 1989; *Ito et al.*, 1990], which is consistent with seismic observations of short-period reflected and converted phases [*Kind and Li*, 2007]. The second bridgmanite formation mechanism is via garnet breakdown, which occurs over a ~100 km interval at the top of the lower mantle. Close to the transition zone–lower mantle boundary, $MgSiO_3$ bridgmanite exsolves, lowering the Mg-majoritic component of the residual garnet. Growth continues as the solution of Al_2O_3 in bridgmanite increases [*Irifune*, 1994; *Nishiyama and Yagi*, 2003; *Irifune and Tsuchiya*, 2007]. The preservation of garnet in the uppermost lower mantle may be responsible for slightly lower velocities than otherwise expected.

Further changes in mineral assemblage for a BSE bulk composition are limited to the base of the mantle. At depths greater than 2600 km bridgmanite may transform to the post-perovskite phase [*Murakami et al.*, 2004; *Irifune and Tsuchiya*, 2007], although the positive and relatively shallow Clausius-Clapeyron slope of this transformation may limit the stability of this phase to only the coldest regions of the lowermost mantle. High thermal gradients may lead to double-crossings of the bridgmanite-postperovskite transition [*Hernlund et al.*, 2005]. Preliminary results indicate that at about 2000 km a relatively $FeSiO_3$-rich phase, termed H phase, may exsolve from bridgmanite, leaving the latter strongly depleted in Fe [*Zhang et al.*, 2014].

18.5 LOWER MANTLE MINERAL COMPOSITIONS

The dominant lower mantle minerals are all solid solutions. Bridgmanite is dominated by the $MgSiO_3$ end member but contains ~10 mole % $FeSiO_3$, $FeAlO_3$ and $AlAlO_3$ end members. Ferropericlase in addition to MgO and FeO contains ~1 wt % SiO_2 and Al_2O_3 and contains the lower mantle compliment of Na_2O. $CaSiO_3$ perovskite is relatively pure but also contains ~1 wt % Al_2O_3, MgO, and FeO; furthermore its very strong affinity for large cations means that it also hosts the lower mantle's compliment of heat-producing elements, U, Th, and K.

Once garnet has transformed to bridgmanite, the proportions of these three minerals should remain relatively constant throughout the lower mantle for a given bulk composition [*Irifune and Tsuchiya*, 2007]. The proportion of $CaSiO_3$ is fixed by the CaO content of the bulk composition and the proportions of bridgmanite and ferropericlase by the remaining SiO_2 and Al_2O_3. The only major variation expected to occur is in the distribution of Fe and Mg between the two dominant minerals. This variation is complicated by the fact that, at least in experiments, bridgmanite contains subequal proportions of both Fe^{2+} and Fe^{3+} cations (*McCammon*, 1997; *Lauterbach et al.*, 2000). Furthermore, Fe^{2+} in both dominant phases and possibly Fe^{3+} in bridgmanite may go through a high spin–low spin transformation as pressures increase, which may also influence the distribution of both Fe components [*Lin et al.*, 2013; *Badro*, 2014]. The smaller volume of low-spin iron (ferric and ferrous) and the redistribution of iron during spin transitions can potentially influence both the densities and seismic wave velocities with depth in the lower mantle. In particular, the distribution of iron will influence both the Fe^{2+} and Fe^{3+} contents of bridgmanite, which have also been shown to influence the elasticity and inferred seismic velocity of this mineral [*Boffa Ballaran et al.*, 2012; *Chantel et al.*, 2012; *Wang et al.*, 2015].

The exchange of Fe^{2+} and Mg between bridgmanite and ferropericlase can be described using the equilibrium

$$\begin{array}{ccccccc} FeO & + & MgSiO_3 & \leftrightarrow & MgO & + & FeSiO_3 \\ \text{ferropericlase} & & \text{bridgmanite} & & \text{ferropericlase} & & \text{bridgmanite} \end{array}$$

(18.1)

and the corresponding exchange coefficient K_D,

$$K_D = \frac{X_{FeSiO_3}^{Bdg} X_{MgO}^{Fper}}{X_{MgSiO_3}^{Bdg} X_{FeO}^{Fper}},$$

(18.2)

where $X_{FeSiO_3}^{Bdg}$ is, for example, the mole fraction of the $FeSiO_3$ end member in bridgmanite, which would equate to the molar cation ratio $Fe^{2+}/(Fe^{2+}+Mg)$. However, because a significant proportion of the Fe in bridgmanite is Fe^{3+} in systems which also contain Al and the majority of studies are not able to analyse the $Fe^{3+}/\Sigma Fe$ ratio, many studies define an apparent Fe-Mg exchange $K_D(app)$ where $X_{FeSiO_3}^{Brid} = (Fe^{2+} + Fe^{3+})/(Fe^{2+} + Fe^{3+} + Mg)$. Figure 18.2 shows determinations for $K_D(app)$ from recent studies as a function of pressure. The experiments are broken into two groups, those that have studied an ultramafic composition in a natural system, most of these have used a pyrolite composition, and experiments performed with

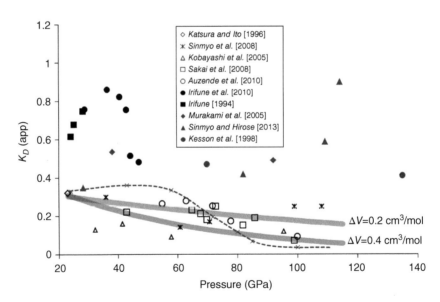

Figure 18.2 Apparent Fe-Mg exchange coefficient (apparent from electron microprobe analyses, i.e., all Fe is assumed to be FeO) plotted as a function of pressure for experiments performed using a mantle olivine starting material (open symbols and crosses) and pyrolite (black filled symbols-from multianvil experiments, filled grey symbols from laser-heated diamond anvil cell experiments). The two curves are thermodynamic calculations employing the model of *Nakajima et al.* [2012] but assuming different volume changes for the exchange equilibrium (18.1). The thickness of each line equates to a temperature variation of ±200°C in the calculation. The dashed grey line is calculated for a mantle olivine bulk composition assuming a high-spin to low-spin Fe^{2+} transition in ferropericlase occurs over a pressure interval between 40 and 90 GPa.

mantle olivine, $(Fe_{0.1},Mg_{0.9})_2SiO_4$, as a starting material. It should be recognized that $K_D(app)$ is expected to vary with bulk Fe concentration, pressure, and temperature [*Nakajima et al.*, 2012] but the expected variations between the studies for a given starting material should be small. As will be discussed later, $K_D(app) \approx K_D$ for samples formed from olivine starting materials, but this is not the case for pyrolite compositions mainly due to the Al content of bridgmanite promoting the presence of Fe^{3+}.

The majority of values of $K_D(app)$ for bridgmanite and periclase produced from olivine starting materials indicate a decrease in $K_D(app)$ with pressure. Studies above 25 GPa are performed in the laser-heated diamond anvil cell and in detail many of the different studies are in poor agreement. Shown for comparison in Figure 18.2 are two curves calculated for a mantle olivine composition using a thermodynamic treatment of the exchange equilibrium (18.1). This model was derived by fitting experimental data collected at 24 GPa as a function of the Fe/(Fe+Mg) ratio to the equation

$$RT \ln K_D = -\Delta G^o_{(18.1)} + W^{Bdg}_{FeMg}\left(2X^{Bdg}_{Fe}-1\right) + W^{Fper}_{FeMg}\left(1-2X^{Fper}_{Fe}\right),$$

(18.3)

where R is the gas constant, W^{Bdg}_{FeMg} and W^{Fper}_{FeMg} are Margules interaction parameters for bridgmanite and ferropericlase that describe nonideal Fe and Mg mixing, and $\Delta G^o_{(18.1)}$ is the standard state Gibbs free energy of equilibrium (18.1) at the pressure and temperature of interest. Values determined for $\Delta G^o_{(1)}$ can be simply described by the expression

$$\Delta G^o_{(18.1)} = \Delta H^\circ - T\Delta S^\circ + P\Delta V^\circ,$$

(18.4)

where $\Delta H^\circ, \Delta S^\circ$ and ΔV° are the standard state enthalpy, entropy, and volume changes of equilibrium (18.1), which are all considered to be independent of pressure and temperature. Equation (18.3) has been fitted to data collected in the $FeO-MgO-SiO_2$ system at 24 GPa, where experiments also contained metallic iron to ensure that $K_D(app) \approx K_D$ [*Nakajima et al.*, 2012]. The equation can then be extrapolated to higher pressures by making assumptions concerning the magnitude of ΔV°, as shown in Figure 18.2. The majority of olivine results are consistent with a ΔV° of 0.2 cm³/mol, although three studies, which are in good agreement at 100 GPa, would require a ΔV° of ~0.4 cm³/mol [*Kobayashi et al.*, 2005; *Sinmyo et al.*, 2008; *Sakai et al.*, 2009]. A ΔV° of 0.2 cm³/mol would be consistent with mineral physics estimates for this volume change based on ambient volume and elasticity data of the end members once the uncertainties are considered [*Xu et al.* 2008]. The mineral physical data and model of *Xu et al.* [2008] actually predict a small but negative value of

ΔV° but a value of 0.2 cm³/mol would still be within the uncertainties. A value of 0.4 cm³/mol is probably too large to be explained by existing volume and elasticity measurements, although many of the thermoelastic properties have not been measured at mantle temperatures. An alternative explanation for data lying along the 0.4 cm³/mol trend is that the Fe^{2+} component in ferropericlase undergoes a high spin–low spin electronic transition [*Burns*, 1993; *Badro et al.*, 2003].

It is relatively straightforward to make a simple estimate for the effect of a ferropericlase Fe^{2+} spin transition on Fe-Mg exchange by using experimental estimates for the proportion of the high-spin and low-spin components as a function of pressure [e.g., *Speziale et al.*, 2007] and by assuming that they mix ideally. If n is the proportion of low-spin Fe^{2+}, then the Gibbs free energy of the FeO component as it crosses the transition can be determined from

$$G_{P,T} = nG^o_{LS} + (1-n)G^o_{HS} + RT\left[n\ln(n)+(1-n)\ln(1-n)\right],$$

(18.5)

where G^o_{LS} and G^o_{HS} are the Gibbs free energies of the low-spin and high-spin FeO components. Using experimental values of n as a function of pressure, the derivative of equation (18.5) with respect to n can be used to determine the difference in Gibbs free energy between high-spin and low-spin FeO, i.e.,

$$G^o_{LS} - G^o_{HS} = RT\ln\left(\frac{1}{n}-1\right).$$

(18.6)

Using these differences at each pressure and values of G^o_{HS} determined using a thermodynamic database [*Xu et al.*, 2008], values of G^o_{LS} can be calculated and used in equation (18.5) to determine the Gibbs free energy of FeO across the transformation. Then $\Delta G^o_{(18.1)}$ can be calculated across the transition using further thermodynamic data for the remaining mineral end Members in equilibrium (18.1) and K_D is then determined using equation (18.3) assuming a $(Fe_{0.1},Mg_{0.9})_2SiO_4$ bulk composition. In Figure 18.2 the results of such a calculation are shown by employing the experimental data for n versus P from *Speziale et al.* [2007], who determined the mixed state transition interval in ferropericlase to be between 40 and 90 GPa. The elasticity data used by *Xu et al.* [2008] actually predict a small negative $\Delta V^\circ_{(18.1)}$ before the spin transition, which results in an initial increase in K_D, but this is reversed as soon as the transition commences. As stated previously, values up to 0.2 cm³/mol would be consistent with the elasticity data within its uncertainties. In general, however, the change in K_D predicted for the spin transition is consistent in magnitude with that seen in some studies and reproduces the low values at 100 GPa shown by three different studies [*Kobayashi et al.*, 2005; *Auzende et al.*, 2008; *Sakai et al.*, 2009]. The shape of the curve can be manipulated by changing the assumptions. The pressure of onset and the interval of the transition will be a major influence, and there is a range in the experimental estimates for both [*Lin et al.*, 2013; *Badro*, 2014]. In the calculation $W_{\text{FeMg}}^{\text{Fper}}$ is held constant across the transition, which is certainly unrealistic. Reducing the magnitude of $W_{\text{FeMg}}^{\text{Fper}}$ as a function of $1 - n$ shifts the curve to lower pressures by ~5 GPa. Interaction terms involving the low-spin Fe^{2+} component have been ignored but are likely to be smaller than $W_{\text{FeMg}}^{\text{Fper}}$ because low-spin Fe should be closer to Mg^{2+} in size. A major prediction of the model, however, is that for mantle olivine bulk compositions between 24 and 100 GPa the ferropericlase Fe/(Fe+Mg) ratio increases from ~0.15 to ~0.19, while that of bridgmanite decreases from ~0.05 to ~0.01. These changes are most likely sufficient to influence calculated velocities and densities for the lower mantle. As detailed below, these changes are probably larger than those taking place in the BSE (pyrolite) composition mantle, where a significant portion of the Fe ends up as Fe^{3+} in bridgmanite. Nevertheless, the loss of Fe^{2+} from bridgmanite may reduce the probability of forming the proposed H phase [*Zhang et al.*, 2014].

Although this simplistic treatment reproduces behavior seen in some experimental data for $(\text{Fe}_{0.1},\text{Mg}_{0.9})_2\text{SiO}_4$ starting compositions, the resulting pressure dependence of the data is not in agreement with all studies. Figure 18.3 shows experimental Fe-Mg partitioning results obtained in the multianvil apparatus by *Tange et al.* [2009] using an olivine composition starting material but with a much higher Fe/(Fe+Mg) ratio than in the studies shown in Figure 18.2. These data were collected between 22 and 43 GPa at 2273 K and are compared with model calculations described above assuming a constant $\Delta V_{(18.1)}^{\circ}$ of 0.4 cm³/mol. As stated previously, a volume change of this magnitude would appear to be inconsistent with the end-member volumes and equation-of-state data [*Xu et al.*, 2008]. On the other hand, an iron spin transition in ferropericlase starting at ~22 GPa is probably also inconsistent with the prediction that higher FeO contents should push this transition to much higher pressures [*Lin et al.*, 2013; *Badro*, 2014]. Such inconsistencies and uncertainties in the pressure range over which the spin transition occurs currently make it difficult to unify the existing experimental data into a single model.

The problems that exist in understanding results in simplified olivine systems are somewhat trivial in comparison to those encountered for studies employing natural pyrolite (i.e., BSE) compositions. As shown in Figure 18.2, pyrolite K_D(app) values are not only higher than for olivine compositions but also are seen to increase up to ~40 GPa and then decrease quite sharply.

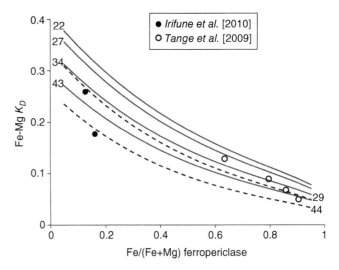

Figure 18.3 Bridgmanite/ferropericlase Fe^{2+}-Mg exchange coefficients determined at 2273 K and between 22 and 43 GPa by *Tange et al.* [2009] in the MgO-FeO-SiO₂ system are compared with (grey) curves calculated at the same conditions using the model of *Nakajima et al.* [2012]. To fit the experimental data the ΔV of the Fe-Mg exchange equilibrium must be assumed to be 0.4 cm³/mol. The Fe^{2+}-Mg K_D has also been estimated for data reported by *Irifune et al.* [2010] from experiments performed on a natural pyrolite composition at 29 GPa and 1873 K and 44 GPa and 2073 K using the reported bridgmanite Fe^{3+}/Σ Fe ratios. These data also fit the same model with the same ΔV (as shown by the dashed lines).

At > 50 GPa there is poor agreement between studies. As shown in Figure 18.4a, the initial excursion to high values of K_D(app) between 22 and 28 GPa is a strong function of the bridgmanite Al content, which increases over this pressure range due to the breakdown of the mineral garnet [*Irifune*, 1994; *Wood and Rubie*, 1996; *Nishiyama and Yagi*, 2003]. The increase in K_D(app) can be mainly attributed to an increase in the bulk Fe content of bridgmanite. Measurements using Mössbauer and electron energy loss spectroscopy have shown that this additional Fe is predominantly Fe^{3+} [*McCammon*, 1997; *Lauterbach et al.*, 2000; *Frost and Langenhorst*, 2002; *McCammon et al.*, 2004]. Figure 18.4b shows the proportion of Fe^{3+} cations measured in bridgmanite as a function of the Al content. The data are shaded to indicate the bulk Fe content. The Fe^{3+} increases with both Al and bulk Fe content and the Al dependence at a given bulk Fe content appears to be nonlinear. Bridgmanite Fe^{3+} contents were determined for samples from 29 and 44 GPa in the study of *Irifune et al.* [2010] and are also shown in Figure 18.4b. These measurements can be used to calculate the Fe^{2+}-Mg exchange coefficient between bridgmanite and ferropericlase, i.e., K_D, which is also plotted in Figure 18.3. The result from 28 GPa is in good agreement with values of K_D determined in Al-free systems at 25 GPa and reproduced by

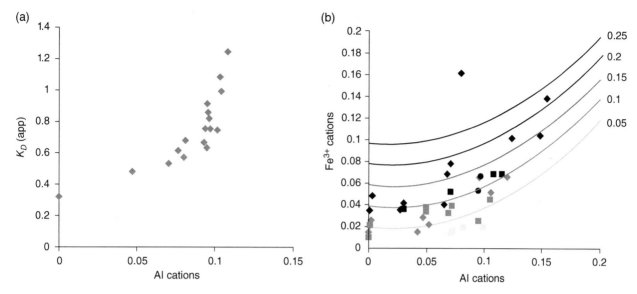

Figure 18.4 (a) Apparent Fe-Mg exchange coefficient between bridgmanite and ferropericlase is shown to be a function of bridgmanite Al content, in cations per 3 oxygen formula units. The data are from experiments performed between 23 and 29 GPa using pyrolitic bulk compositions, except the Al-free data point for which an olivine starting material was employed [*Katsura and Ito*, 1996]. (b) Bridgmanite Fe^{3+} versus Al^{3+} in cation formula units determined from experiments performed at 25 GPa but including the analyses of *Irifune et al.* [2010] from 29 and 44 GPa (circles). Data points are shaded to indicate the bulk Fe content, with the shading following the proportions indicated by the empirical model curves derived by *Nakajima et al.* [2012] that follow the expression $Fe^{3+}=-0.13Al+3.1Al^2+0.9 \Sigma Fe$. Experiments performed in the presence of iron metal or in iron capsules (squares) show similar Fe^{3+} contents to those performed at more oxidizing conditions (diamonds). Experiments performed at the peridotite solidus show lower but still significant Fe^{3+} contents.

the model calculation at this pressure. This supports the argument of *Nakajima et al.* [2012] that the presence of Al influences the Fe^{3+} content of bridgmanite but does not strongly influence Fe^{2+}-Mg exchange between phases. The value determined at 44 GPa indicates that the precipitous drop in K_D(app) that occurs above 40 GPa is caused to a large extent by a more favorable partitioning of Fe^{2+} into ferropericlase. A consequence of the bulk Fe content of bridgmanite decreasing is that the Fe^{3+} content also drops, as expected from the relationships outlined in Figure 18.4b. To model this strong decrease in K_D, a $\Delta V_{(1)}^{\circ}$ of ~0.4 cm³/mol is again required. As stated previously, such a high value is in conflict with mineral physics estimates, although given the consistency with the $\Delta V_{(18.1)}^{\circ}$ made using the data of *Tange et al.* [2009] and some of the diamond anvil cell (DAC) data on olivine, it is hard not to suspect that it is the elasticity data that are in error. The alternative explanation, that this arises from the onset of an Fe spin transition in ferropericlase, seems unlikely as most studies argue for a much higher pressure onset. The start of the transition would need to be at least 10 GPa lower in pressure than that modeled for $(Fe_{0.1},Mg_{0.9})_2SiO_4$ starting compositions in Figure 18.2, and of this transition is at lower pressure than proposed in many studies [*Lin et al.*, 2013; *Badro*, 2014].

The effect of this rapid decrease in K_D(app) is actually smaller than predicted for the spin transition in a mantle olivine bulk composition, because much of the iron remains as Fe^{3+} in bridgmanite. Between 30 and 50 GPa the Fe/(Fe+Mg) ratio of ferropericlase increases from ~0.11 to ~0.15, while bridgmanite decreases from ~0.11 to ~0.08.

Several studies have proposed that Fe^{3+} in bridgmanite may also undergo a high spin–low spin crossover [*Lin et al.*, 2013]. Experiments appear to indicate that Fe^{3+} undergoes a spin crossover when it substitutes onto the smaller octahedrally coordinated Si site (B site). This occurs in Al-free bridgmanite and the spin transition takes place between 15 and 50 GPa [*Catalli et al.*, 2010; *Hsu et al.*, 2011]. In Al-bearing perovskite; however, Fe^{3+} resides almost exclusively on the larger Mg site (A site), meaning that there is little-high spin Fe^{3+} on the B site that could transform. Nevertheless, stabilization of Fe^{3+} on the B site as a result of the transition could drive up the Fe^{3+} occupancy on that site, via the exchange reaction

$$Fe_A^{3+, HS} + Al_B^{3+} \rightarrow Fe_B^{3+, LS} + Al_A^{3+} \qquad (18.7)$$

[*Grocholski et al.*, 2009; *Catalli et al.*, 2010; *Fujino et al.*, 2012]. At low temperatures (≤1300 K), the amount of low-spin Fe^{3+} on the B site appears to be negligible

[*Glazyrin et al.*, 2014]. At higher temperatures, the current data are more ambiguous. Mössbauer analysis of a sample of $Mg_{0.6}Fe_{0.4}Si_{0.63}Al_{0.37}O_3$ bridgmanite before, during, and after laser heating at 1700–1800 K showed no evidence of low-spin Fe^{3+} up to 80 GPa [*Glazyrin et al.*, 2014]. In contrast, Mössbauer and X-ray emission spectroscopy on samples of $Mg_{0.88}Fe_{0.13}Si_{0.88}Al_{0.11}O_3$ and $Mg_{0.85}Fe_{0.15}Si_{0.85}Al_{0.15}O_3$ synthesized at ~2000 K appears to indicate significant low-spin Fe^{3+} after the spin transition [*Catalli et al.*, 2011; *Fujino et al.*, 2012]. Ab initio investigations suggest that temperature plays a key role in stabilizing low-spin Fe^{3+} [*Hsu et al.*, 2012], but it would seem that more work is required to assess the importance of both composition and temperature on Fe^{3+} spin state at lower mantle conditions.

18.6. OXIDATION STATE OF IRON BRIDGMANITE AND ITS IMPLICATIONS

An important aspect of measured bridgmanite Fe^{3+} contents is that they appear to be nominally independent of oxygen fugacity. At upper mantle conditions minerals that can accommodate Fe^{3+} generally have $Fe^{3+}/\Sigma Fe$ ratios that are broadly a function of oxygen fugacity [*O'Neill et al.*, 1993; *Ballhaus et al.*, 1991]. At oxygen fugacities compatible with the existence of metallic iron such minerals have $Fe^{3+}/\Sigma Fe$ ratios that are below detection limits. Aluminum-bearing bridgmanites synthesized in the presence of metallic iron, however, have been shown to have $Fe^{3+}/\Sigma Fe$ ratios similar to those produced at oxygen fugacities buffered by the presence of Re and ReO_2, i.e., >0.5 [*Lauterbach et al.*, 2000; *Frost et al.*, 2004]. The $Re-ReO_2$ buffer has an oxygen fugacity at least 5 log units

higher than Fe metal saturation. There appears to be no barrier to the incorporation of Fe^{3+} in bridgmanite as even Al-free samples have been synthesised containing iron exclusively in this oxidation state [*Hummer and Fei*, 2012]. The high-pressure multianvil assembly is probably not a closed system to oxygen and likely exerts a relatively oxidizing environment due to the presence of adsorbed H_2O within the ceramics of the assembly. For this reason the occurrence of Fe^{3+} in many bridgmanite samples is probably consistent with the ambient oxygen fugacity, which is likely much higher than in the lower mantle. Nevertheless, the presence of iron metal in some of the experiments [*Lauterbach et al.*, 2000; *Frost et al.*, 2004] should ensure that the oxygen fugacity is fixed at the lowest plausible level for the lower mantle as long as equilibrium was achieved. At lower oxygen fugacities than this, iron would be reduced out of bridgmanite. The fact that high $Fe^{3+}/\Sigma Fe$ ratios are still encountered in the bridgmanite samples equilibrated with metallic iron implies that high Fe^{3+} contents most likely also exist in this mineral in the lower mantle, at least at the conditions equivalent to those at which experiments have been performed.

The relationships shown in Figure 18.4 imply that Fe^{3+} is likely stabilized within the bridgmanite structure as a result of an energetically favorable coupled substitution with Al. The Fe^{3+} substitution onto the larger A site is probably charge balanced by Al substitution onto the smaller octahedral B site [*Lauterbach et al.*, 2000; *Frost and Langenhorst*, 2002]. However, as shown in Figure 18.5, this is not the only trivalent substitution mechanism operating in bridgmanite. Most of the samples, particularly at lower total trivalent cation concentrations and including

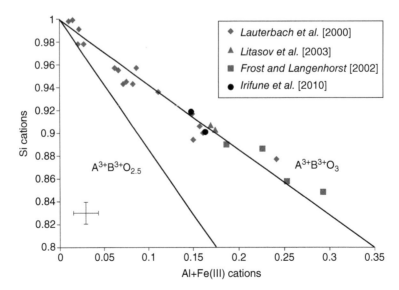

Figure 18.5 Trivalent cation substitution mechanisms in bridgmanite can be assessed through a plot of the proportion of Si cations in the structure, in atoms per 3 oxygen formula units, versus the combined trivalent cation content. The data points are from experiments that employed a range of bulk compositions and oxygen fugacities.

the range of pyrolite samples, do not plot along the compositional curve expected if substitution occurs only through the $FeAlO_3$ component. A proportion of trivalent cations seem to substitute onto the Si site with charge balance provided by the formation of an oxygen vacancy [*Navrotsky*, 1999]. This component appears to be also present in both 29 and 44 GPa experiments of *Irifune et al.* [2010]. Although small, the removal of oxygen from the structure could greatly increase the compressibility of bridgmanite [*Brodholt*, 2000]. Defect perovskites are likely to be restricted to the upper parts of the lower mantle as a result of the large ΔV of the reaction

$$4\left(Mg_{3/4}Al_{1/4}\right)\left(Si_{3/4}Al_{1/4}\right)O_3 + 2MgO = 2MgAlO_{2.5} + 3MgSiO_3.$$

$$(18.8)$$

As bridgmanite contains high levels of Fe^{3+} even in equilibrium with metallic iron, it is probably unavoidable that it contains significant Fe^{3+} in the lower mantle. Experiments suggest that lower mantle bridgmanite should have an $Fe^{3+}/\Sigma Fe$ of ~0.6, constraining the bulk lower mantle $Fe^{3+}/\Sigma Fe$ to be ~0.4. In contrast, estimates for the $Fe^{3+}/\Sigma Fe$ ratio of the upper mantle based on mantle xenoliths are <0.03 [*Canil and O'Neill*, 1996], although recent measurements on MORB glasses may imply slightly higher $Fe^{3+}/\Sigma Fe$ ratios [*Cottrell and Kelley*, 2011; 2013]. If the lower mantle is isochemical with the upper mantle, then the same bulk oxygen concentration can only be maintained if disproportionation of FeO occurs in the lower mantle, i.e.,

$$3FeO_{(bdg/fper)} = Fe_{(metal)} + Fe_2O_{3(bdg)}. \qquad (18.9)$$

To maintain the bulk oxygen concentration would require the formation of approximately 1 wt % metallic iron. The metal would likely contain also Ni, C, S, and other siderophile elements. The very observation that bridgmanite with high Fe^{3+} contents coexists with metallic Fe demonstrates that the process must take place if bridgmanite forms under conditions of insufficient oxygen [*Frost et al.*, 2004; *Frost and McCammon*, 2008]. As material upwells out of the bridgmanite stability field, Fe^{3+} will react with metallic iron to regain a similar Fe^{3+}/Fe^{2+} ratio as the upper mantle. The opposite would occur for downwelling material. The bulk oxygen concentration of the entire mantle would remain a constant, as long as the metallic iron remained locked within the lower mantle assemblage, which is likely as it would be solid throughout much of the mantle. Even upon melting, dihedral angles between metallic liquid and lower mantle assemblages are high throughout most or all of the mantle, limiting connectivity at small metal melt fractions [*Shi et al.*, 2013].

The upper mantle today is more oxidized than it would have been during core formation, when it was in equilibrium with core-forming metallic Fe. Currently the upper mantle is ~5 log units more oxidized than the stability of metallic iron, and measurements on some of the oldest rocks and minerals seem to imply that this oxidation process occurred during or very soon after core formation [*Canil*, 1997; *Delano*, 2001; *Trail et al.*, 2011]. The mechanism by which the oxidation process occurred is important because it would have influenced the nature of volatile species degassing from the mantle. At iron metal saturation degassing species would have been dominated by H_2 and CH_4, whereas the more oxidized species CO_2 and H_2O were important for forming the early atmosphere and hydrosphere.

The formation of Fe^{3+}-bearing bridgmanite in the lower mantle raises a possibility by which this oxidation could have occurred. As the lower mantle likely formed before the end of core formation, it would have formed from material initially poor in Fe^{3+}. Iron metal should, therefore, have formed with bridgmanite. If, toward the end of core formation, some (approximately 10%) of this metallic iron separated to the core, the remaining lower mantle material would contain proportionately more oxygen. Whole-mantle mixing of this oxygen-enriched material would raise the oxidation state of the mantle to the apparent present-day value. Figure 18.6 shows one plausible mechanism by which this could happen, i.e., disproportionated metal is removed from the mantle to the core by core-forming diapirs [*Frost et al.*, 2004; *Frost and McCammon*, 2008].

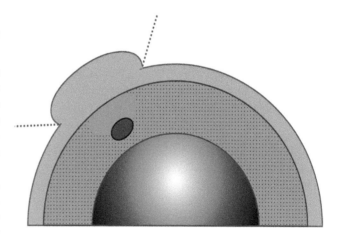

Figure 18.6 Cartoon showing one possible mechanism to oxidize the mantle during accretion and core formation. As the lower mantle forms, Fe metal and Fe^{3+}-rich bridgmanite are produced by disproportionation of FeO. As the cores of accreting planetesimals pass through the lower mantle, some of this metal is entrained and separates to the core. This leaves the lower mantle enriched in oxygen, which after whole-mantle convection raises the oxygen fugacity of the upper mantle. Approximately 10% Fe metal would have to be lost to the core to explain the increase in the redox state of Earth's upper mantle compared to that expected during core formation.

A further possibility would be that the entire lower mantle lost Fe metal to the core, which would have left it oxygen rich in comparison to the upper mantle. This would seem to require that the lower mantle has also remained isolated from the upper mantle because even small amounts of mixing would have continued to raise the upper mantle oxygen fugacity, which appears to have remained constant over geological time [*Canil*, 1997]. It has also been proposed, however, that carbon as diamond or graphite may reduce Fe^{3+} in ascending mantle as it oxidizes to form CO_2 melt components through the reaction

$$C + 2Fe_2O_3 = 4FeO + CO_2, \qquad (18.10)$$

thus masking the $Fe^{3+}/\Sigma Fe$ ratio of the deeper mantle [*O'Neill et al.*, 1993; *Stagno et al.*, 2013]. One quite interesting observation is that OIB magmas are often estimated to have higher CO_2 concentrations than MORB [e.g., *Dixon et al.*, 1997; *Aubaud et al.*, 2005]. Higher carbon concentrations in the OIB source would imply that the source at depth has a much higher $Fe^{3+}/\Sigma Fe$ ratio but that this is reduced as carbon oxidizes to CO_2 during decompression. In this way carbon concentrations in the OIB source might mask a very large initial $Fe^{3+}/\Sigma Fe$ ratio in the lower mantle, which could remove the necessity for the lower mantle to contain metallic iron.

18.7. CONCLUSIONS

Here we have attempted to summarize the arguments for and against large-scale chemical differentiation of the mantle and to examine the variations in mineral chemistry that need to be accounted for when placing constraints on the compositional and thermal structure of the lower mantle using seismic velocity observations. Given the still challenging experimental conditions encompassed by the lower mantle, significant uncertainties remain in experimentally determined phase equilibria and elasticity data. Although a number of mechanisms that may be important in the lower mantle have been recognized, consistent models for the operation of these influences have yet to be developed. In particular, defect structures and spin transitions are rarely considered in thermodynamic models but may have important effects on major element partitioning between phases and on seismic velocities. Disproportionation of ferrous iron may also have a major effect by changing the elastic properties of bridgmanite and by affecting iron partitioning with ferropericlase. It should be possible to employ a self-consistent thermodynamic approach that combines experimental data on both phase equilibria, i.e., chemical partitioning, and equation-of-state measurements into a single model for lower mantle properties. The development of such models will ensure consistency between elasticity models and

phase equilibrium studies and provide an independent technique by which to assess the composition of Earth's deep interior.

ACKNOWLEDGMENTS

An earlier version of this manuscript was improved thanks to the comments of two anonymous reviewers. This work was funded through the support of European Research Council (ERC) Advanced Grant "DEEP" (227893).

REFERENCES

Agee, C. B., and D. Walker (1988), Mass balance and phase density constraints on early differentiation of chondritic mantle, *Earth Planet. Sci. Lett.*, *90*, 144–156.

Albarède, F., and R. D. van der Hilst (2002), Zoned mantle convection, *Philos. Trans. R. Soc. Lond.*, *360*, 2569–2592.

Allegre, C. J., J. P. Poirier, E. Humbler, and A. W. Hofmann (1995), The chemical composition of the Earth, *Earth Planet Sci. Lett.*, *134*, 515–526.

Anderson, D. L. (1968), Chemical inhomogeneity of the mantle, *Earth Planet. Sci. Lett.*, *5*, 89–94.

Anderson, D. L. (1979), Chemical stratification of the mantle, *J. Geophys. Res.*, *84*, 6297–6298.

Andrault, D., N. Bolfan-Casanova, G. Lo Nigro, M. A. Bouhifd, G. Garbarino, and M. Mezouar (2011), Solidus and liquidus profiles of chondritic mantle: Implication for melting of the Earth across its history, *Earth Planet. Sci. Lett.*, *304*, 251–259.

Andrault, D., S. Petitgirard, G. Lo Nigro, J. L. Devidal, G. Veronesi, G. Garbarino, and M. Mezouar (2012), Solid-liquid iron partitioning in Earth's deep mantle, *Nature*, *487*, 354–357.

Antonangeli, D., J. Siebert, J. Badro, D. L. Farber, G. Fiquet, G. Morard, and F. J. Ryerson (2010), Composition of the Earth's inner core from high-pressure sound velocity measurements in Fe-Ni-Si alloys, *Earth Planet. Sci. Lett.*, *295*, 292–296.

Arevalo, R., W. F. McDonough, and M. Luong (2009), The K/U ratio of the silicate Earth: Insights into mantle composition, structure and thermal evolution, *Earth Planet. Sci. Lett.*, *278*: 361–369.

Aubaud, C., P. Francoise, R. Hekinian, and M. Javoy (2005), Degassing of CO_2 and H_2O in submarine lavas from the Society hotspot, *Earth Planet. Sci. Lett.* *235*, 511–527.

Austermann, J., B. T. Kaye, J. X. Mitrovica, and P. Huybers (2014) A statistical analysis of the correlation between large igneous provinces and lower mantle seismic structure, *Geophys. J. Int. 197*, 1–9.

Auzende, A. L., J. Badro, F. J. Ryerson, P. K. Weber, S. J. Fallon, A. Addad, J. Siebert, and G. Fiquet (2008), Element partitioning between magnesium silicate perovskite and ferropericlase: New insights into bulk lower-mantle geochemistry, *Earth. Planet. Sci. Lett.*, *269*, 164–174.

Badro, J. (2014), Spin transitions in mantle minerals., *Annu. Rev. Earth Planet. Sci.*, *42*, 231–248.

Badro, J., G. Fiquet, F. Guyot, J. P. Rueff, V. V. Struzhkin, and G. Vankó (2003), Iron partitioning in Earth's mantle: Toward a deep lower mantle discontinuity, *Science*, *300*, 789–791.

Badro J., A. S. Cote, and J. P. Brodholt (2014), A seismologically consistent compositional model of Earth's core, *PNAS, 132*, 7542–7545.

Ballhaus, C., R. F. Berry, and D. H. Green (1991), High pressure experimental calibration of the olivine-orthopyroxene-spinel oxygen geobarometer: Implications for the oxidation state of the upper mantle., *Contrib. Mineral. Petrol., 107*, 27–40.

Bina C. R., and G. Helffrich (2014), Geophysical constraints upon mantle composition, in *Treatise on Geochemistry, vol. 2*, 2nd ed., edited by H. D. Holland and K. K. Turekian, pp. 41–65,Elsevier, Oxford, doi:10.1016/B978-0-08-095975-7.00202-3.

Birch, F. (1952). Elasticity and constitution of the earth's interior, *J. Geophys. Res., 57*, 227–286.

Boffa Ballaran, T., A. Kurnosov, K. Glazyrin, D. J. Frost, M. Merlini, M. Hanfland, and R. Caracas (2012), Effect of chemistry on the compressibility of silicate perovskite in the lower mantle, *Earth Planet. Sci. Lett., 333–334*, 181–190.

Brodholt, J. P. (2000), Pressure-induced changes in the compression mechanism of aluminous perovskite in the Earth's mantle, *Nature, 407*, 620–622.

Burns, R. G. (1993), *Mineralogical Application of Crystal Field Theory*, Cambridge Univ. Press, Cambridge.

Campbell, I. H., and HStC O'Neill (2012), Evidence against a chondritic Earth, *Nature, 483*, 553–558.

Canil, D. (1997), Vanadium partitioning and the oxidation state of Archean komatiite magmas, *Nature, 389*, 842–845.

Canil, D., and HStC O'Neill (1996), Distribution of ferric iron in some upper-mantle assemblages,. *J. Petrol., 37*, 609–635.

Canup, R. M. (2004), Simulations of a late lunar-forming impact, *Icarus, 168*, 433–456.

Carlson, R. W., E. Garnero, T. M. Harrison, J. Li., M. Manga, W. F. McDonough, S. Mukhopadhyay, B. Romanowicz, D. Rubie, Zhong S., and Williams Q. (2014), How did early Earth become our modern world? *Annu. Rev. Earth Planet. Scie., 42*, 151–178.

Catalli, K., S.-H Shim, V. B. Prakapenka, J. Zhao, W. Struhahn, P. Chow, Y. Xiao, H. Liu, H. Cynn, and W. J. Evans (2010), Spin state of ferric iron in $MgSiO_3$ perovskite and its effect on elastic properties, *Earth Planet. Sci. Lett., 289*, 68–75.

Catalli, K., S. H. Shim, P. Dera, V. B. Prakapenka, J. Zhao, W. Sturhahn, P. Chow, Y. Xiao, H. Cynn, and W. J. Evans (2011), Effects of the Fe^{3+} spin transition on the properties of aluminous perovskite—New insights for lowermantle seismic heterogeneities., *Earth Planet. Sci. Lett., 310*, 293–302.

Chantel, J., D. J. Frost, C. A. McCammon, Z. C. Jing, and Y. B. Wang (2012), Acoustic velocities of pure and iron-bearing magnesium silicate perovskite measured to 25 GPa and 1200 K, *Geophys. Res. Lett., 39*, L19,307, doi:10.1029/2012GL053075.

Christensen, U. R., and D. A. Yuen (1985), Layered convection induced by phase-transitions, *J. Geophys. Res. 90*, 291–300.

Condie, K. C. (1998), Episodic continental growth and supercontinents: A mantle avalanche connection? *Earth Planet. Sci. Lett., 163*, 97–108.

Corgne, A., C. Liebske, B. J. Wood, D. C. Rubie, and D. J. Frost (2005), Silicate perovskite-melt partitioning of trace elements and geochemical signature of a deep perovskitic reservoir, *Geochim. Cosmochim. Acta, 69* (2), 485–496.

Cottaar, S., T. Heister, I. Rose, and C. Unterborn (2014), BurnMan: A lower mantle mineral physics toolkit, *Geochem. Geophys. Geosyst., 15*, 1164–1179.

Cottrell, E., and K. A. Kelley (2011), The oxidation state of Fe in MORB glasses and the oxygen fugacity of the upper mantle, *Earth Planet. Sci. Lett., 305*, 270–282.

Cottrell E., and K. A. Kelley (2013), Redox heterogeneity in mid-ocean ridge basalts as a function of mantle source, *Science, 340*, 1314–1317.

Creager, K. C., and T. H. Jordan (1984), Slab penetration into the lower mantle, *J. Geophys. Res. 89*, 3031–3049.

Dauphas, N., F. Poitrasson, C. Burkhardt, H. Kobayashi, K. Kurosawa (2015), Planetary and meteoritic Mg/Si and image variations inherited from solar nebula chemistry, *Earth Planet. Sci. Lett., 427*, 236–248.

Delano, J. W. (2001), Redox history of the Earth's interior since approximately 3900 Ma: Implications for prebiotic molecules. *Orig. Life Evol. Biospheres J. Int. Soc. Study Orig. Life, 31*, 311–41.

Dixon, J. E., D. A. Clague, P. Wallace, and R. Poreda (1997), Volatiles in alkalic basalts from the north arch volcanic field, Hawaii: Extensive degassing of deep submarine-erupted alkalic series lavas, *J. Petrol., 38*, 911–939.

Dziewonski, A. M., and L. Anderson (1981), Preliminary reference Earth model, *Phys. Earth Planet. Inter., 25*, 297–356.

Frost, D. J., F. Langenhorst (2002), The effect of Al_2O_3 on Fe-Mg partitioning between magnesiowüstite and magnesium silicate perovskite, *Earth Planet. Sci. Lett., 199*, 227–241.

Frost, D. J., and C. A. McCammon (2008), The redox state of Earth's mantle, *Annu. Rev. Earth Planet. Sci. 36*, 389–420.

Frost, D. J., C. Liebske, F. Langenhorst, C. A. McCammon, R. G. Tronnes, and D. C. Rubie (2004), Experimental evidence for the existence of iron-rich metal in the Earth's lower mantle, *Nature, 428*, 409–412.

Fujino, K., D. Nishio-Hamane, Y. Seto, N. Sata, T. Nagai, T. Shinmei, T. Irifune, H. Ishii, N. Hiraoka, Y. Q. Cai, and K.-D. Tsuei (2012), Spin transition of ferric iron in Al-bearing Mg–perovskite up to 200 GPa and its implication for the lower mantle, *Earth Planet. Sci. Lett. 317–318*, 407–412.

Fukao, Y., and O. Masayuki (2013), Subducted slabs stagnant above, penetrating through, and trapped below the 660 km discontinuity, *J. Geophys. Res., 118*, 5920–5938.

Fukao, Y., S., Widiyantoro and M. Obayashi, (2001), Stagnant slabs in the upper and lower mantle transition region, *Rev. Geophy., 39*, 291–323.

Gaherty, J. B., and B. H. Hager (1994) Compositional vs. thermal buoyancy and the evolution of subducted lithosphere, *Geophys. Res. Lett., 21*, 141–144.

Glazyrin, K., T. Boffa Ballaran, D. J. Frost, C. McCammon, A. Kantora, M. Merlinie, M. Hanfland, and L. Dubrovinsky (2014), Magnesium silicate perovskite and effect of iron oxidation state on its bulk sound velocity at the conditions of the lower mantle, *Earth Planet. Sci. Lett., 393*, 182–186.

Grand, S. P. (2002), Mantle shear-wave tomography and the fate of subducted slabs, *Philos. Trans. R. Soc. Lond. Seri. A Math. Phys. Eng. Sci. 360*, 2475–2491.

Grocholski, B., S.-H Shim, W. Sturhahn, J. Zhao, Y. Xiao, and P. C. Chow (2009), Spin and valence states of iron in $(Mg_{0.8}Fe_{0.2})SiO_3$, *Geophys. Res. Lett. 36*, L24,303.

Hofmann, A. W. (2014), Sampling mantle heterogeneity through oceanic basalts: Isotopes and trace elements, in *Treatise on Geochemistry*, 2nd ed., vol. *2*, edited by H. D. Holland and K. K. Turekian) pp. 67–101, Elsevier, Oxford, doi:10.1016/B978-0-08-095975-7.00202-3.

Hsu, H., P. Blaha, M. Cococcioni, and R. M. Wentzcovitch (2011), Spin crossover and hyperfine interactions of ferric iron in $MgSiO_3$ perovskite, *Phys. Rev. Lett. 106*, 118,501.

Hsu, H., Y. G., Yu and R. M. Wentzcovitch, (2012) Spin crossover of iron in aluminous $MgSiO_3$ perovskite and postperovskite, Earth Planet. *Sci. Lett., 359–360*, 34–39.

Hummer, D. R., and Y. Fei (2012), Synthesis and crystal chemistry of Fe^{3+}-bearing $(Mg,Fe^{3+})(Si,Fe^{3+})O_3$ perovskite, *Am. Mineral., 97*, 1915–1921.

Irifune T. (1994), Absence of an aluminous phase in the upper part of the Earth's lower mantle, *Nature, 370*, 131–133.

Irifune, T. and T. Tsuchiya, (2007), Mineralogy of the Earth: Phase transitions and mineralogy of the lower mantle, in *Treatise on Geophysics, vol. 2 edited by G. Schubert*, pp. 33–62, Elsevier, Amsterdam, The Netherlands.

Irifune T., T. Shinmei, C. A. McCammon, N. Miyajima, D. C. Rubie, and D. J. Frost (2010), Iron partitioning and density changes of pyrolite in Earth's lower mantle, *Science 327*, 193–195.

Ito, E., and E. Takahashi (1989), Post-spinel transformations in the system $Mg_2SiO–Fe_2SiO_4$ and some geophysical implications, *J. Geophys. Res., 94*, 10,637–10,646.

Ito, E., M. Akaogi, L. Topor, and A. Navrotsky (1990), Negative pressure-temperature slopes for reactions forming $MgSiO_3$ from calorimetry, *Science, 249*, 1275–1278.

Ito, E., T. Katsura, A. Kubo, and M. Walter (2004), Melting experiments of mantle materials under lower mantle conditions with implication to fractionation in magma ocean, *Phys. Earth Planet. Inter., 143–144*, 397–406.

Jackson, C. R. M., L. B. Ziegler, H. L. Zhang, M. G. Jackson, and D. R. Stegman (2014), A geochemical evaluation of potential magma ocean dynamics using a parameterized model for perovskite crystallization, *Earth. Planet. Sci. Lett., 392*, 154–165.

Jackson, I. (1983) Some geophysical constraints on the chemical composition of the Earth's lower mantle, *Earth. Planet. Sci. Lett. 62*, 91–103.

Jackson, M. G., and A. M. Jellinek (2013), Major and trace element composition of the high He-3/He-4 mantle: Implications for the composition of a nonchondritic Earth, *Geochem. Geophys. Geosyst., 14*, 2954–2976.

Jacobsen S. B., and G. J. Wasserburg (1979), The mean age of mantle and crustal reservoirs. *J. Geophys. Res., 84*, 7411–7427.

Jochum K. P., A. W. Hofmann, E. Ito, H. M. Seufert, and W. M. White (1983), K, U and Th in mid-ocean ridge basalt glasses and heat-production, K/U and K/Rb in the mantle, *Nature, 306*, 431–436.

Kaminsky, F. (2012),. Mineralogy of the lower mantle: A review of "super-deep" mineral inclusions in diamond, *Earth-Sci. Rev., 110*, 127–147.

Kato, T., A. E. Ringwood, and T. Irifune (1988), Experimental determination of element partitioning between silicate perovskites, garnets and liquids: Constraints on early differentiation of the mantle, *Earth Planet. Sci. Lett., 89*, 123–145.

Katsura, T., and E. Ito (1996), Determination of Fe-Mg partitioning between perovskite and magnesiowüstite, *Geophys. Res. Lett., 23*, 2005–2008.

Katsura, T., H. Yamada, T. Shinmei, A. Kubo, S. Ono, M. Kanzaki, A. Yoneda, M. J. Walter, S. Urakawa, E. Ito, K. Funakoshi, and W Utsumi (2003), Post-spinel transition in Mg_2SiO_4 determined by in situ X-ray diffractometry, *Phys. Earth Planet. Inter. 136*, 11–24.

Kind, R., and X. Li (2007), Deep Earth structure—Transition zone and mantle discontinuities, in edited by G. Schubert, B. Romanowicz, A. Dziewonski, *Treatise on Geophy, Vol. 1, Seismology and the Structure of the Earth*, pp. 591–612. Elsevier, Amsterdam.

Kobayashi, Y., T. Kondo, E. Ohtani, N. Hirao, N. Miyajima, T. Yagi, T. Nagase, and T. Kikegawa (2005), Fe-Mg partitioning between $(Mg, Fe)SiO_3$ post-perovskite, perovskite, and magnesiowustite in the Earth's lower mantle, *Geophys. Res. Lett., 32*, L19,301, doi: 10.1029/2005GL023257.

Labrosse, S., J. W. Hernlund, and N. Coltice (2007), A crystallizing dense magma ocean at the base of the Earth's mantle, *Nature, 450*, 866–869.

Larimer, J. W., and E. Anders (1970), Chemical fractionations in meteorites—III. Major element fractionations in chondrites, *Geochim. Cosmochim. Acta, 34*, 367–387.

Lassiter, J. C., and E. H. Hauri (1998), Osmium-isotope variations in Hawaiian lavas: Evidence for recycled oceanic lithosphere in the Hawaiian plume, *Earth Planet. Sci. Lett., 164*, 483–496.

Lauterbach, S., C. A. McCammon, P. van Aken, F. Langenhorst, and F. Seifert (2000), Mossbauer and ELNES spectroscopy of $(Mg,Fe)(Si,Al)O_3$ perovskite: A highly oxidised component of the lower mantle, *Contrib. Mineral. Petrol., 138*, 17–26.

Li, C., and R. D. van der Hilst (2010), Structure of the upper mantle and transition zone beneath Southeast Asia from traveltime tomography, *J. Geophys. Res., 115*, B07308, doi:10.1029/2009JB006882.

Li, X-D, and B. Romanowicz (1996), Global mantle shear velocity model developed using nonlinear asymptotic coupling theory, *J. Geophys. Res., 101*, 22,245–22,272.

Liebske, C., and D. J. Frost (2012), Melting phase relations in the MgO-$MgSiO_3$ system between 16 and 26 GPa: Implications for melting in Earth's deep interior, *Earth Planet. Sci. Lett., 345*, 159–170.

Liebske, C., A. Corgne, D. J. Frost, D. C. Rubie, and B. J. Wood (2005), Compositional effects on element partitioning between Mg-silicate perovskite and silicate melts, *Contrib. Mineral. Petrol., 149*, 113–128.

Lin, J.-F., S. Speziale, Z. Mao, and H. Marquardt (2013), Effects of the electronic spin transitions of iron in lower mantle minerals: Implications for deep mantle geophysics and geochemistry, *Rev. Geophys., 51*, 244–275, doi:10.1002/rog.20010.

Litasov, K., E. Ohtani, F. Langenhorst, H. Yurimoto, T. Kubo, and T. Kondo (2003), Water solubility in Mg-perovskites, and water storage capacity in the lower mantle, *Earth. Planet. Sci. Lett., 211*, 189–203.

Liu, L. G. (1974), Silicate perovskite from phase transformations of pyrope-garnet at high pressure and temperature, *Geophys. Res. Lett.*, *1*, 277–280.

Liu, L. G. (1975) Post-oxide phases of forsterite and enstatite, *Geophys. Res. Lett.*, *2*, 417–419.

Liu, L. G. (1976), The post-spinel phases of forsterite, *Nature*, *262*, 770–772.

McCammon, C. (1997), Perovskite as a possible sink for ferric iron in the lower mantle, *Nature*, *387*, 694–696.

McCammon C. A., S. Lauterbach, F. Seifert, F. Langenhorst, and P. A. van Aken (2004) Iron oxidation state in lower mantle mineral assemblages - I. *Empirical relations derived from high-pressure experiments. Earth Planet. Sci. Lett.* *222*, 435–449.

McDonough, W. F., and S.-S. Sun (1995), The composition of the Earth, *Chem. Geol.*, *120*, 223–253.

Murakami, M., K. Hirose, K. Kawamura, N. Sata, and Y. Ohishi (2004), Post–perovskite phase transition in $MgSiO_3$, *Science*, *304*(5672), 855–858.

Murakami, M., Y. Ohishi, N. Hirao, K. Hirose (2012), A perovskitic lower mantle inferred from high–pressure, high–temperature sound velocity data, *Nature 485*, 90–94.

Myhill, R. (2013), Slab buckling and its effect on the distributions and focal mechanisms of deep–focus earthquakes. *Geophys J. Int.*, *192*, 837–853.

Nakajima, Y., D. J., Frost, D. C. Rubie (2012), Ferrous iron partitioning between magnesium silicate perovskite and ferropericlase and the composition of perovskite in the Earth's lower mantle, *J. Geophys. Res. 117*, B08,201, doi: 10.1029/2012JB009151.

Navrotsky, A (1999), A lesson from ceramics, *Science*, *284*, 1788–1789.

Ni, S., E. Tan, M. Gurnis, and D. Helmberger (2002), Sharp sides to the African superplume, *Science*, *296*, 1850–1852.

Nishiyama N., and T. Yagi (2003), Phase relation and mineral chemistry in pyrolite to 2200 degrees C under the lower mantle pressures and implications for dynamics of mantle plumes, *J. Geophys. Res.*, *108*, 2255 doi:10.1029/2002JB002216.

Nomura, R., H. Ozawa, S. Tateno, K. Hirose, J. Hernlund, S. Muto, H. Ishii, and N. Hiraoka (2011), Spin crossover and ironrich silicate melt in the Earth's deep mantle, *Nature*, *473*, 199–202.

Oliver, J. and B. Isacks (1967), Deep earthquake zones, anomalous structures in upper mantle and lithosphere, *J. Geophys. Res. 72*, 4259–4275.

O'Neill, H.St.C, and H. Palme (1998), Composition of the silicate earth: Implications for accretion and core formation, in *The Earth's Mante-Composition, Structure and Evolution*, edited by I. Jackson, pp. 3–126, Cambridge Univ. Press, Cambridge.

O'Neill, HStC, D. C. Rubie, D. Canil, C. A. Geiger, and C. R. Ross (1993), Ferric iron in the upper mantle and in transition zone assemblages: Implications for relative oxygen fugacities in the mantle, in *Evolution of the Earth and Planets, Geophys. Monogr. 74*, edited by E. Takahashi, R. Jeanloz, and D. C. Rubie, pp. 73–88, AGU, Washington, D.C.

Poirier, J. P. (1994), Light elements in the Earth's outer core—A critical review, *Phys. Earth Planet. Inter.*, *85*, 319–337.

Richter, F., and D. McKenzie (1978), Simple plate models of mantle convection, *J. Geophys.*, *44*, 441–471.

Richter, F. M., and C. E. Johnson (1974), The stability of a chemically layered mantle, *J. Geophys. Res.*, *79*, 1635–1639.

Ringwood, A. E. (1975), *Composition and Petrology of the Earth's Mantle*, McGraw-Hill, New York.

Ringwood, A. E. (1982), Phase transformations and differentiation in subducted lithosphere: Implications for mantle dynamics, basalt petrogenesis, and crustal evolution, *J. Geol. 90*, 611–643.

Ringwood, A. E. (1991), Phase transformations and their bearing on the constitution and dynamics of the mantle, *Geochim. Cosmochim. Acta*, *55*, 2083–2110.

Saikia, A., D. J. Frost, and D. C. Rubie (2008), Splitting of the 520- kilometer seismic discontinuity and chemical heterogeneity in the mantle, *Science*, *319*, 1515–1518.

Sakai, T., E. Ohtani, H. Terasaki, N. Sawada, Y. Kobayashi, M. Miyahara, M. Nishijima, N. Hirao, Y. Ohishi, T. Kikegawa (2009), Fe-Mg partitioning between perovskite and ferropericlase in the lower mantle, *Am. Mineral.*, *94*, 921–925.

Shi, C. Y., L. Zhang, W. Yang, Y. Liu, J. Wang, Y. Meng, J. C. Andrews, and WL Mao (2013), Formation of an interconnected network of iron melt at Earth's lower mantle conditions, *Nature Geosci.*, *6*, 971–975.

Sinmyo, R., and K. Hirose (2013), Iron partitioning in pyrolitic lower mantle, *Phys. Chem. Minerals*, *40*, 107–113.

Sinmyo, R., K. Hirose, D. Nishio-Hamane, Y. Seto, K. Fujino, N. Sata, and Y. Ohishi (2008), Partitioning of iron between perovskite/postperovskite and ferropericlase in the lower mantle, *J. Geophys. Res. 113*, B11,204, doi: 10.1029/2008JB005730.

Sobolev, A. V., A. W. Hofmann, and I. K. Nikogosian (2000), Recycled oceanic crust observed in "ghost plagioclase" within the source of Mauna Loa lavas, *Nature*, *404*, 986–990, doi:10.1038/35010098.

Solheim, L. P., and W. R. Peltier (1994), Avalanche effects in phase transition modulated thermal convection: A model of Earth's mantle, *J. Geophys. Res. 99*, 6997–7018.

Solomatov, V. S., and C. C. Reese (2008), Grain size variations in the Earth's mantle and the evolution of primordial chemical heterogeneities, *J. Geophys. Res.*, *113*, 10.1029/2007JB005319.

Speziale, S., V. E. Lee, S. M. Clark, J. F. Lin, M. P. Pasternak, and R. Jeanloz (2007). Effects of Fe spin transition on the elasticity of (Mg,Fe)O magnesiowüstites and implications for the seismological properties of the Earth's lower mantle, *J. Geophys. Res*, *112*, B10212, doi:10.1029/2006JB004730.

Stagno, V., O. Dickson, C. A. McCammon, and D. J. Frost (2013), The oxidation state of the mantle and the extraction of carbon from Earth's interior, *Nature*, *493*, 84–88. doi:10.1038/nature11679.

Stixrude, L., and R. Jeanloz (2007), Constraints on seismic models from other disciplines—Constraints from mineral physics on seismological models, in *Treatise on Geophysic*, vol. 1: *Seismology and the Structure of the Earth*, edited by G. Schubert, pp. 775–803, Elsevier., Oxford, doi:10.1016/B978-044452748-6/00026-2.

Su, W.-J., R. L. Woodward, and A. M. Dziewonski (1994), Degree 12 model of shear velocity heterogeneity in the mantle, *J. Geophys. Res. Solid Earth*, *99*, 6945–6980.

Tackley, P. J., D. J. Stevenson, G. A. Glatzmaier, and G. Schubert (1993), Effects of an endothermic phase transition at 670 km depth in a spherical model of convection in the Earth's mantle, *Nature*, *361*, 699–704.

Tange, Y., E. Takahashi, Y. Nishihara, and K. I. Funakoshi, N. Sata (2009), Phase relations in the system MgO-FeO-SiO$_2$ to 50 GPa and 2000 degrees C: An application of experimental techniques using multianvil apparatus with sintered diamond anvils, *J. Geophys. Res.*, *114*, B02,214, doi:10.1029/2008JB005891.

Torsvik, T. H., M. A. Smethurst, K. Burke, and B. Steinberger (2008), Long term stability in Deep Mantle structure: Evidence from the ca. 300 Ma Skagerrak-Centered Large Igneous Province (the SCLIP), *Earth Planet. Sci. Lett.*, *267*, 444–452.

Trail, D., E. B. Watson, and N. D. Tailby (2011), The oxidation state of Hadean magmas and implications for early Earth's atmosphere, *Nature*, *480*, 79–82.

Tschauner, O., C. Ma, J. R. Beckett, C. Prescher, V. B. Prakapenka, and G. R. Rossman (2014), Discovery of bridgmanite, the most abundant mineral in Earth, in a shocked meteorite, *Science*, *346*, 1100–1102, doi:10.1126/science.1259369.

van der Hilst, R. D., S. Widiyantoro, and E. R. Engdahl (1997), Evidence for deep mantle circulation from global tomography, *Nature*, *386*, 578–584.

Walter, M. (2003), Melt extraction and compositional variability in mantle lithosphere, in *The Mantle and Core*, (R. W. Carlson), vol. 2, *Treatise on Geochemistry*, edited by H. D. Holland and K. K. Turekian, pp. 363–394, Elsevier-Pergamon, Oxford.

Walter, M. J., E. Nakamura, R. G. Tronnes, and D. J. Frost (2004), Experimental constraints on crystallization differentiation in a deep magma ocean, *Geochim. Cosmochim. Acta*, *68*, 4267–4284.

Wang, X., T. Tsuchiya, and A. Hase (2015), Computational support for a pyrolitic lower mantle containing ferric iron, *Nature Geosci. 8*, 556–559, doi:10.1038/ngeo2458.

Wasson, J. T. and G. W. Kallemeyn (1988), Composition of chondrites, *Philos. Trans. R. Soc. Lond. A*, *325*, 535–544.

Wood, B. J., and D. C. Rubie (1996), The effect of alumina on phase transformations at the 660-km discontinuity from Fe-Mg partitioning experiments, *Science*, *273*, 1522–1524.

Workman, R. K., S. R. Hart (2005), Major and trace element composition of the depleted MORB mantle (DMM), *Earth Planet. Sci. Lett.*, *231*, 53–72.

Xu, W., C. Lithgow-Bertelloni, L. Stixrude, and J. Ritsema (2008), The effect of bulk composition and temperature on mantle seismic structure, *Earth Planet. Sci. Lett.*, *275*, 70, doi:10.1016/j.epsl.2008.08.012.

Zhang, L., Y. Meng, W. Yang, L. Wang, W. L. Mao, Q.-S. Zeng, J. S. Jeong, A. J. Wagner, K. A. Mkhoyan, W. Liu, R. Xu, and H. K. Mao (2014), Disproportionation of (Mg,Fe)SiO$_3$ perovskite in Earth's deep lower mantle, *Science*, *344*, 877–882.

19

Phase Diagrams and Thermodynamics of Lower Mantle Materials

Susannah M. Dorfman

ABSTRACT

Experimental observations of phase equilibria in silicates have provided the key to interpreting seismic discontinuities and wave speeds in Earth's mantle. Recent discoveries, including the post-perovskite phase transition in the deep lower mantle, have provided new constraints on the mantle phase assemblage and the geotherm. However, precise experimental determination of equilibrium chemistry is difficult at high pressures and temperatures. In this chapter experimental and computational approaches to phase equilibria of the lower mantle and their uncertainties are summarized. Experimental results and models are used to describe phase equilibria for a simple system, $(Mg,Fe)SiO_3$, and bulk mantle compositions pyrolite, basalt, and harzburgite. Phase equilibria in these systems are compared to seismic observations.

19.1. INTRODUCTION

Phase equilibria of mantle minerals are fundamentally important to the understanding of seismic observations of Earth's mantle. Discontinuous changes in density and wave speed at multiple depths, including the ~660 km discontinuity that defines the top of the lower mantle [e.g., *Shearer*, 2000] and the D" discontinuity ~250 km above the core-mantle boundary (CMB) [*Wysession et al.*, 1998], have been explained by pressure-induced phase transitions in mantle silicates. Detailed constraints on the effects of pressure, temperature, and composition on the mantle phase assemblage can be used to quantify thermal and chemical heterogeneities in the mantle.

Phase diagrams for high-pressure minerals and the lower mantle phase assemblage are most directly obtained by applying extreme pressures and temperatures to different

compositions and identifying the synthesized phases. In experiments, phases synthesized at mantle pressure-temperature conditions are typically characterized by in situ or ex situ X-ray diffraction or ex situ microscopy [e.g., *Duffy*, 2005]. In addition, multiple ab initio computational methods for predicting stable crystal structures have been applied to minerals at extreme pressures [e.g., *Oganov and Glass*, 2006]. A key to both experimental and theoretical methods is determining whether equilibrium has been reached. This is commonly assessed by reversibility and dependence of the results on run time and starting conditions [*Fyfe*, 1960]. However, the only fundamental measure of equilibrium is the Gibbs free energy, G, which cannot be measured directly and must instead be calculated from thermodynamic properties.

In this chapter experimental and theoretical constraints on phase equilibria in Earth's lower mantle are reviewed. Experimental methods and sources of uncertainty are summarized. The thermodynamic properties of mantle materials are described in the context of modeling equilibria. Recent experimental data and the thermodynamic

Department of Earth and Environmental Sciences, Michigan State University, East Lansing, MI, USA

Deep Earth: Physics and Chemistry of the Lower Mantle and Core, Geophysical Monograph 217, First Edition.
Edited by Hidenori Terasaki and Rebecca A. Fischer.

code developed by *Stixrude and Lithgow-Bertelloni* [2011] are applied to the phase diagram of the dominant lower mantle silicate, $(Mg,Fe)SiO_3$ bridgmanite. This model is also used to produce new phase diagrams for the bulk silicate mantle. Boundaries in these diagrams are compared to constraints from seismology on the depths of mantle discontinuities.

19.2. METHODS FOR DETERMINING PHASE EQUILIBRIA

19.2.1. Experimental Synthesis

Measuring lower mantle phase equilibria in the laboratory depends first on generating extreme pressure-temperature conditions in multianvil presses or the laser-heated diamond anvil cell (LHDAC). Large-volume multianvil presses have provided the most detailed constraints on phase equilibria in the uppermost part of the lower mantle [e.g., *Irifune et al.*, 1998]. Maximum reported pressures reached in multianvil presses are now greater than 100 GPa (2250 km depth) [e.g., *Yamazaki et al.*, 2014]. However, only a few detailed studies of lower mantle phase equilibria using multianvil presses have reached pressures as high as ~50 GPa (1250 km) [e.g., *Tange et al.*, 2009], and the routine maximum is <30 GPa (800 km). At deep lower mantle conditions of up to ~136 GPa and ~3000 K, experiments are performed in the LHDAC.

To reach the conditions of the CMB, the LHDAC presents additional challenges relative to the multianvil press. Calibration of pressure in the LHDAC is based on the equations of state (EOS) of internal pressure standards, which are increasingly uncertain with pressure. For different commonly used pressure standards, existing scales may differ by up to 10% at pressures above 100 GPa [*Dorfman et al.*, 2012]. Recent efforts to cross-compare compression behavior of different calibrant materials have reduced relative differences to <3% [e.g., *Dewaele et al.*, 2004; *Dorfman et al.*, 2012], but which scales provide the most accurate pressures is still debated. Systematic differences in pressure measurement also originate from deviatoric stresses. Stress is applied in the LHDAC along only a single axis. Hydrostaticity of the stress state can be improved by the use of soft pressure-transmitting media and high-temperature annealing [*Angel et al.*, 2007; *Klotz et al.*, 2009; *Dorfman et al.*, 2012]. Due largely to relaxation of deviatoric stresses, measured pressures typically differ by a few gigapascals before and after laser annealing. Laser heating is the only method capable of heating DAC samples to the thousands of Kelvin temperatures of the deep mantle but suffers from instabilities over time, uncertainties in temperature measurement of tens to hundreds of K, and temperature gradients of ~1000 K over a few micrometers [e.g., *Campbell*, 2008]. Temperature stability, measurement,

and homogeneity can be improved by the use of insulating media [e.g., *Boehler and Chopelas*, 1992; *Kiefer and Duffy*, 2005]. Assembly of samples for the LHDAC requires miniaturization to volumes ~3 orders of magnitude smaller than multianvil sample volumes. At ~100 GPa, typical samples in the DAC have diameter <50 μm and thickness ~5 μm. To minimize gradients in pressure and temperature across measurements, analytical probes must be focused to narrower diameters of <10 μm. To isolate individual crystallites formed in typical experiments, submicrometer resolution is required.

In situ characterization of phase assemblages formed in experiments at lower mantle conditions is dominantly performed by synchrotron X-ray diffraction [*Duffy*, 2005]. Synchrotron beamlines dedicated to LHDAC experiments [e.g., *Mezouar et al.*, 2005; *Meng et al.*, 2006; *Prakapenka et al.*, 2008] focus monochromatic X rays to <5 μm for in situ diffraction within the laser-heated spot. From observed crystal lattice spacings the phase assemblage and approximate compositions of each phase can be determined. In situ diffraction is the only method that can be used to find and solve crystal structures of unquenchable or unrecoverable phases formed at high pressure or temperature. For example, perovskite-structured $CaSiO_3$ [*Liu and Ringwood*, 1975] and post-perovskite $MgSiO_3$ [*Murakami et al.*, 2004] both become amorphous upon decompression to ambient conditions. In situ analysis ensures no chemical diffusion or structural changes associated with temperature or pressure quench can influence observations of the phase assemblage.

The relevance of phase assemblages identified in experiments to Earth conditions and time scales depends on the assumption of equilibrium. However, in typical experiments the stability of laser heating and availability of synchrotron X-ray diffraction beam time limit typical run times to less than an hour, which depending on temperature/pressure conditions may be insufficient to reach equilibrium. The most robust constraints on phase equilibria are determined by testing reversibility of phase transitions, i.e., multiple loops of pressure-temperature conditions across boundaries and in situ confirmations of the phase change [*Fyfe*, 1960]. Unfortunately sluggish kinetics of solid-state phase transitions result in hysteresis and make complete experimental reversals essentially impossible for some reactions, particularly those involving disproportionation. Observations of phase mixtures are thus difficult to use to determine the widths of phase boundaries. Phase transition loops can be bounded from above and below by observations of transitions on compression and decompression, respectively. For example, this reasoning has been applied to bracket the post-perovskite phase transition in $(Mg,Fe)SiO_3$ [*Catalli et al.*, 2009]; however, large uncertainties remain. Synthesis of phases from multiple starting structures, such as crystalline versus amorphous samples, provides additional evidence of stability.

Diffraction in the LHDAC suffers from limited sensitivity to chemical composition and sample heterogeneity. X-ray diffraction may not detect trace (less than a few percent) or weakly diffracting (disordered or low-atomic-number) phases. The compositions of observed phases may be determined based on their effects on the crystal lattices of each phase but with low precision. The constraint depends on the magnitude of the effect of composition on structure and accurate previous measurements of the pressure-volume EOS of multiple compositions. For $(Mg,Fe)SiO_3$ bridgmanite, the $Fe/(Mg+Fe)$ (Fe#) can be determined with powder diffraction to about 10% precision under best conditions [Dorfman et al., 2013]. Another major problem for compositional constraints is heterogeneous temperature during laser heating, which causes diffusion of elements along thermal gradients (Soret diffusion) [Campbell et al., 1992; Sinmyo and Hirose, 2010]. In Fe-bearing silicate samples, Fe and Fe-rich phases are known to diffuse away from the sample center during inhomogeneous laser heating. As a result, in situ diffraction at only the sample center may see a composition different from the starting material and miss phases segregated to the edges.

More precise chemical measurements require ex situ analysis. Small, fragile high-pressure samples can be sectioned using the focused ion beam for transmission electron microscopy (TEM) [Wirth, 2009]. TEM has nanometer-scale spatial resolution and ~1% compositional sensitivity suitable for determining compositions of individual crystallites in LHDAC samples. Recent experimenters have applied these techniques to deep lower mantle phase assemblages synthesized in the LHDAC [e.g., Auzende et al., 2008; Hirose et al., 2008]. Compositions of coexisting phases can be used to compute partitioning coefficients and phase boundaries. In addition, TEM analysis has revealed trace phases missed in diffraction experiments such as nanoscale Fe metal droplets [Frost et al., 2004]. For some materials, structures can also be determined in the transmission electron microscope via electron diffraction. However, silicates (especially those formed at high pressure) are sensitive to damage from the electron beam, which can cause changes in both structure and composition [Carrez et al., 2001]. When paired with in situ diffraction for structure determination, ex situ electron microscopy provides particularly powerful constraints on high-pressure phase relations.

19.2.2. Ab Initio Calculations

Ab initio computational modeling is increasingly important for exploring phase equilibria at the extreme conditions of the lower mantle [Gillan et al., 2006]. Ab initio methods based on density functional theory (DFT) [Hohenberg and Kohn, 1964; Kohn and Sham, 1965] can simulate structures and their properties at conditions difficult or impossible to access with experiments. DFT can be used both to test the relative stability of known structures and to search for previously undiscovered stable or metastable structures. For a recent important example from the lower mantle, DFT methods were instrumental in predicting and identifying the $MgSiO_3$ post-perovskite structure [Murakami et al., 2004; Oganov and Ono, 2004]. Since the discovery of the post-perovskite transition, multiple studies have applied DFT to the effects of composition on the stability and structure of post-perovskite phases [e.g., Caracas and Cohen, 2005].

However, DFT results are known to depend on approximations and corrections, particularly the choice of exchange correlation functional. For phases with similar structures and energies, differences between the commonly used local density approximation (LDA) and generalized gradient approximation (GGA) functionals can lead to large differences (tens to hundreds of gigapascals) in calculated phase transition pressures. DFT calculations are thus more reliable for determining trends than absolute phase transition pressures and must be tested by experiments. To predict the effects of nonzero temperatures on phase stability and properties, additional computations are needed of lattice dynamics, which entail additional approximations and uncertainties. The most commonly used approximation, quasi-harmonic lattice vibration, is accurate at temperatures well below melting but begins to fail at higher temperatures relevant to the lower mantle for mantle materials [e.g., Tsuchiya and Wang, 2013]. At higher temperatures, molecular dynamics simulations are required [e.g., Oganov et al., 2001a,b]. Computing power has limited ab initio models to simple compositions. Computational methods are thus best applied to endmembers rather than to complex natural mineral compositions.

19.3. THERMODYNAMICS AND STABILITY

Pressure- and temperature-induced phase transitions are consequences of the different responses of different structures to stress, i.e., their physical properties. Provided with sufficient constraints on physical properties of mantle phases, thermodynamic modeling can thus also predict mantle phase relations. High-pressure, high-temperature diffraction experiments described above are also used to measure volume V, density ρ, and isothermal bulk modulus K_T as a function of pressure, temperature, and composition. Other experimental techniques such as ultrasonic interferometry, Brillouin spectroscopy, and nuclear inelastic scattering can determine both bulk and shear isentropic elastic properties but are limited in pressure and/or composition. For materials at deep lower mantle conditions, DFT calculations provide unique insights and constraints on thermodynamic properties.

Phase equilibria in the mantle have been computed from thermodynamic constraints by numerous previous studies. Here we follow *Stixrude and Lithgow-Bertelloni* [2005, 2011], in which both phase equilibria and physical properties of mantle phases are modeled with a single, self-consistent approach based on fundamental thermodynamics [*Callen*, 1960]. The goal is to evaluate stability based on the Gibbs free energy, $\mathcal{G}(P, T)$, of a system of i phases, each with compositions varying by solutions of j components on k sites. The Gibbs free energy can be expressed as the sum of the chemical potentials μ_i of phases of abundance n_i:

$$\mathcal{G}(P,T,n) = \sum_i n_i \mu_i (P,T,n),$$

$$\mathcal{G}(P,T,n) = \sum_i n_i \left[\mathcal{G}_i(P,T) + \mathcal{G}_{\text{mixing}}(P,T,n) \right].$$

The chemical potential is a function of the physical behavior of each phase as well as chemical mixing. The partial molar Gibbs free energy, $\mathcal{G}_i$, is specified by the EOS of the pure phase as a function of pressure and temperature, and $\mathcal{G}_{\text{mixing}}$ is a combination of the energy of ideal mixing and nonideal mixing. These components of the chemical potential are described separately below and in more detail by *Stixrude and Lithgow-Bertelloni* [2005, 2011].

19.3.1. Energy of Pure Phases

High-pressure EOSs based on Eulerian finite strain theory [*Birch*, 1952] are expressed as functions of volume rather than functions of pressure. For this reason, the Helmholtz free energy, $\mathcal{F}(V, T)$, is more suitable than $\mathcal{G}(P, T)$ for formulating the properties of lower mantle minerals. The Helmholtz free energy can be related to the Gibbs free energy through a partial Legendre transform:

$$\mathcal{G}(P,T) = \mathcal{F}(V,T) + PV.$$

The Helmholtz free energy of each end-member composition can be written as a sum of an ambient part, $\mathcal{F}_0$, a cold, compressed part, $\mathcal{F}_C$, a high-temperature part, $\mathcal{F}_q$, a magnetic part, $\mathcal{F}_m$, and any other significant pressure- or temperature-dependent components. Multiple empirical formulations have been developed to express the dependence of free energy on isothermal compression [*Poirier*, 2000]. The commonly used Birch-Murnaghan EOS [*Birch*, 1952] is developed from the Eulerian finite strain,

$$f = \frac{1}{2} \left[\left(\frac{\rho}{\rho_0} \right)^{2/3} - 1 \right].$$

A Taylor series expansion truncated to the third order gives the following form for $\mathcal{F}_C$:

$$\mathcal{F}_c = \frac{9K_{T0}}{2\rho_0} f^2 + \frac{9K_{T0}}{2\rho_0} \left(K'_{T0} - 4 \right) f^3,$$

The prime indicates the pressure derivative and the subscript zero ambient conditions.

The high-temperature part of the free energy is expressed using the quasi-harmonic approximation for lattice vibrations, the Mie-Grüneisen-Debye approach [*Debye*, 1912; *Grüneisen*, 1912]. The parameter $\mathcal{F}_q$ is a function of temperature, the Boltzmann constant k, the Debye temperature θ, and the Debye free energy function D:

$$\mathcal{F}_q = 9nk\Delta \left[TD \left(\frac{\theta}{T} \right) \right].$$

The characteristic vibrational temperature also varies with strain. The Taylor series expansion has coefficients related to the ambient values of the Grüneisen parameter γ and its pressure derivative q. When truncated to second order, θ varies as $\theta^2 = \theta_0^2 (1 + 6\gamma_0 f + (-6\gamma_0 + 18\gamma_0^2 - 9q_0\gamma_0) f^2)$.

The magnetic part includes the energy of electronic spin degeneracy. The only major mantle element that exhibits electronic spin transitions is iron, for which the spin quantum number S varies in lower mantle phases from 0 to 2.5 [*Lin et al.*, 2013; *Badro*, 2014]. For magnetically disordered phases i with r stoichiometry of Fe, $\mathcal{F}_m = -T\Sigma r_i \ln (2S_i + 1)$. The assumption of magnetic disorder is valid for lower mantle phases due to the high temperatures in Earth's interior.

An additional energy term is required for the SiO_2 phase in the lower mantle. Stishovite undergoes a second-order, or displacive, transition to the $CaCl_2$-type structure at ~50 GPa [*Andrault et al.*, 1998]. To model this change in compression behavior, displacive transition energy $\mathcal{G}_L$ can also be computed as a function of pressure and temperature.

19.3.2. Energy of Mixing

In solid solutions of multiple components, the energy due to mixing is the sum of the energy related to the increase in configurational entropy of a disordered mixture and the energy required to accommodate exchange of atoms of different sizes and chemical properties [*Wood and Nicholls*, 1978; *Helffrich and Wood*, 1989; *Mukhopadhyay et al.*, 1993]. The approximation of ideal mixing assumes only the former and is reasonable for elements with similar size such as Mg^{2+} and Fe^{2+}. Nonideal mixing becomes important for more mismatched elements such as Mg^{2+} and Ca^{2+}.

For each phase, the configurational entropy S_{conf} is related to the total number of combinations, Ω, of numbers of atoms, N_{jk}, of component j on site k, for a total number of components, c. The quantity N_{jk} is the sum $\sum_i s_{ijk} n_i$, where s_{ijk} are the stoichiometric coefficients of each component j on site k in end members i and n_i are the end-member abundances. The number of atoms on site k, N_k, is the sum of N_{jk} over the components:

$$S_{conf} = k \ln \Omega$$

$$= k \ln \prod_k \frac{\left(\sum_j^c N_{jk} \right)!}{\prod_j^c N_{jk}!} \sim R \sum_k \left(N_k \ln N_k - \sum_j^c \left(N_{jk} \ln N_{jk} \right) \right).$$

With this approximation, the Gibbs free energy related to ideal mixing is given by the following:

$$\mathcal{G}_{ideal} = -RT \sum_k \left(\ln N_k \sum_j^c s_{ijk} - \sum_j^c \left(s_{ijk} \ln N_{jk} \right) \right).$$

The nonideal or excess contribution to the energy of mixing accounts for differences in size with parameters d_γ and other chemical interactions with the Margules parameters $W_{\alpha\beta}$ defined for pairs of end members α and β. The proportion of end member α, weighted by d_α, is φ_α, defined as $\varphi_\alpha = n_\alpha d_\alpha / \sum_\gamma n_\gamma d_\gamma$. The excess contribution is then evaluated for each phase as a sum over all pairs of end members:

$$\mathcal{G}_{nonideal} = -\sum_{\beta > \alpha} \frac{2d_i}{d_\alpha + d_\beta} W_{\alpha\beta} \left(\delta_{i\alpha} - \phi_\alpha \right) \left(\delta_{i\beta} - \phi_\beta \right).$$

The Kronecker deltas δ and the sum over terms $\beta > \alpha$ ensure that each pair is counted only once.

19.4. PHASE DIAGRAMS OF LOWER MANTLE MINERALS

The thermodynamic framework above is the basis for the phase equilibrium modeling code HeFESTo [*Stixrude and Lithgow-Bertelloni*, 2011], which is used in this section to generate phase diagrams for mantle phases. The full set of parameters used to model each component of each phase in HeFESTo, defined at ambient conditions denoted with subscript zero, are Helmholtz free energy $\mathcal{F}_0$; pressure-volume EOS parameters V_0, K_0, and K_0'; the Debye temperature θ_0; the Gruneisen parameter γ_0 and its pressure and shear derivatives q_0 and η_{S0}; shear modulus G_0 and its pressure derivative G_0'; and nonideal mixing parameters $W_{\alpha\beta}$ and d_α. The model does not include the effects of spin transitions in Fe-bearing mantle phases [*Lin et al.*, 2013; *Badro*, 2014], disproportionation of ferrous iron to ferric iron and metallic iron [*Frost et al.*, 2004], or multiple post-perovskite phases [*Tschauner et al.*, 2008; *Yamanaka et al.*, 2012; *Zhang et al.*, 2014].

19.4.1. $MgSiO_3$-$FeSiO_3$ System in Lower Mantle

As discussed in previous chapters, the dominant phase of the lower mantle is orthorhombic $GdFeO_3$-type perovskite-structured $(Mg,Fe,Al)(Al,Si)O_3$ bridgmanite. Bridgmanite controls the properties of the lower mantle and therefore its chemistry and properties have been of great interest. Recent experiments have determined that the bridgmanite structure can incorporate up to 75% $FeSiO_3$ at a deep-mantle pressure of ~80 GPa [*Tateno et al.*, 2007; *Dorfman et al.*, 2013], much higher than the previously accepted solubility published in earlier phase diagrams of the $(Mg,Fe)SiO_3$ system [e.g., *Irifune and Tsuchiya*, 2007]. More controversial is the post-perovskite boundary, with different experimental studies reporting opposite Fe partitioning and effect of Fe on the boundary [*Auzende et al.*, 2008; *Hirose et al.*, 2008]. These disagreements result from the increasing challenges associated with accurate characterization of phases formed at the higher pressures of the deep lower mantle.

Recent constraints on the EOS of Fe-bearing bridgmanite and post-perovskite were used to model the phase diagram for $(Mg,Fe)SiO_3$ at lower mantle conditions. The majority of experiments and computational studies have reached consensus on the effect of Fe^{2+} on V and K of bridgmanite [*Dorfman and Duffy*, 2014]. For post-perovskite, the effect of Fe^{2+} on K is relatively controversial, with some experimental studies finding a softening with Fe content while others and most ab initio simulations predicting the opposite [*Dorfman and Duffy*, 2014]. The parameters for $MgSiO_3$ and $FeSiO_3$ end members used in the model here and given in Table 19.1 are fit from EOS parameters for compositions ranging from 0 to 74% $FeSiO_3$ for bridgmanite [*Lundin et al.*, 2008; *Dorfman et al.*, 2013] and post-perovskite [*Guignot et al.*, 2007; *Dorfman et al.*, 2013]. For simplicity, K_0 for post-perovskite was assumed to be invariant with Fe content. Relative to the previous model by *Stixrude and Lithgow-Bertelloni* [2011], V is smaller for $FeSiO_3$ bridgmanite and larger for $FeSiO_3$ post-perovskite. To stabilize $FeSiO_3$ post-perovskite at pressures observed in *Dorfman et al.* [2013], here

Table 19.1 Birch-Murnaghan equation-of-state parameters for bridgmanite and post-perovskite end members used in phase diagram computation.

	MgSiO$_3$ bridgmanite	FeSiO$_3$ bridgmanite	MgSiO$_3$ post-perovskite	FeSiO$_3$ post-perovskite
V_0 (cm^3/mol)	24.441	25.337	24.419	25.806
K_0 (GPa)	253.0	278.0	231.2	231.2
K'_0	4	4	4	4

Note: V_0 and K_0 for bridgmanite were fit from results for 0–74% FeSiO$_3$ from *Lundin et al.* [2008] and *Dorfman et al.* [2013]. V_0 for post-perovskite was fit from MgSiO$_3$ [*Guignot et al.*, 2007] and 74% FeSiO$_3$ [*Dorfman et al.*, 2013] compositions.

the $\mathcal{F}_0$ of FeSiO$_3$ post-perovskite was reduced by 60 kJ/mol relative to the previous model. A difference in this free-energy parameter of 23 kJ/mol was the sole difference in models by *Stixrude and Lithgow-Bertelloni* [2011] for conflicting measurements of partitioning between bridgmanite and post-perovskite phases [*Auzende et al.*, 2008; *Hirose et al.*, 2008]. All other parameters for thermodynamic properties of mantle phases in the model are as optimized by *Stixrude and Lithgow-Bertelloni* [2011].

The modeled solubility of FeSiO$_3$ in bridgmanite at 2000 K (Figure 19.1) increases with pressure to a maximum of ~68% FeSiO$_3$ at 85 GPa. This increase is in better agreement with experimental observations than the previous model, though still lower than the experimental maximum of >75% FeSiO$_3$ [*Tateno et al.*, 2007; *Dorfman et al.*, 2013]. To reconcile the stability of Fe-rich bridgmanite with trends observed at lower pressures and Fe contents, future models may need to incorporate pressure-induced electronic spin transitions in Fe-bearing phases. The phase boundary of the breakdown of pure bridgmanite to a mixture of bridgmanite and oxides is highly sensitive to the difference in V_0 between the Mg and Fe end members, ΔV_0, but relatively insensitive to ΔK_0. In this model, (Mg$_{0.5}$Fe$_{0.5}$)SiO$_3$ bridgmanite is stabilized as a single phase at 65 GPa. A 10% change in ΔV_0 shifts this phase transition pressure by ~40 GPa, whereas a 50% change in ΔK_0 results in a difference of only ~10 GPa. Conveniently, experimental constraints on crystallographic volumes are substantially stronger than those on elastic moduli. Even without explicitly considering electronic spin, a high solubility of FeSiO$_3$ in bridgmanite in the deep lower mantle thus is consistent with both observed elastic properties and direct measurements of phase relations in Fe-rich silicates.

The breadth of the modeled post-perovskite transition remains much narrower than the range observed in most experiments [*Mao et al.*, 2004; *Catalli et al.*, 2009; *Dorfman et al.*, 2013]. Systematically narrow phase transitions were noted in this model by *Stixrude and Lithgow-Bertelloni* [2011] and may be due to bias in the model and/or systematic overestimation of phase transition widths in experiments. The decrease in the transition pressure with Fe content in the model is within the wide range of observations. Additional complexity not

yet modeled in phase equilibria in this system may be due to multiple post-perovskite structures, depending on composition [*Tschauner et al.*, 2008; *Tsuchiya and Tsuchiya*, 2008; *Yamanaka et al.*, 2012; *Zhang et al.*, 2014]. Further work is needed to resolve conflicts between existing studies on the effect of Fe on the post-perovskite transition(s).

19.4.2. Lower Mantle Compositional Models

Multiple compositional models of the lower mantle were reviewed in chapter 12. Here the phase assemblages formed from bulk mantle of pyrolite composition [*McDonough and Sun*, 1995], subducted midocean ridge basalt (MORB) [*Hofmann*, 1988], and depleted harzburgite [*Michael and Bonatti*, 1985] are considered. To first order, these compositions can be described by variation in major element components MgO, SiO$_2$, FeO, Al$_2$O$_3$, CaO, and Na$_2$O. Phase diagrams were modeled for each composition (Figure 19.2) using HeFESTo with the same model parameters described above and an adiabatic temperature profile with a mantle potential temperature of 1600 K [*Stixrude and Lithgow-Bertelloni*, 2011].

The upper boundary of the lower mantle is defined by the reaction from ringwoodite (Mg,Fe)$_2$SiO$_4$ to bridgmanite (Mg,Fe)SiO$_3$ plus ferropericlase (Mg,Fe)O. At slightly higher pressures, increasing with Al-content, majorite garnet also reacts to form bridgmanite, Mg$_3$(Mg,Al,Si)$_2$Si$_3$O$_{12}$ → 3(Mg,Al)(Al,Si)O$_3$. For different bulk mantle compositions (Figure 19.2), the abundances of majorite garnet, ringwoodite, and stishovite in the transition zone vary, resulting in differences in the pressure at which bridgmanite becomes the mantle's dominant phase. The dominance of ringwoodite at transition zone pressures in pyrolite and harzburgite compositions results in the transition to bridgmanite at ~23.5 GPa (660 km). In contrast, the MORB composition in the transition zone is almost entirely garnet. As a result, the completion of the transition in MORB to the lower mantle phase assemblage is delayed to ~30 GPa (~700 km). This broader, higher pressure transition to bridgmanite has been observed in seismology as deepening and doubling of the lower mantle discontinuity in regions with subducting slabs [*Deuss et al.*, 2006].

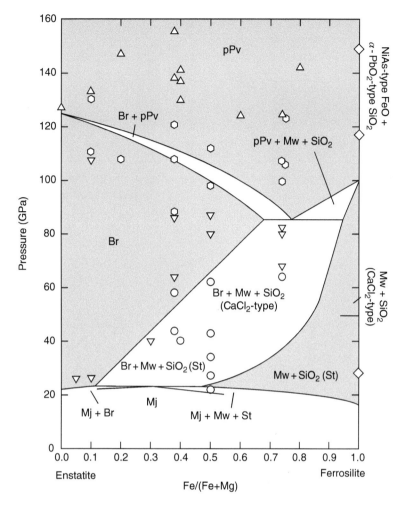

Figure 19.1 Phase diagram of MgSiO₃-FeSiO₃ (enstatite-ferrosilite) compositions at lower mantle pressures and ~2000 K. Phases include majorite (Mj), magnesiowüstite (Mw), stishovite (St), bridgmanite (Br), and post-perovskite (pPv). Phase boundaries shown are computed with HeFESTo code for a constant temperature of 2000 K [*Stixrude and Lithgow-Bertelloni*, 2011]. Experimental data are pressures at which various phase assemblages were observed in previous studies at temperatures of ~1800–2500 K [*Mao et al.*, 2004, 2005, 2006; *Tateno et al.*, 2007; *Catalli et al.*, 2009; *Fujino et al.*, 2009; *Dorfman et al.*, 2013]: oxides (diamonds), bridgmanite plus oxides (circles), bridgmanite (down-pointing triangles), bridgmanite plus post-perovskite (hexagons), post-perovskite (up-pointing triangles).

For the pyrolite composition, the modeled lower mantle phase assemblage comprises 75% bridgmanite, 16% ferropericlase, and 7% Ca-perovskite up to ~120 GPa (Figure 19.2a). As noted by *Stixrude and Lithgow-Bertelloni* [2011], the HeFESTo model for pyrolite predicts 3% calcium-ferrite-type aluminous phase in pyrolite in the lower mantle (Figure 19.2a), which is not observed in experiments [*Irifune*, 1994; *Kesson et al.*, 1998]. This is a limitation of the model, which does not allow solution of Na in ferropericlase or bridgmanite phases due to lack of experimental constraints. With this exception, lower mantle mineralogy predicted by the model agrees well with experimental observations.

A few major differences relative to the pyrolite assemblage are observed in phase assemblages formed from other mantle compositions. Due to higher Ca, Fe, Na, and Si content, MORB in the lower mantle exhibits a modal abundance of bridgmanite only about half that found in pyrolite as well as ~20% each of calcium-ferrite-type aluminous phase, Ca-perovskite and stishovite (Figure 19.2b) [*Ono et al.*, 2001; *Hirose et al.*, 2005; *Ohta et al.*, 2008; *Grocholski et al.*, 2012]. Depleted harzburgite, a mixture of olivine and pyroxenes with very low Ca and Na, transforms in the lower mantle to bridgmanite and ferropericlase associated with only 1% Ca-perovskite and no calcium-ferrite-type phase (Figure 19.2c) [*Irifune*

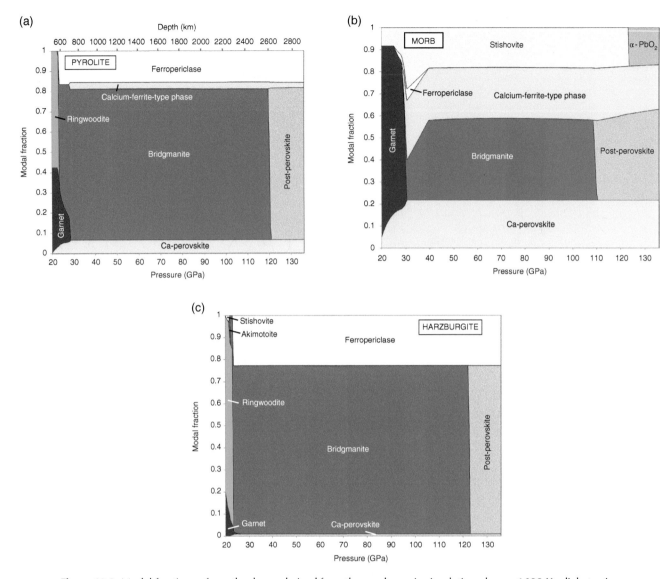

Figure 19.2 Modal fractions of mantle phases derived from thermodynamic simulation along a 1600 K adiabat using HeFESTo [*Stixrude and Lithgow-Bertelloni*, 2011] for (a) pyrolite [*McDonough and Sun*, 1995], (b) MORB [*Hofmann*, 1988], and (c) harzburgite [*Michael and Bonatti*, 1985] compositions. Note that experiments on pyrolite composition do not find a Ca-ferrite-type phase in the lower mantle, and its appearance here is due to limitations in modeling Na incorporation in ferropericlase and bridgmanite [*Stixrude and Lithgow-Bertelloni*, 2011].

and Ringwood, 1987; *Nakagawa et al.*, 2010; *Grocholski et al.*, 2012; *Zhang et al.*, 2013]. These compositions give a sense of the variability expected in phase assemblages in mantle heterogeneities associated with slab subduction.

Above 120 GPa on the 1600 K adiabat (2620 km depth), the bridgmanite component in pyrolite transforms to the post-perovskite phase. This modeled transition pressure is a few gigapascals higher than the 113–116 GPa onset of the post-perovskite transition observed by *Murakami et al.* [2005] and *Ohta et al.* [2008] but substantially lower and narrower than the broad transition observed by

Grocholski et al. [2012] to begin at 139 GPa (below the CMB). The post-perovskite transition is observed in the model at ~10 GPa lower pressures (180 km shallower depth) in MORB relative to pyrolite, in agreement with experimental observations [*Hirose et al.*, 2005; *Ohta et al.*, 2008; *Grocholski et al.*, 2012]. Harzburgite exhibits a post-perovskite transition a few gigapascals higher than the modeled transition in pyrolite. The modeled sharpness of the transition in these compositions is consistent with average observations of the D″ discontinuity, though the depth is too shallow in all compositions [*Wysession et al.*, 1998]. However, the geotherm used to compute these phase diagrams is a simple adiabat without a thermal boundary layer that may be present at the base of the mantle [*Hernlund et al.*, 2005; *Wookey et al.*, 2005]. Higher temperatures in the deep mantle deepen the post-perovskite transition due to its high positive Clapeyron slope [*Catalli et al.*, 2009] and improve consistency with the observed D'' discontinuity.

Differences between phase transitions computed in the pyrolite model and observations of lower mantle discontinuities and properties may be due to uncertainty in both the temperature and composition of the lower mantle. The phase transitions of major mantle phases and assumption of an adiabatic temperature profile are the key constraints on the geotherm. Proposed adiabatic temperature profiles in the mantle differ by hundreds of kelvin [*Brown and Shankland*, 1981; *Ono*, 2008; *Katsura et al.*, 2010], and greater differences are observed for possible superadiabatic profiles due to high thermal conductivity in the deep mantle or a thermal boundary layer at the CMB [*da Silva et al.*, 2000]. Uncertainty in geochemical constraints on the bulk composition of the lower mantle may also be responsible for differences between computed phase diagrams and observed mantle properties. Lower mantle shear wave velocities have been suggested to be more consistent with a higher proportion of bridgmanite relative to pyrolite and thus high Si content [*Murakami et al.*, 2012]. Recently some attempts have been made to produce compositional models for the mantle directly from inversion of seismic data and equations of state of mantle phases [e.g., *Ishii and Tromp*, 2004; *Mattern et al.*, 2005; *Deschamps and Tackley*, 2009]. The success of this approach will require further reduction of uncertainties in properties of the lower mantle phase assemblage, particularly at the highest pressures relevant to the CMB and D'' region.

19.5. CONCLUSIONS

Thermodynamic modeling demonstrates that direct measurements of phase transitions and measurements of physical properties can provide consistent constraints on lower mantle phase equilibria. In the $MgSiO_3$-$FeSiO_3$ binary system, modeling of the stability of the bridgmanite phase is improved by experimental data on physical properties of Fe-rich bridgmanite. However, experimental and theoretical studies have not reached consensus on the compositional dependence of post-perovskite structures and transitions. Improved constraints of the effects of electronic spin transitions on the structures and properties of $(Mg,Fe)O$ and $(Mg,Fe)SiO_3$ phases will help to determine equilibria in this system.

For bulk lower mantle compositions, chemistry and physical properties are increasingly approached as linked problems. Most experiments and modeling agree on phase equilibria of proposed bulk silicate Earth composition, pyrolite. To match phase transition depths and physical properties of the pyrolite composition with seismic discontinuities and wave speeds, superadiabatic temperatures may be required in the lowermost mantle. A combination of constraints from mineral physics experiments, theoretical modeling, and seismic observations will be needed to determine the major element composition of the bulk lower mantle and its heterogeneities.

ACKNOWLEDGMENTS

L. Stixrude provided HeFESTo code and helpful guidance. I thank the editors for the invitation to contribute this chapter and two reviewers for constructive comments that improved the manuscript. I acknowledge support from the Swiss National Science Foundation.

REFERENCES

Andrault, D., G. Fiquet, F. Guyot, and M. Hanfland (1998), Pressure-induced Landau-type transition in stishovite, *Science*, *282*(5389), 720–724, doi:10.1126/science.282.5389.720.

Angel, R. J., M. Bujak, J. Zhao, G. D. Gatta, and S. D. Jacobsen (2007), Effective hydrostatic limits of pressure media for high-pressure crystallographic studies, *J. Appl. Crystallogr.*, *40*(1), 26–32, doi:10.1107/S0021889806045523.

Auzende, A.-L., J. Badro, F. J. Ryerson, P. K. Weber, S. J. Fallon, A. Addad, J. Siebert, and G. Fiquet (2008), Element partitioning between magnesium silicate perovskite and ferropericlase: New insights into bulk lower mantle geochemistry, *Earth Planet. Sci. Lett.*, *269*, 164–174.

Badro, J. (2014), Spin transitions in mantle minerals, *Annu. Rev. Earth Planet. Sci.*, *42*(1), 231–248, doi:10.1146/annurev-earth-042711-105304.

Birch, F. (1952), Elasticity and constitution of the Earth's interior, *J. Geophys. Res.*, *57*(2), 227–286, doi:10.1029/JZ057i002p00227.

Boehler, R., and A. Chopelas (1992), Phase transitions in a 500 Kbar – 3000 K gas apparatus, in *High Pressure Research: Application to Earth and Planetary Sciences*, edited by Y. Syono and M. H. Manghnani, pp. 55–60, Terra Scientific Publ., Tokyo.

Brown, J. M., and T. J. Shankland (1981), Thermodynamic parameters in the Earth as determined from seismic profiles, *Geophys. J. Int.*, *66*(3), 579–596, doi:10.1111/j.1365-246X.1981.tb04891.x.

Callen, H. B. (1960), *Thermodynamics*, Wiley, New York.

Campbell, A. J. (2008), Measurement of temperature distributions across laser heated samples by multispectral imaging radiometry, *Rev. Sci. Instrum.*, *79*(1), 015,108, doi:10.1063/1.2827513.

Campbell, A. J., D. L. Heinz, and A. M. Davis (1992), Material transport in laser-heated diamond anvil cell melting experiments, *Geophys. Res. Lett.*, *19*(10), 1061–1064, doi:10.1029/92GL00972.

Caracas, R., and R. E. Cohen (2005), Effect of chemistry on the stability and elasticity of the perovskite and post-perovskite phases in the MgSiO$_3$-FeSiO$_3$-Al$_2$O$_3$ system and implications for the lowermost mantle, *Geophys. Res. Lett.*, *32*, L16310.

Carrez, P., H. Leroux, P. Cordier, and F. Guyot (2001), Electron-irradiation-induced phase transformation and fractional volatilization in (Mg, Fe)$_2$SiO$_4$ olivine thin films, *Philos. Mag.*, *81*(12), 2823–2840, doi:10.1080/01418610108217167.

Catalli, K., S.-H. Shim, and V. Prakapenka (2009), Thickness and Clapeyron slope of the post-perovskite boundary, *Nature*, *462*(7274), 782–785, doi:10.1038/nature08598.

Da Silva, C. R., R. M. Wentzcovitch, A. Patel, G. D. Price, and S. I. Karato (2000), The composition and geotherm of the lower mantle: Constraints from the elasticity of silicate perovskite, *Phys. Earth Planet. Int.*, *118*(1), 103–109.

Debye, P. (1912), On the theory of specific heats, *The Collected Papers of Peter JW Debye*, Interscience Publishers, New York (book published 1954), 650–696.

Deschamps, F., and P. J. Tackley (2009), Searching for models of thermo-chemical convection that explain probabilistic tomography. II—Influence of physical and compositional parameters, *Phys. Earth Planet. Int.*, *176*(1), 1–18.

Deuss, A., S. A. T. Redfern, K. Chambers, and J. H. Woodhouse (2006), The nature of the 660-kilometer discontinuity in Earth's mantle from global seismic observations of PP precursors, *Science*, *311*, 198–201.

Dewaele, A., P. Loubeyre, and M. Mezouar (2004), Equations of state of six metals above 94 GPa, *Phys. Rev. B*, *70*(9), 094112, doi:10.1103/PhysRevB.70.094112.

Dorfman, S. M., and T. S. Duffy (2014), Effect of Fe-enrichment on seismic properties of perovskite and post-perovskite in the deep lower mantle, *Geophys. J. Int.*, *197*(2), 910–919, doi:10.1093/gji/ggu045.

Dorfman, S. M., V. B. Prakapenka, Y. Meng, and T. S. Duffy (2012), Intercomparison of pressure standards (Au, Pt, Mo, MgO, NaCl and Ne) to 2.5 Mbar, *J. Geophys. Res.*, *117*(B8), B08210, doi:10.1029/2012JB009292.

Dorfman, S. M., Y. Meng, V. B. Prakapenka, and T. S. Duffy (2013), Effects of Fe-enrichment on the equation of state and stability of (Mg,Fe)SiO$_3$ perovskite, *Earth Planet. Sci. Lett.*, *361*, 249–257, doi:10.1016/j.epsl.2012.10.033.

Duffy, T. S. (2005), Synchrotron facilities and the study of the Earth's deep interior, *Rep. Prog. Phys.*, *68*, 1811–1859, doi:10.1088/0034-4885/68/8/R03.

Frost, D. J., C. Liebske, F. Langenhorst, C. A. McCammon, R. G. Trønnes, and D. C. Rubie (2004), Experimental evidence for the existence of iron-rich metal in the Earth's lower mantle, *Nature*, *428*(6981), 409–412, doi:10.1038/nature02413.

Fujino, K., D. Nishio-Hamane, K. Suzuki, H. Izumi, Y. Seto, and T. Nagai (2009), Stability of the perovskite structure and possibility of the transition to the post-perovskite structure in CaSiO$_3$, FeSiO$_3$, MnSiO$_3$ and CoSiO$_3$, *Phys. Earth Planet. Int.*, *177*(3–4), 147–151, doi:10.1016/j.pepi.2009.08.009.

Fyfe, W. (1960), Hydrothermal synthesis and determination of equilibrium between minerals, *J. Geol.*, *68*(5), 553–566.

Gillan, M. J., D. Alfè, J. Brodholt, L. Vočadlo, and G. D. Price (2006), First-principles modelling of Earth and planetary materials at high pressures and temperatures, *Rep. Prog. Phys.*, *69*, 2365–2441.

Grocholski, B., K. Catalli, S.-H. Shim, and V. Prakapenka (2012), Mineralogical effects on the detectability of the postperovskite boundary, *Proc. Natl. Acad. Sci. USA*, *109*(7), 2275–2279, doi:10.1073/pnas.1109204109.

Grüneisen, E. (1912), Theorie des festen Zustandes einatomiger Elemente, *Annalen der Physik*, *344*(12), 257–306.

Guignot, N., D. Andrault, G. Morard, N. Bolfan-Casanova, and M. Mezouar (2007), Thermoelastic properties of post-perovskite phase MgSiO$_3$ determined experimentally at core-mantle boundary P-T conditions, *Earth Planet. Sci. Lett.*, *256*(1–2), 162–168, doi:10.1016/j.epsl.2007.01.025.

Helffrich, G., and B. Wood (1989), Subregular model for multicomponent solutions, *Am. Mineral.*, *74*(9–10), 1016–1022.

Hernlund, J. W., C. Thomas, and P. J. Tackley (2005), A doubling of the post-perovskite phase boundary and structure of the Earth's lowermost mantle, *Nature*, *434*(7035), 882–886, doi:10.1038/nature03472.

Hirose, K., N. Takafuji, N. Sata, and Y. Ohishi (2005), Phase transition and density of subducted MORB crust in the lower mantle, *Earth Planet. Sci. Lett.*, *237*, 239–251.

Hirose, K., N. Takafuji, K. Fujino, S. R. Shieh, and T. S. Duffy (2008), Iron partitioning between perovskite and post-perovskite: A transmission electron microscope study, *Am. Mineral.*, *93*(10), 1678–1681.

Hofmann, A. W. (1988), Chemical differentiation of the Earth: The relationship between mantle, continental crust, and oceanic crust, *Earth Planet. Sci. Lett.*, *90*(3), 297–314, doi:10.1016/0012-821X(88)90132-X.

Hohenberg, P., and W. Kohn (1964), Inhomogeneous electron gas, *Phys. Rev.*, *136*(3B), B864–B871, doi:10.1103/PhysRev.136.B864.

Irifune, T. (1994), Absence of an aluminous phase in the upper part of the Earth's lower mantle, *Nature*, *370*, 131–133.

Irifune, T., and A. E. Ringwood (1987), Phase transformations in a harzburgite composition to 26 GPa: Implications for dynamical behaviour of the subducting slab, *Earth Planet. Sci. Lett.*, *86*(2–4), 365–376, doi:10.1016/0012-821X(87)90233-0.

Irifune, T., and T. Tsuchiya (2007), 2.03—Mineralogy of the Earth—Phase transitions and mineralogy of the lower mantle, in *Treatise on Geophysics*, edited by G. Schubert, pp. 33–62, Elsevier, Amsterdam.

Irifune, T., et al. (1998), The postspinel phase boundary in Mg$_2$SiO$_4$ determined by in situ X-ray diffraction, *Science*, *279*(5357), 1698–1700, doi:10.1126/science.279.5357.1698.

Ishii, M., and J. Tromp (2004), Constraining large-scale mantle heterogeneity using mantle and inner-core sensitive normal modes, *Phys. Earth Planet. Int.*, *146*(1–2), 113–124, doi:16/j.pepi.2003.06.012.

Katsura, T., A. Yoneda, D. Yamazaki, T. Yoshino, and E. Ito (2010), Adiabatic temperature profile in the mantle, *Phys. Earth Planet. Int.*, *183*(1–2), 212–218, doi:10.1016/j.pepi.2010.07.001.

Kesson, S. E., J. D. Fitzgerald, and J. M. G. Shelley (1998), Mineralogy and dynamics of a pyrolite lower mantle, *Nature*, *393*, 252–255.

Kiefer, B., and T. S. Duffy (2005), Finite element simulations of the laser-heated diamond-anvil cell, *J. Appl. Phys.*, *97*(11), 114902, doi:10.1063/1.1906292.

Klotz, S., J.-C. Chervin, P. Munsch, and G. L. Marchand (2009), Hydrostatic limits of 11 pressure transmitting media, *J. Phys. D*, *42*(7), 075413, doi:10.1088/0022-3727/42/7/075413.

Kohn, W., and L. J. Sham (1965), Self-consistent equations including exchange and correlation effects, *Phys. Rev.*, *140*(4A), A1133–A1138, doi:10.1103/PhysRev.140.A1133.

Lin, J.-F., S. Speziale, Z. Mao, and H. Marquardt (2013), Effects of the electronic spin transitions of iron in lower mantle minerals: Implications for deep mantle geophysics and geochemistry, *Rev. Geophys.*, *51*, pp. 244–275, doi:10.1002/rog.20010.

Liu, L.-G., and A. E. Ringwood (1975), Synthesis of a perovskite-type polymorph of $CaSiO_3$, *Earth Planet. Sci. Lett.*, *28*(2), 209–211, doi:10.1016/0012-821X(75)90229-0.

Lundin, S., K. Catalli, J. Santillán, S.-H. Shim, V. B. Prakapenka, M. Kunz, and Y. Meng (2008), Effect of Fe on the equation of state of mantle silicate perovskite over 1 Mbar, *Phys. Earth Planet. Int.*, *168*(1–2), 97–102, doi:10.1016/j.pepi.2008.05.002.

Mao, W. L., G. Shen, V. B. Prakapenka, Y. Meng, A. J. Campbell, D. L. Heinz, J. Shu, R. J. Hemley, and H.-K. Mao (2004), Ferromagnesian postperovskite silicates in the D″ layer of the Earth, *Proc. Natl. Acad. Sci. USA*, *101*(45), 15,867–15,869, doi:10.1073/pnas.0407135101.

Mao, W. L., Y. Meng, G. Shen, V. B. Prakapenka, A. J. Campbell, D. L. Heinz, J. Shu, R. Caracas, R. E. Cohen, and Y. Fei (2005), Iron-rich silicates in the Earth's D″ layer, *Proc. Natl. Acad. Sci. USA*, *102*(28), 9751–9753, doi:10.1073/pnas.0503737102.

Mao, W. L., H. Mao, W. Sturhahn, J. Zhao, V. B. Prakapenka, Y. Meng, J. Shu, Y. Fei, and R. J. Hemley (2006), Iron-rich post-perovskite and the origin of ultralow-velocity zones, *Science*, *312*(5773), 564–565, doi:10.1126/science.1123442.

Mattern, E., J. Matas, Y. Ricard, and J. Bass (2005), Lower mantle composition and temperature from mineral physics and thermodynamic modelling, *Geophys. J. Int.*, *160*(3), 973–990, doi:10.1111/j.1365-246X.2004.02549.x.

McDonough, W. F., and S.-s. Sun (1995), The composition of the Earth, *Chem. Geol.*, *120*(3–4), 223–253, doi:10.1016/0009-2541(94)00140-4.

Meng, Y., G. Shen, and H.-k. Mao (2006), Double-sided laser heating system at HPCAT for in situ x-ray diffraction at high pressures and high temperatures, *J. Phys. Condens. Matter*, *18*(25), 1097, doi:10.1088/0953-8984/18/25/S17.

Mezouar, M., et al. (2005), Development of a new state-of-the-art beamline optimized for monochromatic single-crystal and powder X-ray diffraction under extreme conditions at the ESRF, *J. Synchrotron Radiat.*, *12*(5), 659–664, doi:10.1107/S0909049505023216.

Michael, P. J., and E. Bonatti (1985), Peridotite composition from the North Atlantic: Regional and tectonic variations and implications for partial melting, *Earth Planet. Sci. Lett.*, *73*(1), 91–104, doi:10.1016/0012-821X(85)90037-8.

Mukhopadhyay, B., S. Basu, and M. Holdaway (1993), A discussion of Margules-type formulations for multicomponent solutions with a generalized approach, *Geochim. Cosmochim. Acta*, *57*(2), 277–283, doi:10.1016/0016-7037(93)90430-5.

Murakami, M., K. Hirose, K. Kawamura, N. Sata, and Y. Ohishi (2004), Post-perovskite phase transition in $MgSiO_3$, *Science*, *304*(5672), 855–858, doi:10.1126/science.1095932.

Murakami, M., K. Hirose, N. Sata, and Y. Ohishi (2005), Post-perovskite phase transition and mineral chemistry in the pyrolitic lowermost mantle, *Geophys. Res. Lett.*, *32*, article # L03304, doi:200510.1029/2004GL021956.

Murakami, M., Y. Ohishi, N. Hirao, and K. Hirose (2012), A perovskitic lower mantle inferred from high-pressure,

high-temperature sound velocity data, *Nature*, *485*(7396), 90–94, doi:10.1038/nature11004.

Nakagawa, T., P. J. Tackley, F. Deschamps, and J. A. D. Connolly (2010), The influence of MORB and harzburgite composition on thermo-chemical mantle convection in a 3-D spherical shell with self-consistently calculated mineral physics, *Earth Planet. Sci. Lett.*, *296*(3–4), 403–412, doi:10.1016/j.epsl.2010.05.026.

Oganov, A. R., and C. W. Glass (2006), Crystal structure prediction using ab initio evolutionary techniques: Principles and applications, *J. Chem. Phys.*, *124*(24), 244,704, doi:10.1063/1.2210932.

Oganov, A. R., and S. Ono (2004), Theoretical and experimental evidence for a post-perovskite phase of $MgSiO_3$ in Earth's D″ layer, *Nature*, *430*(6998), 445–448, doi:10.1038/nature02701.

Oganov, A. R., J. P. Brodholt, and G. D. Price (2001a), Ab initio elasticity and thermal equation of state of $MgSiO_3$ perovskite, *Earth Planet. Sci. Lett.*, *184*(3–4), 555–560.

Oganov, A. R., J. P. Brodholt, and G. D. Price (2001b), The elastic constants of $MgSiO_3$ perovskite at pressures and temperatures of the Earth's mantle, *Nature*, *411*(6840), 934–937, doi:10.1038/35082048.

Ohta, K., K. Hirose, T. Lay, N. Sata, and Y. Ohishi (2008), Phase transitions in pyrolite and MORB at lowermost mantle conditions: Implications for a MORB-rich pile above the core–mantle boundary, *Earth Planet. Sci. Lett.*, *267*(1–2), 107–117, doi:10.1016/j.epsl.2007.11.037.

Ono, S. (2008), Experimental constraints on the temperature profile in the lower mantle, *Phys. Earth Planet. Int.*, *170*(3–4), 267–273, doi:10.1016/j.pepi.2008.06.033.

Ono, S., E. Ito, and T. Katsura (2001), Mineralogy of subducted basaltic crust (MORB) from 25 to 37 GPa, and chemical heterogeneity of the lower mantle, *Earth Planet. Sci. Lett.*, *190*(1–2), 57–63, doi:10.1016/S0012-821X(01)00375-2.

Poirier, J.-P. (2000), *Introduction to the Physics of the Earth's Interior*, 2nd ed., Cambridge Univ. Press, Cambridge.

Prakapenka, V. B., A. Kubo, A. Kuznetsov, A. Laskin, O. Shkurikhin, P. Dera, M. L. Rivers, and S. R. Sutton (2008), Advanced flat top laser heating system for high pressure research at GSECARS: Application to the melting behavior of germanium, *High Pressure Res.*, *28*(3), 225–235, doi:10.1080/08957950802050718.

Shearer, P. M. (2000), Upper mantle seismic discontinuities, in *Geophysical Monograph Series*, vol. 117, edited by S. Karato, A. Forte, R. Liebermann, G. Masters, and L. Stixrude, pp. 115–131, AGU, Washington, D. C.

Sinmyo, R., and K. Hirose (2010), The Soret diffusion in laser-heated diamond-anvil cell, *Phys. Earth Planet. Int.*, *180*(3–4), 172–178, doi:16/j.pepi.2009.10.011.

Stixrude, L., and C. Lithgow-Bertelloni (2005), Thermodynamics of mantle minerals—I. Physical properties, *Geophys. J. Int.*, *162*(2), 610–632, doi:10.1111/j.1365-246X.2005.02642.x.

Stixrude, L., and C. Lithgow-Bertelloni (2011), Thermodynamics of mantle minerals—II. Phase equilibria, *Geophys. J. Int.*, *184*(3), 1180–1213, doi:10.1111/j.1365-246X.2010.04890.x.

Tange, Y., E. Takahashi, Y. Nishihara, K. Funakoshi, and N. Sata (2009), Phase relations in the system MgO-FeO-SiO_2 to 50 GPa and 2000°C: An application of experimental techniques using multianvil apparatus with sintered diamond anvils, *J. Geophys. Res.*, *114*, B02214, doi:200910.1029/2008JB005891.

Tateno, S., K. Hirose, N. Sata, and Y. Ohishi (2007), Solubility of FeO in (Mg,Fe)SiO$_3$ perovskite and the post-perovskite phase transition, *Phys. Earth Planet. Int.*, *160*(3–4), 319–325, doi:10.1016/j.pepi.2006.11.010.

Tschauner, O., B. Kiefer, H. Liu, S. Sinogeikin, M. Somayazulu, and S.-N. Luo (2008), Possible structural polymorphism in Al-bearing magnesium silicate post-perovskite, *Am. Mineral.*, *93*(4), 533–539, doi:10.2138/am.2008.2372.

Tsuchiya, J., and T. Tsuchiya (2008), Postperovskite phase equilibria in the MgSiO$_3$–Al$_2$O$_3$ system, *Proc. Natl. Acad. Sci. USA*, *105*(49), 19,160–19,164, doi:10.1073/pnas.0805660105.

Tsuchiya, T., and X. Wang (2013), Ab initio investigation on the high-temperature thermodynamic properties of Fe^{3+}-bearing MgSiO$_3$ perovskite, *J. Geophys. Res. Solid Earth*, *118*(1), 83–91, doi:10.1029/2012JB009696.

Wirth, R. (2009), Focused Ion Beam (FIB) combined with SEM and TEM: Advanced analytical tools for studies of chemical composition, microstructure and crystal structure in geomaterials on a nanometre scale, *Chem. Geol.*, *261*(3–4), 217–229, doi:10.1016/j.chemgeo.2008.05.019.

Wood, B., and J. Nicholls (1978), Thermodynamic properties of reciprocal solid-solutions, *Contrib. Mineral. Petrol.*, *66*(4), 389–400, doi:10.1007/BF00403424.

Wookey, J., S. Stackhouse, J.-M. Kendall, J. Brodholt, and G. D. Price (2005), Efficacy of the post-perovskite phase as an explanation for lowermost-mantle seismic properties, *Nature*, *438*(7070), 1004–1007, doi:10.1038/nature04345.

Wysession, M. E., T. Lay, J. Revenaugh, Q. Williams, E. J. Garnero, R. Jeanloz, and L. H. Kellogg (1998), The D″ discontinuity and its implications, in *The Core-Mantle Boundary Region*, edited by M. Gurnis, M. E. Wysession, E. Knittle, and B. A. Buffett, pp. 273–297, AGU, Washington, D. C.

Yamanaka, T., K. Hirose, W. L. Mao, Y. Meng, P. Ganesh, L. Shulenburger, G. Shen, and R. J. Hemley (2012), Crystal structures of (Mg$_{1-x}$,Fe$_x$)SiO$_3$ postperovskite at high pressures, *Proc. Natl. Acad. Sci. USA*, *109*(4), 1035–1040, doi:10.1073/pnas.1118076108.

Yamazaki, D., E. Ito, T. Yoshino, N. Tsujino, A. Yoneda, X. Guo, F. Xu, Y. Higo, and K. Funakoshi (2014), Over 1 Mbar generation in the Kawai-type multianvil apparatus and its application to compression of (Mg$_{0.92}$Fe$_{0.08}$)SiO$_3$ perovskite and stishovite, *Phys. Earth Planet. Int.*, *228*, 262–267, doi:10.1016/j.pepi.2014.01.013.

Zhang, L., et al. (2014), Disproportionation of (Mg,Fe)SiO$_3$ perovskite in Earth's deep lower mantle, *Science*, *344*(6186), 877–882, doi:10.1126/science.1250274.

Zhang, Y., Y. Wang, Y. Wu, C. R. Bina, Z. Jin, and S. Dong (2013), Phase transitions of harzburgite and buckled slab under eastern China, *Geochem. Geophys. Geosyst.*, *14*(4), 1182–1199, doi:10.1002/ggge.20069.

Part V
Volatiles in Deep Interior

20

Hydrogen in the Earth's Core: Review of the Structural, Elastic, and Thermodynamic Properties of Iron-Hydrogen Alloys

Caitlin A. Murphy

ABSTRACT

This chapter serves as a review of the properties of iron hydrides (FeH_x). At high temperatures and pressures below 60 GPa, FeH_x crystallizes in a face-centered-cubic (fcc) structure, similar to pure iron. However, the incorporation of hydrogen results in a complex melting curve shape at low pressures and a melting point depression of ~800 K relative to fcc Fe. At higher pressures, double hexagonal close-packed (dhcp) FeH remains stable to at least 84 GPa and 2000 K, and its extrapolated sound velocities are faster than the seismically determined values for Earth's core. Based on an inversion of available structural and elastic data for hcp-Fe and dhcp-FeH, an inner core composition of $FeH_{0.14}$ produces reasonable agreement with seismic observations. However, this value is highly uncertain because the properties of Fe and FeH are not well constrained at core conditions, and the influence of hydrogen on the thermal profile of the core must be taken into account. In order to produce more accurate core composition models, suggested future research directions include experiments at higher pressures and temperatures, new techniques for probing the hydrogen content in situ, and exploration of binary- and tertiary-phase diagrams that include iron and hydrogen.

20.1. INTRODUCTION

Earth's core is made up primarily of iron, but experimental constraints on the density of pure iron at core conditions indicate that it is ~8%–10% denser than the liquid outer core and ~3%–5% denser than the solid inner core [e.g., *Dewaele et al.*, 2006]. The incorporation of light elements is often suggested as a means for resolving this so-called core density deficit (CDD), because the resulting alloy would have an expanded volume and reduced average atomic mass relative to pure iron. Constraining the identity and amount of light elements in the core is particularly important because alloying will also affect the sound velocities, conductivity, and melting behavior

of iron and thus has important implications for existing models of the deep Earth. For example, the melting temperature of core materials at the inner core boundary (ICB), where Earth's liquid outer core and solid inner core are in contact, serves as an anchor point for the thermal profile of the core. In turn, this light element component is intimately related to convection-driven models for the generation of Earth's geodynamo and the dynamics of Earth's mantle via heat transport across the core-mantle boundary (CMB).

The core most likely comprises a mixture of light elements, but the standard approach to this underdetermined problem has been to explore the effects of alloying one light element at a time. In that spirit, the focus of this chapter will be on hydrogen, which is the most abundant element in the solar system (~$10^4 \times$ Si). During Earth formation, hydrogen was likely introduced via chondrites

Geophysical Laboratory, Carnegie Institution of Washington, Washington, D.C., USA

Deep Earth: Physics and Chemistry of the Lower Mantle and Core, Geophysical Monograph 217, First Edition.
Edited by Hidenori Terasaki and Rebecca A. Fischer.

and comets in the form of water and in sufficient amounts to explain the entire CDD. However, it is unclear whether hydrogen could have reached and maintained great enough depths to avoid volatilization during the subsequent impacts and heating associated with Earth accretion. Such a discussion is beyond the scope of this review, but qualitative models exist for retaining a significant amount of water within a primitive magma ocean. In addition, the observed degassing of primordial noble gases from mid-ocean ridges and hotspots offers indirect evidence that water may also be stored deep in the Earth [see, e.g., *Williams and Hemley*, 2001].

Regardless of the retention mechanism, it is likely that any interaction between water and metallic iron below a depth of ~80 km would have resulted in the formation of an iron-hydrogen alloy. X-ray diffraction (XRD) experiments have shown that Fe+H_2O mixtures react to form FeO and FeH$_x$ at pressures as low as 2.2 GPa [*Yagi and Hishinuma*, 1995] and up to at least 84 GPa [*Ohtani et al.*, 2005]. Moreover, thermodynamic calculations predict that this reaction will be increasingly favored at high-pressure and high-temperature (*high-PT*) conditions [*Badding et al.*, 1992]; if the same relationship holds for the liquid phase, then any hydride formed at relatively shallow depths should have remained stable during its gravitationally driven descent through the primitive mantle. It is worth noting that such a mechanism does not require hydrogen to be the major light element but instead demonstrates that some hydrogen was likely incorporated in the core.

This chapter will serve as a review of iron hydrides (FeH$_x$) and their role in the core. First, the established *PT* phase diagram for FeH$_x$ and the fundamental material properties of the phases most relevant for Earth's core will be presented (Section 20.2). Next, existing experimental data for the pressure evolution of their structural and elastic properties will be reviewed and synthesized (Section 20.3), and will serve as input parameters for a core composition model (Section 20.4). Finally, gaps in our knowledge of the properties of iron-hydrogen alloys and future research directions that will improve our ability to constrain the amount of hydrogen in Earth's core will be discussed (Section 20.5).

20.2. STABLE PHASES OF IRON-HYDROGEN ALLOYS

The solubility of hydrogen in iron is extremely low (~10^{-5} at %) at ambient conditions, but it increases dramatically at high hydrogen pressures [e.g., *Antonov et al.*, 1980; *Fukai*, 2005]. Under high-*PT* conditions, multiple FeH$_x$ phases are formed, and the corresponding *PT* phase diagram has been constructed from a wide array of experimental techniques: quench texture [e.g., *Okuchi*, 1997, 1998], resistivity [e.g., *Antonov et al.*, 2002], Mössbauer spectroscopy

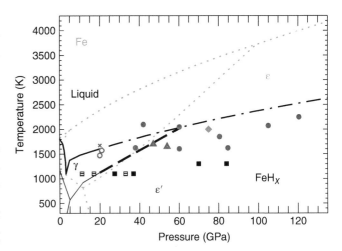

Figure 20.1 Pressure-temperature phase diagrams of FeH$_x$ and Fe. Black solid lines show experimentally determined phase boundaries between bcc (α), fcc (γ), dhcp (ε'), and liquid FeH$_x$ (labeled on the figure); the dashed (dash-dotted) black line reflects the extrapolated ε'-γ phase boundary (γ-FeH$_x$ melting curve), as presented by *Sakamaki et al.* [2009]. Filled (open) symbols show reports of ε' (γ) stability fields: black squares [*Ohtani et al.*, 2005], blue triangles [*Narygina et al.*, 2011], and green diamond [*Saxena et al.*, 2004]; red circles, (Fe,Ni)H$_x$ [*Terasaki et al.*, 2012]. Red x shows the melting point of (Fe,Ni)H$_x$ at ~20 GPa. Grey dotted lines depict representative literature phase boundaries from *Liu and Basset* [1986] and *Anzellini et al.* [2013]. Figure adapted with permission from *Sakamaki et al.* [2009].

[e.g., *Choe et al.*, 1991; *Mao et al.*, 2004; *Narygina et al.*, 2011; *Schneider et al.*, 1991], X-ray magnetic circular dichroism [*Ishimatsu et al.*, 2012], neutron powder diffraction (NPD) [*Antonov et al.*, 1998, 2002], and XRD [e.g., *Badding et al.*, 1991; *Fukai*, 2005; *Hirao et al.*, 2004; *Ohtani et al.*, 2005; *Sakamaki et al.*, 2009; *Saxena et al.*, 2004; *Yagi and Hishinuma*, 1995]. Moreover, the crystal structures of the stable hydrides have been determined in the aforementioned XRD studies and in NPD studies performed on samples that were synthesized under high-*PT* conditions, cooled to liquid nitrogen temperature, and then decompressed in order to retain the hydride at ambient pressure. Products of the latter synthesis technique will be referred to as "quenched samples" for the remainder of this chapter.

The resulting phase diagram for FeH$_x$ is presented with that of pure iron (Fe) [e.g., *Anzellini et al.*, 2013] in Figure 20.1, where one can see that the major features are quite similar. For example, both Fe and FeH$_x$ crystallize in a face-centered-cubic (fcc) structure (γ) at high temperatures and relatively low pressures. However, the dominant high-pressure phase of FeH$_x$ has a double hexagonal close-packed (dhcp) structure (ε'), compared to regular hcp (ε) for pure iron. Because γ-FeH$_x$ and ε'-FeH$_x$ are thought to be the most relevant phases for discussions about hydrogen in Earth's core, their fundamental material properties and how they were determined will be described.

20.2.1. Double Hexagonal Close-Packed FeH

At 300 K, solid iron and liquid hydrogen react to form ε'-FeH$_x$ at ~3.5 GPa and 300 K [e.g., *Badding et al.*, 1991]. Experiments have shown that the kinetics of this reaction are sluggish at both ambient and elevated temperatures, as indicated by the coexistence of α-Fe and ε'-FeH$_x$ in the presence of excess hydrogen up to ~15 GPa [*Choe et al.*, 1991; *Sakamaki et al.*, 2009]. However, the reaction is fully reversible, and the incorporated hydrogen escapes upon decompression at temperatures above ~150 K [*Antonov et al.*, 1980; *Badding et al.*, 1991].

The atomic-scale structure of ε'-FeH$_x$ has been studied in detail with in situ XRD [e.g., *Badding et al.*, 1991] and NPD measurements of quenched samples [e.g., *Antonov et al.*, 2002]. Close-packed iron layers are stacked along the c axis in an ABAC sequence (Figure 20.2), thus producing two inequivalent iron sites ($2a$ and $2c$ in the Wyckoff representation). It is thought that there is a large concentration of faults in this stacking sequence—resulting in an unequal number of $2a$ and $2c$ iron sites—based on a broadening of the (102) diffraction peak [*Antonov et al.*, 2002; *Badding et al.*, 1991; *Hirao et al.*, 2004]. The location of hydrogen

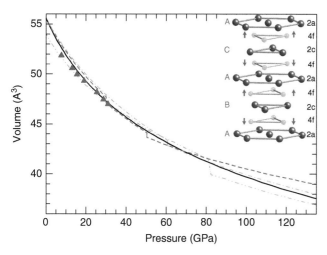

Figure 20.2 High-pressure equations of state for ε'-FeH$_x$. Experimentally determined EOS for ε'-FeH$_x$: solid black line [*Badding et al.*, 1991]; dashed red line, the three compression regimes from *Hirao et al.* [2004]. Triangles reflect preliminary results for our ongoing NPD experiments on FeD$_x$ up to 31 GPa [*Murphy et al.*, 2013b]. Theoretical EOS: brown dash-dotted line [*Elsasser et al.*, 1998], with a FM-NM transition at 60 GPa; grey dash-dot-dotted line [*Isaev et al.*, 2007], with a FM-NM transition near 80 GPa. The inset shows the stacking sequence for ε': grey spheres, close-packed Fe; blue spheres, H in octahedral sites; blue arrows, the suggested vertical displacement of H. Note that the corresponding stacking sequence for ε (γ) is ABAB (ABCABC). Inset figure reproduced with permission from [*Tsumuraya et al.*, 2012].

atoms can only be inferred with XRD, but Rietveld refinements of NPD data for quenched samples suggest that hydrogen atoms occupy all available octahedral interstitial sites ($4f$ in Figure 20.2) [*Antonov et al.*, 1998]. These refinements also suggest a slight (0.06 Å) vertical displacement of hydrogen atoms toward the $2a$ iron site (Figure 20.2), which results in an increased distance between nearest-neighbor hydrogen atoms. Finally, the composition of ε'-FeH$_x$ has been constrained by correlating this structural information with complementary degassing experiments on quenched samples, which showed that ε'-FeH$_x$ is approximately stoichiometric ($x = 1$) [*Schneider et al.*, 1991]. It is important to note that this is the most direct determination of hydrogen content in ε'-FeH$_x$ to date and the foundation for nearly all subsequent estimates of x. In particular, a volume increase per hydrogen atom (v_H) of ~2.6 Å^3/H is often cited as confirmation that a sample is stoichiometric [see e.g., *Fukai*, 2005]: This value arises from a measured difference in unit cell volumes between ε'-FeH$_x$ and ε-Fe of ~10.4 Å^3 at 3.5 GPa [e.g., *Badding et al.*, 1991] and the assumption that $x = 1$.

An important consequence of the aforementioned vertical displacement of hydrogen atoms is that it substantially alters the electron density of states near the Fermi level, thus inducing ferromagnetic (FM) ordering [*Elsasser et al.*, 1998; *Tsumuraya et al.*, 2012]. Mössbauer spectroscopy (MS) experiments have shown that ε'-FeH is indeed FM, with a magnetic moment that is ~80% that of α-Fe [*Choe et al.*, 1991]. In addition, the MS spectra show two overlapping magnetic patterns with similar isomer shifts but different hyperfine fields [*Choe et al.*, 1991; *Narygina et al.*, 2011; *Schneider et al.*, 1991], which is consistent with the presence of two inequivalent iron sites. Finally, the relative intensities of the two magnetic patterns suggest that there may be ~20% more $2c$ iron sites than $2a$ sites [*Schneider et al.*, 1991], which supports the prediction of stacking faults.

20.2.2. Face-Centered Cubic FeH

At high temperatures and pressures below ~60 GPa, γ-FeH$_x$ crystallizes with the same structure as high-temperature γ-Fe, except with hydrogen atoms occupying the octahedral sites [*Narygina et al.*, 2011; *Sakamaki et al.*, 2009; *Skorodumova et al.*, 2004]. There are currently no published estimates for v_H of γ-FeH, so a value of 1.9 Å^3/H is usually adopted from degassing experiments of γ-Fe$_{0.65}$Mn$_{0.29}$Ni$_{0.06}$H$_{0.95}$. Application of this v_H often predicts superstoichiometric γ-FeH$_x$ ($x = 1.1 - 1.2$) [e.g., *Narygina et al.*, 2011; *Sakamaki et al.*, 2009], which implies that either hydrogen atoms occupy some tetrahedral sites or the true v_H is slightly larger.

The α-γ phase boundary (Figure 20.1) is somewhat unclear due to the formation of nonstoichiometric γ-FeH$_x$ at low pressures [*Yagi and Hishinuma*, 1995] and an observed contraction of the γ-FeH$_x$ unit cell with time below 5 GPa. This gradual volume reduction has been interpreted as the formation of superabundant vacancies, which are known to form in other group VI–VIII transition metal hydrides [*Fukai*, 2005]. Above 10 GPa, there have been no reports of superabundant vacancy formation in long-duration heating experiments [*Narygina et al.*, 2011; *Sakamaki et al.*, 2009], which is likely due to the larger $P\Delta V$ term in the vacancy formation enthalpy [*Fukai*, 2005].

The ε′-γ transition for FeH has been explored with in situ XRD experiments, which reveal a phase boundary slope of ~24 K/GPa and no abrupt change in volume [*Sakamaki et al.*, 2009] (Figure 20.2). Theoretical calculations predict that γ-FeH will also be stable above 83 GPa at 300 K, with an intermediate ε-FeH phase stability field between 37 and 83 GPa. According to *Isaev et al.* [2007], the electronic term (total energy) determines the stabilization of ε- and γ-FeH, but such transitions have not been observed experimentally. This discrepancy may be due to the use of ideal crystals with exact stoichiometry in total energy calculations or perhaps the theory does not adequately describe the electronic properties of iron hydride.

Finally, ambient temperature MS measurements of γ-FeH synthesized at 57 GPa and 1700 K reveal that it is either nonmagnetic (NM) or anti-FM, at least under these metastable conditions [*Narygina et al.*, 2011]. By contrast, ab initio calculations predict that γ-FeH will be FM up to ~50 or 60 GPa [*Elsasser et al.*, 1998; *Isaev et al.*, 2007; *Skorodumova et al.*, 2004], where it will undergo a sudden magnetic collapse. This discrepancy has not been discussed in the literature, but if the experimental samples were synthesized within the stability field of the NM phase, then perhaps the corresponding lack of magnetic ordering was quenched to room temperature, along with the crystal structure. Alternatively, it is possible that the magnetic state of γ-FeH is not accurately predicted by ab initio calculations.

20.2.3. Liquid FeH$_x$

The γ-liquid phase boundary for FeH$_x$ has been explored up to 20 GPa, but no experimental constraints exist for the melting behavior of ε′-FeH. Early in situ XRD and quench texture experiments demonstrated that the incorporation of even a small amount of hydrogen ($x \sim 0.3$–0.4) dramatically affects the melting behavior of iron, inducing a complex melting curve shape at low pressures and a melting point depression (ΔT_M) of ~600 K below 10 GPa [e.g., *Okuchi*, 1998; *Yagi and Hishinuma*,

1995]. In addition, *Sakamaki et al.* [2009] determined the melting curve of γ-FeH between 10 and 20 GPa using in situ XRD experiments and found that it has a Clapeyron slope of ~13 K/GPa. Combining this result with the previously described ε′-γ phase boundary from the same study, the authors predict that the γ-ε′-liquid triple point for FeH$_x$ will lie near 60 GPa and 2000 K. Finally, comparison with the melting curve of γ-Fe recently determined using a similar technique [*Anzellini et al.*, 2013] suggests that the incorporation of hydrogen produces $\Delta T_M \sim$ 600–800 K in stoichiometric γ-FeH below 20 GPa (Figure 20.1).

Little is known about the physical properties of liquid FeH$_x$, but quench texture experiments provide some insight into the relative solubility of hydrogen in solid and liquid FeH$_x$. In particular, *Okuchi* [1997, 1998] induced melting in FeH$_x$ by pressurizing and heating the samples in a multianvil press and then rapidly decompressing them in order to trap hydrogen bubbles upon phase separation. Previously molten and solid FeH$_x$ could be distinguished by their textural differences, and *Okuchi* [1997] estimated that the hydrogen concentration (i.e., volume of evacuated bubbles) in molten FeH$_x$ was enriched by 20% relative to solid FeH$_x$ at 7.5 GPa. This result is in qualitative agreement with thermodynamic calculations [*Fukai*, 2005] and suggests that hydrogen would preferentially remain in the liquid outer core during crystallization of the inner core.

20.3. HIGH-PRESSURE PROPERTIES OF FeH

In order to construct accurate models for hydrogen in Earth's core, one must have experimental constraints on the most relevant seismic properties of FeH$_x$, namely its density and sound velocities. Unfortunately, there have been no reports of the elastic properties of γ-FeH, and limited data exist for its structural evolution. The most comprehensive study was performed by *Narygina et al.* [2011], who synthesized γ-FeH at high-PT conditions and measured its pressure-volume (P-V) relationship upon decompression at 300 K. By fitting an equation of state (EOS) to P-V data over their experimental pressure range (54 – 12 GPa), the authors found $K_0 = 99 \pm 5$ GPa and $K_0' = 11.7 \pm 5$, which correspond to the ambient pressure bulk modulus and its pressure derivative, respectively. These experimental EOS parameters indicate that γ-FeH is more compressible than ε-Fe at low pressures ($K_0 = 165$ GPa and $K_0' = 4.97(4)$ from *Dewaele et al.* [2006]), but the former phase will have a larger K at core pressures; in turn, γ-FeH in the core would likely have a larger volume (smaller density) and compressional sound velocity compared to ε-Fe. However, this result could be significantly altered by thermal effects, which were not considered, particularly because the EOS is based on data collected under metastable conditions (300 K) and upon decompression. In addition, ab initio calculations produce a very different result: *Bazhanova et al.* [2012] and *Elsasser et al.* [1998] both predict larger K_0 and much smaller K_0' for NM γ-FeH, resulting in very similar K for γ-FeH and ε-Fe at inner core pressures. Theoretical EOSs for NM iron alloys typically overestimate K_0 relative to experiments, so it is also worth noting that a comparison with calculations for FM γ-FeH indicate it has a *smaller K* than ε-Fe at inner core pressures [*Elsasser et al.*, 1998]. Finally, all of the previously discussed EOSs are based on a third-order Birch-Murnaghan equation; additional work is needed to constrain the second pressure derivatives of K for FeH and Fe, which will be particularly important when extrapolating their elastic properties to core pressures.

The P-V relationship for ε′-FeH has been studied in detail with experiments performed up to 80 GPa at 300 K [*Badding et al.*, 1991; *Hirao et al.*, 2004]. The available compression data show that ε′-FeH is more compressible than ε-Fe at low pressures, and v_H decreases with pressure from 2.4 Å³/H at 6 GPa to 1.7 Å³/H at 70 GPa. As in the case of γ-FeH, theoretical calculations predict larger K_0 and smaller K_0' [*Elsasser et al.*, 1998; *Tsumuraya et al.*, 2012] compared to experiments (Figure 20.2), resulting in similar K for ε-Fe and NM ε′-FeH and a smaller K for FM ε′-FeH at inner core pressures.

The small pressure steps and wide pressure range (≤80 GPa) measured by *Hirao et al.* [2004] allowed them to identify discontinuities in c/a near 30 and 50 GPa (Figure 20.2), resulting in three distinct compression regimes. For example, their EOS for compression data between 50 and 80 GPa indicates that ε′-FeH is distinctly less compressible, with K_0 = 182 ± 45 GPa and K_0' = 8.5±2.9. The origin of this highest-pressure regime is often interpreted through ab initio calculations that predict a FM–NM transition and sudden increase in K at 50–60 GPa [*Elsasser et al.*, 1998; *Tsumuraya et al.*, 2012]. It is worth noting that the more recent study's calculation for the evolution of c/a across the magnetic transition also agrees qualitatively with the observations of *Hirao et al.* [2004].

In contrast to theoretical predictions, multiple lines of experimental evidence indicate that the FM–NM transition occurs near 27 GPa at 300 K. *Mao et al.* [2004] first identified the magnetic collapse in ε′-FeH with MS, which revealed a weakened magnetic field at 22 GPa and a lack of magnetism at 30 GPa. This result was recently reproduced by *Mitsui and Hirao* [2010] and *Narygina et al.* [2011], who place the FM–NM transition between 24.7 and 27.6 GPa. Finally, *Ishimatsu et al.* [2012] found that ε′-FeH is still weakly magnetic at 27.5 GPa based on X-ray magnetic circular dichroism measurements but estimated via extrapolation that the FM state would no longer be stable above 29.5 GPa. The different FM–NM

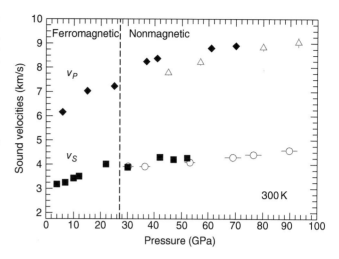

Figure 20.3 Sound velocities of ε′-FeH and ε-Fe. Solid black symbols show the ambient temperature compressional (v_p, diamonds [*Shibazaki et al.*, 2012]) and shear (v_s, squares [*Mao et al.*, 2004]) sound velocities of ε′-FeH. Open symbols show representative v_p (red triangles [*Antonangeli et al.*, 2012]) and v_s (blue circles [*Murphy et al.*, 2013a]) for ε-Fe, measured with the same techniques. Dashed vertical line at ~27 GPa represents the FM–NM transition.

transition pressures predicted by experimental (~27 GPa) and theoretical (~50–60 GPa) studies may be due to temperature effects and/or the use of ideal crystals with exact stoichiometry in total energy calculations or perhaps the existing theoretical treatment of the magnetic behavior of ε′-FeH is not adequate.

The effects of this magnetic collapse are evident not only in the discontinuous pressure-volume (or density, ρ) behavior of ε′-FeH, but also in its compressional (v_p) and shear (v_s) sound velocities. For example, *Shibazaki et al.* [2012] observed a sudden increase and change in slope of $v_p(\rho, 300$ K) between 25 and 37 GPa, as measured by inelastic X-ray scattering (IXS) and in situ XRD experiments up to 70 GPa (Figure 20.3). This corresponds well with an increase in K across the FM-NM transition, but the subsequent smooth evolution of $v_p(\rho)$ between 37 and 70 GPa is seemingly inconsistent with the second proposed discontinuity near 50 GPa [*Hirao et al.*, 2004]. Further, *Mao et al.* [2004] probed the Debye sound velocity of ε′-FeH up to 52 GPa at 300 K using the nuclear resonant inelastic X-ray scattering (NRIXS) technique, which provides a tight constraint on v_s. The authors report discontinuities in their measured sound velocities at 22 and 40 GPa, which they attribute to the magnetic transition and a possible change in hydrogen solubility, respectively. However, based on the synthesis of all aforementioned experimental studies, an alternative interpretation of their $v_s(\rho)$ data set [*Mao et al.*, 2004] is presented here, which involves a single discontinuity between 22 and 30 GPa corresponding to the FM-NM transition (Figure 20.3).

While these low-pressure discontinuities may seem unrelated to discussions about Earth's core, they can have a substantial impact on the extrapolation of $v_p(\rho)$ and $v_s(\rho)$ to core pressures. For example, by applying a Birch's law model to their four highest-pressure $v_p(\rho)$ data points, *Shibazaki et al.* [2012] found v_p (km/s) = 0.71 + 0.92ρ (g/cm³). In turn, this relationship predicts sound velocities for ε'-FeH that are ~10% faster than those of ε-Fe at the density of Earth's ICB [*Antonangeli et al.*, 2012], where a density anchor point is used to remove bias from the different EOSs. A similar fit including all $v_p(\rho)$ data points for ε'-FeH would increase this contrast to ~30% at the same conditions, thus indicating that a proper treatment of the magnetic transition is essential for accurately extending existing data to core conditions. Finally, a linear fit of the four available $v_s(\rho)$ data [*Mao et al.*, 2004] above the magnetic transition produces a Birch's law–like relationship of v_s (km/s) = –0.51 + 0.56ρ (g/cm³), which predicts sound velocities that are ~30% faster than those of ε-Fe at the density of Earth's ICB [*Murphy et al.*, 2013a].

In summary, multiple lines of experimental evidence suggest that a FM-NM transition in ε'-FeH occurs at a pressure near 27 GPa, which is much lower than that predicted by ab initio calculations. This magnetic collapse is likely responsible for the anomalous *P-V* behavior observed by *Hirao et al.* [2004] and produces discontinuities in $v_p(\rho)$ and $v_s(\rho)$ that must be considered when extrapolating to core pressures. Finally, a smooth increase in $v_p(\rho)$ between 37 and 70 GPa is seemingly inconsistent with a second discontinuity in *K* near 50 GPa, which may indicate that the compression data measured by *Hirao et al.* [2004] can be better explained with a single EOS that incorporates a magnetic transition near 30 GPa [e.g., *Chen et al.*, 2012].

20.4. HYDROGEN IN EARTH'S CORE

Based on the aforementioned studies, similar results for ε-Fe, and an assumption that hydrogen is the only light element in Earth's solid inner core, a compositional model will now be constructed. It has already been mentioned that the primary component in the core is thought to be ε-Fe, which is ~3%–5% denser than Earth's solid inner core [e.g., *Dewaele et al.*, 2006]. Moreover, the sound velocities of ε-Fe have been investigated to core pressures at 300 K using the same techniques described in the previous section [e.g., *Antonangeli et al.*, 2012; *Murphy et al.*, 2013a]. Minor discrepancies exist between studies, but there is a general consensus that at 300 K, v_p for ε-Fe at core pressures is similar to that of the core, while v_s for ε-Fe is significantly faster. Finally, a wide array of melting points have been reported for ε-Fe at the pressure of Earth's ICB; here we assign a value of 6000 K, which is consistent with many recent studies [*Anzellini et al.*, 2013].

In the absence of information about the high-*PT* binary-phase diagram for the iron-hydrogen system, it is assumed that hydrogen will be incorporated into the inner core as ε'-FeH. The limited available data largely necessitates this choice, but some evidence exists for the stability of an Fe-FeH mixture at high pressures. In particular, *Badding et al.* [1991] found that in the presence of excess iron, a mixture of iron and hydrogen would form ε'-FeH+Fe at high hydrogen pressures, suggesting a fixed stoichiometry up to at least 62 GPa at 300 K. Moreover, recent density functional theory calculations explored the formation enthalpies of FeH$_x$—which crystallize in hcp structures with hydrogen atoms segregated into layers for *x* < 1—and found that they are slightly less stable than the isochemical mixture of ε'-FeH+Fe [*Bazhanova et al.*, 2012].

20.4.1 Two-Phase Linear Mixing Model

The most common approach for assessing the seismic signature of a two-phase solid in the core is to apply a linear mixing model. In particular, the average density of an ideal solid mixture of ε-Fe and ε'-FeH can be approximated by

$$\rho_\oplus = \chi\rho_{\text{Fe}} + (1 - \chi)\rho_{\text{FeH}}. \quad (20.1)$$

In theory, one can use this equation to solve for the volume fraction of ε-Fe (χ) that is required to match the density of the core ($\rho_\oplus$, taken from [*Dziewonski and Anderson*, 1981]) by applying the aforementioned EOS for ε-Fe and ε'-FeH. However, while the existing compression data for ε'-FeH are largely self-consistent, differences in the reported EOSs are intensified upon extrapolation to densities relevant for the core (Figure 20.2). In particular, the EOS measured by *Badding et al.* [1991] indicates that the core should contain 0.45–1.32 wt % H (20–42.5 at % H), while the EOS from the highest-pressure regime reported by *Hirao et al.* [2004] (50 and 80 GPa) requires only 0.12–0.48 wt % H (6.24–21.09 at % H). Finally, for reference, the EOS reported for γ-FeH by *Narygina et al.* [2011] suggests that 0.08–0.16 wt % H (4.25–8.16 at % H) is consistent with seismic observations.

To improve the accuracy of the predicted χ, one can use the previously discussed sound velocities for ε-Fe and ε'-FeH as an additional constraint in the ideal mixing model:

$$V_\oplus = \frac{V_{\text{Fe}}V_{\text{FeH}}}{(1 - \chi)V_{\text{Fe}} + \chi V_{\text{FeH}}}. \quad (20.2)$$

Extrapolating the existing experimental data for NM ε'-FeH to core densities using $v_p(\rho)$ and $v_s(\rho)$ (Section 20.3), one finds that its v_p and v_s are ~10% and 80% faster (respectively) than the core [*Dziewonski and Anderson*, 1981; *Mao et al.*, 2004; *Shibazaki et al.*, 2012]. At first, this result may seem undesirable, because the experimentally determined v_s for ε-Fe at 300 K is already too fast. However, ab initio calculations for ε-Fe have shown that v_s is significantly reduced at core temperatures (by ~20% at 4000 K and ~40% at 5500 K), while v_p is slightly affected (<10%) [*Vočadlo et al.*, 2009]. In the absence of constraints on the thermal properties of ε′-FeH, the following discussion assumes that they are similar to those predicted for ε-Fe. However, confirmation of this assumption is needed.

Applying all of the aforementioned results for the densities, sound velocities, and thermal properties of ε-Fe and ε′-FeH to equations (20.1) and (20.2), a hydrogen content of ~0.2–0.3 wt % (13–14.3 at %) produces reasonable agreement with seismic observations for the inner core. Considering the aforementioned prediction that hydrogen will preferentially remain in the liquid outer core (Section 20.2.3) [*Okuchi*, 1997], the corresponding total hydrogen content in the core would be ~30 at %. This value may seem unrealistically large based on the amount of hydrogen that would need to be retained during Earth accretion, but recall that other light elements are likely present in the core and would also contribute to the resolution of the CDD. Moreover, this model is temperature dependent, so the effects of hydrogen on the melting behavior of iron must also be considered.

No experimental constraints exist for the melting curve of ε′-FeH, so here a qualitative estimate of the hydrogen-induced ΔT_M based on the extrapolated melting curve of γ-FeH (Figure 20.1) is presented [*Sakamaki et al.*, 2009]. Comparison with the melting curve of ε-Fe reveals that ΔT_M is ~1500 K at the pressure of Earth's CMB (135 GPa) and ~2700 K at the ICB (330 GPa) [*Anzellini et al.*, 2013]. These values are almost certainly overestimates of the true ΔT_M because the melting curve of ε′-FeH should lie above that of γ-FeH. To provide some insight into the magnitude of this overestimate, note that T_M(330 GPa) for ε-Fe is ~500 K higher than that of (hypothetical) γ-Fe. In turn, a more appropriate ΔT_M at the ICB may be ~2000 K, which is still much higher than has been predicted in other light element alloys [e.g., *Sakamaki et al.*, 2009]. However, it is important to restate that this is purely a qualitative exercise, and the main conclusion to draw is that even minor concentrations of hydrogen would significantly influence the thermal profile of the core.

20.4.2. Inherent Uncertainties and Possible Solutions

This model for the potential hydrogen content in Earth's core is rooted in all available data, but it suffers from inherent uncertainties that can only be reduced through new experiments and calculations. First, the ε′-FeH input parameters are derived from data collected below 84 GPa and at 300 K, while the melting behavior has only been constrained up to 20 GPa; significant uncertainties arise from extrapolations that are necessary to apply these experimental results to core conditions. Therefore, it is essential that the density, sound velocities, and melting behavior of iron hydrides be investigated to higher pressures and temperatures with both experimental and theoretical techniques. Alternatively, the existing experimental data could be applied more directly to discussions of smaller terrestrial bodies—such as Mercury, Mars, or the Moon—whose cores lie at lower pressures.

One possible explanation for the limited *PT* space explored thus far is the premature failure of high-pressure experiments due to diffusion of hydrogen into the pressurizing diamonds of a diamond anvil cell (DAC). In an effort to minimize the diamonds' exposure to H_2, a better approach may be to utilize hydrogen sources that do not dissociate until higher pressures. For example, *Terasaki et al.* [2012] recently performed two in situ XRD experiments in laser-heated DACs above 100 GPa and up to 2250 K (Figure 20.1) with natural diaspore (AlOOH) as a hydrogen source. It is possible that the use of diaspore helped to stabilize the sample chamber to such high-*PT* conditions by "trapping" excess hydrogen in the solid source mineral, as evidenced by the observation of diaspore diffraction peaks beyond the hydride formation conditions.

Another fundamental uncertainty in the previously presented model is the assumption that hydrogen would be incorporated into the core as ε′-FeH. Little is known about the binary-phase diagram of an iron-hydrogen mixture at core pressures, which ultimately determines what hydrogen-bearing phase would exist in the core. For example, if the iron-hydrogen system adopts a eutectic phase diagram, then the overall hydrogen content would dictate whether it crystallizes as stoichiometric ε′-FeH (i.e., above the eutectic composition) or as a minor dissolved component in ε-Fe. Moreover, there are currently no experimental constraints on the temperature dependence of hydrogen solubility in iron, so it is possible that a nonstoichiometric ε′-FeH$_x$ would be favored at high-*PT* conditions [e.g., *Bazhanova et al.*, 2012]. Finally, one cannot rule out the stabilization of new phases at unexplored *PT* conditions: Theoretical calculations suggest that stoichiometric γ-FeH and select polyhydrides (FeH$_3$ and FeH$_4$) will be thermodynamically stable at ICB pressures [*Bazhanova et al.*, 2012; *Isaev et al.*, 2007], which could dramatically alter the previous estimate for hydrogen in the core. In particular, the similar size and coordination of Fe and H atoms in the predicted FeH$_3$ and FeH$_4$ structures [*Bazhanova et al.*, 2012] would likely yield very different elastic and thermodynamic properties compared to the hydrides discussed here.

Finally, in order to better constrain the stable phase(s) of FeH$_x$ at core conditions, it is necessary to not only extend existing experiments toward the conditions of Earth's core but also to explore new approaches for determining hydrogen content in situ. High-pressure NPD experiments above 25 GPa are now becoming more widely available [*Boehler et al.*, 2013] and represent a powerful technique for directly exploring the hydrogen sublattice in both FeH and substoichiometric FeH$_x$. However, NPD requires deuterized samples due to the negative scattering cross section of hydrogen, and a clear understanding of isotope effects at high pressures is currently lacking. For example, no explanation has been given for the increased fraction of unreacted α-Fe and metastable ε-FeD$_x$ in quenched deuterized samples [e.g., *Antonov et al.*, 2002]. In addition, high-pressure NPD experiments on hydrogenated and deuterized brucite up to 9 GPa reveal differences in their unit cell volumes and bulk moduli [*Horita et al.*, 2010]. If such effects are present at higher pressures and in FeH$_x$, then one must be careful to distinguish between isotope effects and intrinsic material properties. For example, preliminary refinements of our own high-pressure NPD studies on ε′-FeD$_x$ up to 31 GPa (Figure 20.2) suggest that a minor amount of deuterium may occupy tetrahedral sites [*Murphy et al.*, 2013b]. In order to determine whether this site occupation is the result of isotope effects, complementary XRD experiments on the same material are underway, which will allow for direct comparison with previous EOS studies on hydrogen-bearing samples.

20.5. CONCLUSIONS AND FUTURE DIRECTIONS

This chapter has reviewed the available experimental data for the phase diagram, EOS, and sound velocities of FeH$_x$. The high-temperature phase below 60 GPa is non-magnetic γ-FeH, whose melting curve is complex at low pressures and lies at considerably reduced temperatures relative to that of γ-Fe. The dominant high-pressure phase is ε′-FeH, whose elastic properties are significantly influenced by a ferromagnetic-nonmagnetic transition near 27 GPa. An inversion of available experimental data for ε′-FeH$_x$ and ε-Fe suggests that an iron alloy with 14 at % H is consistent with seismic observations of Earth's solid inner core. However, this prediction is rooted in significant extrapolations and assumptions that must be reduced through higher-pressure and higher-temperature experiments, investigations of the iron-hydrogen binary-phase diagram, and new experimental approaches for constraining the hydrogen content at high-pressure conditions.

Beyond improving the accuracy of a two-phase mixing model, the exploration of multicomponent systems is another important future research direction. The only comprehensive study of this type was published recently by *Terasaki et al.* [2012], who synthesized (Fe,Ni)H$_x$ alloys using the aforementioned diaspore and iron-nickel metal foils with 5% or 10% Ni. The resulting phase behavior for (Fe,Ni)H$_x$ is largely consistent with that of FeH$_x$ (Figure 20.1), but *Terasaki et al.* [2012] report slightly shifted phase stability fields, different P-V data points, and potentially lower hydrogen contents for ε′-(Fe,Ni)H$_x$ compared to ε′-FeH (see Table 2 in *Terasaki et al.* [2012]). Additional experiments and calculations are needed to determine whether these differences reflect experimental scatter or if minor amounts of nickel influence the electronic structure and, in turn, the hydrogen content and structural properties of the alloy.

Finally, the interaction of hydrogen with other light element alloys represents an exciting future research direction. The only available information for such a system is the suggestion that hydrogen and carbon would be incompatible in the core because the reaction of iron carbides and hydrogen to form FeH$_x$ and diamond is favored at high-*PT* conditions [*Narygina et al.*, 2011]. However, hydrogen is likely soluble to some extent in most iron-bearing alloys at high-pressure conditions [*Fukai*, 2005] and could have a substantial influence on their melting behavior and structural properties. From an experimental perspective, an induced melting point depression could be a significant advantage, because it would allow for easier access to the liquid phase of an iron alloy containing hydrogen. In particular, minor amounts of hydrogen could help provide experimental constraints on the liquid outer core by lowering the temperatures necessary for XRD and long-duration inelastic X-ray scattering experiments on liquid iron alloys. Such experiments would be invaluable for constraining the thermal profile of Earth's outer core, which is intimately related to the temperatures and phases found in the lowermost mantle [e.g., *Nomura et al.*, 2014].

ACKNOWLEDGMENTS

I would like to thank R. Hemley, Y. Fei, M. Guthrie, M. Ahart, G. Cody, W. L. Mao, J. Smyth, and two anonymous reviewers for helpful comments and suggestions. EFree, an Energy Frontier Research Center funded by the US Department of Energy (DOE), Office of Science, under Award No. DE-SC0001057, is acknowledged for supporting the author's experiments. Research conducted at the SNS was supported by the Scientific User Facilities division, BES, DOE, under Contract No. DE-AC05-00OR22725 with UT-Battelle, LLC, and salary was supported from the DOE/NNSA under Award No. DE-NA-0002006 to CDAC.

REFERENCES

Antonangeli, D., T. Komabayashi, F. Occelli, E. Borissenko, A. C. Walters, G. Fiquet, and Y. W. Fei (2012), Simultaneous sound velocity and density measurements of Hcp iron up to

93 Gpa and 1100 K: An experimental test of the Birch's law at high temperature, *Earth Planet. Sci. Lett.*, *331*, 210–214.

Antonov, V. E., I. T. Belash, V. F. Degtiareva, E. G. Poniatovskii, and V. I. Shiriaev (1980), Synthesis of iron hydride under high hydrogen pressure, *Doklady Akad. Nauk Sssr*, *252*(6), 1384–1387.

Antonov, V. E., K. Cornell, V. K. Fedotov, A. I. Kolesnikov, E. G. Ponyatovsky, V. I. Shiry, and H. Wipf (1998), Neutron diffraction investigation of the Dhcp and Hcp iron hydrides and deuterides, *J. Alloys Comp.*, *264*(1–2), 214–222.

Antonov, V. E., M. Baier, B. Dorner, V. K. Fedotov, G. Grosse, A. I. Kolesnikov, E. G. Ponyatovsky, G. Schneider, and F. E. Wagner (2002), High-pressure hydrides of iron and its alloys, *J. Phys. Condens. Matter*, *14*(25), 6427–6445.

Anzellini, S., A. Dewaele, M. Mezouar, P. Loubeyre, and G. Morard (2013), Melting of iron at Earth's inner core boundary based on fast X-ray diffraction, *Science*, *340*(6131), 464–466.

Badding, J. V., R. J. Hemley, and H. K. Mao (1991), High-pressure chemistry of hydrogen in metals—In situ study of iron hydride, *Science*, *253*(5018), 421–424.

Badding, J. V., H. K. Mao, and R. J. Hemley (1992), High-pressure crystal structure and equation of state of iron hydride: Implications for the Earth's core, in *High-Pressure Research: Application to Earth and Planetary Sciences*, edited by Y. Syono and M. H. Manghnani, pp. 363–371, AGU, Washington, D.C.

Bazhanova, Z. G., A. R. Oganov, and O. Gianola (2012), Fe-C and Fe-H systems at pressures of the Earth's inner core, *Phys. Uspekhi*, *55*(5), 489–497.

Boehler, R., M. Guthrie, J. J. Molaison, A. M. dos Santos, S. Sinogeikin, S. Machida, N. Pradhan, and C. A. Tulk (2013), Large-volume diamond cells for neutron diffraction above 90gpa, *High Press. Res.*, *33*(3), 546–554.

Chen, B., J. M. Jackson, W. Sturhahn, D. Zhang, J. Zhao, J. K. Wicks, and C. A. Murphy (2012), Spin crossover equation of state and sound velocities of (Mg0.65fe0.35)O ferropericlase to 140 Gpa, *J. Geophys. Res. Solid Earth*, *117*, B08208.

Choe, I., R. Ingalls, J. M. Brown, Y. Satosorensen, and R. Mills (1991), Mossbauer studies of iron hydride at high-pressure, *Phys. Rev. B*, *44*(1), 1–4.

Dewaele, A., P. Loubeyre, F. Occelli, M. Mezouar, P. I. Dorogokupets, and M. Torrent (2006), Quasihydrostatic equation of state of iron above 2 Mbar, *Phys. Rev. Lett.*, *97*(21), 4.

Dziewonski, A. M., and D. L. Anderson (1981), Preliminary reference Earth model, *Phys. Earth Planet. Inter.*, *25*(4), 297–356.

Elsasser, C., J. Zhu, S. G. Louie, B. Meyer, M. Fahnle, and C. T. Chan (1998), Ab initio study of iron and iron hydride: Ii. Structural and magnetic properties of close-packed Fe and Feh, *J. Phys. Condens. Matter*, *10*(23), 5113–5129.

Fukai, Y. (2005), *The Metal-Hydrogen System: Basic Bulk Properties*, 2nd ed., Springer, Berlin.

Hirao, N., T. Kondo, E. Ohtani, K. Takemura, and T. Kikegawa (2004), Compression of iron hydride to 80 Gpa and hydrogen in the Earth's inner core, *Geophys. Res. Lett.*, *31*(6), 4.

Horita, J., A. M. dos Santos, C. A. Tulk, B. C. Chakoumakos, and V. B. Polyakov (2010), High-pressure neutron diffraction study on H-D isotope effects in brucite, *Phys. Chem. Minerals*, *37*(10), 741–749.

Isaev, E. I., N. V. Skorodumova, R. Ahuja, Y. K. Vekilov, and B. Johansson (2007), Dynamical stability of Fe-H in the Earth's mantle and core regions, *Proc. Nat. Acad. Sci. USA*, *104*(22), 9168–9171.

Ishimatsu, N., et al. (2012), Hydrogen-induced modification of the electronic structure and magnetic states in Fe, Co, and Ni monohydrides, *Phys. Rev. B*, *86*(10), 9.

Liu, L.-G., and W. A. Basset (1986), *High-Pressure Phases with Implications for the Earth's Interior*, Oxford Univ. Press, New York.

Mao, W. L., W. Sturhahn, D. L. Heinz, H. K. Mao, J. F. Shu, and R. J. Hemley (2004), Nuclear resonant X-ray scattering of iron hydride at high pressure, *Geophys. Res. Lett.*, *31*(15), 4.

Mitsui, T., and N. Hirao (2010), Ultrahigh-pressure study on the magnetic state of iron hydride using an energy domain synchrotron radiation 57fe Mössbauer spectrometer, *MRS Proceedings*, *1262*, 1262-W06-09.

Murphy, C. A., J. M. Jackson, and W. Sturhahn (2013a), Experimental constraints on the thermodynamics and sound velocities of Hcp-Fe to core pressures, *J. Geophys. Res. Solid Earth*, *118*(5), 1999–2016.

Murphy, C. A., M. Guthrie, R. Boehler, M. Somayazulu, Y. W. Fei, J. J. Molaison, and A. M. dos Santos (2013b), Probing the hydrogen sublattice of Fehx with high-pressure neutron diffraction, edited, Abstract MR12A-08 presented at 2013 Fall Meeting, AGU, San Francisco, Calif.

Narygina, O., L. S. Dubrovinsky, C. A. McCammon, A. Kurnosov, I. Y. Kantor, V. B. Prakapenka, and N. A. Dubrovinskaia (2011), X-ray diffraction and Mossbauer spectroscopy study of Fcc iron hydride Feh at high pressures and implications for the composition of the Earth's core, *Earth Planet. Sci. Lett.*, *307*(3–4), 409–414.

Nomura, R., K. Hirose, K. Uesugi, Y. Ohishi, A. Tsuchiyama, A. Miyake, and Y. Ueno (2014), Low core-mantle boundary temperature inferred from the solidus of pyrolite, *Science*, *343*(6170), 522–525.

Ohtani, E., N. Hirao, T. Kondo, M. Ito, and T. Kikegawa (2005), Iron-water reaction at high pressure and temperature, and hydrogen transport into the core, *Phys. Chem. Minerals*, *32*(1), 77–82.

Okuchi, T. (1997), Hydrogen partitioning into molten iron at high pressure: Implications for Earth's core, *Science*, *278*(5344), 1781–1784.

Okuchi, T. (1998), The melting temperature of iron hydride at high pressures and its implications for the temperature of the Earth's core, *J. Phys. Condens. Matter*, *10*(49), 11,595–11,598.

Sakamaki, K., E. Takahashi, Y. Nakajima, Y. Nishihara, K. Funakoshi, T. Suzuki, and Y. Fukai (2009), Melting phase relation of Fehx up to 20 Gpa: Implication for the temperature of the Earth's core, *Phys. Earth Planet. Inter.*, *174*(1–4), 192–201.

Saxena, S. K., H. P. Liermann, and G. Y. Shen (2004), Formation of iron hydride and high-magnetite at high pressure and temperature, *Phys. Earth Planet. Inter.*, *146*(1–2), 313–317.

Schneider, G., M. Baier, R. Wordel, F. E. Wagner, V. E. Antonov, E. G. Ponyatovsky, Y. Kopilovskii, and E. Makarov (1991),

Mossbauer study of hydrides and deuterides of iron and cobalt, *J. Less-Common Metals*, *172*, 333–342.

Shibazaki, Y., et al. (2012), Sound velocity measurements in Dhcp-Feh up to 70 Gpa with inelastic X-ray scattering: Implications for the composition of the Earth's core, *Earth Planet. Sci. Lett.*, *313*, 79–85.

Skorodumova, N. V., R. Ahuja, and B. Johansson (2004), Influence of hydrogen on the stability of iron phases under pressure, *Geophys. Res. Lett.*, *31*(8), 3.

Terasaki, H., et al. (2012), Stability of Fe-Ni hydride after the reaction between Fe-Ni alloy and hydrous phase (Delta-Alooh) up to 1.2 Mbar: Possibility of H contribution to the core density deficit, *Phys. Earth Planet. Inter.*, *194*, 18–24.

Tsumuraya, T., Y. Matsuura, T. Shishidou, and T. Oguchi (2012), First-principles study on the structural and magnetic properties of iron hydride, *J. Phys. Soc. Jpn.*, *81*(6), 6.

Vočadlo, L., D. P. Dobson, and I. G. Wood (2009), Ab initio calculations of the elasticity of Hcp-Fe as a function of temperature at inner-core pressure, *Earth Planet. Sci. Lett.*, *288*(3–4), 534–538.

Williams, Q., and R. J. Hemley (2001), Hydrogen in the deep Earth, *Annu. Rev. Earth Planet. Sci.*, *29*, 365–418.

Yagi, T., and T. Hishinuma (1995), Iron hydride formed by the reaction of iron, silicate, and water—Implications for the light-element of the Earth's core, *Geophys. Res. Lett.*, *22*(14), 1933–1936.

21

Stability of Hydrous Minerals and Water Reservoirs in the Deep Earth Interior

Eiji Ohtani,[1,2] Yohei Amaike,[1] Seiji Kamada,[1] Itaru Ohira,[1] and Izumi Mashino[1]

ABSTRACT

Water storage capacities in hydrous phases and nominally anhydrous minerals in the deep mantle are summarized. There are several controversies on the water contents in the lower mantle minerals, such as periclase, stishovite, and $MgSiO_3$ perovskite, and we need more studies to clarify the effect of pressure, temperature, and compositions on the H_2O activity of these high-pressure minerals. The transition zone has a water storage capacity of approximately 0.5–1 wt % due to a high water solubility of about 1–3 wt % in wadsleyite and ringwoodite, which are the major constituents in the transition zone. Water has significant effects on the phase boundaries of the phase transformations of the mantle minerals and can explain some topography of the 410 and 660 km seismic discontinuities. Recent discovery of hydrous ringwoodite and phase Egg as inclusions in diamond strongly suggests existence of the hydrous transition zone, a major water reservoir in the Earth's interior. Discovery of a new hydrous phase H, $MgSiO_2(OH)_2$ and its solid solution with isostructural phase δ-AlOOH suggests that water can be stored in this phase in the lower mantle. Water may be transported into the bottom of the lower mantle by this phase in descending slabs.

21.1. INTRODUCTION

The mantle is composed almost entirely of oxide and silicate minerals so small amounts of H included as hydroxyl in the nominally anhydrous as well as nominally hydrous minerals could constitute the bulk of the planet's water. Recent seismic tomography studies indicate that subducting slabs penetrate into the lower mantle and possibly accumulate at the mantle-core boundary [e.g., *Fukao et al.*, 2001; *Grand*, 2002]. These observations suggest that water trapped in the slabs may be transported to the core-mantle boundary (CMB) region. The water transport

capability of different slabs may vary depending on their thermal structures. Hot and young slabs generally transport little water into the deep mantle, whereas colder and older slabs may transport significant amounts of water into the deep mantle, although their water transport capability is still a matter of debate [e.g., *Poli and Schmidt* 2002; *van Keken et al.*, 2011]. Recent estimation of the global water cycle by *van Keken et al.* [2011] suggests that $7–10\times10^{11}$ kg/year of water can penetrate by slab subduction, and two thirds of this water has been lost by dehydration in the slab, whereas one third of bound water in the slab, i.e., 3×10^{11} kg/year, penetrates to depths over 240 km. The deep mantle might contain more water than that estimated by these previous authors. The hydrous 10 Å phase reported by Fumagalli et al. [2001], which is essential to transport water into the transition zone, is sometimes ignored to account for the water transport

[1]*Department of Earth Science, Graduate School of Science, Tohoku University, Sendai, Japan*
[2]*V. S. Sobolev Institute of Geology and Mineralogy, Siberian Branch, Russian Academy of Sciences, Novosibirsk, Russia*

Deep Earth: Physics and Chemistry of the Lower Mantle and Core, Geophysical Monograph 217, First Edition.
Edited by Hidenori Terasaki and Rebecca A. Fischer.
© 2016 American Geophysical Union. Published 2016 by John Wiley & Sons, Inc.

into the transition zone [e.g., *van Keken et al.*, 2011]. Additionally, some amount of water in the deep mantle might be originated from the primitive mantle reservoirs. In spite of these uncertainties, it is very important to consider the potential water reservoir in the deep mantle because water significantly modifies the physical properties of minerals and magmas.

There are several electrical conductivity studies of wadsleyite and ringwoodite to determine the water content in the mantle transition zone, since electrical conductivity is very sensitive to the hydrogen contents in minerals [e.g., *Yoshino et al.*, 2008; *Karato and Dai*, 2009]. These mineral physics data combined with the geophysical observations indicate that there are significant heterogeneities in the water content in the mantle transition zone, and water is localized in the mantle transition zone beneath subduction zones [e.g., *Koyama et al.*, 2006; *Utada et al.*, 2009]. *Khan and Shankland* [2012] argued that the upper mantle is essentially dry, whereas the transition zone is highly heterogeneous in water content; i.e., the transition zone beneath the United States and Northeast China is wet whereas that beneath Europe is dry, supporting earlier conclusions by *Utada et al.* [2009]. There are some studies to determine the water content in the transition zone based on the topography of the 410 and 660 km discontinuities and seismic tomography data [e.g., *Suetsugu et al.*, 2010]. Analyses of the attenuation of seismic shear waves revealed a large low-$Q\mu$ region in the shallow lower mantle beneath eastern Asia, suggesting the possible existence of a wide region of hydrogen enrichment [*Lawrence and Wysession*, 2006].

Additional strong evidence for existence of water in the mantle transition zone and the lower mantle is observed as hydrous mineral inclusions in diamond from the deep mantle. *Wirth et al.* [2007] discovered hydrous phase Egg ($AlSiO_3OH$) in diamond. *Pearson et al.* [2014] discovered hydrous ringwoodite containing 1 wt % water in diamond. These observations from deep-seated diamond crystals imply that the transition zone, where the deep diamond crystals captured the high-pressure minerals as inclusions during their growth, contains a large amount of water indicating the wet nature of the transition zone.

Water plays an important role in the dynamics of Earth's interior because water changes the physical properties of mantle materials, including the melting temperature, viscosity, diffusivity, and strain rates. Hot plumes under wet conditions may be partially molten due to depression of melting temperature [e.g., *Litasov and Ohtani*, 2003; *Ohtani et al.*, 2004]. Water also affects mantle convection and the dynamics of ascending plumes because a trace amount of water lowers the viscosity of the mantle minerals [e.g., *Mei and Kohlstedt*, 2000]. Water also enhances the diffusion of elements and the kinetics of phase transformations [*Kubo et al.*, 1998]. Water affects the position of phase boundaries such as the α-β

transformation and decomposition of ringwoodite into ferroperioclase and Mg-perovskite [e.g., *Higo et al.*, 2001; *Chen et al.*, 2002; *Litasov et al.*, 2005; *Frost and Dolejs*, 2007; *Ghosh et al.*, 2013a,b]. In this work, recent studies on the stability and physical properties of hydrous minerals are reviewed, and potential water reservoirs in the deep mantle are discussed.

21.2. HYDROUS MINERALS AND NOMINALLY ANHYDROUS MINERALS

21.2.1. Water in Nominally Anhydrous Minerals in Transition Zone and Lower Mantle

It has been reported that the transition zone minerals, such as wadsleyite and ringwoodite, can accommodate a large amount of water up to 2–3 wt % in their crystal structures [*Inoue et al.*, 1995; *Koholstedt et al.*, 1996; *Ye et al.*, 2010], and the water solubility of these phases decreases with increasing temperature to 0.1–0.5 wt % along the normal mantle geotherm [*Ohtani et al.*, 2004]. Therefore, the mantle transition zone can be the most important water reservoir in the earth interior. Recent discovery of hydrous ringwoodite containing 1 wt % water as an inclusion in diamond [*Pearson et al.*, 2014] provided direct evidence for the wet mantle transition zone, at least locally, as suggested by several authors [e.g., *Ohtani et al.*, 2004]. Akimotoite ($MgSiO_3$ ilmenite) can exist in the harzburgite layer of subducting slabs in the mantle transition zone. Water solubility in akimotoite was reported to be 350–440 ppm at 19–24 GPa and 1300–1600°C [*Bolfan-Casanova et al.*, 2000].

There are contradictory results on the water solubility of the lower mantle minerals. Periclase and ferropericlase show very limited amount of water solubility; i.e., *Bolfan-Casanova et al.* [2002] reported that they contain 60 ppm water at 25 GPa, whereas *Litasov and Ohtani* [2003] showed the similar amount of water in alumina-bearing periclase and ferropericlase. On the other hand, *Murakami et al.* [2002] reported that periclase contains 2000 ppm water.

There are also contradictory reports on the solubility of water in $MgSiO_3$ perovskite and stishovite. The water contents in these minerals are summarized in Table 21.1. More detailed studies of the silica activity dependency on water solubility in perovskite are needed to explain the discrepancy since the silica activity affects the amount of the oxygen vacancy where hydrogen can be accommodated. A recent first-principles study on water incorporation into $MgSiO_3$ perovskite indicated that water can be partitioned preferentially into ringwoodite compared to perovskite with partition coefficient D_{H_2O} around 10–13, whereas that between $MgSiO_3$ perovskite and periclase is 90, with strong water preference in perovskite [*Hernandez et al.*, 2013]. This theoretical calculation is

Table 21.1 Water content in anhydrous lower mantle minerals.

Minerals	SiO$_2$ (wt %)	TiO$_2$	Al$_2$O$_3$	FeO[a]	MgO	Mg[b]	Pressure synthesis (GPa)	Temperature synthesis (°C)	H$_2$O (ppmwt)	1σ	Reference[c]
MgSiO$_3$ perovskite	60.1				40.36	100	25	1300	104	14	1
Al-Mg perovskite	59.09		2.03		39.77	100	25	1400	101	19	1
Al-Mg perovskite	57.86		4.43		39.55	100	25	1200	1101	156	1
Al-Mg perovskite	55.37		7.16		38.18	100	26	1200	1440	160	1
Al-Fe-Mg perovskite[d]	36.73	2.18	14.5	26.23	20.15	57.8	26	100	110	21	1
Al-Fe-Mg perovskite[d]	38.11	1.91	13.16	24.52	21.37	60.8	26	1200	104	26	1
Al-Fe-Mg perovskite[d]	38.43		16.77	23.4	20.82	61.3	25	1300	47	12	1
Al-Fe-Mg perovskite[e]	52.77		5.5	7.39	34.48	89.3	25	1400	1780	175	1
Al-Fe-Mg perovskite[e]	53.11		5.8	6.09	34.94	91.1	25	1600	1460	130	1
MgSiO$_3$ perovskite	60.1				40.46	100	27	1500	60–70		2
MgSiO$_3$ perovskite	60.1				40.36	100	24	1600	<1		3
Al-Fe-Mg perovskite[e]			2–6	1–4		100	24		<5		3
Al-Fe-Mg perovskite[e]							25.5		1000–4000		4
Ferropericlase							25.5		2000		4
Periclase					100	100	24	1600	2		3
MgSiO$_3$ akimotoite						100	19–24	1300–1600	350–400		3
SiO$_2$ stishovite							10–24		72		3
Stishovite in MORB							10–15		844		5
Stishovite (aluminous)							20	1127	2500		6
CaCl$_2$ type-SiO$_2$ (stishovite)							60	2327	480		7
SiO$_2$ stishovite							10	350–550	1.3+0.2 wt %		8
CaSiO$_3$ perovskite							25.5		0.3–0.4 wt %		4
Mg$_2$SiO$_4$ wadsleyite						100			3.0 wt %		9
Mg$_2$SiO$_4$ ringwoodite						100			2.0 wt %		10
Mg$_2$SiO$_4$ ringwoodite									1.0 wt %		11

[a] Iron calculated as ferrous iron.

[b] $100 \times MgO/(MgO+FeO)$ in mol.

[c] References: 1. *Litasov et al.* [2003]; 2. *Meade et al.* [1994]; 3. *Bolfan-Casanova et al.* [2000]; 4. *Murakami et al.* [2002]; 5. *Chung and Kagi* [2002]; 6. *Litasov et al.* [2007]; 7. *Panero et al.* [2003]; 8. *Spektor et al.* [2011]; 9. *Inoue et al.* [1995]; 10. *Kohlstedt et al.* [1996]; 11. *Peason et al.* [2014].

[d] In MORB.

[e] In peridotite.

consistent with recent experimental results [*Inoue et al.*, 2010a]. There is no report on the water solubility of post-perovskite, $MgSiO_3$, which exists stable at the base of the lower mantle.

The solubility of water in stishovite has been studied experimentally by several authors. The water solubility in pure stishovite was measured to be 72 ppm at 10–24 GPa by *Bolfan-Casanova et al.* [2000]. Recent hydrothermal synthesis experiments with starting materials of silica glass and coesite at 350–550°C near 10 GPa have produced stishovite with significant amounts of H_2O, 1.3 ± 0.2 wt %, in its structure [*Spektor et al.*, 2011]. The primary mechanism for H_2O accommodation is a direct hydrogarnet-like substitution of Si^{4+} by $4H^+$, with protons clustered as hydroxyls around a silicon vacancy [*Spektor et al.*, 2011]. These results strongly indicated that the water solubility in pure stishovite can vary with temperature.

Water contents in stishovite are also strongly related to the aluminum content in this phase by substitution of Si^{4+} by $Al^{3+}+H^+$. *Chung and Kagi* [2002] reported that alumina-bearing stishovite in the midocean ridge basalt (MORB) system contains up to 844 ppm water at 10–15 GPa, whereas *Litasov et al.* [2007] showed that aluminous stishovite can contain more than 2500 ppm water at 20 GPa and above 1400 K. Although *Panero et al.* [2003] reported that aluminous stishovite contains up to 480 ppm water at 60 GPa and above 2600 K, this aluminous silica phase could be the post-stishovite phase, $CaCl_2$-type SiO_2, due to its high-pressure and high-temperature stability [*Lakshtanov et al.*, 2007]. There are no reports on water solubility of post-stishovite phases such as α-PbO_2 type and pyrite type of SiO_2. Further studies are needed for determination of the solubility of water in these phases.

21.2.2. Effect of Water on Phase Boundaries in Mantle Transition Zone and Lower Mantle

Water can affect the pressure of the boundaries of phase transformations in the mantle. The phase boundary between olivine and wadsleyite and that of ringwoodite and Mg-perovskite + periclase shift toward low and high pressures [e.g., *Higo et al.*, 2001; *Litasov et al.*, 2005], respectively, in the presence of H_2O due to the different H_2O solubility in coexisting minerals. The shift of the phase boundaries in Mg_2SiO_4 under wet conditions at different temperatures is schematically shown in Figure 21.1 although we need to quantify the shift of the phase boundaries for given water contents in olivine, wadsleyite, ringwoodite, periclase, and perovskite. Experiments [e.g., *Chen et al.*, 2002; *Litasov and Ohtani*, 2003] showed a clear shift of the olivine-wadsleyite phase transition boundary to a lower pressure by 1–2 GPa if we add water in the

system. The post-spinel transformation boundary in hydrous peridotite shifts to higher pressure by about 0.6 GPa relative to anhydrous peridotite [*Litasov et al.*, 2005] at 1473 K, whereas there is no obvious shift of this boundary at higher temperatures (1773–1873 K). The resulting linear equation for appearance of Mg-perovskite may be expressed as P (GPa) = $-0.002T$ (K) + 26.3 and is applicable for the temperature range 1000–1800 K. However, we should note a possibility that the small shift of the phase boundary at higher temperatures of 1773–1873 K may be caused by the low solubility of water in ringwoodite at high temperature [*Ohtani et al.*, 2004]. *Inoue et al.* [2010b] showed a small shift, approximately 0.8 GPa, of the wadsleyite−ringwoodite phase boundary of $(Mg_{0.9}Fe_{0.1})_2SiO_4$ with 1 wt % toward higher pressure. The shift of the phase boundaries in the mantle minerals such as olivine and garnet under the dry and wet conditions is shown in Figure 21.2.

These data can be used to estimate the water contents in the mantle transition zone from the depths of the 410 and 660 km discontinuities, especially in the regions close to subduction zones. A large elevation of the 410 km discontinuity to 60−70 km in a Izu-Bonin slab was observed by *Collier et al.* [2001]. They argued that the discontinuity can be explained by the equilibrium boundary of the olivine-wadsleyite transformation in the cold slab, and a large elevation of the 410 km discontinuity corresponds to the temperature difference of about 1000°C compared to the surrounding normal mantle. This indicates the temperature of the 410 km discontinuity is around 500°C. Recent studies on the phase transformation kinetics [e.g., *Rubie and Ross*, 1994] indicate that a very low temperature transformation, such as that proposed by *Collier et al.* [2001], would be kinetically inhibited and therefore unlikely to occur in the dry slabs. A small amount of water, about 0.12–0.5 wt %, dramatically enhances the olivine-wadsleyite phase transformation kinetics, which corresponds to a temperature elevation of about 150°C [*Kubo et al.*, 1998; *Ohtani et al.*, 2004]. The elevation of the 410 km discontinuity of the Izu-Bonin slab reported by *Collier et al.* [2001] may be explained as the equilibrium phase boundary under the hydrous conditions. *Koper et al.* [1998] also observed no evidence for depression of the 410 km discontinuity of the coldest Tonga slab originating from a metastable olivine wedge in the slab. In some slabs, on the other hand, a strong depression of the 410 km discontinuity, i.e., the low-velocity anomaly around the 410 km discontinuity, suggests the existence of a metastable olivine wedge in a cold and dry slab descending into the transition zone [*Jiang et al.*, 2008]. This metastable olivine wedge of the cold stabs in the transition zone is not always observed in the slabs, indicating that the slab may also be locally dry and is highly heterogeneous in terms of the water distribution.

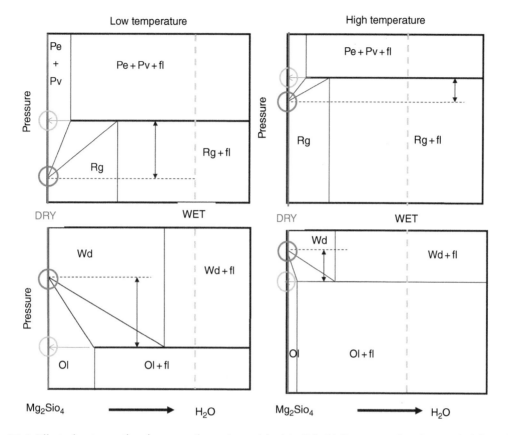

Figure 21.1 Effect of water on the phase transformations of the Mg_2SiO_4-H_2O system at low (~1000 K, left) and high (~2000 K, right) temperatures at high pressure. The shifts of the phase boundaries in Mg_2SiO_4 depend on the solubility of water in wadsleyte and ringdoodite at high pressure. The red line indicates a dry Mg_2SiO_4 composition, and the blue dashed line indicates a water-saturated composition. A red circle indicates the pressure of the transformation under the dry condition, whereas a blue circle indicates the transformation pressure of the water-saturated conditions. Pe, periclase; Pv, bridgmanite; Rg, ringwoodite; fl, fluid water; Ol, olivine; Wd, wadsleyite.

This is consistent with a slow diffusion rate of hydrogen in mantle minerals.

The depression of the 660 km discontinuity may be caused by the hydrated transition zone [*Litasov et al.*, 2005]. Recent in situ X-ray diffraction studies of the phase boundary of the decomposition of ringwoodite [*Katsura et al.*, 2003; *Fei et al.*, 2004] under the dry condition indicate that the slope of the boundary, *dP/dT*, is around –0.4 to −1.3 MPa/K, which is significantly smaller than that determined previously (−2.5 MPa/K [*Irifune et al.*, 1998]). A temperature decrease of 1500 K is needed to account for the depression of the discontinuity of about 40 km observed in some subduction zones [e.g., *Collier et al.*, 2001] indicating unusually cold subducting slabs. *Higo et al.* [2001], *Litasov et al.* [2005], and *Ghosh et al.* [2014] showed that the phase boundary moves toward higher pressure due to the high water content in ringwoodite. Thus, a large depression of the 660 km discontinuity beneath some subduction zones is consistent with the wet subducting slab.

21.2.3. Dense Hydrous Magnesium Silicates and Alphabet Phases

Various high-pressure hydrous minerals have been synthesized at high pressure [e.g., *Ohtani*, 2005]. The high-pressure hydrous phases in the MgO-SiO_2-H_2O system are called dense hydrous magnesium silicates [*Ringwood and Major*, 1967]. Pioneering studies of the stability of high-pressure hydrous phases led to the discovery of several high-pressure alphabet phases such as hydrous phases A, B, C [e.g., *Ringwood and Major*, 1967] and phase D [*Liu et al.*, 1987], although the X-ray data were not sufficient at that time to identify the structures of these phases and thus some confusion was introduced in the identification of the phases. *Pacalo and Parise* [1992] reported a hydrous phase with a composition of $Mg_{10}Si_3O_{14}(OH)_4$ and named it superhydrous phase B. However, later detailed analyses of its crystal structure revealed that this new phase is identical to hydrous phase C reported by *Ringwood and Major* [1967]. *Kanzaki et al.*

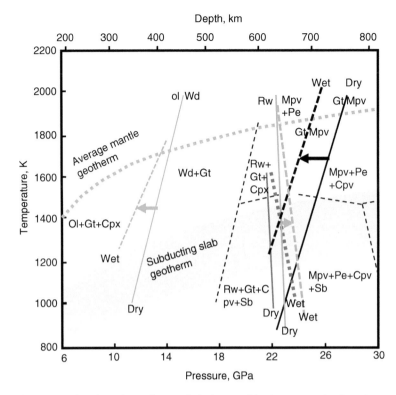

Figure 21.2 Effect of water on the phase boundaries of olivine-wadsleyite, post-spinel, and post-garnet transitions. Blue solid and dashed lines are the boundary of the post-spinel transition in Mg_2SiO_4 under the dry and wet conditions by *Ghosh et al.* [2014]; thick green solid and dashed lines are the same boundary in pyrolite under the dry and wet conditions [*Litasov et al.*, 2005]; thin green solid and dashed lines are the dry and wet boundaries for the olivine-wadsleyite transition [*Litasov and Ohtani*, 2003]; black solid and dashed lines are the dry and wet boundaries for the garnet-perovskite transition [*Sano et al.*, 2006]. Ol, olivine; Gt, garnet; Cpx, clinopyroxene; Wd, wadsleyite; Rw, ringwoodiite; Cpv, Ca-perovskite; Sb, superhydrous phase B; Pe, periclase; Mpv, bridgmanite.

Table 21.2 Definition of the alphabet phases.

Mineral names	Formula	Density (g/cm³)	Mg/Si	H_2O (wt %)	References[a]
Phase A	$Mg_7Si_2O_8(OH)_6$	2.96	3.5	12	1
Phase B	$Mg_{12}Si_4O_{19}(OH)_2$	3.38	3	2.4	1
Superhydrous phase B= phase C	$Mg_{10}Si_3O_{12}(OH)_4$	3.327	3.3	5.8	1,2
Phase D= phase F= phase G	$Mg_{1.14}Si_{1.73}H_{2.81}O_6$	3.5	0.66	14.5–18	3
Phase E	$Mg_{2.3}Si_{1.25}H_{2.4}O_6$	2.88	1.84	11.4	4
Phase H	$MgSiO_2(OH)_2$	3.466	1	15	5,6

[a] References: 1. *Ringwood and Major* [1967]; 2. *Gasparik et al.* [1993]; 3. *Liu* [1987], *Yang et al.* [1997], *Ohtani et al.* [1997]; 4. *Kanzaki* [1991]; 5. *Nishi et al.* [2014]; 6. *Tsuchiya* [2013].

[1991] reported new hydrous phases E and F at around 13–17 GPa. *Ohtani et al.* [1997] and *Kudoh et al.* [1997] reported a new hydrous phase and named it phase G. At the same time, *Yang et al.* [1997] reported a similar hydrous phase and called it hydrous phase D because of its similarity of X-ray diffraction profile with phase D reported by *Liu et al.* [1987]. After detailed crystallographic analyses, we recognized that phase D by *Yang et al.* [1997], phase G by *Ohtani et al.* [1997] and *Kudoh*

et al. [1997], and phase F by *Kanzaki* [1991] are all the same and now are all usually called phase D. The complicated terminology of the alphabet phases are summarized in Table 21.2.

The major hydrous phases in the peridotite layer of the slab change continuously from chlorite to serpentine and to the 10 Å phase [*Fumagalli et al.*, 2001] in the upper mantle. At higher pressures, water can be transported by phase A, the stability field of which overlaps with that of serpentine and/or the 10 Å phase in cold slabs [e.g., *Poli and Schmidt*, 2002]. The stability field of the 10 Å phase is very important for water transport into the deep upper mantle because this phase provides a higher temperature bound for hydrogen transport into the deep mantle. Phase A breaks down to hydroxyl-chondrodite and then to hydroxyl-clinohumite, which are both stable to near-geotherm temperatures at depths of 300–410 km.

Superhydrous phase B (phase C) can also serve as a water reservoir at the base of the transition zone depending on the water content in the slab. This phase is formed by decomposition of hydrous ringwoodite at the base of the transition zone. Superhydrous phase B is stable to the top of the lower mantle at a pressure of approximately 30 GPa [*Ohtani et al.*, 2003]. It decomposes into periclase, Mg-perovskite, and phase D at around 45 GPa [*Shieh et al.*, 1998]. Phase Egg ($AlSiO_3OH$) is also stable in the transition zone and the top of the lower mantle [*Schmidt et al.*, 1998]. This phase was discovered as an inclusion in diamond [*Wirth et al.*, 2007], indicating the existence of the wet transition zone. Phase Egg decomposes to stishovite and phase δ-AlOOH at the top of the lower mantle [*Sano et al.*, 2004].

21.3. EXISTENCE OF NEW HYDROUS PHASE, $MgSiO_2(OH)_2$ (PHASE H) AND ITS RELATION TO δ-ALOOH

21.3.1. Discovery of New Hydrous Phase H

The Mg- and Si-bearing δ-AlOOH phase has been reported by *Suzuki et al.* [2000]. The importance of this phase as a water carrier in the lower mantle has been discussed by *Ohtani* [2005, 2014], *Sano et al.* [2008], and *Terasaki et al.* [2012]. The stability of phase D has been studied by *Shieh et al.* [1998] up to 50 GPa. They suggested a decomposition of phase D to Mg-perovskite, periclase, and unknown phase(s) at around 50 GPa. They suggested that the unknown phase(s) might be a possible high-pressure phase(s) as the decomposition product of phase D. However, detailed crystallographic studies were not possible at that time due to the lack of quality of the X-ray diffraction data. Recently, *Tsuchiya* [2013] predicted by abinitio calculation a new hydrous phase with an $MgSiO_2(OH)_2$ formula which has a similar structure as the hydrous phase δ-AlOOH. More recently, this hydrous phase was synthesized by *Nishi et al.* [2014] using the sintered diamond multianvil press and they named the phase hydrous phase H. They synthesized this phase at pressures from 30 to 50 GPa at 1273 K. They recovered this phase and made a crystallographic analysis [*Bindi et al.*, 2014]. *Nishi et al.* [2014] reported that phase H has an orthorhombic structure with space group $P2_1nm$, which is the same as that of δ-phase AlOOH, whereas *Bindi et al.*

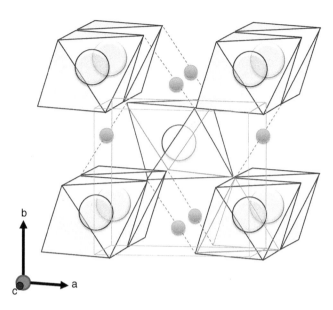

Figure 21.3 Typical structure of δ-phase AlOOH. Large open spheres are aluminum ions; green spheres are hydrogen ions. The structure is similar to $CaCl_2$-type SiO_2.

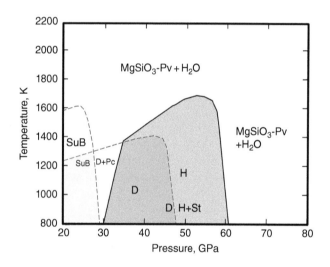

Figure 21.4 Stability field of alphabet phases, including phase H $MgSiO_2(OH)_2$, in the transition zone and lower mantle conditions. SuB, superhydrous phase B; D, phase D; Pc, periclase; H, phase H; $MgSiO_3$-Pv, $MgSiO_3$-bridgmanite.

[2014] reported that it has space group *Pnnm* at an ambient condition. The typical structure of δ-phase AlOOH is shown in Figure 21.3. *Ohtani et al.* [2014] revealed that phase H decomposes at pressure around 60 GPa, and it has a narrow stability field which is consistent with the theoretical calculation by *Tsuchiya* [2013]. The stability field of phase H $MgSiO_2(OH)_2$ in the lower mantle [*Tsuchiya*, 2013; *Nishi et al.*, 2014; *Ohtani et al.*, 2014] is shown in Figure 21.4.

The δ-phase was reported first as a reaction product of pyropic garnet and water at around 20 GPa and 1500 K

[*Suzuki et al.*, 2000], and the δ-AlOOH phase synthesized at that time contained a significant amount of Mg and Si together with Al, and it has a composition $(Mg,Si,Al_2)O_2(OH)_2$. Thus a solid solution between phase H and phase δ-AlOOH, i.e., aluminous phase H, $(Mg,Si,Al_2)O_2(OH)_2$, can exist stably under the transition zone and lower mantle conditions. *Ohira et al.* [2014] reported that this phase coexists with $MgSiO_3$ perouskite (Pv) and post-perouskite (PPv) in the lower mantle.

21.3.2. Stability of Hydrous Phase H, δ-phase AlOOH, and Aluminous Phase H Solid Solution in $MgSiO_2(OH)_2$-AlOOH System

Several hydrous minerals in subducting slabs work as water carriers under the conditions of the upper mantle and transition zone. Some hydrous phases, such as hydrous phase D (=phase F=phase G), superhydrous B (phase C), phase Egg ($AlSiO_3OH$), and hydrous phase H, and nominal anhydrous minerals such as aluminous stishovite in subducting slabs can transport water into the lower mantle. These hydrous minerals, except hydrous aluminous phase H, decompose and dehydrate at pressures around 30–50 GPa and high temperature. Among these high-pressure hydrous minerals, hydrous ringwoodite containing about 1 wt % H_2O and hydrous phase Egg were discovered in nature as inclusions in diamond [*Pearson et al.*, 2014; *Wirth et al.*, 2007]. Existence of these hydrous minerals strongly suggests that the mantle transition zone at least locally is wet, and water can be transported further into the lower mantle by collapse of the stagnant slabs in the transition zone. *Sano et al.* [2008] revealed that δ-AlOOH is stable up to the base of the lower mantle. The stability of hydrous phase δ-AlOOH is shown in Figure 21.5 together with the stability of phase H.

Ohira et al. [2014] reported a reaction of aluminous perovskite and water into alumina-depleted perovskite/post-perovskite and aluminous phase H, Thus, aluminous phase H can coexist with major lower mantle minerals such as Mg-perovskite and post-perovskite, indicating that the hydrous phase is an important water reservoir in the lower mantle. Hydrous aluminous phase H, $(Mg,Si,Al_2)O_2(OH)_2$, can coexist with the alumina-depleted $MgSiO_3$ perovskite along the slab and mantle geotherms, i.e., aluminous phase H, containing 50 mol % of the $MgSiO_2(OH)_2$ component under the lower mantle conditions of approximate 55–87 GPa and 1700–2400 K. Hydrous aluminous phase H coexisting with Mg-post-perovskite at 120 GPa and 2200 K contains 20 mol % of the $MgSiO_2(OH)_2$ component. The stability fields of aluminous phase H solid solutions are shown in Figure 21.5.

Ohira et al. [2014] revealed that aqueous fluid generated by dehydration of the hydrous minerals at the top of the lower mantle can react with aluminous perovskite to

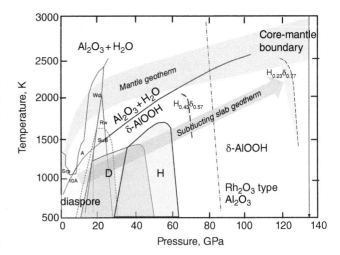

Figure 21.5 Stability of high-pressure hydrous phases under the lower mantle conditions. Srp, serpentine; 10Å, 10 Å phase; A, phase A; Wd, wadsleyite; Rw, ringwoodie; SuB, superhydrous phase B; D, phase D; H, phase H; $H_{0.43}\delta_{0.57}$, phase H-δ solid solution, 0.43 $MgSiO_2(OH)_2$–0.57 AlOOH; $H_{0.23}\delta_{0.77}$, phase H-d solid solution, 0.23 $MgSiO_2(OH)_2$–0.77 AlOOH.

form-alumina depleted perovskite and aluminous phase H. Thus, this phase transports water to the CMB region.

21.3.3. Physical Properties of Hydrous Phase δ and Hydrous Phase H

The effect of hydrogen bond symmetrization on elastic properties has been experimentally studied on ice. *Wolanin et al.* [1997] and *Pruzan et al.* [2003] reported an inflection in compression curve at 66 GPa in H_2O and 84 GPa in D_2O using the Vinet equation of state [*Vinet et al.*, 1986]. This inflection is interpreted as disordering of proton which is a signature of hydrogen bond symmetrization, resulting in increase in bulk modulus from $K=97(4)$ GPa to $K=260(20)$ GPa.

The elastic hardening due to hydrogen bond symmetrization is also observed in other hydrous phases such as hydrous phase D [*Tsuchiya et al.*, 2005, *Hushur et al.*, 2011]. These authors revealed remarkable hardening at around 40 GPa with an increase of the bulk modulus of phase D from $K_0 = 173(2)$ to $K_0 = 212(15)$ GPa [*Hushur et al.*, 2011] due to the hydrogen bonding symmetrization. On the other hand, the study on the equation of state of phase D using a single crystal by *Rosa et al.* [2013] reported $K_T = 151.4(1.2)$ GPa and $K'_T = 4.89(0.08)$ up to 65 GPa at ambient temperature, suggesting the absence of such hardening.

Sano-Furukawa et al. [2009] reported a similar inflection and increase in bulk modulus of δ-AlOOH at around 10 GPa, suggesting hydrogen bond symmetrization at this pressure. The high sound velocity of hydrous phase δ-AlOOH [*Tsuchiya et al.*, 2008] is consistent with its high bulk modulus and is likely to be accounted for by a hydrogen bond symmetrization at high pressure. Phase H, $MgSiO_2(OH)_2$, and aluminous phase H likely have similar high sound velocities because of the similarity of their structures. The elastic hardening of high-pressure hydrous phases due to hydrogen bond symmetrization has important implications for the role of water in the lower mantle. Water stored in aluminous phase H (phase δ–phase H solid solution) in the lower mantle does not decrease the seismic velocity in the lower mantle because of its high bulk modulus and thus its high sound velocity.

Tsuchiya and Tsuchiya [2011] explored the phase transformation of δ-AlOOH by ab inito calculation. They reported that δ-AlOOH transforms to a pyrite type of AlOOH at 170 GPa, which is analogous to the phase transition of InOOH. They also calculated that this pyrite-type phase dissociates into the $CaIrO_3$-type phase of Al_2O_3 and ice X at 300 GPa.

In summary, water can be stored in hydrous and nominally anhydrous minerals in the deep mantle. The mantle transition zone has a high water storage capacity due to high water solubility in its major constituents, wadsleyite and ringwoodite, and stores significant amount of water at least locally. Hydrous phase H, $MgSiO_2(OH)_2$, and its solid solution with isostructural phase δ-AlOOH store water in the lower mantle, and they transport water into the bottom of the lower mantle by slab subduction.

ACKNOWLEDGMENTS

The authors thank A. Suzuki, T. Sakamaki, and M. Murakami for useful discussions on this manuscript. This work was supported by the Grant-in-Aid for Specially Promoted Research from the Japan Society for Promotion of Science (JSPS), Grant Number 22000002, and Ministry of Education and Science of Russian Federation, Project 14.B25.31.0032 (EO). In situ X-ray diffraction study given in this work was conducted under contract at SPring-8 (proposals 2013B0104 and 2014A0104).

REFERENCES

Bindi, L., M. Nishi, J. Tsuchiya, and T. Irifune (2014), Crystal chemistry of dense hydrous magnesium silicates: The structure of phase H, $MgSiH_2O_4$, synthesized at 45 GPa and 1000 °C, *Am. Mineral.*, *99*, 1802–1805.

Bolfan-Casanova, N., H. Keppler, and D. C. Rubie (2000), Water partitioning between nominally anhydrous minerals in the $MgO-SiO_2-H_2O$ system up to 24 GPa: Implications for the distribution of water in the Earth's mantle, *Earth Planet. Sci. Lett.*, *182*, 209–221.

Bolfan-Casanova, N., S. Mackwell, H. Keppler, C. McCammon, and D.C. Rubie (2002), Pressure dependence of H solubility in magnesiowüstite up to 25 GPa: Implications for the storage of water in the Earth's lower mantle, *Geophys. Res. Lett.*, *29*(10), 1449.

Chen, J., T. Inoue, H. Yurimoto, and D. L. Weidner (2002), Effect of water on olivine-wadsleyite phase boundary in the (Mg, Fe)$_2$SiO$_4$ system, *Gophys. Res. Lett.*, *29*, 1875, doi:10.1029/2001GL014429.

Chung, J. I., and H. Kagi (2002), High concentration of water in stishovite in the MORB system, *Geophys. Res. Lett.*, *29*, 16-1-6-4.

Collier, J. D., G. R. Helffrich, and B. J. Wood (2001), Seismic discontinuities and subduction zones, *Phys. Earth Planet. Inter.*, *127*, 35–49.

Fei, Y., J. Van Orman, J. Li, W. van Westrenen, C. Sanloup, W. Minarik, K. Hirose, T. Komabayashi, M. Walter, and K. Funakoshi (2004), Experimentally determined postspinel transformation boundary in Mg_2SiO_4 using MgO as an internal pressure standard and its geophysical implications, *J. Geophys. Res.*, *109*, B02305, doi:10.1029/2003JB002562.

Frost, D. J., and D. Dolejs (2007), Experimental determination of the effect of H_2O on the 410-km seismic discontinuity, *Earth Planet. Sci. Lett.*, *265*, 182–195.

Fukao, Y., S. Widiyantoro, and M. Obayashi (2001), Stagnant slabs in the upper and lower mantle transition region, *Rev Geophys.*, *39*, 291–323.

Fumagalli, P., L. Stixrude, S. Poli, and D. Snyder (2001), The 10Å phase: A high-pressure expandable sheet silicate during subduction of hydrated lithosphere, *Earth Planet. Sci. Lett.*, *186*, 125–141.

Gasparik, T. (1993), The role of volatiles in the transition zone, *J. Geophys. Res.*, *98*, 4287–4300, doi:10.1029/92JB02530.

Ghosh, S., E. Ohtani, K. D. Litasov, A. Suzuki, D. Dobson, and K. Funakoshi (2013a), Effect of water in depleted mantle on post-spinel transition and implication for 660 km seismic discontinuity, *Earth Planet. Sci. Lett.*, *371–372*, 103–111.

Ghosh, S., E. Ohtani, K. D. Litasov, A. Suzuki, D. Dobson, and K. Funakoshi (2013b), Corrigendum to "Effect of water in depleted mantle on post-spinel transition and implication for 660 km seismic discontinuity" [*Earth Planet. Sci. Lett.*, *371–372*, 103–111], *Earth Planet. Sci. Lett.*, *382*, 85–86.

Ghosh, S., and M. W. Schmidt (2014), Melting of phase D in the lower mantle and implications for recycling and storage of H_2O in the deep mantle, *Geochim. Cosmochim. Acta*, *145*, 72–88.

Grand, S. P. (2002), Mantle shear-wave tomography and the fate of subducted slabs, *Philos. Trans. R. Soc. Lond. A*, *360*, 2475–2491.

Hernandez, E. R., D. Alfe, and J. Brodholt (2013), The incorporation of water into lower-mantle perovskites: A first-principles study, *Earth Planet. Sci. Lett.*, *364*, 37–43.

Higo, Y., T. Inoue, T. Irifune, and H. Yurimoto (2001), Effect of water on the spinel-postspinel transformation in Mg_2SiO_4, *Geophys. Res. Lett.*, *28*, 3505–3508.

Hushur, A., M. H. Manghnani, J. R. Smyth, Q. Williams, E. Hellebrand, D. Lonappan, Y. Ye, P. Dera, and D. J. Frost (2011), Hydrogen bond symmetrization and equation of state of phase D, *J. Geophys. Res.*, *116*, B6, doi:10.1029/2010JB008087.

Inoue, T., H. Yurimoto, and Y. Kudoh (1995), Hydrous modified spinel, $Mg_{1.75}SiH_{0.5}O_4$: A new water reservoir in the mantle transition region, *Geophys. Res. Lett.*, *22*(2), 117–120.

Inoue, T., T. Wada, R. Sasaki, and H. Yurimoto (2010a), Water partitioning in the Earth's mantle, *Phys. Earth Planet. Inter.*, *183*, 245–251.

Inoue, T., T. Ueda, Y. Tanimoto, A. Yamada, and T. Irifune (2010b), The effect of water on the high-pressure phase boundaries in the system Mg_2SiO_4–Fe_2SiO_4, *J. Phys. Conf. Ser.*, *215*, Art. No. 012101.

Irifune, T., N. Nishiyama, K. Kuroda, T. Inoue, M. Isshiki, W. Utsumi, K. Funakoshi, S. Urakawa, T. Uchida, T. Katsura, and O. Ohtaka (1998), The postspinel phase boundary in Mg_2SiO_4 determined by *in situ* X-ray diffraction, *Science*, *279*, 1698–1700.

Jiang, G., D. Zhao, and G. Zhang (2008), Seismic evidence for a metastable olivine wedge in the subducting Pacific slab under Japan Sea, *Earth Planet. Sci. Lett.*, *270*, (3–4), 300–307.

Kanzaki, M. (1991), Stability of hydrous magnesium silicates in the mantle transition zone, *Phys. Earth Planet. Inter.*, *66*: 307–312.

Karato, S., and L. Dai (2009), Comments on "Electrical conductivity of wadsleyite as a function of temperature and water content" by Manthilake *et al.*, *Phys. Earth Planet. Inter.*, *174*, 19–21.

Katsura, T., H. Yamada, T. Shinmei, A. Kubo, S. Ono, M. Kanzaki, A. Yoneda, M. J. Walter, S. Urakawa, E. Ito, K. Funakoshi, and W. Utsumi (2003), Post-spinel transition in Mg_2SiO_4 determined by in situ X-ray diffractometry, *Phys. Earth Planet. Inter.*, *136*, 11–24.

Khan, A., and T. J. Shankland (2012), A geophysical perspective on mantle water content and melting: Inverting electromagnetic sounding data using laboratory-based electrical conductivity profiles, *Earth Planet. Scie. Lett.*, *317–318*, 27–43.

Kohlstedt, D. L., H. Keppler, and D. C. Rubie (1996), Solubility of water in the α, β, γ phases of $(Mg,Fe)_2SiO_4$, *Contrib. Mineral. Petrol.*, *123*, 345–357.

Koper, K. D., D. A. Wiens, L. M. Dorman, J. A. Hildebrand, and S. C. Webb (1998), Modeling the Tonga slab: Can travel time data resolve a metastable olivine wedge? *J. Geophys. Res.*, *103*, 30,079–30,100.

Koyama, T., H. Shimizu, H. Utada, M. Ichiki, E. Ohtani, and R. Hae (2006), Water content in the mantle transition zone beneath the North Pacific derived from the electrical conductivity anomaly, *Geophys. Monogr. Ser. 168*, 171–179, doi:10.1029/168GM13m.

Kubo, T., E. Ohtani, T. Kato, T. Shimmei, and K. Fujino (1998), Effect of water on the α–β transformation kinetics in San Carlos olivine, *Science*, *281*, 85–87.

Kudoh, Y., T. Nagase, H. Mizobata, E. Ohtani, M. Sasaki, and M. Tanaka (1997), Structure and crystal chemistry of phase G, a new hydrous magnesium silicate synthesized at 22 GPa and 1050°C, *Geophys. Res. Lett.*, *24*, 1051–1105.

Lakshtanov, D. L., S. V. Sinogeikin, K. D. Litasov, V. B. Prakapenka, H. Hellwi, J. Wang, C. Sanches-Valle, J-P. Perrillat, B. Chen, M. Somayazulu, J. Li, E. Ohtani, and D.J. Bass (2007), The post-stishovite phase transition in hydrous alumina-bearing SiO_2 in the lower mantle of the earth, *Proc. Natl. Acad. Sci*, *104*(34), 13,588–13,590, doi:10.1073/pnas.0706113104.

Lawrence, J. F., and M. E. Wysession (2006), Seismic evidence for subduction-transported water in the lower mantle, in Earth's Deep Water Cycle, *Geophys. Monogr. Ser.*, *168*, 251–261, doi:10.1029/168GM19.

Litasov, K. D. and E. Ohtani (2003), Stability of various hydrous phases in CMAS pyrolite–H_2O system up to 25 GPa, *Phys. Chem. Minerals*, *30*, 147–156.

Litasov, K. D., E. Ohtani, F. Langenhorst, H. Yurimoto, T. Kubo, and T. Kondo (2003), Water solubility in Mg-perovskites and water storage capacity in the lower mantle, *Earth Planet. Sci. Lett.*, *211*, 189–203.

Litasov, K. D., E. Ohtani, A. Sano, A. Suzuki, and K. Funakoshi (2005). In situ X-ray diffraction study of post-spinel transformation in a peridotite mantle implication to the 660-km discontinuity, *Earth Planet. Sci. Lett.*, *238*, 311–328.

Litasov, K. D., H. Kagi, A. Shatzky, E. Ohtani, D. L. Lakshtanov, J. D. Bass, and E. Ito (2007), High hydrogen solubility in Al-rich stishovite and water transport in the lower mantle, *Earth Planet. Sci. Lett.*, *262*, 620–634.

Liu, L. (1987), Effects of H_2O in the phase behavior of the forsterite–enstatite system at high pressures and temperatures and implications for the Earth, *Phys Earth Planet. Inter.*, *49*: 142–167.

Meade, C., R. Jeanloz, J.A. Reffner, and E. Ito (1994), Synchrotron infrared sbsorbance measurements of hydrogen in $MgSiO_3$ perovskite, *Science*, *264*, 1558–1560, doi:10.1126/science.264.5165.1558.

Mei, S., and D. L. Kohlstedt (2000), Influence of water on plastic deformation of olivine aggregates: 1. Diffusion creep regime, *J. Geophys. Res. Solid Earth*, *105* (B9), 21,457–21,469, doi:10.1029/2000JB900179.

Murakami, M., K. Hirose, H. Yurimoto, S. Nakashima, and F. Takafuji (2002), Water in the Earth's lower mantle, *Science*, *295*, 1885–1887.

Nishi, M., T. Irifune, J. Tsuchiya, Y. Tange, Y. Nishihara, K. Fujino, and Y. Higo (2014), Stability of hydrous silicate at high pressures and water transport to the deep lower mantle, *Nature Geosci.*, *7*, 224–227, doi:10.1038/ngeo2074.

Ohira, I., E. Ohtani, T. Sakai, M. Miyahara, N. Hirao, Y. Ohishi, and M. Nishijima (2014), Stability of a hydrous δ-phase, $AlOOH$-$MgSiO_2(OH)_2$, and a mechanism for water transport into the base of lower mantle, *Earth Planet. Sci. Lett.*, *401*, 12–17.

Ohtani, E. (2005), Water in the mantle, *Elements*, *1*, 25–30.

Ohtani, E., H. Mizobata, Y. Kudoh, T. Nagase, H. Arashi, H. Yurimoto, and I. Miyagi (1997), A new hydrous silicate, a water reservoir, in the upper part of the lower mantle, *Geophys. Res. Lett.*, *24*, 1047–1050.

Ohtani, E., M. Toma, T. Kubo, T. Kondo, and T. Kikegawa (2003), In situ X-ray observation of decomposition of super hydrous phase B at high pressure and temperature, *Geophys. Res. Lett.*, *30*, 1029, doi:10.1029/2002GL015549.

Ohtani, E., K. D. Litasov, T. Hosoya, T. Kubo, and T. Kondo (2004), Water transport into the deep mantle and formation of a hydrous transition zone, *Phys. Earth Planet. Inter.*, *143–144*, 255–269.

Ohtani, E., Y. Amaike, S. Kamada, T. Sakamaki, and N. Hirao (2014), Stability of hydrous phase H $MgSiO_4H_2$ under lower mantle conditions, *Geophys. Res. Lett.*, *41*, 8283–8287, doi:10.1002/2014GL061690.

Pacalo, R. E. G., and J. B. Parise (1992), Crystal structure of superhydrous B, a hydrous magnesium silicate synthesized at 1400 °C and 20 GPa, *Am. Mineral.*, *77*, 681–684.

Panero, W. R., L. R. Benedetti, and R. Jeanloz (2003), Transport of water into the lower mantle: Role of stishovite, *J. Geophys. Res.*, *108*, 2039, doi:10.1029/2002JB002053.

Pearson, D. G., F. E. Brenker, F. Nestola, J. McNeill, L. Nasdala, M. T. Hutchison, S. Matveev, K. Mather, G. Silversmit, S. Schmitz, B. Vekemans, and L. Vincze (2014), Hydrous mantle transition zone indicated by ringwoodite included within diamond, *Nature*, *507* (7491), 221, doi:10.1038/nature13080.

Poli, S., and M. W. Schmidt (2002), Petrology of subducted slabs, *Annw. Rev. Earth Planet. Sci.*, *30*, 207–235.

Pruzan, P. J., J. C. Chervin, B. Wolanin, M. Canny, M. Gautier, and M. Hanfland (2003), Phase diagram of ice in the VII-VIII-X domain. Vibrational and structural data for strongly compressed ice III, *J. Raman Spectrosc.*, *34*, 591.

Ringwood, A. E., and A. Major (1967), High–pressure reconnaissance investigations in the system Mg_2SiO_4-MgO-H_2O, *Earth Planet. Sci. Lett.*, *2*, 130–133.

Rosa, A. D., M. Mezouar, G. Garbarino, P. Bouvier, S. Ghosh, A. Rohrbach, and C. Sanchez-Valle (2013), Single-crystal equation of state of phase D to lower mantle pressures and the effect of hydration on the buoyancy of deep subducted slabs, *J. Geophys. Res.*, *118*, 6124–6133.

Rubie, D. C., and C. R. Ross II (1994), Kinetics of olivine-spinel transformation in subducting lithosphere: Experimental constraints and implications on deep slab processes, *Phys. Earth Planet. Inter*, *86*, 223–241.

Sano, A., E. Ohtani, T. Kubo, and K. Funakoshi (2004), In situ X-ray observation of decomposition of hydrous aluminum silicate $AlSiO_3OH$ and aluminum oxide hydroxide δ-$AlOOH$ at high pressure and temperature, *J. Phys. Chem. Solids*, *65*, 1547–1554.

Sano, A., E. Ohtani, K. D. Litasov, T. Kubo, T. Hosoya, K. Funakoshi, and T. Kikegawa, (2006), Effect of water on garnet-perovskite transformation in MORB and implications for penetrating slab into the lower mantle, *Phys. Earth Planet. Inter.*, *159* (1–2), 118–126.

Sano, A., E. Ohtani, T. Kondo, N. Hirao, T. Sakai, N. Sata, Y. Ohishi, and T. Kikegawa (2008), Aluminous hydrous mineral δ-$AlOOH$ as a carrier of hydrogen into the core-mantle boundary, *Geophys. Res. Lett.*, *35*, L03,303.

Sano-Furukawa, A., H. Kagi, T. Nagai, S. Nakano, S. Fukur, D. Ushijima, R. Iizuka, E. Ohtani, and T. Yagi (2009), Change in compressibility of δ-$AlOOH$ and δ-$AlOOD$ at high pressure: A study of isotope effect and hydrogen-bond symmetrization, *Am. Mineral.*, *94*, (8–9), 1255–1261.

Schmidt, M. W., L. W. Finger, R. J. Angel, and R. E. Dinnebier (1998), Synthesis, crystal structure, and phase relations of $AlSiO_3(OH)$, a high-pressure hydrous phase, *Am. Mineral.*, *83*, 881–888.

Shieh, S. R., H. K. Mao, R. J. Hemley, and L. C. Ming (1998), Decomposition of phase D in the lower mantle and the fate of dense hydrous silicates in subducting slabs, *Earth Planet. Sci. Lett.*, *159*, 13–23.

Spektor, K., J. Nylen, E. Stoyanov, A. Navrotsky, R. L. Hervig, K. Leinenweber, G. P. Holland, and U. Häussermann (2011), Ultrahydrous stishovite from high-pressure hydrothermal treatment of SiO_2, *Proc. Natl. Acad. Sci.*, *108* (52), 20,918–20,922, doi:10.1073/pnas.1117152108.

Suetsugu, D., T. Inoue, M. Obayashi, A. Yamada, H. Shiobara, H. Sugioka, A. Ito, T. Kanazawa, H. Kawakatsu, A. Shito, and Y. Fukao (2010), Depths of the 410-km and 660-km discontinuities in and around the stagnant slab beneath the Philippine Sea: Is water stored in the stagnant slab?, *Phys. Earth Planet. Inter.*, *183*, 270–279.

Suzuki, A., E. Ohtani, and T. Kamada (2000), A new hydrous phase δ-$AlOOH$ synthesized at 21GPa and 1000°C, *Phys. Chem. Minerals*, *27*, 689–693.

Terasaki, H., E. Ohtani, T. Sakai, S. Kamada, H. Asanuma, Y Shibazaki, N. Hirao, N. Sata, Y. Ohishi, T. Sakamaki, A. Suzuki, and K. Funakoshi (2012), Stability of Fe–Ni hydride after the reaction between Fe-Ni alloy and hydrous phase (δ-$AlOOH$) up to 1.2 Mbar: Possibility of H contribution to the core density deficit, *Phys. Earth Planet. Inter.*, *194–195*, 18–24.

Tsuchiya, J. (2013), First principles prediction of a new high-pressure phase of dense hydrous magnesium silicates in the lower mantle, *Geophys. Res. Lett.*, *40*, 4570–4573, doi:10.1002/grl.50875.

Tsuchiya, J., and T. Tsuchiya (2011), First-principles prediction of a high-pressure hydrous phase of AlOOH, *Phys. Rev. B*, *83*, 054,115–054,1120, doi:10.1103/PhysRevB.83.054115.

Tsuchiya, J., T. Tsuchiya, and S. Tsuneyuki (2005), First-principles study of hydrogen bond symmetrization of phase D under high pressuren *Am. Mineral.*, *90*, 44–49.

Tsuchiya, J., T. Tsuchiya, and R. M. Wentzcovitch (2008), Vibrational properties of δ-$AlOOH$ under pressure, *Am. Mineral.*, *93*, 477–482.

Utada, H., T, Koyama, M. Obayashi, and Y. Fukao (2009), A joint interpretation of electromagnetic and seismic tomography models suggests the mantle transition zone below Europe is dry, *Earth Planet. Sci. Lett.*, *281*, 249–257.

van Keken, P. E., B. R. Hacker, E. M. Syracuse, and G. A. Abers (2011), Subduction factory: 4. Depth-dependent flux of H_2O from subducting slabs worldwide, *J. Geophys. Res.*, *116*, B01,401, doi:10.1029/2010JB007922.

Vinet, P. J., J. Feerante, J. R. Smith, and J. H. Rose (1986), A universal equation of state of solid, *J. Phys. C Solid State*, *19*, L467.

Wirth, R., C. Vollmer, F. Brenker, S. Matsyuk, and F. Kaminsky (2007), Inclusions of nanocrystalline hydrous aluminium silicate "Phase Egg" in superdeep diamonds from Juina (Mato Grosso State, Brazil), *Earth Planet. Sci. Lett.*, *259*, 384–399.

Wolanin, E., Ph. J. C. Pruzan, J. C. Chervin, D. Gauthier, D. Hausermann, and M (1997). Hanfland, Equation of state of ice VII up to 106 GPa, *Phys. Rev. B*, *56*, 5781.

Yang, H., C. T. Prewitt, and D. J. Frost (1997), Crystal structure of the dense hydrous magnesium silicate, phase D, *Am. Mineral.*, *82*, 651–654.

Ye, Y., J. R. Smyth, A. Hushur, D. Lonappan, M. H. Manghnani, P. Dera, and D. J. Frost (2010), Crystal structure of hydrous wadsleyite with 2.8% H_2O and compressibility to 60 GPa, *Am. Mineral.*, *95*, 1765–1772.

Yoshino, T., G. Manthilake, T. Matsuzaki, and T. Katsura (2008), Dry mantle transition zone inferred from the conductivity of wadsleyite and rngwoodite, *Nature*, *451*, 326–329.

22

Carbon in the Core

Bin Chen[1] and Jie Li[2]

ABSTRACT

Carbon is one of the principal candidates for the light elements in Earth's core. The content and chemical bonding environments of carbon in the core are essential for understanding the nature and dynamics of the core. Cosmochemical and geochemical studies suggested that the depletion of carbon in the bulk silicate Earth (BSE) in comparison with the Sun and carbonaceous chondrites might have resulted from the evaporative loss of volatiles to the space and the core formation. The segregation of a carbon-containing core may also lead to the observed enrichment of ^{13}C in the BSE. Furthermore, petrological experiments indicate that carbon is fairly soluble in iron-nickel liquids under high pressures, allowing the descending liquids to carry carbon to the core. We focus our discussion on mineral physics investigations of density and sound velocities of candidate iron carbides and assess the role of carbon in accounting for the core density deficit and sound velocity discrepancies. Based on exisiting data, the seismological models of the inner core could be adequately explained by the presence of carbides. The carbon-rich core composition model makes the core the largest carbon reservoirs in Earth's interior and could have significant implications for the deep carbon cycle.

22.1. INTRODUCTION

Carbon occupies a rather unique position in the periodic table of elements. As the skeleton for almost all known life on Earth, it plays a central and indispensable role in the existence of life and habitable environment on Earth's surface. Engaging in the extraordinary range of chemical bonding environments, carbon can form numerous materials with distinct physical and chemical properties—from diamond, the hardest mineral found in nature, to graphite, one of the softest materials, and to steel, having

revolutionized human society. The global carbon cycle is generally referred to as the exchange and cycle of carbon in subsurface environments of Earth. These parts of the carbon cycle have been extensively studied from multi-disciplinary efforts in the last decades. In contrast, our knowledge of the carbon cycle in the deep Earth is relatively limited, although the deep interior may contain orders of magnitude more carbon than is present at the surface. The nature and extent of carbon reservoirs in Earth's deep interior are central aspects of the carbon cycle in the deep Earth. Of the many carbon reservoirs in Earth's deep interior, the core may be the largest one [e.g., *Wood*, 1993; *Dasgupta and Walker*, 2008; *Wood et al.*, 2013]. The mineralogy of carbon-bearing compounds, its concentration and chemical bonding environments, in the core is thus of significant importance to understanding the carbon reservoirs and cycle in the deep Earth.

[1]*Hawaii Institute of Geophysics and Planetology, School of Ocean and Earth Science and Technology, University of Hawai'i at Mānoa, Honolulu, Hawaii, USA*

[2]*Department of Earth and Environmental Sciences, University of Michigan, Ann Arbor, Michigan, USA*

Deep Earth: Physics and Chemistry of the Lower Mantle and Core, Geophysical Monograph 217, First Edition.
Edited by Hidenori Terasaki and Rebecca A. Fischer.

Since the discovery of Earth's core in 1906 [*Oldham*, 1906], tremendous efforts have been expended to constrain the core composition from seismological, geochemical, and cosmochemical constraints as well as high-pressure and high-temperature investigations on candidate materials for the core. The core, consisting of a liquid outer core and a solid inner core, is predominantly iron (Fe) diluted with ~5–8% nickel (Ni) [*McDonough and Sun*, 1995] and considerable amount of lighter elements [*Birch*, 1952, 1964; *Ahrens*, 1980; *Poirier*, 1994]. Almost unmistakably, the Fe-Ni in the core is alloyed with nearly all other elements to a certain degree [*Stevenson*, 1981; *Jeanloz*, 1990]. Notwithstanding, it is pivotal to identify the most abundant alloying lighter constituents in the core, as the nature of the light alloying elements is crucial for understanding the origin and dynamics of the core [*Newsom and Sims*, 1991].

Sulfur (S), silicon (Si), carbon (C), oxygen (O), and hydrogen (H) have been suggested as leading candidates for the alloying light elements in the core [*Anderson*, 1977, 1982; *Stevenson*, 1981; *Jeanloz*, 1990; *Li and Fei*, 2014], as required to explain the density deficit and sound velocity discrepancies between Fe and reference Earth models such as the preliminary reference Earth model (PREM) [*Dziewonski and Anderson*, 1981]. The estimated density deficit varies between 6% and 10% relative to liquid Fe for the outer core [e.g., *Anderson and Ahrens*, 1994; *Anderson and Isaak*, 2002] and between 1% and 3% relative to solid Fe for the inner core [*Shearer and Masters*, 1990; *Stixrude et al.*, 1997]. With respect to the sound velocity discrepancies, the measured compressional (P-wave) velocity (v_p) of liquid Fe by shockwave experiments is lower than that of the outer core [*Anderson and Ahrens*, 1994], and both v_p and the S-wave velocity (v_s) of solid Fe are significantly higher than those of the inner core even after correcting for the anticipated velocity depression from high temperatures [*Mao et al.*, 1999; *Badro et al.*, 2007; *Antonangeli et al.*, 2012; *Murphy et al.*, 2013; *Ohtani et al.*, 2013].

To assess the candidacy of certain light alloying element(s) in the core, a simple but effective approach is to compare the density and sound velocities of candidate Fe–lightelement alloys or liquids with seismically determined values of the core. Establishing a self-consistent model and thermochemical state of the core requires accurate and precise phase relation, equation-of-state (EOS), and sound velocities of various Fe-rich alloys under the high-pressure and high-temperature conditions found in the core. Of the many candidate light elements considered for the principal light element in the core, carbon has been suggested as one of the most plausible elements, largely due to its cosmochemical abundance, occurrence of Fe carbide phases in meteorites, and high

affinity and solubility to Fe-Ni liquids during core-mantle differentiation [*Wood*, 1993; *Wood et al.*, 2013].

The remainder of this chapter on carbon in the core is outlined as follows. We first briefly review cosmochemical and geochemical constraints on carbon in the deep Earth from the compositions of meteorite and terrestrial samples, followed by experimental results on the solubility of carbon in Fe-Ni liquids and the processes of transporting carbon to the core during the coreformation differentiation in the early history of Earth. The chapter will focus on experimental investigations of the density and sound velocities of candidate Fe carbides and assess the role of carbon in accounting for the density deficit and sound velocity discrepancies of the inner core. Finally we will discuss the carbon inventory in Earth's core and the implications for the deep carbon cycle in Earth's interior.

22.2. COSMOCHEMICAL AND GEOCHEMICAL CONSTRAINTS ON EARTH'S CARBON BUDGET

The concentration of carbon in the bulk Earth is poorly constrained because the volatile inventories of Earth's building blocks are uncertain and the extent of volatile loss during the planet's history of accretion and evolution is a subject of active research [*Hirschmann*, 2012; *Halliday*, 2013; *Wood et al.*, 2013; e.g. *Carlson et al.*, 2014]. Current models of planet formation postulate that Earth accreted from planetesimals by energetic impacts, some of which probably led to global magma oceans and outgassing of volatiles, as well as blow-off of an atmosphere that likely contained a considerable amount of carbon [e.g., *Chambers*, 2004]. Despite considerable uncertainties, a number of constraints have been derived from investigating the elemental and isotopic compositions of terrestrial and meteoritic samples.

Compared with the Sun and carbonaceous chondrites, Earth is certainly depleted in carbon (Figure 22.1) [*McDonough and Sun*, 1995]. In the Sun, carbon is the fourth most abundant element, after hydrogen, helium, and oxygen [*Anders and Grevesse*, 1989]. The CI carbonaceous chondrites, named after Ivuna meteorites, contain 3.2–3.5 wt% carbon in the form of graphite and organic carbon, which are the most volatilerich among primitive carbonaceous chondrites [*Anders et al.*, 1964]. From carbon-to-argon ratios of basalts and the carbon concentration in the atmosphere, hydrosphere, and crust, the carbon concentration of bulk silicate Earth (BSE) has been estimated at 120 ppm with a factor of uncertainty [*Zhang and Zindler*, 1993; *McDonough and Sun*, 1995]. The C/Si atomic ratio of the BSE is four orders of magnitude smaller than that of the Sun, at about 7, which is 10 times that of the CI chondrites, at 0.7 [*McDonough and Sun*, 1995; *Lodders*, 2003]. This depletion of carbon with

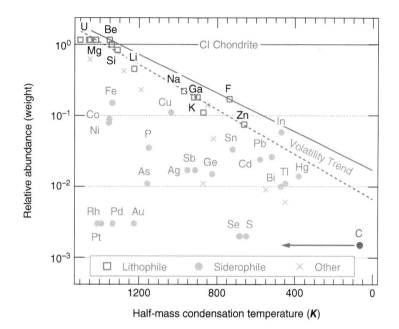

Figure 22.1 Carbon content in the BSE [adapted from *McDonough and Sun*, 1995]. On the plot of relative abundance of elements (normalized to CI chondrites and Mg) v. their half-mass condensation temperatures the volatile lithophile elements (open squares) form a broad linear trend known as the "volatility trend (VT)" with an upper bound defined by F and Be (solid line) and a lower bound defined by eight elements Mg, Si, Li, Na, Ga, K, F, Zn (dashed line). Siderophile elements (solid dots) fall below the VT, reflecting their incorporation into the core. The estimated half-mass condensation temperature of carbon ranges between 40 K [*Lodders*, 2003] and 500 K [*Wasson*, 1985]. To estimate the bulk Earth carbon abundance, we used the upper bound of the volatility trend.

respect to silicon probably resulted from a combination of evaporative volatile loss and core formation. In the silicate Earth, moderately volatile and lithophile elements such as sodium, potassium, and fluorine are depleted to different degrees depending on their volatility, forming a nearly linear correlation on a plot of logarithmic relative abundance versus half-mass condensation temperature, known as the volatility trend (Figure 22.1) [*McDonough and Sun*, 1995]. If we assume that highly volatile elements follow the same trend, then BSE carbon content can be estimated at 2% of the CI chondrite value, or ~600 ppm, using a half-mass condensation temperature of 40 K [*Lodders*, 2003], or 8% of the CI chondrite value, which is 0.2 wt% using 500 K [*Wasson*, 1985]. By mass balance calculation, these estimates correspond to 0.2–0.6 wt% carbon in the core, assuming 120 ppm in the BSE [*McDonough and Sun*, 1995].

The $^{107}Ag/^{109}Ag$ ratio of the silicate Earth has provided an important constraint on the planet's history of volatile accretion [*Carlson et al.*, 2010]. The stable isotope ^{107}Ag is a decay product of short-lived radionuclide ^{107}Pd with a half-life of 6.5 Ma. If all the Ag were delivered prior to the main phase, the core segregation by about 30 Ma when the Earth was 90% accreted, then a subchondritic

$^{107}Ag/^{109}Ag$ would be expected because of strong depletion of siderophile element Pd in the silicate Earth. The $^{107}Ag/^{109}Ag$ ratio of the silicate Earth was found to resemble that of CI chondrite, suggesting that a significant portion of the Earth's volatiles, including carbon, were delivered during the late stage of the accretion in the form of volatile rich materials. Using the C/S mass ratio of carbonaceous chondrite (0.65) and the sulfur content of 0.6 wt% in the bulk Earth [*McDonough and Sun*, 1995], the carbon concentration of the core is estimated at 1.2 wt%. This is considered an upper limit because the C/S ratio of carbonaceous chondrite is the highest among the potential building blocks [*Jarosewich*, 1990].

The carbon isotopic compositions of mantle xenoliths show a bimodal distribution in $\delta^{13}C$ values with one peak at −5‰, which was thought to represent the signature of the primitive mantle [*Deines*, 2002]. In contrast, the $\delta^{13}C$ values of chondrites, Vesta, and Mars are considerably lighter at −20 ± 5‰. The more negative value was interpreted as representing the signature of the bulk Earth, and the enrichment of ^{13}C in the BSE can be explained by the segregation of an Fe-rich core containing carbon [e.g., *Satish-Kumar et al.*, 2011]. To generate the observed fractionation in ^{13}C, the carbon content of the core was inferred to be 0.5–2 wt% carbon for an estimate mantle abundance of 120 ppm and as much as 15 wt% if the mantle abundance is 756 ± 300 ppm [*Marty*, 2012; *Dasgupta*, 2013].

22.3. CARBON CONTENT OF CORE FROM SOLUBILITY AND PARTITIONING EXPERIMENTS

During the accretion and early differentiation of Earth, what is undoubtedly expected to occur is the redistribution of carbon among different reservoirs—space, the Earth's surface, and its deep interiors. The formation of the metallic core is considered one of the most significant events in Earth's early history, giving rise to the mass redistribution of elements in Earth's interiors. On the premise that the proto-Earth formed from chondritic building blocks, a considerable amount of carbon could have entered the metallic core during the processes that liquid metallic Fe alloys segregated from the silicate part of the planet to form the Fe-rich core. In addition to the cosmochemical and geochemical constraints as discussed in the previous section, the solubility of carbon in metallic liquids and partitioning of carbon between metallic liquids and silicates during core formation need to be studied to understand carbon in the core.

The question of whether carbon can readily dissolve in metallic liquids has received much attention in the past decades. Some recent experimental studies focused on the equilibrium partitioning between the silicate and metallic melts [*Hirayama et al.*, 1993; *Wood*, 1993; *Dasgupta and Walker*, 2008]. At ambient pressure, carbon strongly partitions into Fe liquids compared to silicate, and the carbon solubility in Fe liquids at 1673 K was found to be ~5 wt% and decrease with increasing Ni content, reaching a minimum of ~2 wt% carbon in Fe-Ni liquids with 80 wt% Ni [*Tsymbulov and Tsemekhman*, 2001]. At 2 GPa, carbon was found to be readily soluble in Fe liquids, with the solubility reaching 5.8 wt% at 2273 K, 6.7 wt% at 2473 K, and 7.4 wt% at 2683 K, whereas adding 5.2 wt% Ni to Fe liquids only slightly decreases the solubility of carbon by ~0.2–0.3 wt% [*Dasgupta and Walker*, 2008]. More recently, the carbon solubility at 2273 K in molten Fe was found to reach 8.5 wt% at 5 GPa and 7 wt% at 10–14 GPa [*Nakajima et al.*, 2009]. To date, there is however no carbon solubility measurement at magma ocean conditions (~50 GPa) [*Li and Agee*, 1996], which is required to further understand the carbon distribution between mantle and core. Nevertheless, for shallow magma ocean scenario during Earth's early history and based on the low-pressure carbon partitioning and solubility data, carbon could readily enter the core during core formation [*Dasgupta and Walker*, 2008]. If the high solubility of carbon remains under the deep magma ocean conditions, the descending Fe liquids during the formation of the core might be able to transport considerable amounts of carbon to the core.

22.4. MINERAL PHYSICS CONSTRAINTS ON CARBON IN CORE

Mineral physicists have reported scores of static and dynamic studies for determining high-pressure density and sound velocities of numerous solid Fe–light-element alloys as candidates for the inner core. In contrast, for most of the core—the liquid outer core—such investigations and discussions are still significantly lacking, largely due to the huge gap in the *P-T* ranges between experimental and actual core conditions and the very limited experimental data the thermoelastic properties of Fe-rich liquids. Only a handful of studies have involved the structure, density, or sound velocities of Fe, Fe-Si, Fe-S, and Fe-C liquids [*Anderson and Ahrens*, 1994; *Sanloup et al.*, 2000, 2011; *Chen et al.*, 2005; *Yu and Secco*, 2008; *Terasaki et al.*, 2010; *Huang et al.*, 2011; *Tateyama et al.*, 2011; *Morard et al.*, 2013; *Jing et al.*, 2014]. The candidacy of certain light elements in the core is often assessed by its competence in accounting for the core density deficit and explaining the sound velocity discrepancies. For practical reasons, such comparisons are often made for the solid inner core, rarely for the outer core, except for a few shock wave experiments [e.g., *Huang et al.*, 2011].

The thermochemistry and high-pressure phase relations in the Fe-C system under high pressures are essential for understanding the role of carbon in the core. Despite limited data of the phase relations in the Fe-C system at high pressures, existing experimental investigations consistently indicated the formation of a new Fe carbide compound, Fe_7C_3 from the incongruent melting of Fe_3C at pressures >6 GPa [*Tsuzuki et al.*, 1984; *Lord et al.*, 2009; *Nakajima et al.*, 2009; *Bazhanova et al.*, 2012; *Fei and Brosh*, 2014]. The liquidus field of Fe_7C_3 at high temperature was found to exist at least up to 25 GPa [*Nakajima et al.*, 2009; *Fei and Brosh*, 2014]. Based on knowledge of the thermochemistry of the Fe-C binary system at moderate pressures, *Fei and Brosh* [2014] established a comprehensive thermodynamic model for the Fe-C system and provided predications of the phase relations at 136 and 330 GPa. The predicted phase diagram showed that Fe_7C_3 might be the liquidus phase in the Fe-C system under inner core conditions and implied that Fe_7C_3 would replace Fe_3C as the crystalline phase solidifying from the carbon-containing liquid outer core. A recent theoretical study at 0 K suggested that Fe_2C might be the most stable Fe carbide phase at the inner core pressures [*Bazhanova et al.*, 2012], the stability of which, however, awaits further experimental verification.

22.4.1. Density Deficit Considerations

The phase stability, thermal EOS, and sound velocities of relevant Fe carbides and Fe-C liquids at high pressures have recently been investigated to test the carbon-rich

Table 22.1 Existing studies on the Fe-C system.

Source	Composition	P-T range	Method	Results
Density				
Jimbo et al. [1993]	Fe–(0–4 wt%) C (liquid)	0 GPa, 1523–1823 K	Sessile drop profile	ρ (g/cm^3) = [7.10–0.0732 C (wt%)]–[8.28–0.874 C (wt%)]× 10^{-4}[T (K)–1823]
Terasaki et al. [2010]	Fe$_3$C (liquid)	⩽10 GPa, 1973 K	X-ray absorption	K_0 = 50±7 GPa at 1973 K
Sanloup et al. [2011]	Fe–5.7 wt% C (liquid)	⩽7.8 GPa, 2273 K	X-ray absorption	P<5.4 GPa: similar to liquid Fe$_3$C [*Terasaki et al.*, 2010]; P> 5.4 GPa: higher compressibility possibly due to structural change in liquid
Scott et al. [2001]	Fe$_3$C (solid)	⩽73 GPa, 300 K	PXD	K_0 = 175 ± 4 GPa, K' = 5.2 ± 0.3
Li et al. [2002]	Fe$_3$C (solid)	⩽30 GPa, 300 K	PXD	K_0 = 174 ± 6 GPa, K' = 4.8 ± 0.8
Sata et al. [2010]	Fe$_3$C (solid)	⩽187 GPa, 300 K	PXD	K_0 = 290 ± 13 GPa, K' = 3.76 ± 0.18
Ono and Mibe [2010]	Fe$_3$C (solid)	⩽67 GPa, 300 K	PXD	Ferromagnetic phase (P⩽35 GPa): K_0 = 167 GPa, K' = 6.7
Litasov et al. [2013]	Fe$_3$C (solid)	⩽31 GPa, 1473 K	PXD	Paramagnetic phase: K_0 = 192 ± 3 GPa, K' = 4.5 ± 0.1, γ_0 = 2.09 ± 0.04, θ_0 = 490 ± 120 K, q = –0.1 ± 0.3
Prescher et al. [2012]	Fe$_3$C (solid)	⩽50 GPa, 300 K	SXD	K_0 = 161 ± 2 GPa, K' = 5.9 ± 0.2
Nakajima et al. [2011]	Fe$_7$C$_3$ (solid)	⩽72 GPa, 300–1973 K	PXD	Paramagnetic phase: K_0 = 253 ± 7 GPa, K' = 3.6 ± 0.2, γ_0 = 2.57 ± 0.05, θ_0 = 920 ± 140 K, q = 2.2 ± 0.5
Chen et al. [2012]	Fe$_7$C$_3$ (solid)	⩽170 GPa, 300 K	SXD	Nonmagnetic phase: K_0 = 307 ± 6 GPa, K' = 3.2 ± 0.1
Sound velocity				
Gao et al. [2008]	Fe$_3$C (solid)	⩽50 GPa, 300 K	NRIXS	v_p (m/s) = –3990 + 1290 ρ (g/cm^3); v_s (m/s) = 1450 + 240 ρ (g/cm^3)
Gao et al. [2011]	Fe$_3$C (solid)	⩽50 GPa, 300–1450 K	NRIXS	High-T v_s does not follow Birch's law
Fiquet et al. [2009]	Fe$_3$C (solid)	⩽68 GPa, 300 K	IXS	v_p (m/s) = –8671 + 1900ρ (g/cm^3)
Chen et al. [2014]	Fe$_7$C$_3$ (solid)	⩽154 GPa, 300 K	NRIXS	Nonmagnetic phase (>70 GPa): v_p (m/s) = 2160 + 660ρ (g/cm^3), v_s (m/s) = 843 + 242ρ (g/cm^3)

Note: NIRXS, nuclear resonant inelastic X-ray scattering; PXD, powder X-ray diffraction; SXD, single-crystal X-ray diffraction; IXS, inelastic X-ray scattering.

core composition model (see Table 22.1). In the past decade, many high-pressure experiments on the compressibility of Fe carbides, Fe$_3$C and Fe$_7$C$_3$, have been conducted to test the hypothesis of Fe$_3$C as an inner core component, focusing on whether it and its mixture with Fe could account for the density deficit of the inner core [e.g.,

Scott et al., 2001; *Li et al.*, 2002; *Ono and Mibe*, 2010; *Sata et al.*, 2010; *Nakajima et al.*, 2011; *Chen et al.*, 2012].

Under experimentally investigated P-T conditions up to core pressures, no structural change has been found in Fe$_3$C. *Li et al.* [2002] and *Scott et al.* [2001] determined the EOS of Fe$_3$C at 300 K up to 30 and 73 GPa, respectively, and found no phase transition up to the highest pressures studied. Similarly, no structural change in Fe_3C was observed from more recent X-ray diffraction measurements up to 187 GPa [*Sata et al.*, 2010]. Despite that no structural change was observed in Fe_3C, magnetic transitions in Fe_3C may have significant effect on its thermoelastic properties [*Vočadlo et al.*, 2002]. Previous high-pressure studies on magnetic transitions in Fe_3C from the ferromagnetic (FM) to paramagnetic (PM) or nonmagnetic (NM) state often gave contradictory results in the transition pressures. First-principles calculations predicted a pressure-induced magnetic transition at ~60 GPa and 0 K and found the NM phase less compressible than the low-pressure magnetic counterpart [*Vočadlo et al.*, 2002]. Within experimental uncertainties, earlier X-ray diffraction measurements at 300 K did not show any discontinuity in the compression curve of Fe_3C up to 73 GPa [*Scott et al.*, 2001; *Li et al.*, 2002]. Nevertheless, recent experimental investigations for Fe_3C at 300 K confirmed the occurrence of magnetic transitions but showed a wide range of transition pressures of, i.e., ~5 GPa by synchrotron Mössbauer spectroscopy (SMS) [*Gao et al.*, 2008], ~10 GPa by X-ray magnetic circular dichroism (XMCD) [*Duman et al.*, 2005], ~25 GPa by X-ray emission spectroscopy [*Lin et al.*, 2004], ~55 GPa as indicated by a significant reduction in *b* axis from X-ray diffraction measurements [*Ono and Mibe*, 2010], 68–83 GPa from the observed anomalous behaviors of the inelastic X-ray scattering dispersion curve [*Fiquet et al.*, 2009], ~8–10 GPa for the FM-PM transition and 22 GPa for the spin transition from a Mössbauer spectroscopic study [*Prescher et al.*, 2012], and 7–9 GPa for the FM-NM transition from a thermal EOS study to 31 GPa and 1473 K [*Litasov et al.*, 2013] (Figure 22.2). The magnetic transition at <10 GPa is probably an FM-PM transition, as the FM-PM transition pressures determined from different studies are quite consistent, ranging from ~5 to 10 GPa [*Duman et al.*, 2005; *Gao et al.*, 2008; *Prescher et al.*, 2012; *Litasov et al.*, 2013]. However, experimental studies on the PM-NM transition or spin transition in Fe_3C showed inconsistent results: The PM-NM transition pressures from different studies spanned a large range of 22 to >68 GPa. Therefore, the nature and occurrence of the PM-NM transition in Fe_3C still await further experimental investigations.

Due to the nonnegligible effect from magnetic transitions on the thermoelastic properties [*Vočadlo et al.*, 2002], the extrapolation of density of low-pressure data to inner core pressures may be misleading. Indeed, *Ono et al.* [2010] found significant volume change in Fe_3C at 55 GPa and attributed the change to an FM-NM transition (Figure 22.2). According to the magnetic diagram, Fe_3C should be well within the NM (low-spin) stability field under inner core pressures of 329–364 GPa. Therefore, the density of Fe_3C should be determined

from the EOS of the NM phase. *Litasov et al.* [2013] determined the thermal EOS of Fe_3C up to 31 GPa and 1473 K, extrapolated the density of NM-Fe_3C to inner core pressures using their thermal EOS parameters for the PM phase combined with the compressibility data of the NM phase at 50–187 GPa from *Sata et al.* [2010], and then suggested that 2–3 wt% carbon in the inner core can explain the density of the inner core (Figure 22.2).

An Fe_7C_3 inner core composition model has also been tested by comparing the density of Fe_7C_3 with that of the inner core [*Nakajima et al.*, 2011; *Chen et al.*, 2012]. No structural change was observed in an experimental investigation of Fe_7C_3 up to 71.5 GPa and 1973 K [*Nakajima et al.*, 2011]. Likewise, recent single-crystal X-ray diffraction measurements up to 167 GPa at 300 K and synchrotron Mössbauer studies suggested no structural change in Fe_7C_3 [*Chen et al.*, 2012]. Similar to Fe_3C, the magnetic transitions in Fe_7C_3 may have significant effects on its thermoelastic properties [*Mookherjee et al.*, 2011]. Ab initio calculations suggested that Fe_7C_3 undergoes a high spin to low spin transition near 67 GPa and 0 K with associated elastic softening [*Mookherjee et al.*, 2011]. Experimental investigations of the magnetic transitions showed inconsistent results. A pressure-induced FM-PM transition at 18 GPa in Fe_7C_3 was inferred from anomalies in isothermal compression and thermal expansion behavior [*Nakajima et al.*, 2011]. On the contrary, *Chen et al.* [2012] revealed two magnetic transitions in Fe_7C_3 at 300 K up to 167 GPa, first of which corresponds to an FM-PM transition and manifests as the disappearance of fast oscillations in the synchrotron Mössbauer spectra of Fe_7C_3 at 5.5–7.5 GPa. Near 53 GPa, Fe_7C_3 softens after the second discontinuity, presumably caused by a PM-NM (high-spin to low-spin) transition [*Chen et al.*, 2012], later confirmed by X-ray emission spectroscopy (XES) measurements [*Chen et al.*, 2014].

On the basis of the EOS of the NM (low-spin) Fe_7C_3 [*Chen et al.*, 2012] and existing thermoelastic parameters [*Mao et al.*, 1990; *Seagle et al.*, 2006; *Nakajima et al.*, 2011], the volume fraction of Fe_7C_3 with Fe in the inner core is estimated at 62–95% in order to match the inner core density, corresponding to a carbon content of 5.1–8.0 wt%. Likewise, *Nakajima et al.* [2011] suggested that and Fe_7C_3-dominant inner core composition model provides a good explanation for the density of the inner core. An Fe_7C_3 inner core model would make the core by far the largest reservoir of carbon in Earth, accounting for more than 90% of the planet's total carbon budget [*Dasgupta and Hirschmann*, 2010].

22.4.2. Sound Velocity Discrepancies Considerations

In addition to the phase stability and density of light-element-bearing Fe-rich alloys, their sound velocities could further provide critical constraints on the core

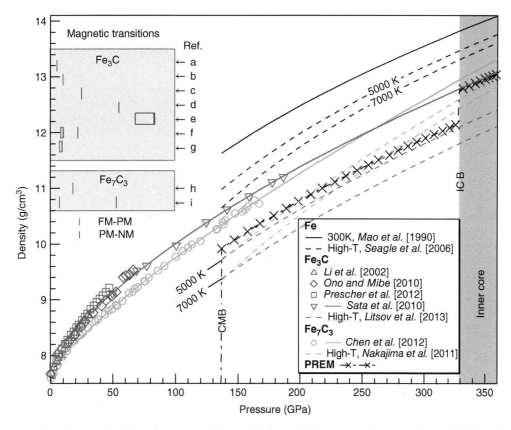

Figure 22.2 Density as a function of pressure of iron carbides at high pressures in comparison with Fe and PREM. Density of hcpFe: at 300 K, black curve [*Mao et al.*, 1990]; at 5000 and 7000 K with Mie-Grüneisen-Debye EOS parameters from *Seagle et al.* [2006], black dashed curve. Density of Fe_3C at 300 K: red upper triangles [*Li et al.*, 2002]; red down triangles [*Sata et al.*, 2010]; blue diamonds, showing an abrupt density jump at 55 GPa likely due to the highspin–lowspin transition in Fe_3C [*Ono and Mibe*, 2010]; red curve, EOS fit to the experimental data at 300 K above 50 GPa [*Sata et al.*, 2010]; red dashed curves, calculated high-temperature density from thermal parameters of the PM phase by *Litasov et al.* [2013]. Density of Fe_7C_3 at 300 K: green circles, which shows an anomalous compressibility behavior change near 53 GPa [*Chen et al.*, 2012], possibly due to the spin transition at 50 GPa as revealed by XES measurements [*Chen et al.*, 2014]; green curve, EOS fit to the experimental data for NM-Fe_7C_3 phase above 53 GPa; green dashed curves, calculated high-temperature density from thermal parameters of the PM phase by *Nakajima et al.* [2011]. PREM [*Dziewonski and Anderson*, 1981], black crosses with dash-dotted curve. The top-left inset depicts experimentally determined or inferred magnetic transition pressures of Fe_3C and Fe_7C_3 at 300 K. Fe_3C: (a) 5 GPa [*Gao et al.*, 2008]; (b) 10 GPa [*Duman et al.*, 2005]; (c) 25 GPa [*Lin et al.*, 2004]; (d) ~55 GPa [*Ono and Mibe*, 2010]; (e) 68–83 GPa [*Fiquet et al.*, 2009]; (f) 8–10 GPa for FM-PM transition (depicted as red bars) and 22 GPa for spin transition [*Prescher et al.*, 2012]; (g) 7–9 GPa for FM-PM transition [*Litasov et al.*, 2013]. Fe_7C_3: (h) 18 GPa [*Nakajima et al.*, 2011]; (i) 7–7.5 GPa for FM-PM transition and 53 GPa for PM-NM transition (depicted as blue bars) [*Chen et al.*, 2012, 2014].

composition. In particular, the v_S of the inner core (3.5–3.7 km/s) is lower than that of Fe and Fe-Ni alloys at corresponding pressures at 300 K [*Mao et al.*, 1999, 2001; *Lin et al.*, 2003, 2005; *Antonangeli et al.*, 2004] by 1–3 km/s (Figure 22.3). The mismatch may reflect the effect of temperature and/or the presence of light elements such as carbon. Numerous efforts have been made to determine the sound velocities of Fe carbides, as a candidate component for the inner core, under high pressures (Table 22.1 and Figure 22.3) [*Gao et al.*, 2008, 2009, 2011; *Fiquet et al.*, 2009; *Chen et al.*, 2014]. To date, there is no

experimental data on the sound velocities of Fe-C liquids at high pressures (Table 22.1). Therefore, in this chapter, we only focus our discussion on the sound velocities of Fe carbides with implications onto the carbon in the inner core.

Previous studies on the sound velocities of Fe_3C by nuclear resonant inelastic X-ray scattering (NRIXS) measurements up to 50 GPa at 300 K showed that the addition of carbon to Fe reduces v_S by ~1 km/s at relevant pressures, thus bringing the v_S closer to seismic values [*Gao et al.*, 2008] (Figure 22.3). A similar conclusion was

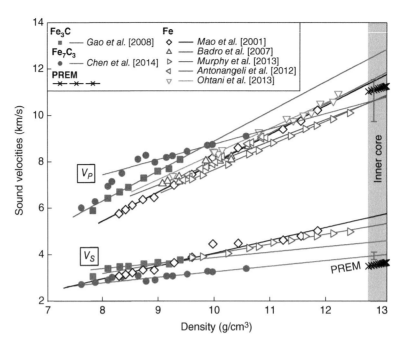

Figure 22.3 Sound velocities of iron and iron carbides as a function of density. v_p or v_s of Fe at 300 K: black diamonds [*Mao et al.*, 2001], blue up-pointing triangles [*Badro et al.*, 2007], purple right-pointing triangles [*Murphy et al.*, 2013], magenta left-pointing triangles [*Antonangeli et al.*, 2012], brown down-pointing triangles [*Ohtani et al.*, 2013]; v_p and v_s of Fe_3C at 300 K: red solid squares [*Gao et al.*, 2008]; v_p and v_s of Fe_7C_3 at 300 K: green solid circles [*Chen et al.*, 2014]. Green error bars represent 90% confidence intervals for the linear fitting to NM-Fe_7C_3 data. PREM [*Dziewonski and Anderson*, 1981], black crosses.

reached by *Fiquet et al.* [2009], who measured the v_p of Fe_3C up to 68 GPa by inelastic X-ray scattering (IXS) and found that Fe_3C has relatively low v_p among other Fe alloys. The effects of temperature on the sound velocities of Fe_3C were also investigated: From the NRIXS experiments of Fe_3C up to 47 GPa and 1450 K, v_s decreases with increasing temperature in a nonlinear fashion and its relationship with density deviates from Birch's law behavior towards smaller values [*Gao et al.*, 2011], supporting Fe_3C as a major component of the inner core.

Magnetic transitions in Fe_3C may have significant effects on the sound velocities [*Lin et al.*, 2004; *Gao et al.*, 2008]. It is found that v_p of Fe_3C significantly softens at pressures >68 GPa [*Fiquet et al.*, 2009], possibly corresponding to a predicted magnetic transition from the PM to the NM state at ~60 GPa and 0 K by theoretical calculations [*Vočadlo et al.*, 2002]. The FM-PM transition as revealed by the SMS experiments may have an apparent effect on both v_p and v_s of Fe_3C: The sound velocities of the FM phase fall below the linear trend of the NM phase [*Gao et al.*, 2008]. Nevertheless, due to the limited pressure range investigated, there is no clear evidence of the effect of the PM-NM transition on Fe_3C's v_p and v_s.

Recent NRIXS measurements on the sound velocities of Fe_7C_3 revealed significant shear softening in Fe_7C_3 at 40–50 GPa, coincident with the spin transition at 50 GPa,

as revealed by XES measurements [*Chen et al.*, 2014]. Accompanying the spin transition are pronounced reductions in both the absolute value of the bulk sound velocity v_D and the rate at which it increases with density. Because v_D weighs heavily toward v_s ($3/v_D^3 = 1/v_P^3 + 2/v_S^3$, and hence $v_D \sim 0.84\ v_S$ if $v_S \sim 1/2\ v_P$), similar reductions in v_S (see Figure 22.3) and shear modulus G occurred in the same pressure range, indicating shear softening [*Chen et al.*, 2014]. Extrapolating the exceptionally small pressure dependence of the low-spin Fe_7C_3 to the inner core pressure of 330–360 GPa, a v_S at 300 K that is only ~14% above the observed values at the inner core boundary pressures was determined [*Chen et al.*, 2014]. In comparison, pure Fe exceeds the observation by more than 65% [*Mao et al.*, 2001; *Antonangeli et al.*, 2012; *Murphy et al.*, 2013; *Ohtani et al.*, 2013; *Badro et al.*, 2014]. The FM-PM transition at 7.0–7.5 GPa [*Chen et al.*, 2012; 2014], however, does not demonstrate a significant effect on sound velocities. As suggested by the experimental data, the presence of Fe_7C_3 in the inner core could lower its v_S sufficiently low to a level comparable to the PREM values [*Chen et al.*, 2014]. The actual volume fraction of Fe_7C_3 is, however, dependent on the effect of high temperature on sound velocities, which ranges from ~20% to 100%, corresponding to ~2–8 wt% carbon in the inner core. The presence of Fe_7C_3 in the inner core, likely in a large volume fraction explaining the anomalously low v_S of the inner core, could make the core potentially the largest carbon reservoir in Earth's deep interior.

22.5. CONCLUDING REMARKS AND IMPLICATIONS

A critical and practical test for any compositional model for the core is that it must reproduce the density and sound velocities constrained by seismic and other geophysical observations such as the PREM model [*Dziewonski and Anderson*, 1981]. This chapter has reviewed carbon as a light element in the core from the perspectives of cosmochemical, geochemical, petrological, and mineral physics considerations. As an abundant element in the chondritic building blocks of the proto-Earth, carbon is also readily soluble in Fe-Ni liquids under the magma ocean conditions, indicating a considerable amount of carbon could have descended together with Fe-rich liquids to the core. This process might also enrich the core with ^{12}C on the basis of the existing experimental investigations of carbon isotope fractionation between Fe carbide liquid and graphite/diamond at high pressures, as manifested by the observed relatively high δ ^{13}C values of the BSE in comparison with those of Mars, Vesta, and carbonaceous chondrites. Mineral physics studies on the density and sound velocities of Fe carbides provide critical constraints on the concentration of carbon in the inner core, by matching the density and sound velocities of a Fe-carbide-containing inner core composition model with seismological models. Both Fe carbide phases, Fe_3C and Fe_7C_3, are FM at 1 bar and undergo pressure-induced magnetic transitions. The FM-PM transitions of both Fe carbides most likely occur at <10 GPa, with the exception of the study by *Nakajima et al.* [2011] indicating an FM-PM transition pressure of 18 GPa for Fe_7C_3. For Fe_3C, a wide range of PM-NM (or highspin-lowspin) transition pressures have been proposed, from 22 to >68 GPa (Figure 22.2). For Fe_7C_3, the PM-NM transition has been proposed to occur at ~53 GPa (Figure 22.2), comparable to the predicted transition pressure of ~67 GPa [*Mookherjee et al.*, 2011]. The magnetic transitions in the Fe-carbide phases, particularly the PM-NM transition, have significant effects on their densities and sound velocities. Anomalous behaviors in the compressibility and sound velocities at high pressures have been observed for both Fe_3C and Fe_7C_3 due to the PM-NM transitions at ~50 GPa (Figure 22.2). An inner core composition model containing Fe carbide, Fe_3C or Fe_7C_3, has been proposed to account for the density deficit and sound velocity discrepancies of the inner core, thus making it unnecessary to invoke partial melting in the inner core to explain the anomalously low shear v_S of the inner core, as suggested previously [*Singh*, 2000]. A totally solid inner core may have unique seismic anisotropy and attenuation and viscosity, which affects the evolution and dynamics of the inner core [*Jeanloz and Wenk*, 1988; *Deguen et al.*, 2013].

Other light elements, such as sulfur, may also exist in the core. For example, for the Fe–C–S outer core composition model, Fe carbides would be the liquidus phase at core pressures and would form the solid inner core, even for carbon concentration as low as 0.3 wt% in the outer core [*Wood*, 1993]. During the solidification of solid Fe and/or Fe carbides from the carbon-containing liquid outer core, carbon would become more depleted in the outer core, which could enhance the transport of carbon from the mantle to the core. Such transport might occur rapidly through grain boundary diffusion and is facilitated by solid-state convection in the mantle [*Hayden and Watson*, 2007]. This process could, in turn, have profound effects on the outgassing and recycling of CO_2 in Earth's interior [*Zhang and Zindler*, 1993].

ACKNOWLEDGMENTS

The authors thank the editors and two anonymous reviewers for their constructive reviews. Chen acknowledges the support from the National Science Foundation (NSF) grants EAR-1440005 and EAR-1555388. Li acknowledges the support from NSF grants AST-1344133, EAR-1219891, EAR-1023729, CDAC DOE CI JL 2009-05246, and University of Michigan APSF and Crosby grants. This is contribution no. 2063 from the Hawai'i Institute of Geophysics and Planetology and contribution no. 9281 from the School of Ocean and Earth Science and Technology, University of Hawai'i at Mānoa, Honolulu.

REFERENCES

Ahrens, T. J. (1980), Dynamic compression of Earth materials, *Science*, 207, 1035–1041.

Anders, E., and N. Grevesse (1989), Abundances of the elements: Meteoritic and solar, *Geochim. Cosmochim. Acta*, 53(1), 197–214, doi:10.1016/0016-7037(89)90286-X.

Anders, E., E. R. Dufresne, R. Hayatsu, A. Cavaillé, A. Dufresne, and F. W. Fitch (1964), Contaminated meteorite, *Science*, 146(3648), 1157–1161, doi:10.1126/science.146. 3648.1157.

Anderson, D. L. (1977), Composition of the mantle and core, *Annu. Rev. Earth Planet. Sci.*, 5, 179–202.

Anderson, O. L. (1982), The Earth's core and the phase diagram of iron, *Philos. Trans. R. Soc. Lond. A*, 306, 21–35.

Anderson, O. L., and D. G. Isaak (2002), Another look at the core density deficit of Earth's outer core, *Phys. Earth Planet. Inter.*, 131, 19–27.

Anderson, W. W., and T. J. Ahrens (1994), An equation of state for liquid iron and implications for the Earth's core, *J. Geophys. Res.*, 99, 4273–4284.

Antonangeli, D., F. Occelli, H. Requardt, J. Badro, G. Fiquet, and M. Krisch (2004), Elastic anisotropy in textured hcp-iron to 112 GPa from sound wave propagation measurements, *Earth Planet. Sci. Lett.*, *225*(1), 243–251, doi:10.1016/j.epsl.2004.06.004.

Antonangeli, D., T. Komabayashi, F. Occelli, E. Borissenko, A. C. Walters, G. Fiquet, and Y. Fei (2012), Simultaneous sound velocity and density measurements of hcp iron up to 93 GPa and 1100 K: An experimental test of the Birch's law at high temperature, *Earth Planet. Sci. Lett.*, *331*, 210–214, doi:10.1016/j.epsl.2012.03.024.

Badro, J., G. Fiquet, F. Guyot, E. Gregoryanz, F. Occelli, D. Antonangeli, and M. D'Astuto (2007), Effect of light elements on the sound velocities in solid iron: Implications for the composition of Earth's core, *Earth Planet. Sci. Lett.*, *254*(1–2), 233–238, doi:10.1016/j.epsl.2006.11.025.

Badro, J., A. S. Cote, and J. P. Brodholt (2014), A seismologically consistent compositional model of Earth's core, *Proc. Natl. Acad. Sci. USA*, *111*(21), 7542–7545, doi:10.1073/pnas.1316708111.

Bazhanova, Z. G., A. R. Oganov, and O. Gianola (2012), Fe-C and Fe-H systems at pressures of the Earth's inner core, *Phys. Usp.*, *55*(5), 489–497, doi:10.3367/UFNe.0182.201205c.0521.

Birch, F. (1952), Elasticity and constitution of the Earth's interior, *J. Geophys. Res.*, *57*(2), 227–286.

Birch, F. (1964), Density and composition of mantle and core, *J. Geophys. Res.: Solid Earth (1978–2012)*, *69*(20), 4377–4388, doi:10.1029/JZ069i020p04377.

Carlson, R. W., M. F. Horan, T. D. Mock, and E. H. Hauri (2010), Heterogeneous accretion and the moderately volatile element budget of Earth, *Science*, *328*(5980), 884–887, doi:10.1126/science.1186239.

Carlson, R. W., et al. (2014), How did early Earth become our modern world? *Annu. Rev. of Earth Planet. Sci.*, *42*, 151–178, doi:10.1146/annurev-earth-060313-055016.

Chambers, J. E. (2004), Planetary accretion in the inner Solar System, *Earth Planet. Sci. Lett.*, *223*(3), 241–252, doi:10.1016/j.epsl.2004.04.031.

Chen, B., L. Gao, B. Lavina, P. Dera, E. E. Alp, J. Zhao, and J. Li (2012), Magneto-elastic coupling in compressed Fe_7C_3 supports carbon in Earth's inner core, *Geophys. Res. Lett.*, *39*(18), L18,301, doi:10.1029/2012GL052875.

Chen, B., Z. Li, M. Y. Hu, J. Zhao, E. E. Alp, Y. Xiao, P. Chow, and J. Li (2014), Hidden carbon in Earth's inner core revealed by shear softening in dense Fe_7C_3, *Proc. Natl. Acad. Sci. USA*, *111*(50), 17,755–17,758, doi:10.1073/pnas.1411154111.

Chen, J., D. J. Weidner, L. Wang, M. T. Vaughan, and C. E. Young (2005), Density measurements of molten materials at high pressure using synchrotron X-ray radiography: Melting volume of FeS, in *Advances in High-Pressure Techniques for Geophysical Applications*, edited by J. Chen, Y. Wang, T. S. Duffy, G. Shen, and L. F. Dobrzhinetskaya, pp. 185–194. Elsevier B. V., Amsterdam, Netherlands.

Dasgupta, R. (2013), Ingassing, storage, and outgassing of terrestrial carbon through geologic time, *Rev. Minera Geochem.*, *75*(1), 183–229.

Dasgupta, R., and M. M. Hirschmann (2010), The deep carbon cycle and melting in Earth's interior, *Earth Planet. Sci. Lett.*, *298*(1–2), 1–13, doi:10.1016/j.epsl.2010.06.039.

Dasgupta, R., and D. Walker (2008), Carbon solubility in core melts in a shallow magma ocean environment and distribution of carbon between the Earth's core and the mantle, *Geochim. Cosmochim. Acta*, *72*, 4627–4641, doi:10.1016/j.gca.2008.06.023.

Deguen, R., T. Alboussière, and P. Cardin (2013), Thermal convection in Earth's inner core with phase change at its boundary, *Geophys. J. Int.*, *194*(3), 1310–1334, doi:10.1093/gji/ggt202.

Deines, P. (2002), The carbon isotope geochemistry of mantle xenoliths, *Earth Sci. Rev.*, *58*(3–4), 247–278, doi:10.1016/S0012-8252(02)00064-8.

Duman, E., M. Acet, E. Wassermann, J. Itié, F. Baudelet, O. Mathon, and S. Pascarelli (2005), Magnetic instabilities in Fe_3C cementite particles observed with Fe K-edge X-ray circular dichroism under pressure, *Phys. Rev. Lett.*, *94*(7), 075502, doi:10.1103/PhysRevLett.94.075502.

Dziewonski, A. M., and D. L. Anderson (1981), Preliminary reference Earth model, *Phys. Earth Planet. Inter.*, *25*, 297–356.

Fei, Y., and E. Brosh (2014), Experimental study and thermodynamic calculations of phase relations in the Fe–C system at high pressure, *Earth Planet. Sci. Lett.*, *408*(C), 155–162, doi:10.1016/j.epsl.2014.09.044.

Fiquet, G., J. Badro, E. Gregoryanz, Y. Fei, and F. Occelli (2009), Sound velocity in iron carbide (Fe_3C) at high pressure: Implications for the carbon content of the Earth's inner core, *Phys. Earth Planet. Inter.*, *172*(1–2), 125–129, doi:10.1016/j.pepi.2008.05.016.

Gao, L., et al., (2008), Pressure-induced magnetic transition and sound velocities of Fe_3C: Implications for carbon in the Earth's inner core, *Geophys. Res. Lett.*, *35*, L17,306, doi:10.1029/2008GL034817.

Gao, L., B. Chen, M. Lerche, E. E. Alp, W. Sturhahn, J. Zhao, and J. Li (2009), Sound velocities of compressed Fe_3C from simultaneous synchrotron X-ray diffraction and nuclear resonant scattering measurements, *J. Sync. Rad.*, *16*(6), 714–722, doi:10.1107/S0909049509033731.

Gao, L., B. Chen, J. Zhao, E. E. Alp, W. Sturhahn, and J. Li (2011), Effect of temperature on sound velocities of compressed Fe_3C, a candidate component of the Earth's inner core, *Earth Planet. Sci. Lett.*, *309*(3–4), 213–220, doi:10.1016/j.epsl.2011.06.037.

Halliday, A. N. (2013), The origins of volatiles in the terrestrial planets, *Geochim. Cosmochim. Acta*, *105*, 146–171, doi:10.1016/j.gca.2012.11.015.

Hayden, L. A., and E. B. Watson (2007), A diffusion mechanism for core-mantle interaction, *Nature*, *450*(7170), 709–711, doi:10.1038/nature06380.

Hirayama, Y., T. Fujii, and K. Kurita (1993), The melting relation of the system, iron and carbon at high pressure and its bearing on the early stage of the Earth, *Geophys. Res. Lett.*, *20*(1), 2095–2098, doi:10.1029/93GL02131.

Hirschmann, M. M. (2012), Magma ocean influence on early atmosphere mass and composition, *Earth Planet. Sci. Lett.*, *341–344*, 48–57, doi:10.1016/j.epsl.2012.06.015.

Huang, H., Y. Fei, L. Cai, F. Jing, X. Hu, H. Xie, L. Zhang, and Z. Gong (2011), Evidence for an oxygen-depleted liquid outer core of the Earth, *Nature*, *479*(7374), 513–516, doi:10.1038/nature10621.

Jarosewich, E. (1990), Chemical analyses of meteorites — A compilation of stony and iron meteorite analyses, *Meteoritics*, *25*, 323–337.

Jeanloz, R. (1990), The nature of the Earth's core, *Annu. Rev. Earth Planet. Sci.*, *18*, 357–386.

Jeanloz, R., and H. R. Wenk (1988), Convection and anisotropy of the inner core, *Geophys. Res. Lett. 15*(1), 72–75.

Jing, Z., Y. Wang, Y. Kono, T. Yu, T. Sakamaki, C. Park, M. L. Rivers, S. R. Sutton, and G. Shen (2014), Sound velocity of Fe–S liquids at high pressure: Implications for the Moon's molten outer core, *Earth Planet. Sci. Lett.*, *396*, 78–87, doi:10.1016/j.epsl.2014.04.015.

Li, J., and C. B. Agee (1996), Geochemistry of mantle-core differentiation at high pressure, *Nature*, *381*, 686–689.

Li, J., and Y. Fei (2014), 3.15—Experimental constraints on core composition, in *Treatise on Geochemistry*, 2nd ed., edited by H. D. K. Turekian, pp. 527–557, Elsevier, Oxford.

Li, J., H. K. Mao, Y. Fei, E. Gregoryanz, M. Eremets, and C.-S. Zha (2002), Compression of Fe₃C to 30 GPa at room temperature, *Phys. Chem. Minerals*, *19*(3), 166–169, doi:10.1007/s00269-001-0224-4.

Lin, J.-F., V. V. Struzhkin, W. Sturhahn, E. Huang, J.-C. Zhao, M. Y. Hu, E. E. Alp, H.-K. Mao, N. Z. Boctor, and R. J. Hemley (2003), Sound velocities of iron-nickel and iron-silicon alloys at high pressures, *Geophys. Res. Lett.*, *30*(21), 2112, doi:10.1029/2003GL018405.

Lin, J.-F., V. V. Struzhkin, H.-K. Mao, R. J. Hemley, P. Chow, M. Y. Hu, and J. Li (2004), Magnetic transition in compressed Fe₃C from x-ray emission spectroscopy, *Phys. Rev. B*, *70*(21), 212405.

Lin, J.-F., W. Sturhahn, J.-C. Zhao, G. Shen, H.-K. Mao, and R. J. Hemley (2005), Sound velocities of hot dense iron: Birch's law revisited, *Science*, *308*, 1892–1894, doi:10.1126/science.1111724.

Litasov, K. D., I. S. Sharygin, P. I. Dorogokupets, A. Shatskiy, P. N. Gavryushkin, T. S. Sokolova, E. Ohtani, J. Li, and K. Funakoshi (2013), Thermal equation of state and thermodynamic properties of iron carbide Fe₃C to 31 GPa and 1473 K, *J. Geophys. Res. Solid Earth*, *118*(1), 5274–5284, doi:10.1002/2013JB010270.

Lodders, K. (2003), Solar system abundances and condensation temperatures of the elements, *Astrophys. J.*, *591*(2), 1220–1247, doi:10.1086/375492.

Lord, O., M. Walter, R. Dasgupta, D. Walker, and S. M. Clark (2009), Melting in the Fe-C system to 70 GPa, *Earth Planet. Sci. Lett.*, *284*, 157–167, doi:10.1016/j.epsl.2009.04.017.

Mao, H. K., Y. Wu, L. C. Chen, and J. F. Shu (1990), Static compression of iron to 300 GPa and Fe₀.₈Ni₀.₂ alloy to 260 GPa: Implications for compositions of the core, *J. Geophys. Res.*, *95*(B13), 21,737–21,742, doi:10.1029/jb095ib13p21737.

Mao, H. K., J. F. Shu, G. Y. Shen, R. J. Hemley, B. S. Li, and A. K. Singh (1999), Elasticity and rheology of iron above 220 GPa and the nature of the Earth's inner core, *Nature*, *399*, 741–743.

Mao, H. K., J. Xu, V. V. Struzhkin, J. Shu, R. J. Hemley, W. Sturhahn, M. Y. Hu, E. E. Alp, L. Vočadlo, and D. Alfé (2001), Phonon density of states of iron up to 153 gigapascals, *Science*, *292*(5518), 914–916.

Marty, B. (2012), The origins and concentrations of water, carbon, nitrogen and noble gases on Earth, *Earth Planet. Sci. Lett.*, *313–314*, 56–66, doi:10.1016/j.epsl.2011.10.040.

McDonough, W. F., and S.-S. Sun (1995), The composition of the Earth, *Chem. Geol.*, *120*, 223–253, doi:10.1016/0009-2541(94)00140-4.

Mookherjee, M., Y. Nakajima, G. Steinle-Neumann, K. Glazyrin, X. Wu, L. S. Dubrovinsky, C. McCammon, and A. I. Chumakov (2011), High-pressure behavior of iron carbide (Fe₇C₃) at inner core conditions, *116*(B4), B04201, doi:10.1029/2010JB007819.

Morard, G., J. Siebert, D. Andrault, N. Guignot, G. Garbarino, F. Guyot, and D. Antonangeli (2013), The Earth's core composition from high pressure density measurements of liquid iron alloys, *Earth Planet. Sci. Lett.*, *373*, 169–178, doi:10.1016/j.epsl.2013.04.040.

Murphy, C. A., J. M. Jackson, and W. Sturhahn (2013), Experimental constraints on the thermodynamics and sound velocities of hcp-Fe to core pressures, *J. Geophys. Res. Solid Earth*, *118*(5), 1999–2016, doi:10.1002/jgrb.50166.

Nakajima, Y., E. Takahashi, T. Suzuki, and K.-I. Funakoshi (2009), "Carbon in the core" revisited, *Phys. Earth Planet. Inter.*, *174*(1–4), 202–211, doi:10.1016/j.pepi.2008.05.014.

Nakajima, Y., E. Takahashi, N. Sata, Y. Nishihara, K. Hirose, K. Funakoshi, and Y. Ohishi (2011), Thermoelastic property and high-pressure stability of Fe₇C₃: Implication for iron-carbide in the Earth's core, *Am. Mineral.*, *96*(7), 1158–1165, doi:10.2138/am.2011.3703.

Newsom, H. E., and K. W. W. Sims (1991), Core formation during early accretion of the Earth, *Science*, *252*(5008), 926–933, doi:10.1126/science.252.5008.926.

Ohtani, E., Y. Shibazaki, T. Sakai, K. Mibe, H. Fukui, S. Kamada, T. Sakamaki, Y. Seto, S. Tsutsui, and A. Q. R. Baron (2013), Sound velocity of hexagonal close-packed iron up to core pressures, *Geophys. Res. Lett.*, *40*(1), 5089–5094, doi:10.1002/grl.50992.

Oldham, R. D. (1906), Constitution of the interior of the *Earth as revealed by earthquakes*, *Quarterly J. Geological Society*, *62*, 456–475.

Ono, S., and K. Mibe (2010), Magnetic transition of iron carbide at high pressures, *Phys. Earth Planet. Inter.*, *180*(1–2), 1–6.

Poirier, J.-P. (1994), Light elements in the Earth's outer core: A critical review, *Phys. Earth Planet. Inter.*, *85*(3–4), 319–337, doi:10.1016/0031-9201(94)90120-1.

Prescher, C., L. S. Dubrovinsky, C. McCammon, K. Glazyrin, Y. Nakajima, A. Kantor, M. Merlini, and M. Hanfland (2012), Structurally hidden magnetic transitions in Fe₃C at high pressures, *Phys. Rev. B*, *85*(1), 140,402, doi:10.1103/PhysRevB.85.140402.

Sanloup, C., F. Guyot, P. Gillet, G. Fiquet, M. Mezouar, and I. Martinez (2000), Density measurements of liquid Fe-S alloys at high-pressure, *Geophys. Res. Lett.*, *27*(6), 811–814, doi:10.1029/1999GL008431.

Sanloup, C., W. van Westrenen, R. Dasgupta, H. Maynard-Casely, and J. P. Perrillat (2011), Compressibility change in iron-rich melt and implications for core formation models, *Earth Planet. Sci. Lett.*, *306*(1–2), 118–122, doi:10.1016/j.epsl.2011.03.039.

Sata, N., K. Hirose, G. Shen, Y. Nakajima, Y. Ohishi, and N. Hirao (2010), Compression of FeSi, Fe₃C, Fe₀.₉₅O, and FeS under the core pressures and implication for light element in the Earth's core, *115*(B9), B09204, doi:10.1029/2009JB006975.

Satish-Kumar, M., H. So, T. Yoshino, M. Kato, and Y. Hiroi (2011), Experimental determination of carbon isotope fractionation between iron carbide melt and carbon: ¹²C-enriched carbon in the Earth's core? *Earth Planet. Sci. Lett.*, *310*(3–4), 340–348, doi:10.1016/j.epsl.2011.08.008.

Scott, H. P., Q. Williams, and E. Knittle (2001), Stability and equation of state of Fe$_3$C to 73 GPa: Implications for carbon in the Earth's core, *Geophys. Res. Lett.*, *28*(9), 1875–1878.

Seagle, C. T., A. J. Campbell, D. L. Heinz, and G. Shen (2006), Thermal equation of state of Fe$_3$S and implications for sulfur in Earth's core, *111*, B06,209, doi:10.1029/2005JB004091.

Shearer, P., and G. Masters (1990), The density and shear velocity contrasts at the inner core boundary, *Geophys. J. Int.*, *102*, 408–491.

Singh, S. C. (2000), On the presence of liquid in Earth's inner core, *Science*, *287*(5462), 2471–2474, doi:10.1126/science.287. 5462.2471.

Stevenson, D. J. (1981), Models of the Earth's core, *Science*, *214*, 611–619.

Stixrude, L., E. Wasserman, and R. E. Cohen (1997), Composition and temperature of Earth's inner core, *J. Geophys. Res. Solid Earth*, *102*, 24,729–24,740.

Tateyama, R., E. Ohtani, H. Terasaki, K. Nishida, Y. Shibazaki, A. Suzuki, and T. Kikegawa (2011), Density measurements of liquid Fe-Si alloys at high pressure using the sink-float method, *Phys. Chem. Minerals*, *38*(1), 801–807, doi:10.1007/s00269-011-0452-1.

Terasaki, H., K. Nishida, Y. Shibazaki, T. Sakamaki, A. Suzuki, E. Ohtani, and T. Kikegawa (2010), Density measurement of Fe$_3$C liquid using X-ray absorption image up to 10 GPa and effect of light elements on compressibility of liquid iron, *J. Geophys. Res. Solid Earth*, *115*(B), B06207, doi:10.1029/2009JB006905.

Tsuzuki, A., S. Sago, S. I. Hirano, and S. Naka (1984), High temperature and pressure preparation and properties of iron carbides Fe$_7$C$_3$ and Fe$_3$C, *J. Mater. Sci.*, *19*(8), 2513–2518.

Tsymbulov, L. B., and L. S. Tsemekhman (2001), Solubility of carbon in sulfide melts of the system Fe-Ni-S, *Russ. J. Appl. Chem*, *74*(6), 925–929.

Vočadlo, L., J. Brodholt, D. P. Dobson, K. S. Knight, W. G. Marshall, G. D. Price, and I. G. Wood (2002), The effect of ferromagnetism on the equation of state of Fe$_3$C studied by first-principles calculations, *Earth Planet. Sci. Lett.*, *203*(1), 567–575, doi:10.1016/S0012-821X(02)00839-7.

Wasson, J. T. (1985), *Meteorites: Their Record of Early Solar-System History*, W. H. Freeman and Co. New York, 274p.

Wood, B. J. (1993), Carbon in the core, *Earth Planet. Sci. Lett.*, *117*, 593–607.

Wood, B. J., J. Li, and A. Shahar (2013), Carbon in the core: Its influence on the properties of core and mantle, *Rev. in Mineral Geochem.*, *75*(1), 231–250, doi:10.2138/rmg.2013.75.8.

Yu, X., and R. A. Secco (2008), Equation of state of liquid Fe–17 wt%Si to 12 GPa, *High Press. Res.*, *28*(1), 19–28, doi:10.1080/08957950701882138.

Zhang, Y., and A. Zindler (1993), Distribution and evolution of carbon and nitrogen in Earth, *Earth Planet. Sci. Lett.*, *117*(3–4), 331–345, doi:10.1016/0012-821X(93)90088-Q.

INDEX

Ab initio mineral physics data
  CMB heat flow
    across CMB, 25–26
    Fe incorporation effects on, 24–25, 25t
    MgO, 22–23, 23f
    MgPv, 23–24, 24f, 25f
    radiative contribution to k, 24–25, 25t
    thermal conductivity of LM phases, 22
  core and lower mantle temperature based on, 13–26
  methods for
    Ab initio molecular dynamics method, 14
    density functional theory, 14
    EoS from molecular dynamics, 15–16
    lattice thermal conductivity, 17
    phonon and crystal thermodynamics, 14–15
    thermoelasticity, 16–17
  temperature profiles for
    core adiabat, 19, 19f, 19t
    mantle adiabat, 17–19, 18f, 18t
  thermal structure in D″ layer, 19–22, 21f
Ab initio techniques
  determining phase equilibria with, 243
  electron-electron scattering computed with, 48
  Fe melting in, 5
Absorption coefficient, 37–38
Accretion models
  combined core-mantle differentiation model and, 186–88, 187f
  linked to siderophile elements
    accretion with predictive D(metal/silicate) expressions, 170–71, 171f
    continuous accretion and heterogeneous accretion, 170
    continuous core formation, 170
  oxidation state during accretion in, 172
Adams–Williamson equation, 150
Adiabatic temperature profiles
  core adiabat, 9, 9f, 19, 19f, 19t
  liquid Fe, 19t
  mantle adiabat, 17–19, 18f, 18t
  MgPv, 18t
Ag/Ag ratio, 279
AK135 model, 114
  core sound speed in, 116, 116f
δ-AlOOH, 79–80
  new hydrous phase H in relation to, 271–73, 271f, 272f
  physical properties of, 272–73

  stability in $MgSiO_2(OH)_2$-AlOOH system of, 272, 272f
  typical structure of, 271f
Alphabet phases, 269–71, 270t
Aluminous phase H solid solution, 272, 272f
Aluminum, studies of iron melting curve with, 102–3
Aluminum-rich phases, crystal structures, 76–77, 77f
Analytical methods in plastic deformation
  average polycrystalline properties, 91
  individual deformation mechanisms, 91
  numerical modeling, 91
  numerical modeling of defect interaction, 89
  numerical modeling of defect properties, 91–92
  PNG method, 92
  PN model, 92
  single crystals to polycrystals, 92
  transmission electron microscopy, 91
  X-ray line profile analysis, 91
Anisotropy
  inner core, 114, 116–17, 117f, 124–25
  layering with, 117f
Average polycrystalline properties, 91

Banded-iron formations (BIFs), 218
Basal magma ocean (BMO), 50
BCC. *See* Body-centered cubic
BIFs. *See* Banded-iron formations
Birch-Murnaghan (BM) EOS
  density-pressure relationship for hcp-Fe fit with, 122f
  outer core with, 130, 131t, 132f, 135
  phase diagram computation with, 246t
Bloch-Grüneisen formula, 48
BM EOS. *See* Birch-Murnaghan EOS
BMO. *See* Basal magma ocean
Body-centered cubic (BCC), 58, 58f
Boltzmann's transport equation, 17
Born-Oppenheimer energy surface, 14
Bridgmanite. *See* Perovskite
Brown-Shankland model, 18
Bulk modulus
  δ-AlOOH, 272–73
  FeH, 258
  H_2O, 272
  isothermal, 16, 243
  Liquid Fe, 130
  Liquid Fe-S, Fe-Si, Fe-C, 130, 131t
  Phase D, 79, 272

Deep Earth: Physics and Chemistry of the Lower Mantle and Core, Geophysical Monograph 217, First Edition.
Edited by Hidenori Terasaki and Rebecca A. Fischer.
© 2016 American Geophysical Union. Published 2016 by John Wiley & Sons, Inc.